Instructor's Solutions Manual

University Physics

Ninth Edition

Volume 2

Instructor's Solutions Manual

Young/Freedman

University Physics

Ninth Edition
Volume 2

Peter A. Morse
Arizona Western College

Addison-Wesley Publishing Company
Reading, Massachusetts • Menlo Park, California • New York • Don Mills, Ontario
Harlow, United Kingdom • Amsterdam • Bonn • Sydney • Singapore • Tokyo
Madrid • San Juan • Milan • Paris

ISBN 0-201-83310-7
1 2 3 4 5 6 7 8 9 10--VG--0099989796

Preface

This instructor's solution manual to Young and Freedman's *University Physics (9th Edition)* contains the solutions to all problems in the second volume of the text's extended version. That is, it covers Chapter 22 through Chapter 46. I wish to acknowledge the contributions of both Craig Watkins, who provided all answers to the even numbered questions for Chapter 39 through Chapter 46, and A. Lewis Ford, whose help came in many forms: suggestions for solutions on sticky problems, proofing on draft solutions, and his Student's Solution Manual which was an invaluable cross-check on many problems.

As all practising physicists know, there may be many ways to solve a given problem, and I have attempted to present the most straight-forward method that came to mind for each particular question. I do not claim divine insight, however, and would be pleased to hear from anyone using this manual who believes the suggested solution is overly cumbersome and that a more elegant path is available.

For most questions I list fundamental constants to just three significant figures and have taken care to list the answers with the appropriate number of significant figures. I have used the common and reasonable technique of carrying all digits from early calculations in a calculator's memory for use in the subsequent parts of a problem. To carry out a calculation, which may involve a number of steps, and to arrive at an answer that differs from the solution in the "back of the book" by one or even two units in the least significant figure is definitely *not* something to be concerned about. This should *not* be allowed to overshadow the point of the problem.

I also wish to thank Roger Freedman for keeping me in mind and putting forward my name for this manual – he has been, and still is, a great inspiration and role model for me. Thanks also to Jennifer Albanese at Addison-Wesley for much assistance throughout the completion of this work. Finally, the greatest thanks and love go to my wife Lori for infinite patience and support, and my children Henry and Gabriella for putting up with my too frequent absences.

P.A.M.
pmorse@awc.cc.az.us

Department of Physics
Arizona Western College
P.O. Box 929
Yuma, AZ 85366

UNIVERSITY PHYSICS, Ninth edition, Vol. II, Instructor's Solutions Manual

Table of Contents

Chapter 22: Electric Charge and Electric Field

22-1: $3.00 \text{ mol } = 3.00 \times 6.02 \times 10^{23} \text{ H atoms} = 1.81 \times 10^{24}$

$\text{charge } = 1.81 \times 10^{24} \times 1.60 \times 10^{-19}\text{C} = 2.89 \times 10^{5} \text{ C}.$

22-2: Mass of gold = 15.2 g and the atomic weight of gold is 197 g/mol.

So the number of atoms= $N_A \times \text{mol} = (6.02 \times 10^{23}) \times \left(\dfrac{15.2\text{g}}{197\text{g / mol}}\right) = 4.64 \times 10^{22}.$

(a) $n_p = 79 \times 4.64 \times 10^{22} = 3.67 \times 10^{24}$

$q = n_p \times 1.60 \times 10^{-19}\text{C} = 5.87 \times 10^{5}\text{C}$

(b) $n_e = n_p = 3.67 \times 10^{24}.$

22-3: $m_{\text{lead}} = 10.0 \text{ g}$ and charge $= -2.5 \times 10^{-9}$ C

(a) $n_e = \dfrac{-2.5 \times 10^{-9}\text{ C}}{-1.6 \times 10^{-19}\text{ C}} = 1.56 \times 10^{10}.$

(b) $n_{\text{lead}} = N_A \times \dfrac{10.0}{207} = 2.91 \times 10^{22}$ and $\dfrac{n_e}{n_{\text{lead}}} = 5.36 \times 10^{-13}.$

22-4: current = 20,000 C/s and t = 100μs = 10^{-4} s

$Q = It = 2.00\text{C}$

$n_e = \dfrac{Q}{1.60 \times 10^{-19}\text{C}} = 1.25 \times 10^{19}.$

22-5: The mass is primarily protons and neutrons of m= 1.67×10^{-27} kg, so:

$n_{\text{p and n}} = \dfrac{70.0\text{kg}}{1.67 \times 10^{-27}\text{kg}} = 4.19 \times 10^{28}$

About one half are protons, so $n_p = 2.10 \times 10^{28} = n_e$ and the charge on the electrons is given by: $Q = (1.60 \times 10^{-19}\text{C}) \times (2.10 \times 10^{28}) = 3.35 \times 10^{9}\text{C}.$

22-6: (a) $F = \dfrac{1}{4\pi\varepsilon}\dfrac{kq_1q_2}{r^2} \Rightarrow 0.5\text{N} = \dfrac{1}{4\pi\varepsilon_o}\dfrac{(0.600 \times 10^{-6}\text{ C})q_2}{(0.25 \text{ m})^2}$

$\Rightarrow q_2 = +5.79 \times 10^{-6}\text{C}.$

(b) $F = 0.5$ N, and is attractive.

22-7: Since the charges are equal in sign the force is repulsive and of magnitude:

$$F = \frac{kq^2}{r^2} = \frac{(8.99 \times 10^{9} \text{ Nm}^2 / \text{C}^2)(4.00 \times 10^{-6} \text{ C})^2}{(0.500 \text{ m})^2} = 0.576 \text{ N}$$

22-8: First find the total charge on the spheres:

$$F = \frac{1}{4\pi\varepsilon_o}\frac{q^2}{r^2} \Rightarrow q = \sqrt{4\pi\varepsilon_o F r^2} = \sqrt{4\pi\varepsilon_o (2.3\times10^{-22})(0.3)^2} = 4.80\times10^{-17}\text{C}$$

And therefore, the total number of electrons required is

$n = q / e = 4.80\times10^{-17}\text{C} / 1.60\times10^{-19}\text{C} = 300.$

22-9: (a) Using Coulomb's Law for equal charges, we find:

$$F = 0.150\text{ N} = \frac{1}{4\pi\varepsilon_o}\frac{q^2}{(0.300\text{ m})^2} \Rightarrow q = \sqrt{1.50\times10^{-12}} = 1.22\times10^{-6}\text{ C.}$$

(b) When one charge is three times the other, we have:

$$F = 0.150\text{ N} = \frac{1}{4\pi\varepsilon_o}\frac{3q^2}{(0.300\text{ m})^2} \Rightarrow q = \sqrt{0.50\times10^{-12}\text{ C}^2} = 7.07\times10^{-7}\text{ C}$$

So one charge is 7.07×10^{-7} C, and the other is 2.12×10^{-6} C.

22-10: (a) The total number of electrons on each sphere equals the number of protons.

$$n_e = n_p = 29\times N_A\times\frac{0.0400\text{ kg}}{0.0635\text{ kg/mol}} = 1.10\times10^{25}.$$

(b) For a force of 1.00×10^4 N to act between the spheres,

$$F = 10^4\text{ N} = \frac{1}{4\pi\varepsilon_o}\frac{q^2}{r^2} \Rightarrow q = \sqrt{4\pi\varepsilon_o(10^4\text{ N})(2.00\text{ m})^2} = 2.11\times10^{-3}\text{ C.}$$

$\Rightarrow n_e' = q/e = 1.32\times10^{+16}.$

(c) n_e' is 1.20×10^{-9} of the total number.

22-11: The force of gravity must equal the electric force.

$$mg = \frac{1}{4\pi\varepsilon_o}\frac{q^2}{r^2} \Rightarrow r^2 = \frac{1}{4\pi\varepsilon_o}\frac{(1.60\times10^{-19}\text{ C})^2}{(9.11\times10^{-31}\text{ kg})(9.8\text{ m/s})} = 25.8\text{ m}^2 \Rightarrow r = 5.08\text{ m.}$$

22-12: $\vec{F} = \vec{F}_1 + \vec{F}_2$ and $F = F_1 - F_2$ so,

$$F = \frac{1}{4\pi\varepsilon_o}q_3\left(\frac{4.00\times10^{-9}\text{ C}}{(0.300\text{ m})^2} - \frac{5.00\times10^{-9}\text{ C}}{(0.200\text{ m})^2}\right) = 5.80\times10^{-6}\text{ N to the left.}$$

22-13: $\vec{F} = \vec{F}_1 + \vec{F}_2$ and $F = F_2 - F_1$ so,

$$F = \frac{1}{4\pi\varepsilon_o}(5.00\times10^{-9}\text{ C})\left(\frac{1.50\times10^{-9}\text{ C}}{(0.400\text{ m})^2} - \frac{3.60\times10^{-9}\text{ C}}{(1.00\text{ m})^2}\right) = 2.60\times10^{-7}\text{ N upwards.}$$

22-14: We only need the y-components, and each charge contributes equally.

$$F = \frac{1}{4\pi\varepsilon_o}\frac{(2.0\times10^{-6}\text{ C})(4\times10^{-6}\text{ C})}{(0.500\text{ m})^2}\sin\alpha = 0.173\text{ N (since }\sin\alpha = 0.6).$$

Therefore, the total force is $2F = 0.35$ N, downward.

22-15:

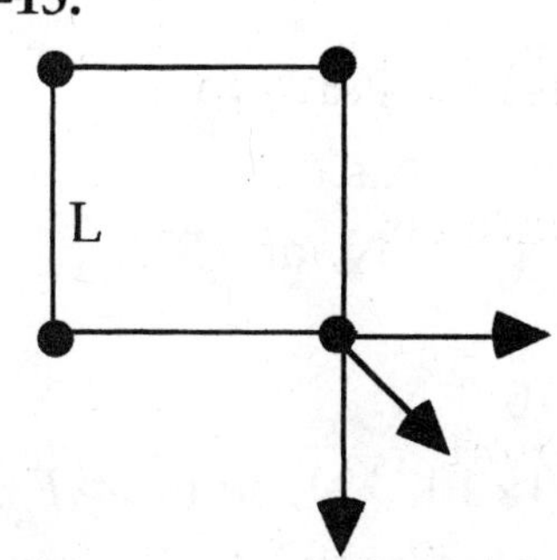

(b) $F_x = \frac{1}{4\pi\varepsilon_o}\frac{q^2}{2L^2} + \sqrt{2}\frac{1}{4\pi\varepsilon_o}\frac{q^2}{L^2} = (1 + 2\sqrt{2})\frac{1}{4\pi\varepsilon_o}\frac{q^2}{2L^2}$ at and angle of 45° below the positive x-axis.

22-16:

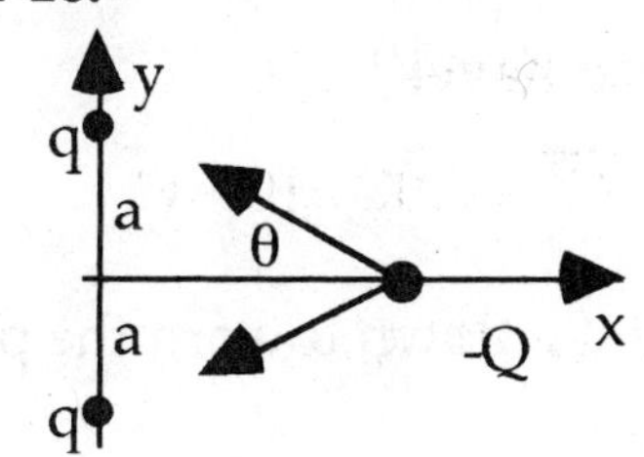

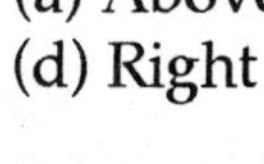
(a) Above
(d) Right

(b) $F_x = -2\frac{1}{4\pi\varepsilon_o}\frac{kqQ}{(a^2 + x^2)}\cos\theta = \frac{1}{4\pi\varepsilon_o}\frac{-2kqQx}{(a^2 + x^2)^{3/2}}$, $F_y = 0$

(c) At $x = 0$, $F = 0$.

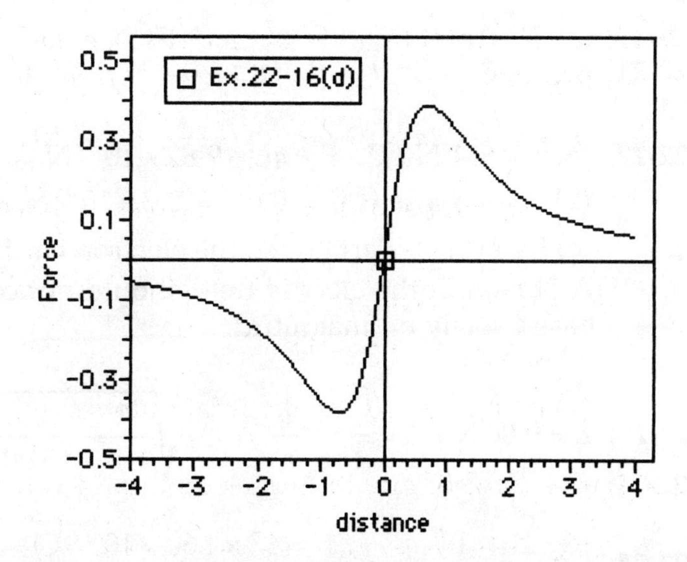

22-17:

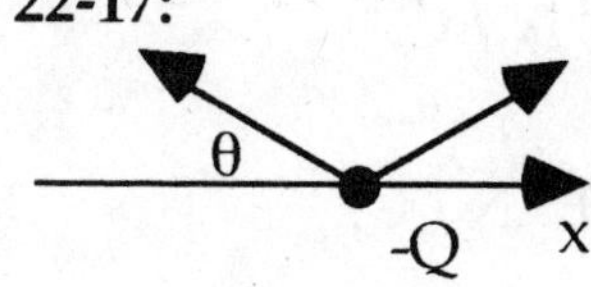

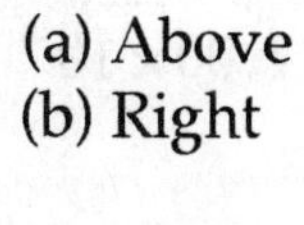
(a) Above
(b) Right

(b) $F_x = 0$, $F_y = 2\frac{1}{4\pi\varepsilon_o}\frac{qQ}{(a^2 + x^2)}\sin\theta = \frac{1}{4\pi\varepsilon_o}\frac{2qQa}{(a^2 + x^2)^{3/2}}$.

(c) At $x = 0$, $F = \frac{1}{4\pi\varepsilon_o}\frac{2qQ}{a^2}$.

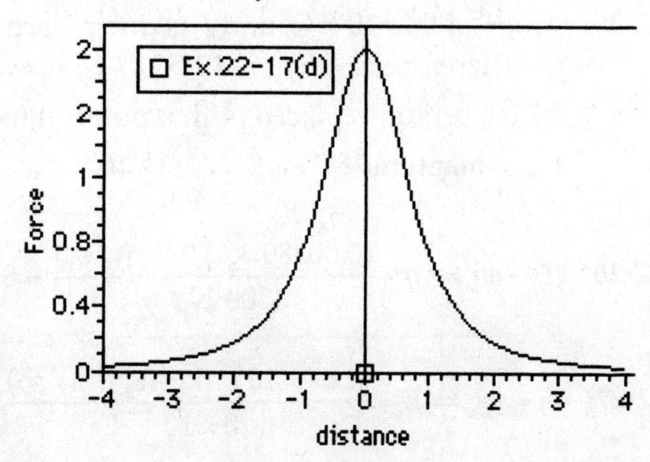

22-18: (a) $\vec{E} = -11\ \mathrm{N/C}\,\hat{i} + 14\ \mathrm{N/C}\,\hat{j}$, so $E = \sqrt{(-11)^2 + (14)^2} = 17.8\ \mathrm{N/C}$.
$\theta = \tan^{-1}(-14/11) = -51.8°$, so $\theta = 128°$ from the x-axis.
(b) $\vec{F} = \vec{E}q$ so $F = (17.8\ \mathrm{N/C})(3.0 \times 10^{-9}\mathrm{C}) = 5.34 \times 10^{-8}\ \mathrm{N}$, at -52°.

22-19: (a) $F_g = m_e g = (9.11 \times 10^{-31}\ \mathrm{kg})(9.8\ \mathrm{m/s^2}) = 8.93 \times 10^{-30}\ \mathrm{N}$.
$F_e = eE = (1.60 \times 10^{-19}\ \mathrm{C})(1.00 \times 10^4\ \mathrm{N/C}) = 1.60 \times 10^{-15}\ \mathrm{N}$, so $F_e >> F_g$.
(b) $E = 10^4\ \mathrm{N/C} \Rightarrow F_e = 1.6 \times 10^{-15}\ \mathrm{N} = mg \Rightarrow m = 1.63 \times 10^{-16}\ \mathrm{kg}$
$\Rightarrow m = 1.79 \times 10^{14}\ m_e$.
(c) No.

22-20: $E = \dfrac{1}{4\pi\varepsilon_o}\dfrac{q}{r^2} = \dfrac{1}{4\pi\varepsilon_o}\dfrac{(5.00 \times 10^{-6}\ \mathrm{C})}{(0.400\ \mathrm{m})^2} = 2.81 \times 10^5\ \mathrm{N/C}$, down toward the particle.

22-21: $\hat{r} = \cos\theta\,\hat{i} + \sin\theta\,\hat{j} = 0.868\,\hat{i} + 0.496\,\hat{j}$; $\varphi = \tan^{-1}(0.8/1.4) = 30°$.

22-22: (a) $E = 614\ \mathrm{N/C}$, $F = qE = 9.82 \times 10^{-17}\mathrm{N}$.
(b) $F = e^2 / 4\pi\varepsilon_o(1.0 \times 10^{-10})^2 = 2.3 \times 10^{-8}\ \mathrm{N}$.
(c) Part (b) >> Part (a), so the electron hardly notices the electric field.
A person in the electric field should notice nothing if physiological effects are based solely on magnitude.

22-23: $E = 5.00\ \mathrm{N/C} = \dfrac{1}{4\pi\varepsilon_o}\dfrac{q}{r^2} \Rightarrow r = \sqrt{\dfrac{1}{4\pi\varepsilon_o}\dfrac{(6.00 \times 10^{-9}\ \mathrm{C})}{(5.00\ \mathrm{N/C})}} = 3.29\ \mathrm{m}$.

22-24: (a) $E = \dfrac{1}{4\pi\varepsilon_o}\dfrac{q}{r^2} = \dfrac{1}{4\pi\varepsilon_o}\dfrac{(47 \times 1.60 \times 10^{-19}\ \mathrm{C})}{(6.00 \times 10^{-10}\ \mathrm{m})^2} = 1.88 \times 10^{11}\ \mathrm{N/C}$.
(b) $E_{\mathrm{proton}} = \dfrac{1}{4\pi\varepsilon_o}\dfrac{q}{r^2} = \dfrac{1}{4\pi\varepsilon_o}\dfrac{(1.60 \times 10^{-19}\ \mathrm{C})}{(5.28 \times 10^{-11}\ \mathrm{m})^2} = 5.16 \times 10^{11}\ \mathrm{N/C}$.

22-25: (a) $q = -4.00 \times 10^{-9}\ \mathrm{C}$, and F is downward with magnitude $5.00 \times 10^{-8}\ \mathrm{N}$
Therefore, $E = F/q = 12.5\ \mathrm{N/C}$, upward.
(b) If a proton is placed at that point, it feels an upward force of magnitude $F = qE = 2.00 \times 10^{-18}\ \mathrm{N}$.

22-26: $qE = mg \Rightarrow q = \dfrac{(0.00380\ \mathrm{kg})(9.8\ \mathrm{m/s^2})}{4500\ \mathrm{N/C}} = 8.28 \times 10^{-6}\ \mathrm{C}$

22-27: $qE = mg \Rightarrow E = \dfrac{(1.67 \times 10^{-27}\ \mathrm{kg})(9.8\ \mathrm{m/s^2})}{1.60 \times 10^{-19}\ \mathrm{C}} = 1.02 \times 10^{-7}\ \mathrm{N/C}$.

22-28: (a) The electric field of the Earth points toward the ground, so a NEGATIVE charge will hover above the surface.

$$mg = qE \Rightarrow q = -\frac{(75.0\ \text{kg})(9.8\ \text{m/s}^2)}{150\ \text{N/C}} = -4.90\ \text{C}.$$

(b) $F = \frac{1}{4\pi\varepsilon_o}\frac{q^2}{r^2} = \frac{1}{4\pi\varepsilon_o}\frac{(4.90\ \text{C})^2}{(50.0\ \text{m})^2} = 8.64\times 10^7\ \text{N}$. The magnitude of the charge is too great for practical use.

22-29: (a) Passing between the charged plates the electron feels a force upwards, and just misses the top plate. The distance it travels in the y-direction is 0.500 m.

time of flight $= t = \frac{0.0200\ \text{m}}{4.00\times 10^6\ \text{m/s}} = 5.00\times 10^{-9}\ \text{s}$ and initial y-velocity is zero.

Now, $y = v_{oy}t + \frac{1}{2}at^2$ so $0.005 = \frac{1}{2}a(5.00\times 10^{-9})$ and $a = 4.00\times 10^{14}\ \text{m/s}^2$. But also

$$a = \frac{F}{m} = \frac{eE}{m_e} \Rightarrow E = \frac{(9.11\times 10^{-31}\ \text{kg})(4.00\times 10^{14}\text{m/s}^2)}{1.60\times 10^{-19}\ \text{C}} = 2280\ \text{N/C}.$$

(b) Since the proton is more massive, it will accelerate less, and NOT hit the plates. To find the vertical displacement when it exits the plates, we use the kinematic equations again:

$$y = \frac{1}{2}at^2 = \frac{1}{2}\frac{eE}{m_p}(5.00\times 10^{-9}\ \text{s})^2 = 2.73\times 10^{-6}\ \text{m}.$$

(c) As mentioned in (b), the proton will not hit one of the plates because although the electric force felt by the proton is the same as the electron felt, a smaller acceleration results for the more massive proton.

22-30: (a) $x = \frac{1}{2}at^2 = \frac{1}{2}\frac{eE}{m_e}t^2 \Rightarrow E = \frac{2(0.0160\ \text{m})(9.11\times 10^{-31}\ \text{kg})}{(1.60\times 10^{-19}\ \text{C})(1.50\times 10^{-8}\ \text{s})^2} = 810\ \text{N/C}.$

(b) $v = v_0 + at = \frac{eE}{m_e}t = 2.13\times 10^6\ \text{m/s}.$

22-31: Point charge q_1 (-6.00nC) is at the origin and q_2 (4.00 nC) is at x=0.800 m.

(a) At $x = 0.200$ m, $E = \frac{k|q_1|}{(0.200\ \text{m})^2} + \frac{k|q_2|}{(0.600\ \text{m})^2} = 1450\ \text{N/C}$ left.

(b) At $x = 1.20$ m, $E = \frac{k|q_2|}{(0.400\ \text{m})^2} - \frac{k|q_1|}{(1.20\ \text{m})^2} = 187\ \text{N/C}$ right.

(c) At $x = -0.200$ m, $E = \frac{k|q_1|}{(0.200\ \text{m})^2} - \frac{k|q_2|}{(1.00\ \text{m})^2} = 1310\ \text{N/C}$ right.

22-32: Point charges q_1 (1.00nC) and q_2 (3.00 nC) are separated by x=1.20 m.

The electric field is zero when $E_1 = E_2 \Rightarrow \frac{kq_1}{r_1^2} = \frac{kq_2}{(1.20 - r_1)^2}$

$$\Rightarrow 9r_1^2 - 18(1.20)r_1 + 9(1.20)^2 = 27r_1^2 \quad \Rightarrow 2r_1^2 + 2.40r_1 - 1.44 = 0.$$

$\Rightarrow r_1 = 0.439$ m or -1.64 m. But the question asks for the point on the line between the two charges where the fields cancel, so $r_1 = 0.439$ m.

22-33: Two positive charges, q, are on the x-axis, a distance a from the origin.
(a) Half way between them, $E = 0$.

(b) At any position x, $E = \begin{cases} \frac{1}{4\pi\varepsilon_o}\left(\frac{q}{(a+x)^2} - \frac{q}{(a-x)^2}\right), |x| < a \\ \frac{1}{4\pi\varepsilon_o}\left(\frac{q}{(a+x)^2} + \frac{q}{(a-x)^2}\right), |x| > a \end{cases}$

(c) See below, left.

22-34: A positive and negative charge, of equal magnitude q, are on the x-axis, a distance a from the origin.

(a) Half way between them, $E = \frac{1}{4\pi\varepsilon_o}\frac{2q}{a^2}$, to the left.

(b) At any position x, $E = \begin{cases} \frac{1}{4\pi\varepsilon_o}\left(\frac{-q}{(a+x)^2} - \frac{q}{(a-x)^2}\right), |x| < a \\ \frac{1}{4\pi\varepsilon_o}\left(\frac{-q}{(a+x)^2} + \frac{q}{(a-x)^2}\right), x > a \\ \frac{1}{4\pi\varepsilon_o}\left(\frac{q}{(a+x)^2} - \frac{q}{(a-x)^2}\right), x < -a \end{cases}$ with "+" to the right.

This is graphed below, on the right.

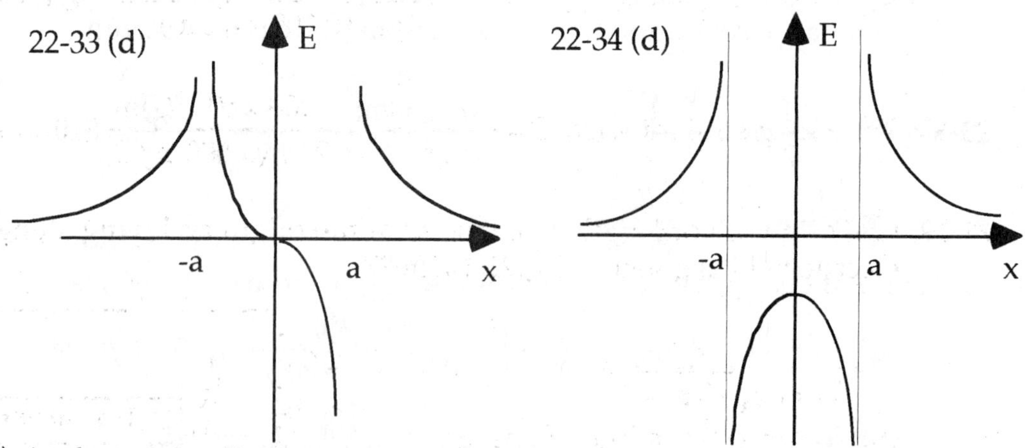

22-35: (a) At the origin, $E = 0$.
(b) At $x = 0.200$ m, $y = 0$:

$$E = \frac{1}{4\pi\varepsilon_o}(2.00 \times 10^{-8}\ \text{C})\left(\frac{1}{(0.100\ \text{m})^2} + \frac{1}{(0.300\ \text{m})^2}\right)\hat{i} = 20{,}000\hat{i}\ \text{N/C}.$$

(c) At $x = 0.100$ m, $y = 0.150$ m:

$$E = \frac{1}{4\pi\varepsilon_o}(2.00 \times 10^{-8}\ \text{C})\left(\frac{1}{(0.150\ \text{m})^2}\hat{j} + \frac{1}{(0.250\ \text{m})^2}\frac{0.20}{0.25}\hat{i} + \frac{1}{(0.250\ \text{m})^2}\frac{0.15}{0.25}\hat{j}\right)$$

$\Rightarrow E = (2304\hat{\imath} + 9730\hat{\jmath})$ N/C $\Rightarrow E = 10{,}000$ N/C and $\theta = 76.7°$ from the x-axis.

(d) $x = 0, y = 0.100$ m: $E = \dfrac{1}{4\pi\varepsilon_o}\dfrac{2(2.00\times10^{-8}\text{ C})}{(0.100\times\sqrt{2}\text{ m})^2} = 18{,}000\hat{\jmath}$ N/C

22-36: Calculate in vector form the electric field for each charge, and add them.

$$\vec{E}_{-} = \frac{1}{4\pi\varepsilon_o}\frac{(6.00\times10^{-9}\text{ C})}{(0.800\text{ m})^2}\hat{\imath} = 84.4\hat{\imath}\text{ N/C}$$

$$\vec{E}_{+} = -\frac{1}{4\pi\varepsilon_o}(4.00\times10^{-9}\text{ C})\left(\frac{1}{(1.00\text{ m})^2}(0.8)\hat{\imath} - \frac{1}{(1.00\text{ m})^2}(0.6)\hat{\jmath}\right) = -28.8\hat{\imath} - 21.6\hat{\jmath}\text{ N/C}$$

$$\Rightarrow E = \sqrt{(21.6)^2 + (55.5)^2} = 59.6\text{ N/C, at } \theta = \tan^{-1}(21.6/55.6) = 21.2°.$$

22-37: (a) At the origin, $E = -\dfrac{1}{4\pi\varepsilon_o}\dfrac{2(2.00\times10^{-8}\text{ C})}{(0.100\text{ m})^2}\hat{\imath} = 36{,}000\hat{\imath}$ N/C..

(b) At $x = 0.200$ m, $y = 0$:

$$E = \frac{1}{4\pi\varepsilon_o}(2.00\times10^{-8}\text{ C})\left(\frac{1}{(0.100\text{ m})^2} - \frac{1}{(0.300\text{ m})^2}\right)\hat{\imath} = 16{,}000\hat{\imath}\text{ N/C}.$$

(c) At $x = 0.100$ m, $y = 0.150$ m:

$$E = \frac{1}{4\pi\varepsilon_o}(2.00\times10^{-8}\text{ C})\left(\frac{1}{(0.150\text{ m})^2}\hat{\jmath} - \frac{1}{(0.250\text{ m})^2}\frac{0.20}{0.25}\hat{\imath} - \frac{1}{(0.250\text{ m})^2}\frac{0.15}{0.25}\hat{\jmath}\right)$$

$\Rightarrow E = (6272\hat{\imath} - 2304\hat{\jmath})$ N/C $\Rightarrow E = 6680$ N/C and $\theta = 110°$ from the x-axis.

(d) $x = 0, y = 0.100$ m: $E_y = 0, E = -\dfrac{1}{4\pi\varepsilon_o}\dfrac{2(2.00\times10^{-8}\text{ C})}{(0.100\times\sqrt{2}\text{ m})^2} = -18{,}000\hat{\imath}$ N/C

22-38: For a long straight wire, $E = \dfrac{\lambda}{2\pi\varepsilon_0 r} \Rightarrow r = \dfrac{3.00\times10^{-10}\text{C/m}}{2\pi\varepsilon_0(0.500\text{ N/C})} = 10.8$ m.

22-39: (a) For a wire of length 2a centered at the origin and lying along the y-axis, the electric field is given by Eq.22-10, P687:

$$\vec{E} = \frac{1}{2\pi\varepsilon_0}\frac{\lambda}{x\sqrt{x^2/a^2+1}}\hat{\imath}$$

(b) For an infinite line of charge:

$$\vec{E} = \frac{\lambda}{2\pi\varepsilon_0 x}\hat{\imath}$$

Graphs of electric field versus position for both are shown at right

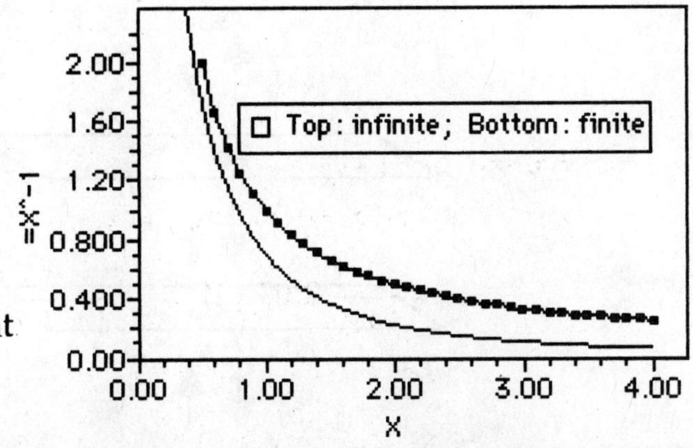

Top: infinite; Bottom: finite
y = x^-1
X
0.00 1.00 2.00 3.00 4.00
0.00 0.400 0.800 1.20 1.60 2.00

22-40: For a ring of charge, the electric field is given by Eq.22-8, p686:

(a) $\vec{E} = \frac{1}{4\pi\varepsilon_0}\frac{Qx}{(x^2+a^2)^{3/2}}\hat{i}$ so with $Q = 8.40\times10^{-6}$ C, $a = 0.250$ cm and $x = 0.50$ m

$\Rightarrow \vec{E} = 3.02\times10^5\hat{i}\,\text{N/C}.$

(b) $\vec{F}_{\text{on ring}} = -\vec{F}_{\text{on q}} = -q\vec{E} = -(-2.50\times10^{-6}\text{ C})(3.02\times10^5\hat{i}\,\text{N/C}) = 0.755\hat{i}$ N.

22-41: For a uniformly charged disk, the electric field is given by Eq.22-11, p688:

$$\vec{E} = \frac{\sigma}{2\varepsilon_0}\left(1-\frac{1}{\sqrt{R^2/x^2+1}}\right)\hat{i}$$

The x-component of the electric field is shown below.

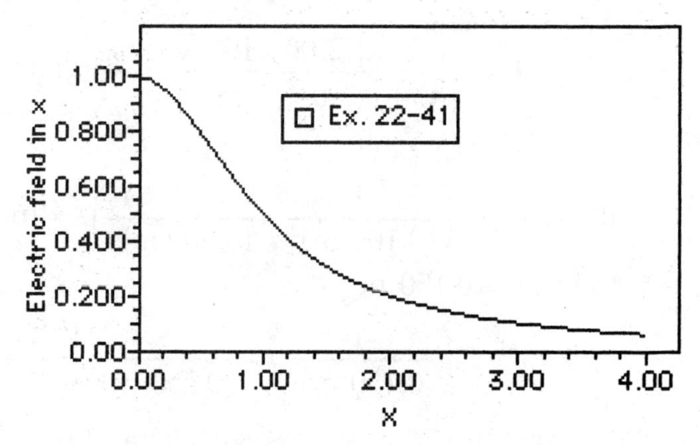

22-42: The earth's electric field is 150 N/C, directly downward. So,

$$E = 150 = \frac{\sigma}{2\varepsilon_0} \Rightarrow \sigma = 300\,\varepsilon_0 = 2.66\times10^{-9}\,\text{C/m}^2,$$ and is negative.

22-43: For an infinite sheet, $E = \frac{\sigma}{2\varepsilon_0} = 3.00 \Rightarrow \sigma = 6\varepsilon_0 = 5.31\times10^{-11}\text{C/m}^2$.

22-44: By superposition we can add the electric fields from two parallel sheets of charge.

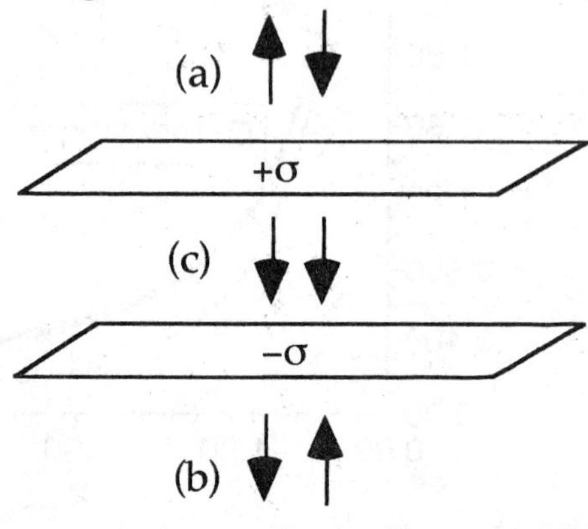

(a) $E = 0$.

(b) $E = 0$.

(c) $E = 2\frac{\sigma}{2\varepsilon_0} = \frac{\sigma}{\varepsilon_0}$, directed downward.

22-45:

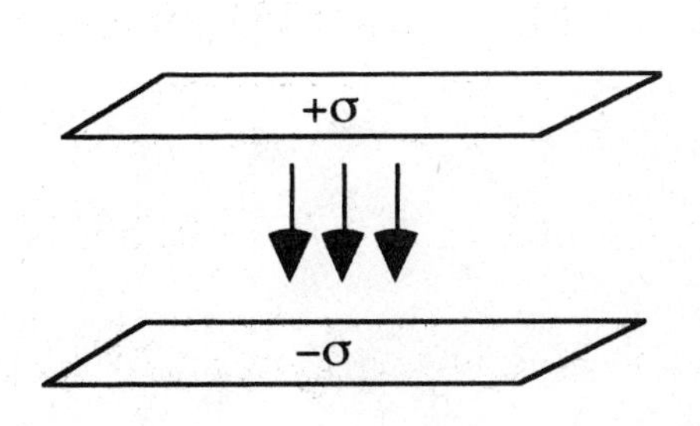

22-46:
The field appears like that of a point charge a long way from the disk and an infinte plane close to the disk's center. The field is symmetrical on right and left (not shown).

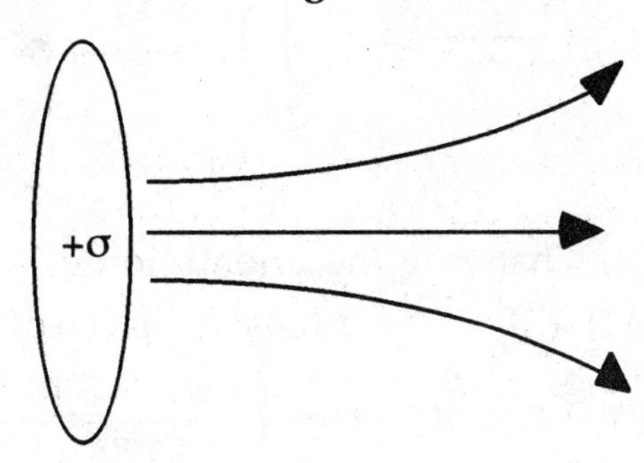

22-47: An infinite line of charge has a radial field in the plane through the wire, and constant in the plane of the wire, mirror-imaged about the wire:

Cross-section through the wire:

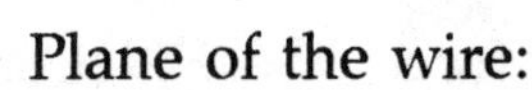

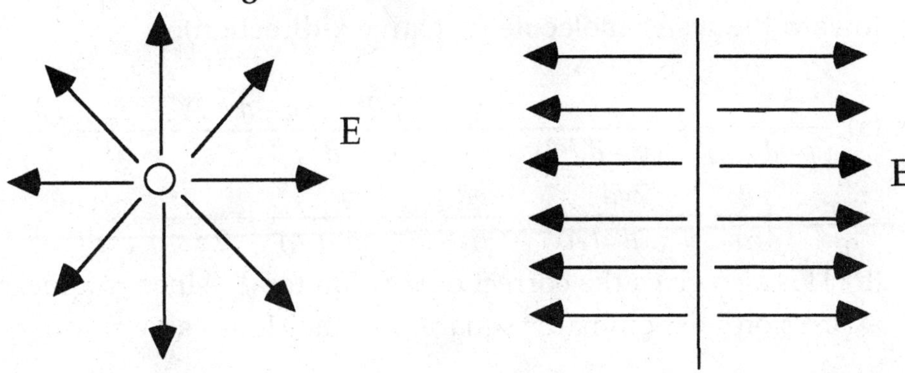

22-48: (a) Since field lines pass from positive charges and toward negative charges, we can deduce that the top charge is positive, middle is negative, and bottom is positive.
(b) The electric field is the smallest on the horizontal line through the middle charge, at two positions on either side where the field lines are least dense. Here the y-components of the field are cancelled between the positive charges and the negative charge cancels the x-component of the field from the two positive charges.

22-49: (a) $p = qd \Rightarrow d = p/q = (8.2\times10^{-12}\ \text{C}\cdot\text{m})/(3.5\times10^{-9}\ \text{C}) = 0.234\ \text{cm}.$
(b) If $\vec{E}$ is at 35°, and the torque $\tau = pE\sin\phi$, then:

$$E = \frac{\tau}{p\sin\phi} = \frac{7.0\times10^{-9}\ \text{N}\cdot\text{m}}{(8.2\times10^{-12}\ \text{C}\cdot\text{m})\sin 35°} = 1500\ \text{N/C}.$$

22-50: (a) $d = p/q = (8.9\times10^{-30}\ \text{C}\cdot\text{m})/(1.6\times10^{-19}\ \text{C}) = 5.56\times10^{-11}\ \text{m}.$
(b) $\tau_{max} = pE = (8.9\times10^{-30}\ \text{C}\cdot\text{m})(5.0\times10^{4}\ \text{N/C}) = 4.45\times10^{-25}\ \text{N}\cdot\text{m}.$

Maximum torque:

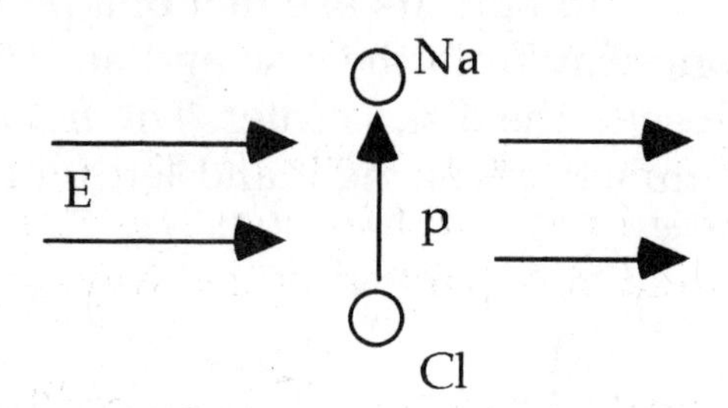

22-51: (a) Changing the orientation of a dipole from parallel to perpendicular yields:
$\Delta U = U_f - U_i = pE\cos 90° - pE\cos 0° = -(5.0\times 10^{-30}\ \mathrm{C\cdot m})(2.0\times 10^{5}\ \mathrm{N/C}) = -10^{-24}\ \mathrm{J}.$

(b) $\frac{3}{2}kT = 10^{-24}\ \mathrm{J} \Rightarrow T = \frac{2(10^{-24}\ \mathrm{J})}{3(1.38\times 10^{-23}\ \mathrm{J/K})} = 0.048\ \mathrm{K}.$

22-52: $E_{\text{dipole}}(x) = \frac{p}{2\pi\varepsilon_o x^3} \Rightarrow E_{\text{dipole}}(5.00\times 10^{-8}\ \mathrm{m}) = \frac{6.17\times 10^{-30}\ \mathrm{C\cdot m}}{2\pi\varepsilon_o(5.00\times 10^{-8}\ \mathrm{m})^3} = 888\ \mathrm{N/C}.$

The electric force $F = qE = (1.60\times 10^{-19}\ \mathrm{C})(888\ \mathrm{N/C}) = 1.42\times 10^{-16}\ \mathrm{N}$ and is toward the water molecule (negative x-direction).

22-53: (a) $\frac{1}{(y-d/2)^2} - \frac{1}{(y+d/2)^2} = \frac{(y+d/2)^2-(y-d/2)^2}{(y^2-d^2/4)^2} = \frac{2yd}{(y^2-d^2/4)^2}$

$\Rightarrow E_y = \frac{q}{4\pi\varepsilon_o}\frac{2yd}{(y^2-d^2/4)^2} = \frac{qd}{2\pi\varepsilon_o}\frac{y}{(y^2-d^2/4)^2} \approx \frac{p}{2\pi\varepsilon_o y^3}.$

(b) This also gives the correct expression for E_y since y appears in the full expression's denominator squared, so the signs carry through correctly.

22-54: (a) $F_3 = 6.00\times 10^{-4}\ \mathrm{N} = \frac{kq_1q_3}{r_{13}^2} + \left|\frac{kq_2q_3}{r_{23}^2}\right| = kq_3\left(\frac{(6.00\times 10^{-9}\ \mathrm{C})}{(0.300\ \mathrm{m})^2} + \frac{(4.00\times 10^{-9}\ \mathrm{C})}{(-0.200\ \mathrm{m})^2}\right)$

$\Rightarrow q_3 = \frac{6.00\times 10^{-4}\ \mathrm{N}}{(1500\ \mathrm{N/C})} = 40.0\ \mathrm{nC}.$

(b) The force acts on the middle charge to the left.

(c) The force equal zero if the two forces from the other charges cancel. Because of the magnitude and size of the charges, this can only occur to the left of the negative charge q_2. Then:

$F_{13} = F_{23} \Rightarrow \frac{kq_1}{(0.300-x)^2} = \frac{kq_2}{(-0.200-x)^2}$ where x is the distance from the origin.

Solving for x we find: $2x^2 + 0.48x - 0.12 = 0 \Rightarrow x = -0.393$ m. The other value of x was to the right of the origin and is not allowed.

22-55: (a) 0.100 mol NaCl $\Rightarrow$ $m_{\mathrm{Na}} = (0.100\ \mathrm{mol})(22.99\ \mathrm{g/mol}) = 2.30\ \mathrm{g}$

$\Rightarrow m_{\mathrm{Cl}} = (0.100\ \mathrm{mol})(35.45\ \mathrm{g/mol}) = 3.55\ \mathrm{g}$

Also the number of ions is $(0.100\ \mathrm{mol})N_A = 6.02\times 10^{22}$ so the charge is:

$q = (6.02 \times 10^{22})(1.60 \times 10^{-19}\ \text{C}) = 9630\ \text{C}$. The force between two such charges is:

$$F = \frac{1}{4\pi\varepsilon_0}\frac{q^2}{r^2} = \frac{1}{4\pi\varepsilon_0}\frac{(9630)^2}{(0.0200\ \text{m})^2} = 2.09 \times 10^{21}\ \text{N}.$$

(b) $a = F/m = (2.09 \times 10^{21}\ \text{N})/(3.55 \times 10^{-3}\ \text{kg}) = 5.89 \times 10^{23}\text{m/s}^2$.

(c) With such a large force between them, it does not seem reasonable to think the sodium and chlorine ions could be separated in this way.

22-56: Examining the forces: $\sum F_x = T\sin\theta - F_e = 0$ and $\sum F_y = T\cos\theta - mg = 0$.

So $\dfrac{mg\sin\theta}{\cos\theta} = F_e = \dfrac{kq^2}{d^2}$ But $\tan\theta \approx \dfrac{d}{2L} \Rightarrow d^3 = \dfrac{2kq^2L}{mg} \Rightarrow d = \left(\dfrac{q^2L}{2\pi\varepsilon_0 mg}\right)^{1/3}$.

22-57:

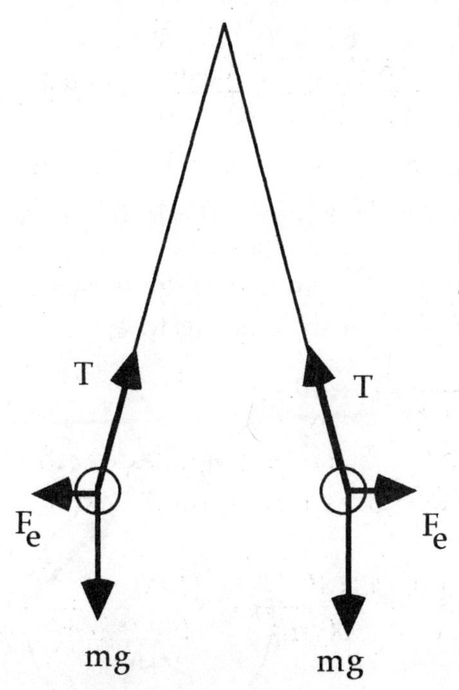

(b) Using the same force analysis as **22-56**, we find:

$q^2 = 4\pi\varepsilon_0 d^2 mg\tan\theta$ and $d = 2\sin 20°$

$\Rightarrow q = \sqrt{4\pi\varepsilon_0(2\sin 20)^2\tan 20°(0.0120\ \text{kg})(9.80\ \text{m/s}^2)}$

$\Rightarrow q = 1.49 \times 10^{-6}\ \text{C}$.

(c) Shorten the strings to 0.400 m and the new angle is:

22-58: (a) Free body diagram as in **22-57**. Each charge still feels equal and opposite electric forces.

(b) $T = mg/\cos 20° = 0.0834\ \text{N}$ so $F_e = T\sin 20° = 0.0285\ \text{N} = \dfrac{kq_1q_2}{r_1^2}$.

(Note: $r_1 = 2(0.500\ \text{m})\sin 20° = 0.342\ \text{m}$.)

(c) The charges must be of the same sign, since they are repelling.

(d) The charges on the spheres are made equal by connecting them with a wire, but we still have $F_e = mg\tan\theta = 0.0453\ \text{N} = \dfrac{1}{4\pi\varepsilon_0}\dfrac{Q^2}{r_2^2}$ where $Q = \dfrac{q_1 + q_2}{2}$.

But the separation r_2 is known: $r_2 = 2(0.500\ \text{m})\sin 30° = 0.500\ \text{m}$. Hence:

$Q = \frac{q_1 + q_2}{2} = \sqrt{4\pi\varepsilon_0 F_e r_2{}^2} = 1.12 \times 10^{-6}$ C. This equation, along with that from part (b) gives us two equations in q_1 and q_2.
$q_1 + q_2 = 2.24 \times 10^{-6}$ C and $q_1 q_2 = 3.70 \times 10^{-13}$ C^2. By elimination, substitution and after solving the resulting quadratic equation, we find:
$q_1 = 2.06 \times 10^{-6}$ C and $q_2 = 1.80 \times 10^{-7}$ C.

22-59: (a) With the mass of the book about 1.0 kg, most of which is protons and neutrons, we find: $\#\text{protons} = \frac{1}{2}(1.0 \text{ kg}) / (1.67 \times 10^{-27} \text{ kg}) = 3.0 \times 10^{26}$. Thus the charge difference present if the electron's charge was 99.999 % of the proton's is $\Delta q = (3.0 \times 10^{26})(0.00001)(1.6 \times 10^{-19} \text{ C}) = 480$ C.
(b) $F = k(\Delta q)^2 / r^2 = k(480 \text{ C})^2 / (5.0 \text{ m})^2 = 8.3 \times 10^{13}$ N - repulsive. The acceleration $a = F / m = (8.3 \times 10^{13} \text{ N}) / (1 \text{ kg}) = 8.3 \times 10^{13} \text{ m / s}^2$.
(b) Thus even the slightest charge imbalance in matter would lead to explosive repulsion!

22-60: (a) 30.0 g carbon $\Rightarrow \frac{30.0 \text{ g}}{12.0 \text{ g / mol}} = 2.5$ mol carbon $\Rightarrow 6(2.5) = 15.0$ mol electrons
$\Rightarrow q = (15.0)N_A(1.60 \times 10^{-19} \text{ C}) = 1.45 \times 10^6$ C. This much charge is placed at the earth's poles, (negative at north, positive at south) leading to a force:

$$F = \frac{1}{4\pi\varepsilon_0}\frac{q^2}{(2R_{\text{earth}})^2} = \frac{1}{4\pi\varepsilon_0}\frac{(1.45 \times 10^6 \text{ C})^2}{(1.276 \times 10^7 \text{ m})^2} = 1.15 \times 10^8 \text{ N}.$$

(b) A positive charge at the equator of the same magnitude as above will feel a force in the south-to-north direction, perpendicular to the earth's surface:

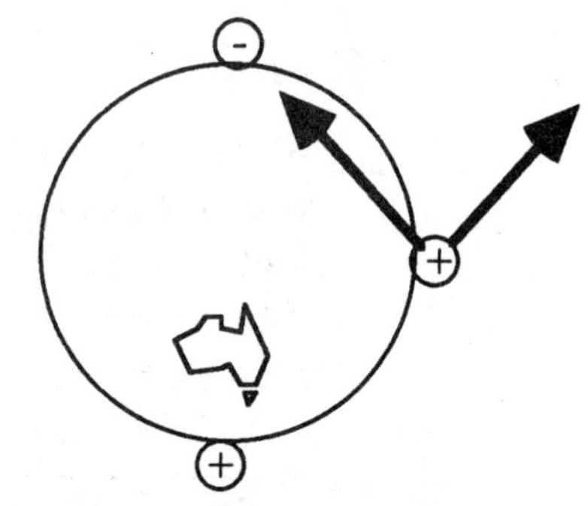

$$F = 2\frac{1}{4\pi\varepsilon_o}\frac{q^2}{(\sqrt{2}R_{\text{earth}})^2}\sin 45°$$

$$\Rightarrow F = \frac{1}{4\pi\varepsilon_o}\frac{4}{\sqrt{2}}\frac{(1.45 \times 10^6 \text{ C})^2}{(1.276 \times 10^7 \text{ m})^2} = 3.25 \times 10^8 \text{ N}.$$

22-61: (a) $F_q = \frac{1}{4\pi\varepsilon_o}\frac{qQ}{(a+x)^2} - \frac{1}{4\pi\varepsilon_o}\frac{qQ}{(a-x)^2} = \frac{1}{4\pi\varepsilon_o}\frac{qQ}{a^2}\left(\frac{1}{(1+x/a)^2} - \frac{1}{(1-x/a)^2}\right)$

$$\Rightarrow F_q \approx \frac{1}{4\pi\varepsilon_o}\frac{qQ}{a^2}(1 - 2\frac{x}{a}\ldots-(1+2\frac{x}{a}\cdots)) = \frac{1}{4\pi\varepsilon_o}\frac{qQ}{a^2}\left(-4\frac{x}{a}\right) = -\left(\frac{qQ}{\pi\varepsilon_o a^3}\right)x.$$

But this is the equation of a simple harmonic oscillator, so:

$$\omega = 2\pi f = \sqrt{\frac{qQ}{\pi\varepsilon_o a^3}} \Rightarrow f = \frac{1}{2\pi}\sqrt{\frac{qQ}{\pi\varepsilon_o a^3}} = \sqrt{\frac{kqQ}{\pi^2 a^3}}.$$

(b) If the charge was placed on the y-axis there would be no restoring force if q and Q had the same sign. It would move straight out from the origin along the y-axis, since the x-components of force would cancel.

22-62: (a) $\vec{F}_{13} = -\left|\frac{1}{4\pi\varepsilon_o}\frac{q_1 q_3}{r_{13}^2}\right|\sin\theta\,\hat{i} - \left|\frac{1}{4\pi\varepsilon_o}\frac{q_1 q_3}{r_{13}^2}\right|\cos\theta\,\hat{j}$

$$\Rightarrow \vec{F}_{13} = -\frac{1}{4\pi\varepsilon_o}\frac{(3.00\text{ nC})(6.00\text{ nC})}{((9.00+25.0)\times 10^{-4}\text{ m})}\frac{3}{\sqrt{34}}\hat{i} - \frac{1}{4\pi\varepsilon_o}\frac{(3.00\text{ nC})(6.00\text{ nC})}{((9.00+25.0)\times 10^{-4}\text{ m})}\frac{5}{\sqrt{34}}\hat{j}$$

$\Rightarrow \vec{F}_{13} = -(2.45\times 10^{-5}\text{ N})\,\hat{i} - (4.08\times 10^{-5}\text{ N})\,\hat{j}.$

Similarly for the force from the other charge:

$$\vec{F}_{23} = +\frac{1}{4\pi\varepsilon_o}\frac{q_2 q_3}{r_{23}^2}\hat{i} = \frac{1}{4\pi\varepsilon_o}\frac{(2.00\text{ nC})(6.00\text{ nC})}{(0.0300\text{ m})^2} = (1.20\times 10^{-4}\text{N})\,\hat{i}$$

Therefore the two force components are:

$$F_x = 9.55\times 10^{-5}\text{ N and } F_y = -4.08\times 10^{-5}\text{ N}$$

(b) Thus, $F = \sqrt{F_x^2 + F_y^2} = \sqrt{(9.55\times 10^{-5}\text{ N})^2 + (4.08\times 10^{-5}\text{ N})^2} = 1.04\times 10^{-4}\text{ N}$, and the angle is $\theta = \arctan(F_y / F_x) = 23.0°$, below the axis.

22-63: (a) $F = +\frac{1}{4\pi\varepsilon_o}\frac{q(3q)}{(L/\sqrt{2})^2} = \frac{1}{4\pi\varepsilon_o}\frac{6q^2}{L^2}$, toward the lower left charge.

The other two forces are equal and opposite.

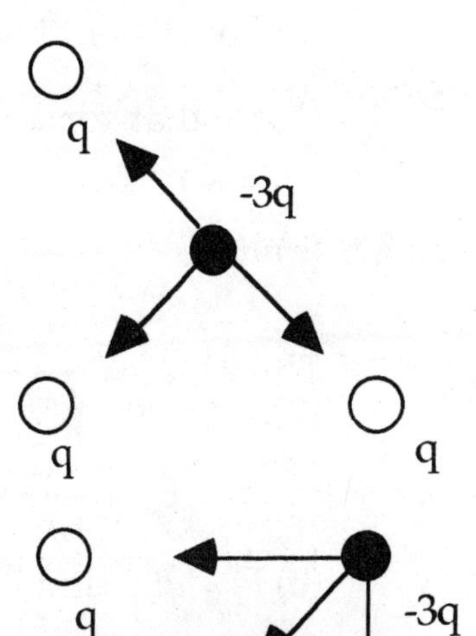

(b) The upper left charge and lower right charge have equal magnitude forces at right angles to each other, resulting in a total force of twice the force of one, directed toward the lower left charge. So, all the forces sum to:

$$F = \frac{\sqrt{2}}{4\pi\varepsilon_o}\left(\frac{q(3q)}{L^2} + \frac{q(3q)}{(\sqrt{2}L)^2}\right) = \frac{q^2}{4\pi\varepsilon_o L^2}\left(3\sqrt{2} + \frac{3}{2}\right)\text{ N}.$$

22-64: (a) $E(p) = \frac{1}{4\pi\varepsilon_o}\left(\frac{q}{(y-a)^2} + \frac{q}{(y+a)^2} - \frac{2q}{y^2}\right).$

(b) $E(p) = \frac{1}{4\pi\varepsilon_o}\frac{q}{y^2}\left((1-a/y)^{-2} + (1+a/y)^{-2} - 2\right)$. Using the binomial expansion:

$$\Rightarrow E(p) \approx \frac{1}{4\pi\varepsilon_o}\frac{q}{y^2}\left(1 + \frac{2a}{y} + \frac{3a^2}{y^2} + \ldots + 1 - \frac{2a}{y} + \frac{3a^2}{y^2} + \ldots - 2\right) = \frac{1}{4\pi\varepsilon_o}\frac{6qa^2}{y^4}.$$

Note that a point charge drops off like $\frac{1}{y^2}$ and a dipole like $\frac{1}{y^3}$.

22-65: (a) The field is all in the x-direction (the rest cancels). From the +q charges:

$$E = \frac{1}{4\pi\varepsilon_0}\frac{q}{a^2+x^2} \Rightarrow E_x = \frac{1}{4\pi\varepsilon_0}\frac{q}{a^2+x^2}\frac{x}{\sqrt{a^2+x^2}} = \frac{1}{4\pi\varepsilon_0}\frac{qx}{(a^2+x^2)^{3/2}}.$$

(Each +q contributes this.) From the -2q:

$$E = -\frac{1}{4\pi\varepsilon_0}\frac{2q}{x^2} \Rightarrow E_{\text{total}} = \left(\frac{qx}{(a^2+x^2)^{3/2}} - \frac{2q}{x^2}\right) = \frac{1}{4\pi\varepsilon_0}\frac{2q}{x^2}\left((a^2/x^2+1)^{-3/2}-1\right).$$

(b) $E_{\text{total}} \approx \frac{1}{4\pi\varepsilon_0}\frac{2q}{x^2}\left(1-\frac{3a^2}{2x^2}+\ldots-1\right) = \frac{1}{4\pi\varepsilon_0}\frac{3qa^2}{x^4}$, for $x >> a$.

Note that a point charge drops off like $\frac{1}{x^2}$ and a dipole like $\frac{1}{x^3}$.

22-66: (a) On the x-axis:

$$dE_x = \frac{1}{4\pi\varepsilon_0}\frac{dq}{(a+r)^2} \Rightarrow E_x = \frac{1}{4\pi\varepsilon_0}\int_0^a \frac{Qdx}{a(a+r-x)^2} = \frac{1}{4\pi\varepsilon_0}\frac{Q}{a}\left(\frac{1}{r}-\frac{1}{a+r}\right). \text{ And } E_y = 0.$$

(b) If $a+r=x$, then $E = \frac{1}{4\pi\varepsilon_0}\frac{Q}{a}\left(\frac{1}{x-a}-\frac{1}{x}\right) \Rightarrow \vec{F} = q\vec{E} = \frac{1}{4\pi\varepsilon_0}\frac{qQ}{a}\left(\frac{1}{x-a}-\frac{1}{x}\right)\hat{i}.$

(c) For $x >> a$, $F = \frac{kqQ}{a}\left((1-a/x)^{-1}-1\right) = \frac{kqQ}{ax}(1+a/x+\ldots-1) \approx \frac{kqQ}{x^2} \approx \frac{1}{4\pi\varepsilon_0}\frac{qQ}{r^2}.$

(Note that for $x >> a$, $r = x-a \approx x$.)

22-67: (a) $dE = \frac{k\,dq}{(x^2+y^2)} = \frac{kQ\,dy}{a(x^2+y^2)}$ with $dE_x = \frac{kQx\,dy}{a(x^2+y^2)^{3/2}}$ and $dE_y = \frac{kQy\,dy}{a(x^2+y^2)^{3/2}}$

Thus: $E_x = \frac{1}{4\pi\varepsilon_0}\frac{Qx}{a}\int_0^a \frac{dy}{(x^2+y^2)^{3/2}} = \frac{1}{4\pi\varepsilon_0}\frac{Qx}{a}\frac{1}{(x^2+a^2)^{1/2}}$

$$E_y = \frac{1}{4\pi\varepsilon_0}\frac{Q}{a}\int_0^a \frac{ydy}{(x^2+y^2)^{3/2}} = \frac{1}{4\pi\varepsilon_0}\frac{Q}{a}\left(\frac{1}{x}-\frac{1}{(x^2+a^2)^{1/2}}\right)$$

(b) $F_x = qE_x$ and $F_y = qE_y$ where E_x and E_y are given in (a).

(c) For $x >> a$, $F_y = \frac{1}{4\pi\varepsilon_0}\frac{qQ}{ax}\left(1-(1+a^2/x^2)^{-1/2}\right) \approx \frac{1}{4\pi\varepsilon_0}\frac{qQ}{ax}\frac{a^2}{2x^2} = \frac{1}{4\pi\varepsilon_0}\frac{qQa}{2x^3}.$

22-68: (a) From Eq. (22-9), $\vec{E} = \frac{1}{4\pi\varepsilon_0}\frac{Q}{x\sqrt{x^2+a^2}}\hat{i}$

$$\Rightarrow \vec{E} = \frac{1}{4\pi\varepsilon_0}\frac{(5.00\times10^{-9}\text{ C})}{(4.00\times10^{-3}\text{ m})\sqrt{(4.00\times10^{-3}\text{ m})^2+(0.0200\text{ m})^2}}\hat{i} = (5.52\times10^5\text{ N/C})\hat{i}.$$

(b) The electric field is less than that at the same distance from an infinite line of charge ($E_\infty = \frac{1}{4\pi\varepsilon_0}\frac{2\lambda}{x} = \frac{1}{4\pi\varepsilon_0}\frac{2Q}{x2a} = 5.63\times10^5$ N/C). This is because in the approximation, the terms left off were positive.

(c) For a 1% difference, we need the next highest term in the expansion that was left off to be less than 0.01:

$$\frac{x^2}{2a^2} < 0.01 \Rightarrow x < a\sqrt{2(0.01)} = 0.0200\sqrt{2(0.01)} \Rightarrow x < 0.0028\text{ m}.$$

22-69: (a) From Eq. (22-9), $\vec{E} = \frac{1}{4\pi\varepsilon_o} \frac{Q}{x\sqrt{x^2 + a^2}} \hat{i}$

$$\Rightarrow \vec{E} = \frac{1}{4\pi\varepsilon_o} \frac{(5.00 \times 10^{-9}\ \text{C})}{(0.100\ \text{m})\sqrt{(0.100\ \text{m})^2 + (0.0200\ \text{m})^2}} = (4410\ \text{N/C})\ \hat{i}.$$

(b) The electric field is less than that at the same distance from a point charge (4500 N / C). This is because in the approximation, terms were added.)

(c) For a 1% difference, we need the next highest term in the expansion that was left off to be less than 0.01:

$$\frac{a^2}{2x^2} \approx 0.01 \Rightarrow x \approx a\sqrt{1/(2(0.01))} = 0.0200\sqrt{1/0.02} \Rightarrow x \approx 0.141\ \text{m}.$$

22-70: (a) On the axis,

$$E = \frac{\sigma}{2\varepsilon_o}\left[1 - \left(\frac{R^2}{x^2} + 1\right)^{-1/2}\right] = \frac{5.00\ \text{nC/m}^2}{2\varepsilon_o}\left[1 - \left(\frac{(0.020\ \text{m})^2}{(0.0020\ \text{m})^2} + 1\right)^{-1/2}\right]$$

$$\Rightarrow E = 254\ \text{N/C}.$$

(b) The electric field is less than that of an infinte sheet $E_\infty = \frac{\sigma}{2\varepsilon_o} = 282\ \text{N/C}$.

(c) Finite disk electric field can be expanded using the binomial theorem since the expansion terms are small: $\Rightarrow E \approx \frac{\sigma}{2\varepsilon_o}\left[1 - \frac{x}{R} + \frac{x^3}{2R^3} - \ldots\right]$ So the difference between the infinite sheet and finite disk goes like $\frac{x}{R}$. Thus:

$$\Delta E(x = 0.20\ \text{cm}) \approx 0.2/2 = 0.1 \equiv 10\% \text{ and } \Delta E(x = 0.40\ \text{cm}) \approx 0.4/2 = 0.2 \equiv 20\%.$$

22-71: (a) As in **22-70**: $E = \frac{\sigma}{2\varepsilon_o}\left[1 - \left(\frac{R^2}{x^2} + 1\right)^{-1/2}\right] = \frac{5.00\ \text{nC}}{2\varepsilon_o}\left[1 - \left(\frac{(0.0200\ \text{m})^2}{(0.200\ \text{m})^2} + 1\right)^{-1/2}\right]$

$$\Rightarrow E = 1.40\ \text{N/C}$$

(b) $x >> R$, $E = \frac{\sigma}{2\varepsilon_o}\left[1 - (1 - R^2/2x^2 + 3R^4/8x^4 - \ldots)\right] \approx \frac{\sigma}{2\varepsilon_o}\frac{R^2}{2x^2} = \frac{\sigma\pi R^2}{4\pi\varepsilon_o x^2} = \frac{Q}{4\pi\varepsilon_o x^2}$.

(c) The electric field of (a) is less than that of the point charge since the correction term that was omitted was negative.

(d) The difference between the point charge and finite disk goes like $\frac{3R^2}{4x^2}$. So:

$$\Delta E(x = 20\ \text{cm}) \approx \frac{3(0.02)^2}{4(0.2)^2} = 0.0075 \equiv 0.75\%\ ;\ \Delta E(x = 10\ \text{cm}) \approx \frac{3(0.02)^2}{4(0.1)^2} = 0.03 \equiv 3\%.$$

22-72: (a) $f(x) = f(-x)$: $\int_{-a}^{a} f(x)dx = \int_{-a}^{0} f(x)dx + \int_{0}^{a} f(x)dx = \int_{0}^{-a} f(-x)d(-x) + \int_{0}^{a} f(x)dx$

Now replace $-x$ with y: $\Rightarrow \int_{-a}^{a} f(x)dx = \int_{0}^{a} f(y)d(y) + \int_{0}^{a} f(x)dx = 2\int_{0}^{a} f(x)dx$.

(b) $g(x) = -g(-x)$: $\int_{-a}^{a} g(x)dx = \int_{-a}^{0} g(x)dx + \int_{0}^{a} g(x)dx = -\int_{0}^{-a} -g(-x)(-d(-x)) + \int_{0}^{a} g(x)dx$

Now replace $-x$ with y: $\Rightarrow \int_{-a}^{a} g(x)dx = -\int_{0}^{a} g(y)d(y) + \int_{0}^{a} g(x)dx = 0$.

(c) The integrand in E_y for example 22-11 is odd, so $E_y = 0$.

22-73: The forces must balance out so:

$$F_e = 10mg \Rightarrow qE = 10mg \Rightarrow q = \frac{10mg}{E} = \frac{10(3.0\times10^{-12}\text{ kg})(9.8\text{ m/s}^2)}{2.00\times10^{-5}\text{ N/C}} = 1.47\times10^{-15}\text{ C}$$

22-74: (a) The y-components of the electric field cancel, and the x-components from both charges, as given in **Pr. (22-67)** is:

$$E_x = \frac{1}{4\pi\varepsilon_0}\frac{-2Q}{a}\left(\frac{1}{y} - \frac{1}{(y^2+a^2)^{1/2}}\right) \Rightarrow \vec{F} = \frac{1}{4\pi\varepsilon_0}\frac{-2Qq}{a}\left(\frac{1}{y} - \frac{1}{(y^2+a^2)^{1/2}}\right)\hat{i}.$$

If $y >> a$, $\vec{F} \approx \frac{1}{4\pi\varepsilon_0}\frac{-2Qq}{ay}\left(1-(1-a^2/2y^2+\ldots)\right)\hat{i} = -\frac{1}{4\pi\varepsilon_0}\frac{Qqa}{y^3}$.

(b) If the point charge is now on the x-axis, the two charged parts of the rods provide different forces, though still along the x-axis (see **Pr. (22-66)**).

$$\vec{F}_+ = q\vec{E}_+ = \frac{1}{4\pi\varepsilon_0}\frac{Qq}{a}\left(\frac{1}{x-a} - \frac{1}{x}\right)\hat{i} \text{ and } \vec{F}_- = q\vec{E}_- = -\frac{1}{4\pi\varepsilon_0}\frac{Qq}{a}\left(\frac{1}{x} - \frac{1}{x+a}\right)\hat{i}$$

So, $\vec{F} = \vec{F}_+ + \vec{F}_- = \frac{1}{4\pi\varepsilon_0}\frac{Qq}{a}\left(\frac{1}{x-a} - \frac{2}{x} + \frac{1}{x+a}\right)\hat{i}$

For $x >> a$, $\vec{F} \approx \frac{1}{4\pi\varepsilon_0}\frac{Qq}{ax}\left((1+\frac{a}{x}+\frac{a^2}{x^2}+\ldots)-2+(1-\frac{a}{x}+\frac{a^2}{x^2}-\ldots)\right)\hat{i} = \frac{1}{4\pi\varepsilon_0}\frac{2Qqa}{x^3}\hat{i}$.

22-75: (a) $F = qE = (-1.60\times10^{-19}\text{ C})(400\text{ N/C}) = -6.40\times10^{-17}\text{ N}$.

And $a_y = F/m = (-6.4\times10^{-17}\text{ N})/(9.11\times10^{-31}\text{ kg}) = -7.03\times10^{13}\text{ m/s}^2$. Then,

$v_y^2 = v_{yo}^2 + 2a\Delta y \Rightarrow 0 = (3.00\times10^6\sin 30°)^2 - 2(7.03\times10^{13})\Delta y \Rightarrow \Delta y = 1.07\times10^{-8}$ m.

(b) Time to top,

$v_y = v_{yo} + at \Rightarrow t = -v_{yo}/a = (1.50\times10^6\text{ m/s})/(7.03\times10^{13}\text{ m/s}^2) = 2.13\times10^{-8}$s.

$t_{total} = 2t = 4.27\times10^{-8}\text{ s} \Rightarrow x = v_{xo}t_{total} = ((3\times10^6\cos30°)\text{ m/s})(4.27\times10^{-8}\text{ s}) = 0.11$ m

22-76: First, $E = \frac{\sigma}{2\varepsilon_0} = \frac{2.50\times10^{-5}\text{ C/m}^2}{2\varepsilon_0} = 1.41\times10^6\text{ N/m}$.

Then, we must sum the forces in x and y directions, yielding two equations:

$qE = T\sin\theta$ and $mg = T\cos\theta$. So: $\tan\theta = \frac{qE}{mg}$

$$\Rightarrow \theta = \arctan\left(\frac{(3.00\times10^{-10}\text{ C})(1.41\times10^6\text{ N/C})}{(6.00\times10^{-4}\text{ kg})(9.80\text{ m/s}^2)}\right) = 4.12°.$$

22-77: (a) $E = 50\ \mathrm{N/C} = \left|\frac{1}{4\pi\varepsilon_o}\frac{q_1}{r_1^2}\right| + \left|\frac{1}{4\pi\varepsilon_o}\frac{q_2}{r_2^2}\right| = \frac{1}{4\pi\varepsilon_o}\left(\left|\frac{q_1}{r_1^2}\right| + \left|\frac{q_2}{r_2^2}\right|\right) \Rightarrow q_2 = r_2^2\left(4\pi\varepsilon_o E - \left|\frac{q_1}{r_1^2}\right|\right)$

$$\Rightarrow q_2 = (0.600\ \mathrm{m})^2\left(4\pi\varepsilon_o 50.0\ \mathrm{N/C} - \frac{(4.00\times10^{-6}\ \mathrm{C})}{(1.20\ \mathrm{m})^2}\right) = 1.00\times10^{-9}\ \mathrm{C}.$$

(b) $E = -50\ \mathrm{N/C} = \left|\frac{1}{4\pi\varepsilon_o}\frac{q_1}{r_1^2}\right| + \left|\frac{1}{4\pi\varepsilon_o}\frac{q_2}{r_2^2}\right| = \frac{1}{4\pi\varepsilon_o}\left(\left|\frac{q_1}{r_1^2}\right| + \left|\frac{q_2}{r_2^2}\right|\right) \Rightarrow q_2 = r_2^2\left(\frac{-50}{k} - \left|\frac{q_1}{r_1^2}\right|\right)$

$$\Rightarrow q_2 = (0.600\ \mathrm{m})^2\left(4\pi\varepsilon_o(-50.0) - \frac{(4.00\times10^{-6}\ \mathrm{C})}{(1.20\ \mathrm{m})^2}\right) = -3.00\times10^{-9}\ \mathrm{C}.$$

22-78: First, the mass of the drop:

$$m = \rho V = (1000\ \mathrm{kg/m^3})\left(\frac{4\pi(15.0\times10^{-6}\ \mathrm{m})^3}{3}\right) = 1.41\times10^{-11}\ \mathrm{kg}.$$

Next, the time of flight: $t = D/v = 0.02/20 = 0.00100\ \mathrm{s}$ and the acceleration:

$$d = \frac{1}{2}at^2 \Rightarrow a = \frac{2d}{t^2} = \frac{2(3.00\times10^{-4}\ \mathrm{m})}{(0.001\ \mathrm{s})^2} = 600\ \mathrm{m/s^2}.\ \text{So:}$$

$$a = F/m = qE/m \Rightarrow q = ma/E = \frac{(1.41\times10^{-11}\ \mathrm{kg})(600\ \mathrm{m/s^2})}{8.00\times10^4\ \mathrm{N/C}} = 1.06\times10^{-13}\ \mathrm{C}.$$

22-79: $E = 18.0\ \mathrm{N/C} = \frac{k(16.0\ \mathrm{nC})}{(3.00\ \mathrm{m})^2} + \frac{k(12.0\ \mathrm{nC})}{(2.00\ \mathrm{m})^2} + \frac{kq}{(6.00\ \mathrm{m})^2}$

$$\Rightarrow q = 36.0\ \mathrm{m^2}\left(\frac{18}{k} - \frac{1.60\times10^{-8}\ \mathrm{C}}{81.0\ \mathrm{m^2}} - \frac{1.20\times10^{-8}\ \mathrm{C}}{4.00\ \mathrm{m^2}}\right) = -4.30\times10^{-8}\ \mathrm{C} = -43.0\ \mathrm{nC}.$$

22-80: The electric field in the x-direction cancels from the left and right halves of the semi-circle. The remaining y-component points in the negative y-direction. The charge per unit length of the semi-circle is:

$$\lambda = \frac{Q}{\pi a}\ \text{and}\ dE = \frac{k\lambda\, dl}{a^2} = \frac{k\lambda\, d\theta}{a}\ \text{but}\ dE_y = dE\sin\theta = \frac{k\lambda \sin\theta\, d\theta}{a}.$$

$$\text{So, } E_y = \frac{2k\lambda}{a}\int_0^{\pi/2}\sin\theta\, d\theta = \frac{2k\lambda}{a}\left[-\cos\theta\right]_0^{\pi/2} = \frac{2k\lambda}{a} = \frac{2kQ}{\pi a^2},\ \text{downwards.}$$

22-81: $E_x = E_y$ and the magnitude of the field in the y-direction must be half of that obtained **22-80,** so $E_x = E_y = \frac{1}{4\pi\varepsilon_o}\frac{\lambda}{a}$. But the total field is the vector sum of these components:

$$\therefore E = \sqrt{E_x^2 + E_y^2} = \sqrt{2}\frac{1}{4\pi\varepsilon_o}\frac{\lambda}{a} = 2\sqrt{2}\frac{1}{4\pi\varepsilon_o}\frac{Q}{\pi a^2}$$

22-82: (a) $E_x = E_y$, and $E_x = 2E_{\text{length of wire } a,\ \text{charge } Q} = 2\frac{1}{4\pi\varepsilon_o}\frac{Q}{x\sqrt{x^2 + a^2}}$, where $x = \frac{a}{2}$.

$$\Rightarrow E_x = \frac{Q}{\pi\varepsilon_o a^2\sqrt{5/4}} = \frac{2Q}{\pi\varepsilon_o a^2\sqrt{5}}, \text{ in } +\hat{i},\ E_y = \frac{2Q}{\pi\varepsilon_o a^2\sqrt{5}}, \text{ in } -\hat{j}.$$

(b) If all edges of the square had equal charge, the electric fields would cancel by symmetry at the center of the square.

22-83: (a) $E(P) = -\frac{|\sigma_1|}{2\varepsilon_o} - \frac{|\sigma_2|}{2\varepsilon_o} + \frac{|\sigma_3|}{2\varepsilon_o} = -\frac{0.0200\ \text{C/m}^2}{2\varepsilon_o} - \frac{0.0100\ \text{C/m}^2}{2\varepsilon_o} + \frac{0.0200\ \text{C/m}^2}{2\varepsilon_o}$

$$\Rightarrow E(P) = \frac{0.0100\ \text{C/m}^2}{2\varepsilon_o} = 5.65\times10^8\ \text{N/C, in the } -x\text{-direction.}$$

(b) $E(R) = +\frac{|\sigma_1|}{2\varepsilon_o} - \frac{|\sigma_2|}{2\varepsilon_o} + \frac{|\sigma_3|}{2\varepsilon_o} = +\frac{0.0200\ \text{C/m}^2}{2\varepsilon_o} - \frac{0.0100\ \text{C/m}^2}{2\varepsilon_o} + \frac{0.0200\ \text{C/m}^2}{2\varepsilon_o}$

$$\Rightarrow E(R) = \frac{0.0300\ \text{C/m}^2}{2\varepsilon_o} = 1.69\times10^9\ \text{N/C, in the } +x\text{-direction.}$$

(c) $E(S) = +\frac{|\sigma_1|}{2\varepsilon_o} + \frac{|\sigma_2|}{2\varepsilon_o} + \frac{|\sigma_3|}{2\varepsilon_o} = +\frac{0.0200\ \text{C/m}^2}{2\varepsilon_o} + \frac{0.0100\ \text{C/m}^2}{2\varepsilon_o} + \frac{0.0200\ \text{C/m}^2}{2\varepsilon_o}$

$$\Rightarrow E(S) = \frac{0.0500\ \text{C/m}^2}{2\varepsilon_o} = 2.82\times10^9\ \text{N/C, in the } +x\text{-direction.}$$

(d) $E(T) = +\frac{|\sigma_1|}{2\varepsilon_o} + \frac{|\sigma_2|}{2\varepsilon_o} - \frac{|\sigma_3|}{2\varepsilon_o} = +\frac{0.0200\ \text{C/m}^2}{2\varepsilon_o} + \frac{0.0100\ \text{C/m}^2}{2\varepsilon_o} - \frac{0.0200\ \text{C/m}^2}{2\varepsilon_o}$

$$\Rightarrow E(S) = \frac{0.0100\ \text{C/m}^2}{2\varepsilon_o} = 5.65\times10^8\ \text{N/C, in the } +x\text{-direction.}$$

22-84: $\frac{F_{\text{on I}}}{A} = \frac{qE_{\text{at I}}}{A} = \sigma_1\left(\frac{-|\sigma_2| + |\sigma_3|}{2\varepsilon_o}\right) = \frac{2.00\times10^{-4}\ \text{C/m}^2}{2\varepsilon_o} = +1.13\times10^7\ \text{N/m}.$

$$\frac{F_{\text{on II}}}{A} = \frac{qE_{\text{at II}}}{A} = \sigma_2\left(\frac{+|\sigma_1| + |\sigma_3|}{2\varepsilon_o}\right) = \frac{4.00\times10^{-4}\ \text{C/m}^2}{2\varepsilon_o} = +2.26\times10^7\ \text{N/m}$$

$$\frac{F_{\text{on III}}}{A} = \frac{qE_{\text{at III}}}{A} = \sigma_3\left(\frac{+|\sigma_1| + |\sigma_2|}{2\varepsilon_o}\right) = \frac{-6.00\times10^{-4}\ \text{C/m}^2}{2\varepsilon_o} = -3.39\times10^7\ \text{N/m}$$

(Note that "+" means toward the right, and "-" is toward the left.)

22-85: By inspection the fields in the different regions are as shown below:

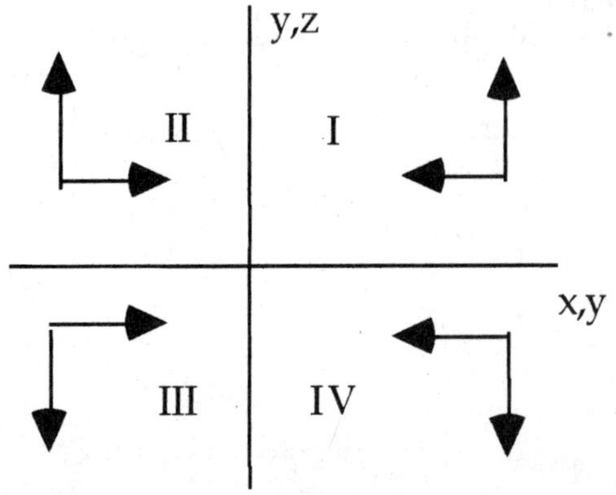

y,z
II I
III IV
x,y

$$E_I = \left(\frac{\sqrt{2}\sigma}{2\varepsilon_o}\right)(-\hat{i} + \hat{k}),\quad E_{II} = \left(\frac{\sqrt{2}\sigma}{2\varepsilon_o}\right)(+\hat{i} + \hat{k})$$

$$E_{III} = \left(\frac{\sqrt{2}\sigma}{2\varepsilon_o}\right)(+\hat{i} - \hat{k}),\quad E_{IV} = \left(\frac{\sqrt{2}\sigma}{2\varepsilon_o}\right)(-\hat{i} - \hat{k})$$

$$\therefore E = \left(\frac{\sqrt{2}\sigma}{2\varepsilon_o}\right)\left(-\frac{|x|}{x}\hat{i} + \frac{|z|}{z}\hat{k}\right).$$

22-86: (a) $Q = A\sigma = \pi(R_2^{\ 2} - R_1^{\ 2})\sigma$

(b) Recall the electric field of a disk, Eq. (22-11): $E = \dfrac{\sigma}{2\varepsilon_o}\left[1 - 1/\sqrt{(R/x)^2 + 1}\right]$.

So, $E(x) = \dfrac{\sigma}{2\varepsilon_o}\left(\left[1 - 1/\sqrt{(R_2/x)^2 + 1}\right] - \left[1 - 1/\sqrt{(R_1/x)^2 + 1}\right]\right)\dfrac{|x|}{x}\hat{i}$

$\Rightarrow E(x) = \dfrac{\sigma}{2\varepsilon_o}\left(1/\sqrt{(R_2/x)^2 + 1} - 1/\sqrt{(R_1/x)^2 + 1}\right)\dfrac{|x|}{x}\hat{i}$

(c) Note that $1/\sqrt{(R_1/x)^2 + 1} = \dfrac{x}{R_1}(1 + (x/R_1)^2)^{-1/2} \approx \dfrac{x}{R_1}(1 - \dfrac{(x/R_1)^2}{2} + \ldots)$

$\Rightarrow E(x) = \dfrac{\sigma}{2\varepsilon_o}\left(\dfrac{x}{R_1} - \dfrac{x}{R_2}\right)\dfrac{x}{|x|}\hat{i} = \dfrac{\sigma}{2\varepsilon_o}\left(\dfrac{1}{R_1} - \dfrac{1}{R_2}\right)\dfrac{x^2}{|x|}\hat{i}$, and sufficiently close means that $(x/R_1)^2 << 1$.

(d) $F = qE(x) = -\dfrac{q\sigma}{2\varepsilon_o}\left(\dfrac{1}{R_1} - \dfrac{1}{R_2}\right)x = m\ddot{x} \Rightarrow f = \dfrac{\omega}{2\pi} = \dfrac{1}{2\pi}\sqrt{\dfrac{q\sigma}{2\varepsilon_o m}\left(\dfrac{1}{R_1} - \dfrac{1}{R_2}\right)}$

22-87: (a) The four possible force diagrams are:

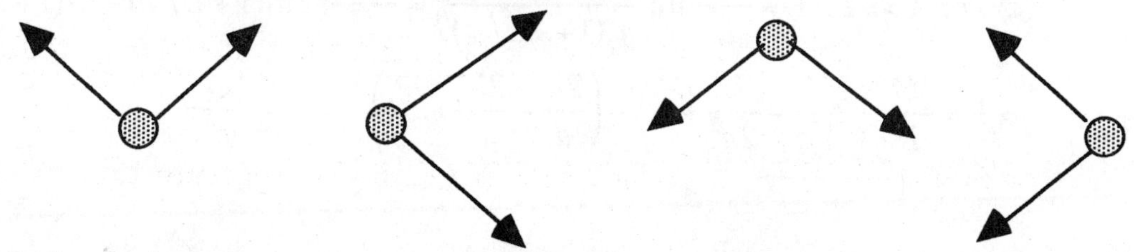

Only the last picture can result in an electric field in the -x-direction.
(b) $q_1 = -3.00\ \mu C$, $q_3 = +4.00\ \mu C$, and $q_2 > 0$.

(c) $E_y = 0 = \dfrac{1}{4\pi\varepsilon_o}\dfrac{q_1}{(0.0400\ \text{m})^2}\sin\theta_1 - \dfrac{1}{4\pi\varepsilon_o}\dfrac{q_2}{(0.0300\ \text{m})^2}\sin\theta_2$

$\Rightarrow q_2 = \dfrac{9}{16}q_1\dfrac{\sin\theta_1}{\sin\theta_2} = \dfrac{9}{16}q_1\dfrac{3/5}{4/5} = \dfrac{27}{64}q_1 = 1.27\ \mu C.$

(d) $F_3 = q_3E_x = q_3\dfrac{1}{4\pi\varepsilon_o}\left(\dfrac{q_1}{0.0016}\dfrac{4}{5} + \dfrac{q_2}{0.0009}\dfrac{3}{5}\right) = 84.3\ \text{N}$

22-88: (a) The four possible force diagrams are:

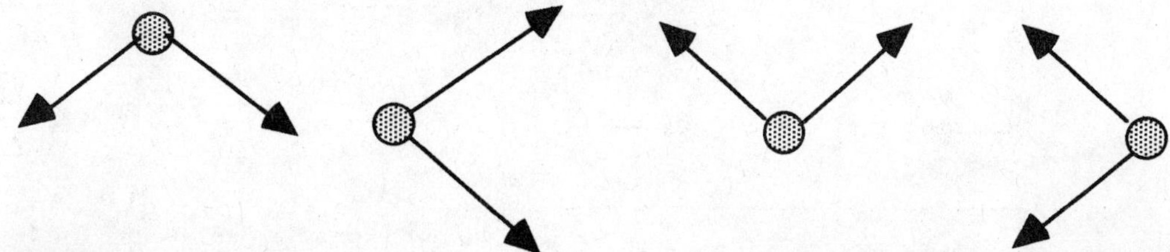

The first diagram is the only one in which the electric field must point in the negative y-direction.
(b) $q_1 = -3.00\ \mu C$, and $q_2 < 0$.

(c) $E_x = 0 = \frac{kq_1}{(0.050\text{ m})^2}\frac{5}{13} - \frac{kq_2}{(0.120\text{ m})^2}\frac{12}{13} \Rightarrow \frac{kq_2}{(0.120\text{ m})^2} = \frac{kq_1}{(0.050\text{ m})^2}\frac{5}{12}$

$E = E_y = \frac{kq_1}{(0.050\text{ m})^2}\frac{12}{13} + \frac{kq_2}{(0.120\text{ m})^2}\frac{5}{13} = \frac{kq_1}{(0.05\text{ m})^2}\left(\frac{12}{13} + \left(\frac{5}{12}\right)\left(\frac{5}{13}\right)\right)$

$\Rightarrow E = E_y = 1.17 \times 10^7 \text{ N/C}.$

22-89: (a) For a rod in general of length L, $E = \frac{kQ}{L}\left(\frac{1}{r} - \frac{1}{L+r}\right)$ and here $r = x + \frac{a}{2}$.

So, $E_{\text{left rod}} = \frac{kQ}{L}\left(\frac{1}{x+a/2} - \frac{1}{L+x+a/2}\right) = \frac{2kQ}{L}\left(\frac{1}{2x+a} - \frac{1}{2L+2x+a}\right).$

(b) $dF = dq\, E \Rightarrow F = \int E\, dq = \int_{a/2}^{L+a/2} \frac{EQ}{L}\, dx = \frac{2kQ^2}{L^2}\int_{a/2}^{L+a/2}\left(\frac{1}{x+a/2} - \frac{1}{L+x+a/2}\right)dx$

$\Rightarrow F = \frac{2kQ^2}{L^2}\frac{1}{2}\left(\left[\ln(a+2x)\right]_{a/2}^{L+a/2} - \left[\ln(2L+2x+a)\right]_{a/2}^{L+a/2}\right)$

$\Rightarrow F = \frac{kQ^2}{L^2}\ln\left(\left(\frac{a+2L+a}{2a}\right)\left(\frac{2L+2a}{4L+2a}\right)\right) = \frac{kQ^2}{L^2}\ln\left(\frac{(a+L)^2}{a(a+2L)}\right).$

(c) For $a >> L$: $F = \frac{kQ^2}{L^2}\ln\left(\frac{a^2(1+L/a)^2}{a^2(1+2L/a)}\right) = \frac{kQ^2}{L^2}(2\ln(1+L/a) - \ln(1+2L/a))$

$\Rightarrow F \approx \frac{kQ^2}{L^2}\left(2\left(\frac{L}{a} - \frac{L^2}{2a^2} + \ldots\right) - \left(\frac{2L}{a} - \frac{2L^2}{a^2} + \ldots\right)\right) \Rightarrow F \approx \frac{kQ^2}{a^2}.$

Chapter 23: Gauss's Law

23-1: (a) $\Phi = \vec{E}\cdot\vec{A} = (12.0\ \text{N/C})(0.500\ \text{m}^2)\cos 70° = 2.05\ \text{Nm}^2/\text{C}.$
(b) As long as the sheet is flat, its shape does not matter.
(c) The maximum flux occurs at an angle $\phi = 0°$ between the sheet and field.
(d) The minimum flux occurs at an angle $\phi = 90°$ between the sheet and field.

23-2: $\Phi = \vec{E}\cdot\vec{A} = (75.0\ \text{N/C})(0.240\ \text{m}^2)\cos 70° = 6.16\ \text{Nm}^2/\text{C}.$

23-3: (a) $\Phi = \vec{E}\cdot\vec{A} = \dfrac{\lambda}{2\pi\varepsilon_o r}(2\pi r l) = \dfrac{\lambda l}{\varepsilon_o} = \dfrac{(5.00\times10^{-6}\ \text{C})(0.300\ \text{m})}{\varepsilon_o} = 1.69\times10^5\ \text{Nm}^2/\text{C}.$
(b) We would get the same flux as in (a) if the cylinder's radius was made larger - the field lines must still pass through the surface.
(c) If the length was increased to $l = 0.900\ \text{m}$, the flux would increase by a factor of three: $\Phi = 5.08\times10^5\ \text{Nm}^2/\text{C}.$

23-4: (a) $\Phi = \vec{E}\cdot\vec{A} = EA\cos\phi = (5.00\times10^3\ \text{N/C})(0.0800\ \text{m})^2\cos 53.1° = 19.2\ \text{Nm}^2/\text{C}.$
(b) The total flux through the cube must be zero, any flux entering the cube must also leave it.

23-5: (a) Given that $\vec{E} = (2.50\ \text{N/C})\,\hat{i} - (4.20\ \text{N/C})\,\hat{j}$, edge length $L = 0.200\ \text{m}$, and
$\hat{n}_{S_1} = -\hat{j} \Rightarrow \Phi_1 = \vec{E}\cdot\hat{n}_{S_1} = +4.20L^2\ \text{Nm}^2/\text{C} = 0.168\ \text{Nm}^2/\text{C}.$
$\hat{n}_{S_2} = +\hat{k} \Rightarrow \Phi_2 = \vec{E}\cdot\hat{n}_{S_2} = 0.$
$\hat{n}_{S_3} = +\hat{j} \Rightarrow \Phi_3 = \vec{E}\cdot\hat{n}_{S_3} = -4.20L^2\ \text{Nm}^2/\text{C} = -0.168\ \text{Nm}^2/\text{C}.$
$\hat{n}_{S_4} = -\hat{k} \Rightarrow \Phi_4 = \vec{E}\cdot\hat{n}_{S_4} = 0.$
$\hat{n}_{S_5} = +\hat{i} \Rightarrow \Phi_5 = \vec{E}\cdot\hat{n}_{S_5} = +2.50L^2\ \text{Nm}^2/\text{C} = 0.100\ \text{Nm}^2/\text{C}.$
$\hat{n}_{S_6} = -\hat{i} \Rightarrow \Phi_6 = \vec{E}\cdot\hat{n}_{S_6} = -2.50L^2\ \text{Nm}^2/\text{C} = -0.100\ \text{Nm}^2/\text{C}.$

23-6: $\Phi = q/\varepsilon_o \Rightarrow q = \varepsilon_o\Phi = \varepsilon_o(4.90\ \text{Nm}^2/\text{C}) = 4.34\times10^{-11}\ \text{C}.$

23-7: $\Phi = q/\varepsilon_o = (4.80\times10^{-6}\ \text{C})/\varepsilon_o = 5.42\times10^5\ \text{Nm}^2/\text{C}.$

23-8: (a) $\Phi_{S_1} = q_1/\varepsilon_o = (2.50\times10^{-9}\ \text{C})/\varepsilon_o = 282\ \text{Nm}^2/\text{C}.$
(b) $\Phi_{S_2} = q_2/\varepsilon_o = (-4.00\times10^{-9}\ \text{C})/\varepsilon_o = -452\ \text{Nm}^2/\text{C}.$
(c) $\Phi_{S_3} = (q_1+q_2)/\varepsilon_o = ((2.50-4.00)\times10^{-9}\ \text{C})/\varepsilon_o = -169\ \text{Nm}^2/\text{C}.$
(d) $\Phi_{S_4} = (q_1+q_3)/\varepsilon_o = ((2.50+6.40)\times10^{-9}\ \text{C})/\varepsilon_o = 1010\ \text{Nm}^2/\text{C}.$
(e) $\Phi_{S_5} = (q_1+q_2+q_3)/\varepsilon_o = ((2.50-4.00+6.40)\times10^{-9}\ \text{C})/\varepsilon_o = 554\ \text{Nm}^2/\text{C}.$
(f) All that matters for Gauss's Law is the total amount of charge enclosed by the surface, not its distribution within the surface.

23-9: $\Phi_{6\text{ sides}} = q/\varepsilon_o = (3.60\times10^{-9}\text{ C})/\varepsilon_o = 407\text{ Nm}^2/\text{C}$. But the box is symmetrical, so for one side, the flux is: $\Phi_{1\text{ side}} = 67.8\text{ Nm}^2/\text{C}$.

23-10: (a) Only the charge at the origin is enclosed in a sphere of radius 0.500 m.
$\Phi = q/\varepsilon_o = (5.00\times10^{-6}\text{ C})/\varepsilon_o = 5.65\times10^5\text{ Nm}^2/\text{C}$.
(b) Both charges are enclosed if the sphere's radius is 1.50 m.
$\Phi = (q_1+q_2)/\varepsilon_o = ((5.00-3.00)\times10^{-6}\text{ C})/\varepsilon_o = 2.26\times10^5\text{ Nm}^2/\text{C}$.
(c) Again, both charges are enclosed, so the flux is the same as in (b):
$\Phi = 2.26\times10^5\text{ Nm}^2/\text{C}$.

23-11: (a) Since $\vec{E}$ is uniform, the flux through a closed surface must be zero. That is:
$\Phi = \oint \vec{E}\cdot d\vec{A} = \dfrac{q}{\varepsilon_o} = \dfrac{1}{\varepsilon_o}\int\rho\,dV = 0 \Rightarrow \int\rho\,dV = 0$. But because we can choose any volume we want, ρ must be zero if the integral equals zero.
(b) If there is no charge in a region of space, that does NOT mean that the electric field is uniform. Consider a closed volume close to, but not including, a point charge. The field diverges there, but there is no charge in that region.

23-12: (a) If $\rho > 0$ and uniform, then q inside any closed surface greater than zero.
$\Rightarrow \Phi > 0 \Rightarrow \oint \vec{E}\cdot d\vec{A} > 0$ and so the electric field cannot be uniform.
(b) However, inside a small bubble of zero density within the material with density ρ, the field CAN be uniform. All that is important is that there be zero flux through the surface of the bubble (since it encloses no charge). (See Exercise 23-47.)

23-13: $E = \dfrac{1}{4\pi\varepsilon_o}\dfrac{q}{r^2} \Rightarrow q = 4\pi\varepsilon_o E r^2 = 4\pi\varepsilon_o(1400\text{ N/C})(0.220\text{ m})^2 = 7.53\times10^{-9}\text{ C}$.

So the number of electrons is: $n_e = \dfrac{7.53\times10^{-9}\text{ C}}{1.60\times10^{-19}\text{ C}} = 4.71\times10^{10}$.

23-14: (a) $E(r = 0.6+0.1\text{ m}) = \dfrac{1}{4\pi\varepsilon_o}\dfrac{q}{r^2} = \dfrac{1}{4\pi\varepsilon_o}\dfrac{(1.50\times10^{-10}\text{ C})}{(0.700\text{ m})^2} = 2.76\text{ N/C}$.
(b) Inside the metal sphere, there is no charge, so a point inside the sphere has no electric field.

23-15: $E = \dfrac{1}{4\pi\varepsilon_o}\dfrac{q}{r^2} \Rightarrow r = \sqrt{\dfrac{1}{4\pi\varepsilon_o}\dfrac{q}{E}} = \sqrt{\dfrac{1}{4\pi\varepsilon_o}\dfrac{(1.20\times10^{-6}\text{ C})}{614\text{ N/C}}} = 1.33\text{ m}$.

23-16: (a) $\Phi = EA = q/\varepsilon_o \Rightarrow q = \varepsilon_o EA = \varepsilon_o(2.00\times10^5\text{ N/C})(0.0610\text{ m}^2) = 1.08\times10^{-7}\text{ C}$.
(b) Double the surface area: $q = \varepsilon_o(2.00\times10^5\text{ N/C})(0.122\text{ m}^2) = 2.16\times10^{-7}\text{ C}$.

23-17: (a) $\sigma = \frac{Q}{A} = \frac{Q}{2\pi RL} \Rightarrow \frac{Q}{L} = \sigma 2\pi R = \lambda.$

(b) $\oint \vec{E}\cdot d\vec{A} = E(2\pi rL) = \frac{Q}{\varepsilon_o} = \frac{\sigma 2\pi RL}{\varepsilon_o} \Rightarrow E = \frac{\sigma R}{2r}.$

(c) But from (a), $\lambda = \sigma 2\pi R$, so $E = \frac{\lambda}{2\pi\varepsilon_o r}.$

23-18: (a) Negative charge is attracted to the inner surface of the conductor by the inner charge. Its magnitude is the same of the inner charge: $q = -5.00$ nC.
(b) On the outer surface the charge is a combination of the net charge on the conductor and the charge "left behind" when the -5.00 nC moved to the inner surface: $q = (7.00 + 5.00)\text{ nC} = 12.0\text{ nC}.$

23-19: S_2 and S_3 enclose no charge, so the flux is zero, and electric field outside the plates is zero.
For between the plates, S_1 shows that: $EA = q/\varepsilon_o = \sigma A/\varepsilon_o \Rightarrow E = \sigma/\varepsilon_o.$

23-20: (a) At a distance of 0.1 mm from the center, the sheet appears "infinite", so:

$$\oint \vec{E}\cdot d\vec{A} = E2A = \frac{q}{\varepsilon_o} \Rightarrow E = \frac{q}{2\varepsilon_o A} = \frac{2.50\times10^{-9}\text{ C}}{2\varepsilon_o(9.00\text{ m}^2)} = 15.7\text{ N/C}.$$

(b) At a distance of 100 m from the center, the sheet looks like a point, so:

$$E \approx \frac{1}{4\pi\varepsilon_o}\frac{q}{r^2} = \frac{1}{4\pi\varepsilon_o}\frac{(2.50\times10^{-9}\text{ C})}{(100\text{ m})^2} = 2.25\times10^{-3}\text{ N/C}.$$

(c) There would be no difference if the sheet was a conductor since the charge would automatically spread out evenly, if we neglect edge effects.

23-21: To find the charge enclosed, we need the flux through the parallelepiped:

$$\Phi_1 = AE_1\cos 60° = (0.0500\text{ m})(0.0600\text{ m})(3.50\times10^4\text{ N/C})(0.5) = 52.5\text{ N}\cdot\text{m}^2/\text{C}$$

$$\Phi_2 = AE_2\cos 120° = (0.0500\text{ m})(0.0600\text{ m})(6.00\times10^4\text{ N/C})(0.5) = -90.0\text{ N}\cdot\text{m}^2/\text{C}$$

So the total flux is $\Phi = \Phi_1 + \Phi_2 = (52.5\ -90.0)\text{ N}\cdot\text{m}^2/\text{C} = -37.5\text{ N}\cdot\text{m}^2/\text{C}$, and

$$q = \Phi\varepsilon_o = (-37.5\text{ N}\cdot\text{m}^2/\text{C})\varepsilon_o = -3.32\times10^{-10}\text{ C}.$$

(b) There must be a net charge (negative) in the parallelepiped since there is a net flux flowing into the surface.

23-22: Given $\vec{E} = (3.00\text{ N/C}\cdot\text{m})x\,\hat{i} + (4.00\text{ N/C}\cdot\text{m})y\,\hat{j}$, edge length $L = 0.200$ m, and

$\hat{n}_{S_1} = -\hat{j} \Rightarrow \Phi_1 = \vec{E}\cdot\hat{n}_{S_1}A = 0.$ $\quad \hat{n}_{S_2} = +\hat{k} \Rightarrow \Phi_2 = \vec{E}\cdot\hat{n}_{S_2}A = 0.$

$\hat{n}_{S_3} = +\hat{j} \Rightarrow \Phi_3 = \vec{E}\cdot\hat{n}_{S_3}A = +4.00L^3\text{ N}\cdot\text{m}^2/\text{C} = 0.0320\text{ N}\cdot\text{m}^2/\text{C}.$

$\hat{n}_{S_4} = -\hat{k} \Rightarrow \Phi_4 = \vec{E}\cdot\hat{n}_{S_4}A = 0.$

$\hat{n}_{S_5} = +\hat{i} \Rightarrow \Phi_5 = \vec{E}\cdot\hat{n}_{S_5}A = +3.00L^3\text{ N}\cdot\text{m}^2/\text{C} = 0.0240\text{ N}\cdot\text{m}^2/\text{C}.$

$\hat{n}_{S_6} = -\hat{i} \Rightarrow \Phi_6 = \vec{E}\cdot\hat{n}_{S_6}A = 0.$

(b) Total flux: $\Phi = \Phi_3 + \Phi_5 = 0.0560\text{ N}\cdot\text{m}^2/\text{C} \Rightarrow q = \Phi\varepsilon_o = 4.96\times10^{-13}\text{ C}.$

23-23:

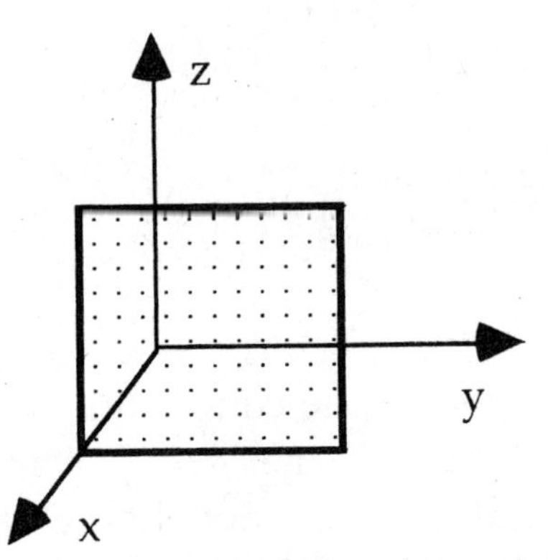

(b) Imagine a charge q at the center of a cube of edge length 2L. Then: $\Phi = q / \varepsilon_o$

Here the square is one 24th of the surface area of the imaginary cube, so it intercepts 1/24 of the flux. That is, $\Phi = q / 24\varepsilon_o$.

23-24: (a) $\Phi = EA = (250\ \text{N/C})(6.0\ \text{m}^2) = 1500\ \text{N} \cdot \text{m}^2/\text{C}$.
(b) Since the field is parallel to the surface, $\Phi = 0$.
(c) Choose the gaussian surface to equal the volume's surface. Then:
$1500 + EA = q / \varepsilon_o \Rightarrow E = \frac{1}{6.0\ \text{m}^2}(3.66 \times 10^{-8}\ \text{C} / \varepsilon_o - 1500) = 440\ \text{N/C}$, in the positive x-direction.
(d) Since the field is uniform, it must in part be due to external charges, because such a field is only possible for an infinitely extended distribution.

23-25: (a) The sphere acts as a point charge on an external charge, so:
$F = qE = \frac{1}{4\pi\varepsilon_0}\frac{qQ}{r^2}$, radially outward.
(b) If the point charge was inside the sphere (where there is no electric field) it would feel zero force.

23-26: (a) For $r < a$, $E = 0$, since no charge is enclosed.
For $a < r < b$, $E = \frac{1}{4\pi\varepsilon_0}\frac{q}{r^2}$, since there is +q inside a radius r.
For $b < r < c$, $E = 0$, since now the -q cancels the inner +q.
For $r > c$, $E = \frac{1}{4\pi\varepsilon_0}\frac{q}{r^2}$, since again the total charge enclosed is +q.

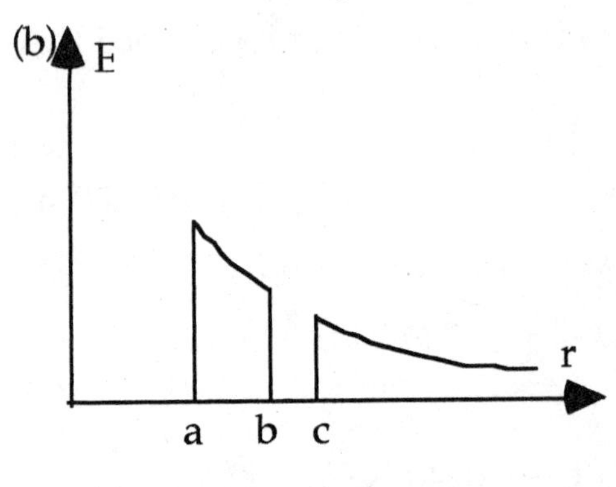

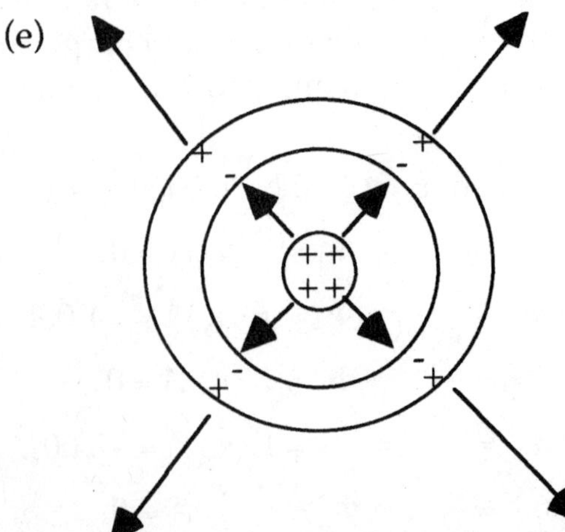

(c) Charge on inner shell surface is -q. (d) Charge on outer shell surface is +q.

23-27: (a) $r < R,\ E = 0$, since no charge is enclosed.

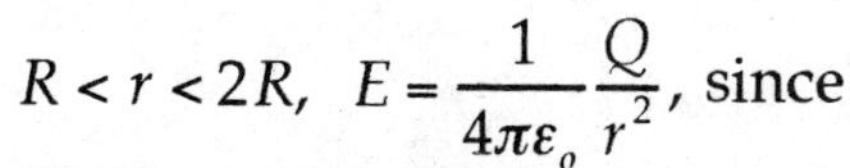

$R < r < 2R,\ E = \dfrac{1}{4\pi\varepsilon_0}\dfrac{Q}{r^2}$, since charge enclosed is Q.

$r > 2R,\ E = \dfrac{1}{4\pi\varepsilon_0}\dfrac{2Q}{r^2}$, since charge enclosed is 2Q.

(b)

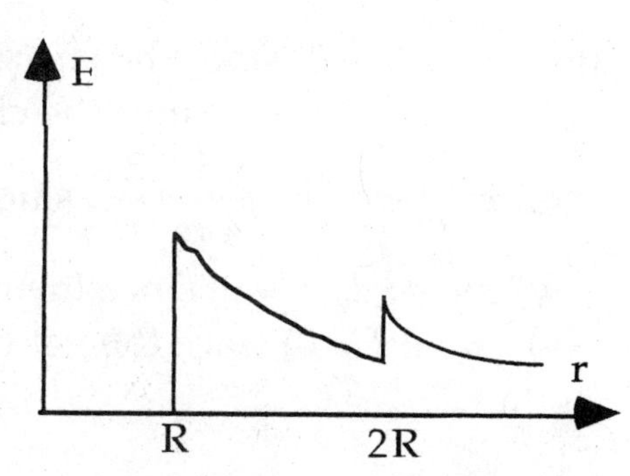

23-28: (a) $r < a,\ E = \dfrac{1}{4\pi\varepsilon_0}\dfrac{Q}{r^2}$, since the charge enclosed is Q.

$a < r < b,\ E = 0$, since the -Q on the inner surface of the shell cancels the +Q at the center of the sphere.

$r > b,\ E = -\dfrac{1}{4\pi\varepsilon_0}\dfrac{2Q}{r^2}$, since the total enclosed charge is -2Q.

(b) The surface charge density on inner surface: $\sigma = -\dfrac{Q}{4\pi a^2}$.

(c) The surface charge density on the outer surface: $\sigma = -\dfrac{2Q}{4\pi b^2}$.

(d)

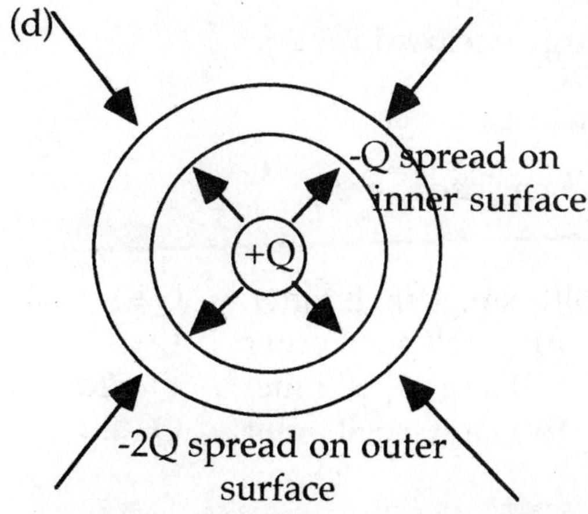

(e)

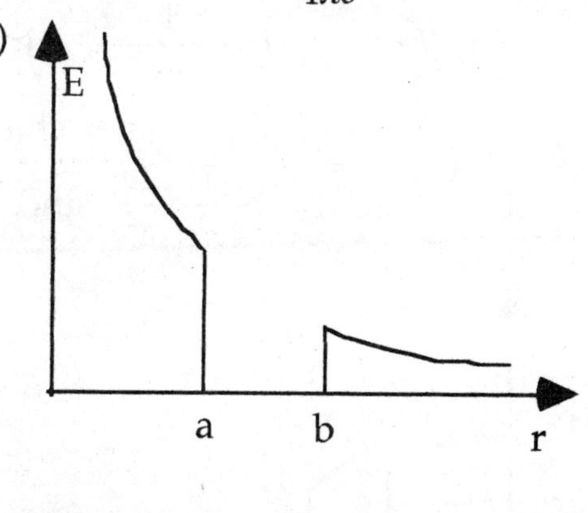

23-29: (a)(i) $r < a,\ E = 0$, since Q = 0.

(ii) $a < r < b,\ E = 0$, since Q = 0.

(iii) $b < r < c,\ E = \dfrac{1}{4\pi\varepsilon_0}\dfrac{2q}{r^2}$, since Q = +2q.

(iv) $c < r < d,\ E = 0$, since Q = 0.

(v) $r > d,\ E = \dfrac{1}{4\pi\varepsilon_0}\dfrac{6q}{r^2}$, since Q = +6q.

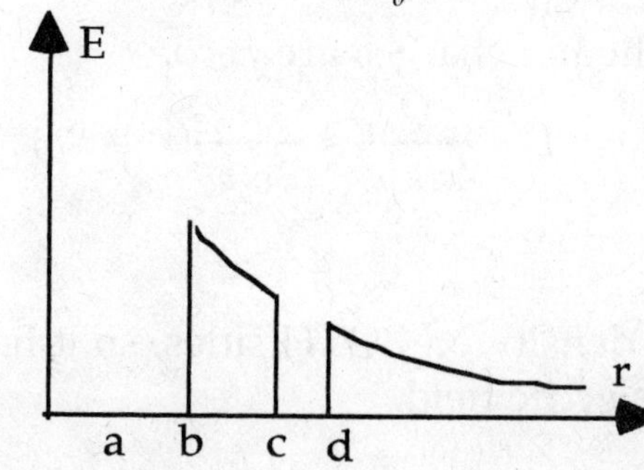

(b)(i) small shell inner: Q=0
(ii) small shell outer: Q=+2q
(iii) large shell inner: Q=-2q
(iv) large shell outer: Q=+6q

23-30: (a)(i) $r < a,\ E = 0$, since the charge enclosed is zero.

(ii) $a < r < b,\ E = 0$, since the charge enclosed is zero.

(iii) $b < r < c,\ E = \dfrac{1}{4\pi\varepsilon_0}\dfrac{2q}{r^2}$, since charge enclosed is +2q.

(iv) $c < r < d,\ E = 0$, since the net charge enclosed is zero.

(v) $r > d,\ E = 0$, since the net charge enclosed is zero.

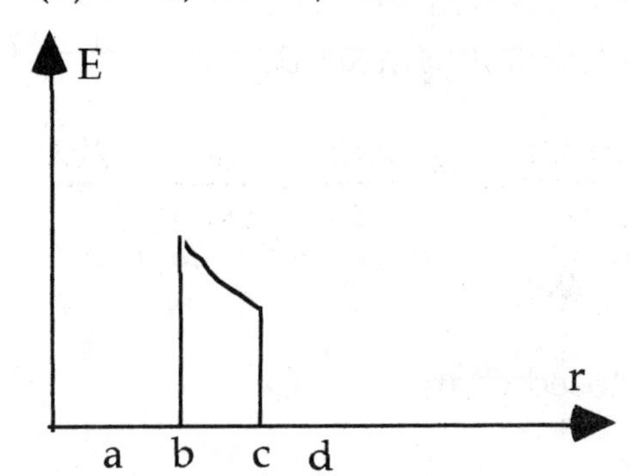

(b)(i) small shell inner: Q=0
(ii) small shell outer: Q=+2q
(iii) large shell inner: Q=-2q
(iv) large shell outer: Q=0

23-31: (a)(i) $r < a,\ E = 0$, since charge enclosed is zero.

(ii) $a < r < b,\ E = 0$, since charge enclosed is zero.

(iii) $b < r < c,\ E = \dfrac{1}{4\pi\varepsilon_0}\dfrac{2q}{r^2}$, since charge enclosed is +2q.

(iv) $c < r < d,\ E = 0$, since charge enclosed is zero.

(v) $r > d,\ E = -\dfrac{1}{4\pi\varepsilon_0}\dfrac{2q}{r^2}$, since charge enclosed is -2q.

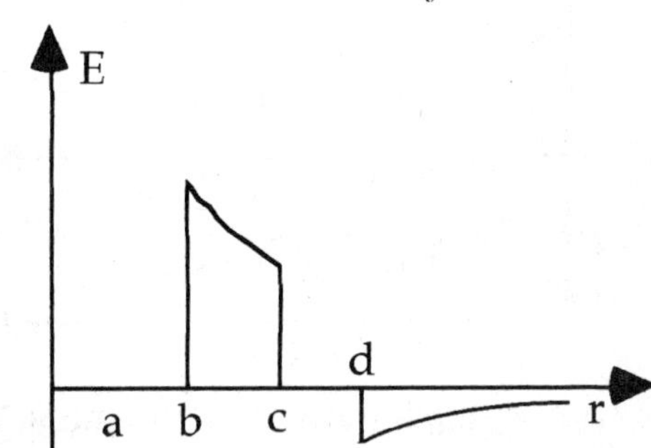

(b)(i) small shell inner: Q=0
(ii) small shell outer: Q=+2q
(iii) large shell inner: Q=-2q
(iv) large shell outer: Q=-2q

23-32: (a) We need: $-Q = \dfrac{4\pi\rho}{3}((2R)^3 - R^3) \Rightarrow Q = \dfrac{-28\pi\rho R^3}{3} \Rightarrow \rho = -\dfrac{3Q}{28\pi R^3}$.

(b) $r < R,\ E = 0$ and $r > 2R,\ E = 0$, since the net charges are zero.

$R < r < 2R,\ \Phi = E(4\pi r^2) = \dfrac{Q}{\varepsilon_0} + \dfrac{4\pi\rho}{3\varepsilon_0}(r^3 - R^3) \Rightarrow E = \dfrac{Q}{4\pi\varepsilon_0 r^2} + \dfrac{\rho}{3\varepsilon_0 r^2}(r^3 - R^3)$,

radially outward.

23-33: (a) The conductor has the surface charge density on BOTH sides, so it has twice the enclosed charge and twice the electric field.

(b) We have a conductor with surface charge density σ on both sides. thus the electric field outside the plate is $\Phi = E(2A) = (2\sigma A)/\varepsilon_0 \Rightarrow E = \sigma/\varepsilon_0$. To find the field inside the conductor use a gaussian surface that has one face inside the conductor, and one outside. Then:
$\Phi = E_{out}A + E_{in}A = (\sigma A)/\varepsilon_0$ but $E_{out} = \sigma/\varepsilon_0 \Rightarrow E_{in}A = 0 \Rightarrow E_{in} = 0$.

23-34: (a) $r < R,\ E(2\pi rl) = \dfrac{q}{\varepsilon_0} = \dfrac{\rho\pi r^2 l}{\varepsilon_0} \Rightarrow E = \dfrac{\rho r}{2\varepsilon_0}$, radially outward.

(b) $r > R,$ and $\lambda = \rho\pi R^2,\ E(2\pi rl) = \dfrac{q}{\varepsilon_0} = \dfrac{\rho\pi R^2 l}{\varepsilon_0} \Rightarrow E = \dfrac{\rho R^2}{2\varepsilon_0 r} = \dfrac{\lambda}{2\pi\varepsilon_0 r} = \dfrac{2k\lambda}{r}$.

(c) $r = R,$ the electric field for BOTH regions is $E = \dfrac{\rho R}{2\varepsilon_0}$, so they are consistent.

(d)

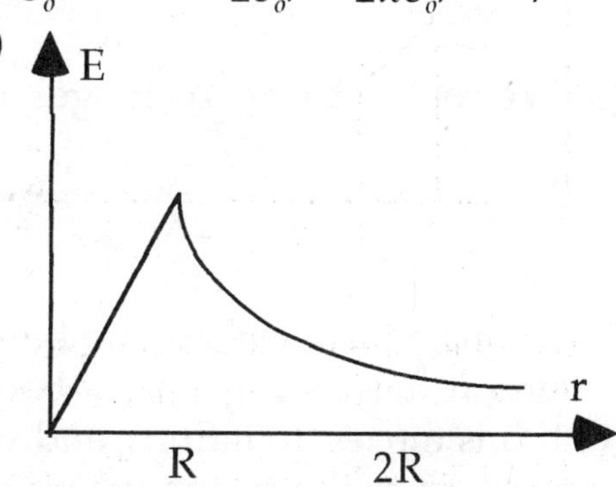

23-35: (a) $a < r < b,\ E = \dfrac{1}{4\pi\varepsilon_0}\dfrac{2\lambda}{r}$, as in **23-34(b)**.

(b) $r > c,\ E = \dfrac{1}{4\pi\varepsilon_0}\dfrac{2\lambda}{r}$, since again the charge enclosed is the same as in part (a).

(c)

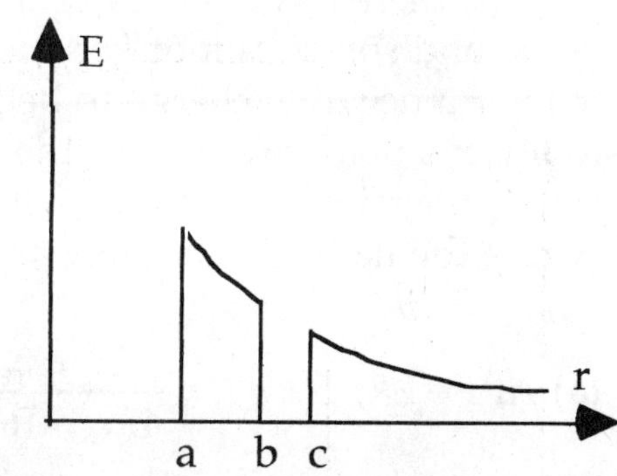

(d) The inner and outer surfaces of the outer cylinder must have the same amount of charge on them: $\lambda l = -\lambda_{inner} l \Rightarrow \lambda_{inner} = -\lambda$, and $\lambda_{outer} = \lambda$.

23-36: (a) (i) $r < a,\ E(2\pi rl) = \dfrac{q}{\varepsilon_0} = \dfrac{\alpha l}{\varepsilon_0} \Rightarrow E = \dfrac{\alpha}{2\pi\varepsilon_0 r}$.

(ii) $a < r < b$, there is no net charge enclosed, so the electric field is zero.

(iii) $r > b,\ E(2\pi rl) = \dfrac{q}{\varepsilon_0} = \dfrac{2\alpha l}{\varepsilon_0} \Rightarrow E = \dfrac{\alpha}{\pi\varepsilon_0 r}$.

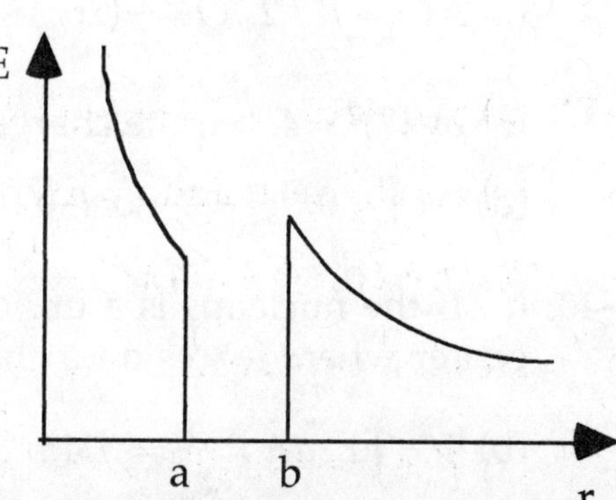

(b) (i) Inner charge per unit length is $-\alpha$. (ii) Outer charge per length is $+2\alpha$.

23-37: (a)(i) $r < a,\ E(2\pi rl) = \frac{q}{\varepsilon_0} = \frac{\alpha l}{\varepsilon_0} \Rightarrow E = \frac{\alpha}{2\pi\varepsilon_0 r}$.

(ii) $a < r < b$, there is no net charge enclosed, so the electric field is zero.

(iii) $r > b$, there is no net charge enclosed, so the electric field is zero.

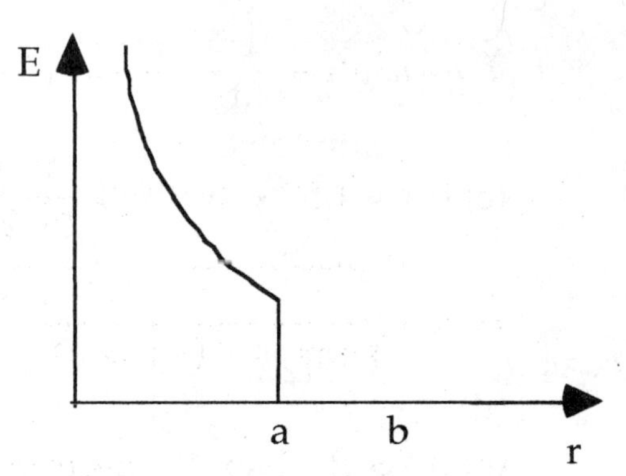

(b) (i) Inner charge per unit length is -α. (ii) Outer charge per length is ZERO.

23–38: (a) We could place two charges +Q on either side of the charge +q:

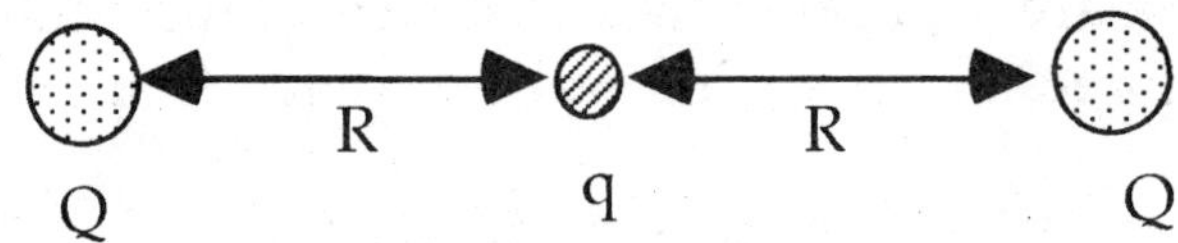

(b) In order for the charge to be stable, the electric field in a neighbourhood around it must always point back to the equilibrium position.

(c) If q is moved to infinity and we require there to be an electric field always pointing in to the region where q had been, we could draw a small gaussian surface there. We would find that we need a negative flux into the surface. That is, there has to be a negative charge in that region. However there is none, and so we cannot get such a stable equilibrium.

(d) For a negative charge to be in stable equilibrium, we need the electric field to always point away from the charge position. The argument in (c) carries through again, this time infering that a positive charge must be in the space where the negative charge was if stable equilibrium is to be attained.

23-39: (a) At $r = 2R,\ F = q_e E = \frac{1}{4\pi\varepsilon_0}\frac{q_e q_{Fe}}{4R^2} = \frac{1}{4\pi\varepsilon_0}\frac{(26)(1.6\times10^{-19}\ \text{C})}{4(4.6\times10^{-15}\ \text{m})^2} = 71\ \text{N}$.

So: $a = F/m = 71\ \text{N}/9.11\times10^{-31}\ \text{kg} = 7.8\times10^{31}\ \text{m/s}^2$.

(b) At $r = R,\ a = 4a_{(a)} = 3.1\times10^{32}\ \text{m/s}^2$.

(c) At $r = R/2,\ Q = \frac{1}{8}(26e)$ so with radius decreasing by 2, the acceleration goes up by $(2)^2 = 4$, but the charge decreased by 8, so $a = \frac{4}{8}a_{(b)} = 1.6\times10^{32}\ \text{m/s}^2$.

(d) At $r = 0,\ Q = 0$, so $F = 0$.

23-40: (a) If the nucleaus is a uniform positively charged sphere, it is only at it's very center where forces on a charge would balance or cancel.

(b) $\Phi = \oint \vec{E}\cdot d\vec{A} = \frac{q}{\varepsilon_0} \Rightarrow E4\pi r^2 = \frac{e}{\varepsilon_0}\left(\frac{r^3}{R^3}\right) \Rightarrow E = \frac{er}{4\pi\varepsilon_0 R^3}$

$\Rightarrow F = qE = -\frac{1}{4\pi\varepsilon_0}\frac{e^2 r}{R^3}$. So from the simple harmonic motion equation:

$$F = -m\omega^2 r = -\frac{1}{4\pi\varepsilon_0}\frac{e^2 r}{R^3} \Rightarrow \omega = \sqrt{\frac{1}{4\pi\varepsilon_0}\frac{e^2}{mR^3}} \Rightarrow f = \frac{1}{2\pi}\sqrt{\frac{1}{4\pi\varepsilon_0}\frac{e^2}{mR^3}}.$$

(c) If $f = 4.57 \times 10^{14}$ Hz $= \frac{1}{2\pi}\sqrt{\frac{1}{4\pi\varepsilon_0}\frac{e^2}{mR^3}}$

$$\Rightarrow R = \sqrt[3]{\frac{1}{4\pi\varepsilon_0}\frac{(1.60 \times 10^{-19}\ \text{C})^2}{4\pi^2(9.11 \times 10^{-31}\ \text{kg})(4.57 \times 10^{14}\ \text{Hz})^2}} = 3.13 \times 10^{-10}\ \text{m}.$$

(d) $r_{\text{actual}} / r_{\text{Thompson}} \approx 1$

(e) If $r > R$ then the electron would still oscillate, (be attracted), but not undergo simple harmonic motion, because for $r > R$, $F \propto 1/r^2$, and is not linear.

23-41: The electrons are separated by a distance 2d, and the amount of the positive nucleus's charge that is within radius d is all that exerts a force on the electron. So: $F_e = \frac{ke^2}{(2d)^2} = F_{\text{nucleus}} = 2ke^2\frac{d}{R^3} \Rightarrow d^3 = R^3/8 \Rightarrow d = R/2.$

23-42: (a) $Q(r) = Q - \int \rho\, dV = Q - \frac{Q}{\pi a_0^3}\iiint e^{-2r/a_0} r^2 dr \sin\theta d\theta d\phi = Q - \frac{4Q}{a_0^3}\int_0^r x^2 e^{-2x/a_0} dx$

$$\Rightarrow Q(r) = Q - \frac{4Qe^{-\alpha r}}{a_0^3\alpha^3}\left(2e^{\alpha r} - \alpha^2 r^2 - 2\alpha r - 2\right).$$

Note if $r \to \infty$ and $\alpha = 2/a_0$, $Q(r) \to Q - \frac{4Q}{a_0^3}\left(\frac{2}{\alpha^3}\right) = Q - Q = 0.$

(b) The electric field is radially outward, and has magnitude:

$$\Rightarrow E = \frac{kQ}{r^2}\left(1 - \frac{e^{-2r/a_0}}{2}(2e^{2r/a_0} - 4r^2/a_0^2 - 4r/a_0 - 2)\right).$$

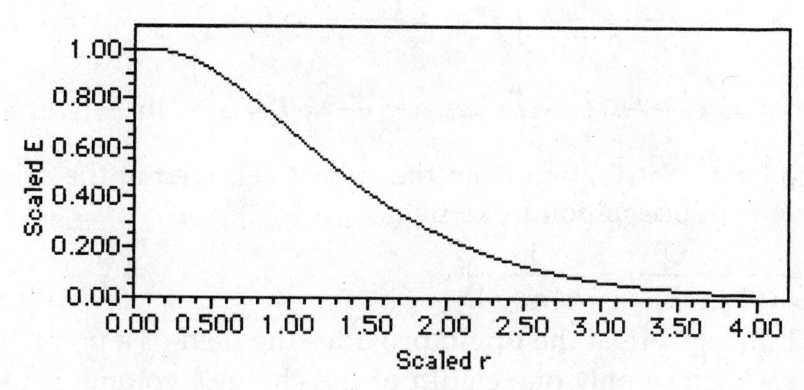

23-43: (a) The total charge: $q = 4\pi\int_0^R \rho_o(1 - r/R)r^2 dr = 4\pi\left[\int_0^R r^2\, dr - \int_0^R r^3/R\, dr\right]$

$$\Rightarrow q = 4\pi\rho_o\left[R^3/3 - R^3/4\right] = \frac{4\pi R^3\rho_o}{12} = \frac{4\pi R^3}{12}\frac{3Q}{\pi R^3} = Q.$$

(b) $r \geq R$, all the charge Q is enclosed, and: $\Phi = E(4\pi r^2) = Q / \varepsilon_o \Rightarrow E = \dfrac{1}{4\pi\varepsilon_o}\dfrac{Q}{r^2}$, the same as a point charge.

(c) $r \leq R$, then $Q(r) = q(r^3 / R^3)$.

Also, $Q(r) = 4\pi \int_0^R \rho_o (1 - r / R) r^2 dr = 4\pi \rho_o (\dfrac{r^3}{3} - \dfrac{r^4}{4R})$.

$\Rightarrow E(r) = \dfrac{12kQ}{r^2}\left(\dfrac{1}{3}\dfrac{r^3}{R^3} - \dfrac{1}{4}\dfrac{r^4}{R^4}\right) = \dfrac{kQ}{r^2}\left(\dfrac{4r^3}{R^3} - \dfrac{3r^4}{R^4}\right)$, for $r \leq R$.

(d) At $r = R$, both (b) and (c) yield the same result $E = \dfrac{1}{4\pi\varepsilon_o}\dfrac{Q}{R^2}$.

23-44: (a) The electric field of the slab must be zero by symmetry. There is no preferred direction in the y-z plane, so the electric field can only point in the x-direction. But at the origin in the x-direction, neither the positive nor negative directions should be singled out as special, and so the field must be zero.

(b) Use a Gaussian surface that has one face of area A on in the y-z plane at x = 0, and the other face at a general value x. Then:

$x \leq d$: $\Phi = EA = \dfrac{Q_{\text{encl}}}{\varepsilon_o} = \dfrac{\rho A x}{\varepsilon_o} \Rightarrow E = \dfrac{\rho x}{\varepsilon_o}$, with direction given by $\dfrac{x}{|x|}\hat{i}$.

Note that E is zero at x = 0.

Now outside the slab, the enclosed charge is constant with x:

$x \geq d$: $\Phi = EA = \dfrac{Q_{\text{encl}}}{\varepsilon_o} = \dfrac{\rho A d}{\varepsilon_o} \Rightarrow E = \dfrac{\rho d}{\varepsilon_o}$, again with direction given by $\dfrac{x}{|x|}\hat{i}$.

23-45: (a) Again, E is zero at x = 0, by symmetry arguments.

(b) $x \leq d$: $\Phi = EA = \dfrac{Q_{\text{encl}}}{\varepsilon_o} = \dfrac{\rho_o A}{\varepsilon_o d^2}\int_0^x x^2\,dx = \dfrac{\rho_o A x^3}{3\varepsilon_o d^2} \Rightarrow E = \dfrac{\rho_o x^3}{3\varepsilon_o d^2}$, in $\dfrac{x}{|x|}\hat{i}$ direction.

(c) $x \geq d$: $\Phi = EA = \dfrac{Q_{\text{encl}}}{\varepsilon_o} = \dfrac{\rho_o A}{\varepsilon_o d^2}\int_0^d x^2\,dx = \dfrac{\rho_o A d}{3\varepsilon_o} \Rightarrow E = \dfrac{\rho_o d}{3\varepsilon_o}$, in $\dfrac{x}{|x|}\hat{i}$ direction.

23-46: (a) $x = 0$: no field contribution from the sphere centered at the origin, and the other sphere produces a point-like field:

$\vec{E}(x = 0) = -\dfrac{1}{4\pi\varepsilon_o}\dfrac{Q}{(2R)^2}\hat{i} = -\dfrac{1}{4\pi\varepsilon_o}\dfrac{Q}{4R^2}\hat{i}$.

(b) $x = R / 2$: the sphere at the origin provides the field of a point charge of charge $q = Q / 8$ since only one eighth of the charge's volume is included. So:

$\vec{E}(x = R / 2) = \dfrac{1}{4\pi\varepsilon_o}\left(\dfrac{(Q / 8)}{(R / 2)^2} - \dfrac{Q}{(3R / 2)^2}\right)\hat{i} = \dfrac{1}{4\pi\varepsilon_o}\dfrac{Q}{R^2}(1 / 2 - 4 / 9)\hat{i} = \dfrac{1}{4\pi\varepsilon_o}\dfrac{Q}{18R^2}\hat{i}$.

(c) $x = R$: the two electric fields cancel, so $\vec{E} = 0$.

(d) $x = 3R$: now both spheres contribute fields pointing to the right:

$\vec{E}(x = 3R) = \dfrac{1}{4\pi\varepsilon_o}\left(\dfrac{Q}{(3R)^2} + \dfrac{Q}{R^2}\right)\hat{i} = \dfrac{1}{4\pi\varepsilon_o}\dfrac{10Q}{9R^2}\hat{i}$.

23-47: (a) For a sphere NOT at the coordinate origin:

$\vec{r}' = \vec{r} - \vec{b} \Rightarrow \Phi = 4\pi r'^2 E = \frac{Q_{encl}}{\varepsilon_o} = \frac{\rho}{\varepsilon_o}\frac{4\pi r'^3}{3} \Rightarrow E = \frac{\rho r'}{3\varepsilon_o}$, in the $\hat{r}'$-direction.

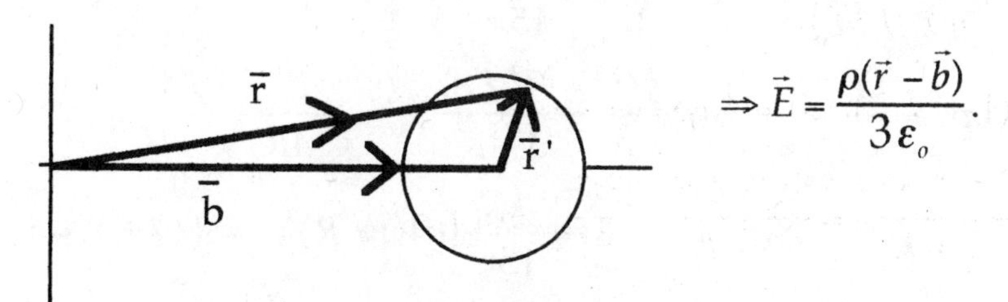

$$\Rightarrow \vec{E} = \frac{\rho(\vec{r} - \vec{b})}{3\varepsilon_o}.$$

(b) The electric field inside a hole in a charged insulating sphere is:

$\vec{E}_{hole} = \vec{E}_{sphere} - \vec{E}_{(a)} = \frac{\rho\vec{r}}{3\varepsilon_o} - \frac{\rho(\vec{r} - \vec{b})}{3\varepsilon_o} = \frac{\rho\vec{b}}{3\varepsilon_o}$. Note that $\vec{E}$ is uniform.

23-48: Using the technique of **23-47**, we first find the field of a cylinder off-axis, then the electric field in a hole in a cylinder is the difference between two electric fields - that of a solid cylinder on-axis, and one off-axis.

$\vec{r}' = \vec{r} - \vec{b} \Rightarrow \Phi = 2\pi r' l E = \frac{Q_{encl}}{\varepsilon_o} = \frac{\rho}{\varepsilon_o} l\pi r'^2 \Rightarrow E = \frac{\rho r'}{2\varepsilon_o} \Rightarrow \vec{E} = \frac{\rho(\vec{r} - \vec{b})}{2\varepsilon_o}.$

$\vec{E}_{hole} = \vec{E}_{cylinder} - \vec{E}_{above} = \frac{\rho\vec{r}}{2\varepsilon_o} - \frac{\rho(\vec{r} - \vec{b})}{2\varepsilon_o} = \frac{\rho\vec{b}}{2\varepsilon_o}$. Note that $\vec{E}$ is uniform.

23-49: (a) $\Phi_g = \oint \vec{g}.d\vec{A} = -Gm\oint \frac{r^2 \sin\theta \, dr d\theta d\phi}{r^2} = -4\pi Gm$.

For any closed surface, mass OUTSIDE the surface contributes zero to the flux passing through the surface. Thus the formula above holds for any situation where m is the mass enclosed by the Gaussian surface.

That is: $\Phi_g = \oint \vec{g}.d\vec{A} = -4\pi GM_{encl}$.

23-50: (a) $\Phi_g = g4\pi r^2 = -4\pi GM \Rightarrow g = -\frac{GM}{r^2}$, which is the same as for a point mass.

(b) Inside a hollow shell, the $M_{encl} = 0$, so $g = 0$.

(c) Inside a uniform spherical mass:

$\Phi_g = g4\pi r^2 = -4\pi GM_{encl} = -4\pi G\left(M\frac{r^3}{R^3}\right) \Rightarrow g = -\frac{GMr}{R^3}$, which is linear in r.

23-51: (a) The charge enclosed: $Q = Q_i + Q_o$, where $Q_i = \alpha\frac{4\pi(R/2)^3}{3} = \frac{\alpha\pi R^3}{6}$, and

$Q_o = 4\pi(2\alpha)\int_{R/2}^{R}(r^2 - r^3/R)dr = 8\alpha\pi\left(\frac{(R^3 - R^3/8)}{3} - \frac{(R^4 - R^4/16)}{4R}\right) = \frac{11\alpha\pi R^3}{24}.$

$$\Rightarrow Q = \frac{15\alpha\pi R^3}{24} \Rightarrow \alpha = \frac{8Q}{5\pi R^3} = \frac{8(4.00\times10^{-17}\ \text{C})}{5\pi(1.50\times10^{-14}\ \text{m})^3} = 6.04\times10^{24}\ \text{C/m}^3.$$

(b) $r \le R/2$: $\Phi = E4\pi r^2 = \frac{\alpha 4\pi r^3}{3\varepsilon_0} \Rightarrow E = \frac{\alpha r}{3\varepsilon_0} = \frac{8Qr}{15\pi\varepsilon_0 R^3}$.

$R/2 \le r \le R$: $\Phi = E4\pi r^2 = \frac{Q_i}{\varepsilon_0} + \frac{1}{\varepsilon_0}\left(8\alpha\pi\left(\frac{(r^3 - R^3/8)}{3} - \frac{(r^4 - R^4/16)}{4R}\right)\right)$

$$\Rightarrow E = \frac{\alpha\pi R^3}{24\varepsilon_0(4\pi r^2)}\left(64(r/R)^3 - 48(r/R)^4 - 1\right) = \frac{kQ}{15r^2}\left(64(r/R)^3 - 48(r/R)^4 - 1\right).$$

$r \ge R$: $E = \frac{Q}{4\pi\varepsilon_0 r^2}$, since all charge is enclosed.

(c) $\frac{Q_i}{Q} = \frac{(4Q/15\pi)}{Q} = \frac{4}{15\pi} = 0.0849.$

(d) $r \le R/2$: $F = -eE = -\frac{8eQ}{15\pi\varepsilon_0 R^3}r$, so the force depends upon displacement to the first power, and we have simple harmonic motion.

(e) $F = -\omega^2 r \Rightarrow \omega = \sqrt{\frac{8eQ}{15\pi\varepsilon_0 R^3}} \Rightarrow T = \frac{2\pi}{\omega} = 2\pi\sqrt{\frac{15\pi\varepsilon_0 R^3}{8eQ}}$

$$\Rightarrow T = 2\pi\sqrt{\frac{15\pi\varepsilon_0(1.50\times10^{-14}\ \text{m})^3}{8(1.60\times10^{-19}\ \text{C})(4.00\times10^{-17}\ \text{C})}} = 3.29\times10^{-8}\ \text{s}.$$

(f) If the amplitude of oscillation is greater than R/2, the force is no longer linear in r, and is thus no longer simple harmonic.

23-52: (a) Charge enclosed:

$Q = Q_i + Q_o$ where $Q_i = 4\pi\int_0^{R/2}\frac{3\alpha r^3}{2R}dr = \frac{6\pi\alpha}{R}\frac{1}{4}\frac{R^4}{16} = \frac{3}{32}\pi\alpha R^3.$

and $Q_o = 4\pi\alpha\int_{R/2}^{R}(1-(r/R)^2)r^2\,dr = 4\pi\alpha R^3\left(\frac{7}{24} - \frac{31}{160}\right) = \frac{47}{120}\pi\alpha R^3.$

Therefore, $Q = \left(\frac{3}{32} + \frac{47}{120}\right)\pi\alpha R^3 = \frac{233}{480}\pi\alpha R^3$

$$= \frac{233}{480}\pi(3.00\times10^{11}\ \text{C/m}^3)(2.00\times10^{-10}\ \text{m})^3 = 3.66\times10^{-18}\ \text{C}.$$

(b) $r \le R/2$: $\Phi = E4\pi r^2 = \frac{4\pi}{\varepsilon_0}\int_0^r\frac{3\alpha r^3}{2R}dr = \frac{3\pi\alpha r^4}{2\varepsilon_0 R} \Rightarrow E = \frac{6\alpha r^2}{16\varepsilon_0 R} = \frac{180Qr^2}{233\pi\varepsilon_0 R^4}$.

$R/2 \le r \le R$: $\Phi = E4\pi r^2 = \frac{Q_i}{\varepsilon_0} + \frac{4\pi\alpha}{\varepsilon_0}\int_{R/2}^{r}(1-(r/R)^2)r^2\,dr$

$$= \frac{Q_i}{\varepsilon_0} + \frac{4\pi\alpha}{\varepsilon_0}\left(\frac{r^3}{3} - \frac{R^3}{24} - \frac{r^5}{5R^2} + \frac{R^3}{160}\right) = \frac{3}{128}\frac{4\pi\alpha R^3}{\varepsilon_0} + \frac{4\pi\alpha R^3}{\varepsilon_0}\left(\frac{1}{3}\left(\frac{r}{R}\right)^3 - \frac{1}{5}\left(\frac{r}{R}\right)^5 - \frac{17}{480}\right)$$

$$\Rightarrow E = \frac{480Q}{233\pi\varepsilon_0 r^2}\left(\frac{1}{3}\left(\frac{r}{R}\right)^3 - \frac{1}{5}\left(\frac{r}{R}\right)^5 - \frac{17}{480} + \frac{3}{128}\right).$$

$r \geq R$: $E = \frac{Q}{4\pi\varepsilon_o r^2}$, since all charge is enclosed.

(c) The fraction of Q between $R/2 \leq r \leq R$: $\frac{Q_o}{Q} = \frac{47}{120}\frac{480}{233} = 0.807$.

(d) $E(r = R/2) = \frac{180}{233}\frac{Q}{4\pi\varepsilon_o R^2}$, using either of the electric field expressions above, evaluated at r = R/2.

(e) The force an electron would feel never is proportional to $-r$ which is necessary for simple harmonic oscillations. It is oscillatory since the force is always attractive, but it has the wrong power of r to be simple harmonic.

Chapter 24: Electric Potential

24-1: (a) $U = \frac{kQq}{r} = \frac{k(9.10\times10^{-6}\ \text{C})(-0.420\times10^{-6}\ \text{C})}{0.960\ \text{m}} = -0.0358\ \text{J}.$

(b) $K_f = K_i + U_i - U_f = -0.0358\ \text{J} - \frac{k(9.10\times10^{-6}\ \text{C})(-0.420\times10^{-6}\ \text{C})}{0.240\ \text{m}} = 0.1075\ \text{J}$

$\Rightarrow K_f = 0.1075\ \text{J} = \frac{1}{2}mv^2 \Rightarrow v = \sqrt{\frac{2(0.1075\ \text{J})}{3.20\times10^{-4}\ \text{kg}}} = 25.9\ \text{m/s}.$

24-2: $U = 0.500\ \text{J} = \frac{kq_1q_2}{r} \Rightarrow r = \frac{k(2.30\times10^{-6}\ \text{C})(7.20\times10^{-6}\ \text{C})}{0.500\ \text{J}} = 0.298\ \text{m}.$

24-3: $\Delta U = kq_1q_2\left(\frac{1}{r_2} - \frac{1}{r_1}\right) = k(-5.80\ \mu\text{C})(4.30\ \mu\text{C})\left(\frac{1}{0.380\ \text{m}} - \frac{1}{0.260\ \text{m}}\right) = 0.272\ \text{J}$

$\Rightarrow W = -\Delta U = -0.272\ \text{J}.$

24-4: $W = 4.20\times10^{-8}\ \text{J} = -\Delta U = U_i - U_f \Rightarrow U_f = -4.20\times10^{-8}\ \text{J} - 6.40\times10^{-8}\ \text{J} = -1.06\times10^{-7}\ \text{J}$

24-5: (a) $E_i = K_i + U_i = \frac{1}{2}(.00200\ \text{kg})(22.0\ \text{m/s})^2 + \frac{k(3.00\times10^{-6}\ \text{C})(7.50\times10^{-6}\ \text{C})}{0.800\ \text{m}} = 0.737\ \text{J}$

$\Rightarrow E_f = \frac{1}{2}mv^2 + \frac{kq_1q_2}{r} \Rightarrow v = \sqrt{\frac{2(0.737\ \text{J} - 0.405\ \text{J})}{0.00200\ \text{kg}}} = 18.2\ \text{m/s}.$

(b) At the closest point, the velocity is zero:

$\Rightarrow 0.737\ \text{J} = \frac{kq_1q_2}{r} \Rightarrow r = \frac{k(3.00\times10^{-6}\ \text{C})(7.50\times10^{-6}\ \text{C})}{0.737\ \text{J}} = 0.275\ \text{m}.$

24-6: From **Ex.24-1**, the initial energy E_i can be calculated:

$E_i = K_i + U_i = \frac{1}{2}(9.11\times10^{-31}\ \text{kg})(3.00\times10^{6}\ \text{m/s})^2 + \frac{k(-1.60\times10^{-19}\ \text{C})(3.20\times10^{-19}\ \text{C})}{10^{-10}\ \text{m}}$

$\Rightarrow E_i = -5.03\times10^{-19}\ \text{J}.$

When velocity equals zero, all energy is electric potential energy, so:

$-5.03\times10^{-19}\ \text{J} = -\frac{k2e^2}{r} \Rightarrow r = 9.15\times10^{-10}\ \text{m}.$

24-7: Since the work done is zero, the sum of the work to bring in the two equal charges q must equal the work done in bringing in charge Q.

$W_{qq} = W_{qQ} \Rightarrow -\frac{kq^2}{d} = \frac{2kqQ}{d} \Rightarrow Q = -\frac{q}{2}.$

24-8: $U = \frac{kq^2}{1.00\ \text{m}} + \frac{2kq^2}{1.00\ \text{m}} = 3kq^2 = 3k(8.40\times10^{-7}\ \text{C})^2 = 0.0191\ \text{J}.$

24-9: (a) $U = k\left(\frac{q_1q_2}{r_{12}} + \frac{q_1q_3}{r_{13}} + \frac{q_2q_3}{r_{23}}\right)$

$= k\left(\frac{(2.00 \text{ nC})(-3.00 \text{ nC})}{(0.200 \text{ m})} + \frac{(2.00 \text{ nC})(5.00 \text{ nC})}{(0.100 \text{ m})} + \frac{(-3.00 \text{ nC})(5.00 \text{ nC})}{(0.100 \text{ m})}\right)$

$= -7.20 \times 10^{-7} \text{ J}.$

(b) If $U = 0$, $0 = k\left(\frac{q_1q_2}{r_{12}} + \frac{q_1q_3}{x} + \frac{q_2q_3}{r_{12} - x}\right)$. So solving for x we find:

$0 = -30 + \frac{10}{x} - \frac{15}{0.2 - x} \Rightarrow 30x^2 - 31x + 2 = 0 \Rightarrow x = 0.0691 \text{ m}, \ 0.964 \text{ m}.$

Therefore x = 0.0691 m since it is the only value between the two charges.

24-10: $V = \frac{kq}{r} \Rightarrow q = \frac{rV}{k} = \frac{(0.750 \text{ m})(48.0 \text{ V})}{k} = 4.00 \times 10^{-9} \text{ C}.$

24-11: (a) $V = \frac{kq}{r} \Rightarrow r = \frac{kq}{V} = \frac{k(6.00 \times 10^{-11} \text{ C})}{90.0 \text{ V}} = 6.00 \times 10^{-3} \text{ m}.$

(b) $V = \frac{kq}{r} \Rightarrow r = \frac{kq}{V} = \frac{k(6.00 \times 10^{-11} \text{ C})}{30.0 \text{ V}} = 0.0180 \text{ m}.$

24-12: (a) $W = -\Delta U = qEd = \Delta K = 2.50 \times 10^{-6} \text{ J}.$

(b) The initial point was at a higher potential than the latter since any positive charge, when free to move, will move from greater to lesser potential.

$\Delta V = \Delta U / q = (2.50 \times 10^{-6} \text{ J}) / (4.30 \text{ nC}) = 581 \text{ V}.$

(c) $qEd = 2.50 \times 10^{-6} \text{ J} \Rightarrow E = \frac{2.50 \times 10^{-6} \text{ J}}{(4.30 \text{ nC})(0.05 \text{ m})} = 11600 \text{ N/C}.$

24-13: (a) Point b has a higher potential since it is "upstream" from where the positive charge moves.

(b) $E = \frac{V}{d} = \frac{730 \text{ V}}{0.400 \text{ m}} = 1830 \text{ N/C}.$

(c) $W = -\Delta U = q\Delta V = (-0.20 \times 10^{-6} \text{ C})(730 \text{ V}) = -1.46 \times 10^{-7} \text{ J}.$

24-14: (a) $E = \frac{V}{d} \Rightarrow d = \frac{V}{E} = \frac{452 \text{ V}}{226 \text{ N/C}} = 2.00 \text{ m}.$

(b) $V = \frac{kq}{d} \Rightarrow q = \frac{Vd}{k} = \frac{(452 \text{ V})(2.00 \text{ m})}{k} = 1.00 \times 10^{-7} \text{ V}.$

(c) The electric field is directed away from q since it is a positive charge.

24-15: (a) Work done is zero since the motion is along an equipotential, perpendicular to the electric field.

(b) $W = qEd = -1.24 \times 10^{-3} \text{ J}.$

(c) $W = qEd = (37.0 \text{ nC})E(2.60 \ \cos 45°) = 3.40 \times 10^{-3} \text{ J}$

24-16: Initial energy equals final energy: $E_i = E_f \Rightarrow -\frac{keq_1}{r_{1i}} - \frac{keq_2}{r_{2i}} = -\frac{keq_1}{r_{1f}} - \frac{keq_2}{r_{2f}} + \frac{1}{2}mv_f^{\,2}$

$$\Rightarrow k(-1.60\times10^{-19}\ \text{C})\left(\frac{(-1.13\times10^{-9}\ \text{C})}{0.01\ \text{m}} + \frac{(3.87\times10^{-9}\ \text{C})}{0.04\ \text{m}}\right)$$

$$= k(-1.60\times10^{-19}\ \text{C})\left(\frac{(-1.13\times10^{-9}\ \text{C})}{0.04\ \text{m}} + \frac{(3.87\times10^{-9}\ \text{C})}{0.01\ \text{m}}\right) + \frac{1}{2}(9.11\times10^{-31}\ \text{kg})v^2$$

$$\Rightarrow v = \sqrt{\frac{2}{9.11\times10^{-31}\ \text{kg}}(2.34\times10^{-17}\text{J} + 5.16\times10^{-16}\text{J})} = 3.44\times10^{7}\ \text{m/s}.$$

24-17: (a) At A: $V_A = k\left(\frac{q_1}{r_1} + \frac{q_2}{r_2}\right) = k\left(\frac{6.80\times10^{-9}\ \text{C}}{0.05\ \text{m}} + \frac{-5.10\times10^{-9}\ \text{C}}{0.05\ \text{m}}\right) = 306\ \text{V}.$

(b) At B: $V_B = k\left(\frac{q_1}{r_1} + \frac{q_2}{r_2}\right) = k\left(\frac{6.80\times10^{-9}\ \text{C}}{0.08\ \text{m}} + \frac{-5.10\times10^{-9}\ \text{C}}{0.06\ \text{m}}\right) = 0.$

(c) $W = q\Delta V = (2.50\times10^{-9}\ \text{C})(306\ \text{V}) = 7.65\times10^{-7}\ \text{J}.$

24-18: (a)

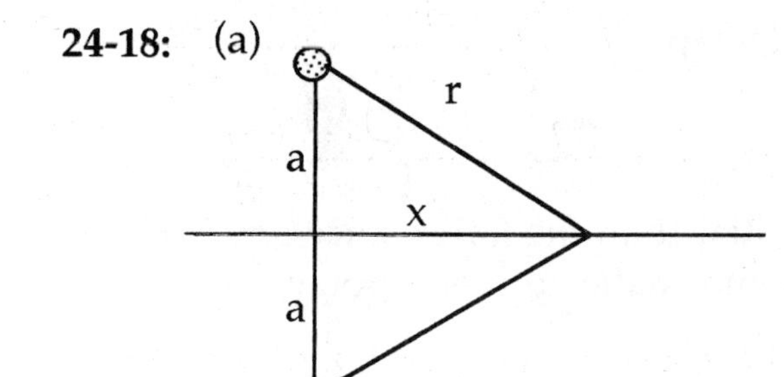

(b) $V = 2\frac{1}{4\pi\varepsilon_o}\frac{q}{a}.$

(c) Looking at the diagram in (a): $V(x) = 2\frac{1}{4\pi\varepsilon_o}\frac{q}{r} = 2\frac{1}{4\pi\varepsilon_o}\frac{q}{\sqrt{a^2+x^2}}.$

(e) When $x >> a$, $V = \frac{1}{4\pi\varepsilon_o}\frac{2q}{x}$, just like a point charge of charge +2q.

(d)

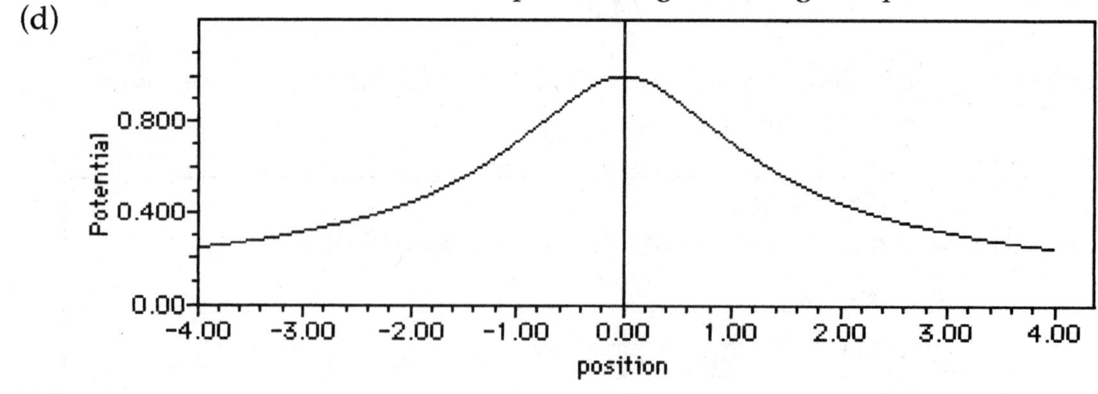

24-19:

(a)

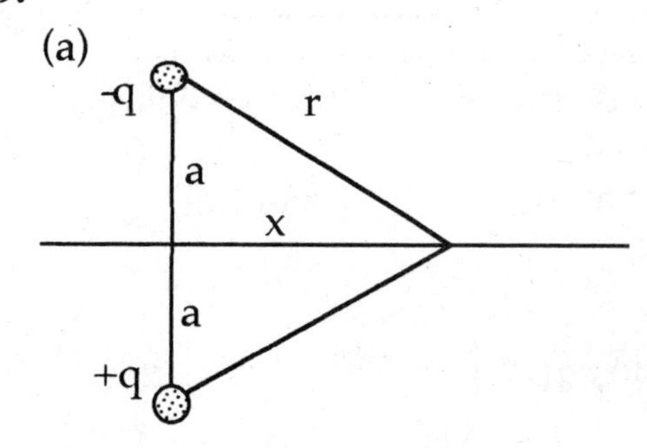

(b) $V_x = \frac{kq}{r} + \frac{k(-q)}{r} = 0.$

(c) The potential along the x-axis is always zero, so a graph would be flat.

(e) If the two charges are interchanged, then the results of (b) and (c) still hold. The potential is zero.

24-20:

(a)

$$|y| < a:\ V = \frac{kq}{(a+y)} - \frac{kq}{(a-y)} = \frac{2kqy}{y^2 - a^2}.$$

$$y > a:\ V = \frac{kq}{(a+y)} - \frac{kq}{(y-a)} = \frac{-2kqa}{y^2 - a^2}.$$

$$y < -a:\ V = \frac{-kq}{(a+y)} - \frac{kq}{(-y+a)} = \frac{2kqa}{y^2 - a^2}$$

(c) $y >> a:\ V = \dfrac{kq}{(a+y)} - \dfrac{kq}{(y-a)} = \dfrac{-2kqa}{y^2}$

(b)

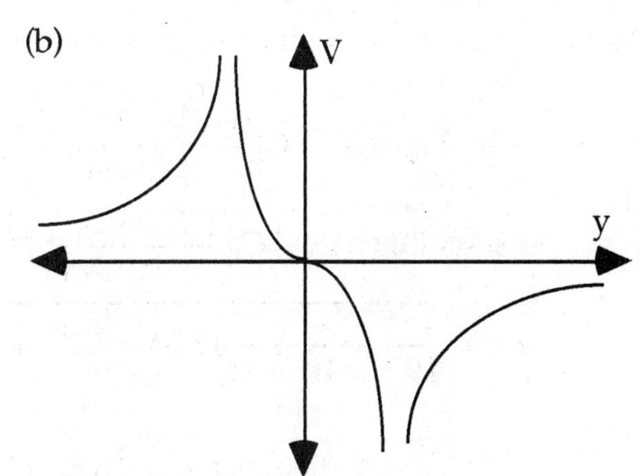

(d) If the charges are interchanged,
then the signs on the answers above become swapped.

24-21:

(a)

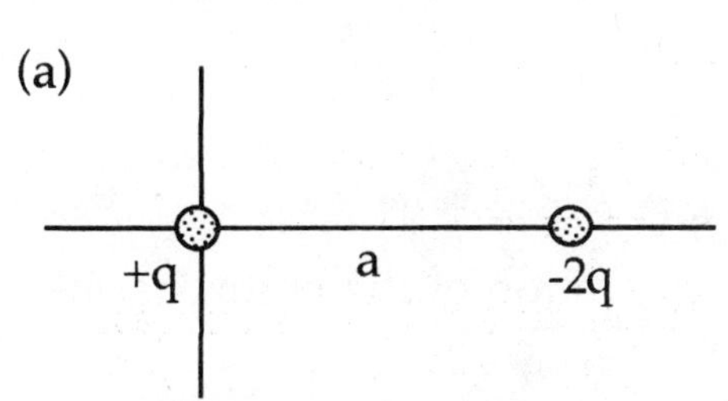

(b) $x > a:\ V = \dfrac{kq}{x} - \dfrac{2kq}{x-a} = \dfrac{-kq(x+a)}{x(x-a)}.$

$0 < x < a:\ V = \dfrac{kq}{x} - \dfrac{2kq}{a-x} = \dfrac{kq(3x-a)}{x(x-a)}$

$x < 0:\ V = \dfrac{-kq}{x} + \dfrac{2kq}{x-a} = \dfrac{kq(x+a)}{x(x-a)}$

(d) The potential is zero at $x = -a$ and $a/3$

(e) For $x >> a:\ V \approx \dfrac{-kqx}{x^2} = \dfrac{-kq}{x}$, which is the same as the potential of a point charge -q . (Note: the two charges must be added with the correct sign.)

24-22: (a) $V = \dfrac{kq}{|y|} - \dfrac{2kq}{r} = kq\left(\dfrac{1}{|y|} - \dfrac{2}{\sqrt{a^2+y^2}}\right).$

(b)

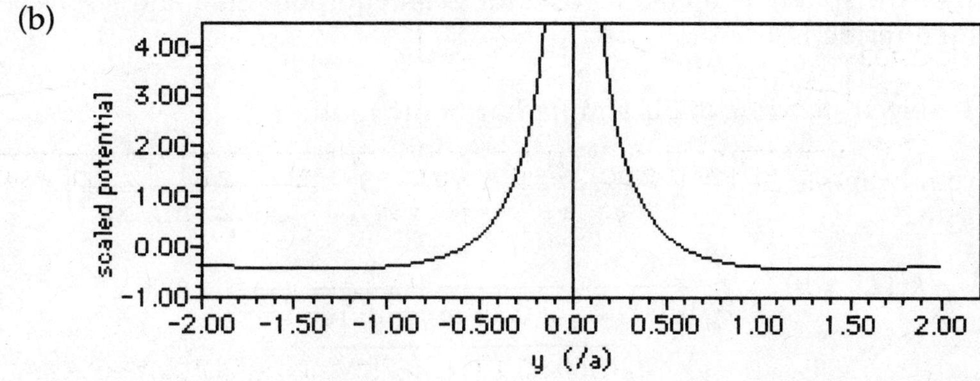

(c) $V = 0$, when $y^2 = \dfrac{a^2+y^2}{4} \Rightarrow 3y^2 = a^2 \Rightarrow y = \pm\dfrac{a}{\sqrt{3}}.$

(d) $y >> a:\ \ V \approx kq\left(\dfrac{1}{y} - \dfrac{2}{y}\right) = -\dfrac{kq}{y}$, which is the potential of a point charge -q.

24-23: $W = -\Delta U = -Vq = (360\ \text{V})(1.60 \times 10^{-19}\ \text{C}) = 5.76 \times 10^{-17}\ \text{J}$. But also:

$$W = \Delta K = \frac{1}{2}mv^2 \Rightarrow v = \sqrt{\frac{2(5.76 \times 10^{-17}\ \text{J})}{9.11 \times 10^{-31}\ \text{kg}}} = 1.12 \times 10^7\ \text{m/s}.$$

24-24: (a) $V = Ed = (670\ \text{N/C})(5.2 \times 10^{-2}\ \text{m}) = 34.8\ \text{V}$.
(b) The higher potential is at the positive sheet.
(c) $E = \dfrac{\sigma}{\varepsilon_o} \Rightarrow \sigma = \varepsilon_o(670\ \text{N/C}) = 5.93 \times 10^{-9}\ \text{C/m}^2$.

24-25: (a) $E = \dfrac{V}{d} = \dfrac{500\ \text{V}}{0.0850\ \text{m}} = 5880\ \text{N/C}$.
(b) $F = Eq = (5880\ \text{N/C})(2.40 \times 10^{-9}\ \text{C}) = 1.41 \times 10^{-5}\ \text{N}$.
(c) $W = Fd = (1.41 \times 10^{-5}\ \text{N})(0.0850\ \text{m}) = 1.20 \times 10^{-6}\ \text{J}$.
(d) $\Delta U = Vq = (500\ \text{N})(2.40 \times 10^{-9}\ \text{C}) = 1.20 \times 10^{-6}\ \text{J}$.

24-26: (a) $E = \dfrac{\sigma}{\varepsilon_o} = \dfrac{3.50 \times 10^{-9}\ \text{C/m}^2}{\varepsilon_o} = 395\ \text{N/C}$

$\Rightarrow V = Ed = (395\ \text{N/C})(0.0450\ \text{m}) = 17.8\ \text{V}$.
(b) The electric field stays the same if the separation of the plates doubles, while the potential between the plates doubles.

24-27: (a) $E = \dfrac{V}{d} \Rightarrow d = \dfrac{V}{E} = \dfrac{3750\ \text{V}}{3.00 \times 10^6\ \text{N/C}} = 1.25 \times 10^{-3}\ \text{m}$.
(b) $E = \dfrac{\sigma}{\varepsilon_o} \Rightarrow \sigma = \varepsilon_o(3.00 \times 10^{-6}\ \text{N/C}) = 2.66 \times 10^{-5}\ \text{C/m}^2$.

24-28: (a) $V = \dfrac{kq}{r} = \dfrac{k(2.60 \times 10^{-9}\ \text{C})}{0.200\ \text{m}} = 117\ \text{V}$.
(b) Since the sphere is metal, its interior is an equipotential, and so the potential inside is 117 V.

24-29: (a) The electron will exhibit simple harmonic motion.
(b) From Example 24-11, $V = \dfrac{kQ}{\sqrt{x^2 + a^2}} \Rightarrow \Delta V = kQ\left(\dfrac{1}{a} - \dfrac{1}{\sqrt{(0.25\ \text{m})^2 + a^2}}\right)$

$$\Rightarrow \Delta V = k(12.0 \times 10^{-9}\ \text{C})\left(\frac{1}{0.100\ \text{m}} - \frac{1}{\sqrt{(0.25\ \text{m})^2 + (0.100\ \text{m})^2}}\right) = 678\ \text{V}$$

But $W = q\Delta V = \dfrac{1}{2}mv^2 \Rightarrow v = \sqrt{\dfrac{2(1.60 \times 10^{-19}\ \text{C})(678\ \text{V})}{9.11 \times 10^{-31}\ \text{kg}}} = 1.54 \times 10^7\ \text{m/s}$.

24-30: Energy is conserved:

$$\frac{1}{2}mv^2 = q\Delta V \Rightarrow \Delta V = \frac{(1.67\times10^{-27}\ \text{kg})(2500\ \text{m/s})^2}{2(1.60\times10^{-19}\ \text{C})} = 0.0326\ \text{V}.$$

But: $\Delta V = \frac{\lambda}{2\pi\varepsilon_o}\ln(r_o/r) \Rightarrow r_o = r\exp\left(\frac{2\pi\varepsilon_o\Delta V}{\lambda}\right) \Rightarrow r = r_o\exp\left(-\frac{2\pi\varepsilon_o\Delta V}{\lambda}\right)$

$$\Rightarrow r = (0.180\ \text{m})\exp\left(-\frac{2\pi\varepsilon_o(0.0326\ \text{V})}{4.00\times10^{-12}\ \text{C/m}}\right) = 0.114\ \text{m}.$$

24-31:

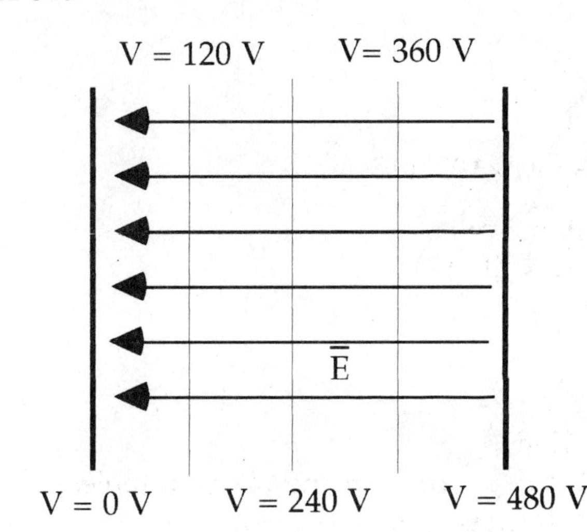

(a) Equipotentials and electric field lines of two large parallel plates are shown at left.

(b) The electric field lines and the equipotential lines are mutually perpendicular.

24-32: (a) (i) $r < r_a$: $V = \frac{kq}{r_a} - \frac{kq}{r_b} = kq\left(\frac{1}{r_a} - \frac{1}{r_b}\right)$.

(ii) $r_a < r < r_b$: $V = \frac{kq}{r} - \frac{kq}{r_b} = kq\left(\frac{1}{r} - \frac{1}{r_b}\right)$.

(iii) $r > r_b$: $V = 0$, since outside a sphere the potential is the same as for a point charge. Therefore we have the identical potential to two oppositely charged point charges at the same location. These potentials cancel.

(b) $V_a = \frac{1}{4\pi\varepsilon_o}\left(\frac{q}{r_a} - \frac{q}{r_b}\right)$ and $V_b = 0 \Rightarrow V_{ab} = \frac{1}{4\pi\varepsilon_o}q\left(\frac{1}{r_a} - \frac{1}{r_b}\right)$.

(c) $r_a < r < r_b$: $E = -\frac{\partial V}{\partial r} = -\frac{q}{4\pi\varepsilon_o}\frac{\partial}{\partial r}\left(\frac{1}{r} - \frac{1}{r_b}\right) = +\frac{1}{4\pi\varepsilon_o}\frac{q}{r^2} = \frac{V_{ab}}{\left(\frac{1}{r_a}-\frac{1}{r_b}\right)}\frac{1}{r^2}$.

(d) From Equation 24-23: $E = 0$, since V is zero outside the spheres.

(e) If the outer charge is different, then outside the outer sphere the potential is no longer zero but is $V = \frac{1}{4\pi\varepsilon_o}\frac{q}{r} - \frac{1}{4\pi\varepsilon_o}\frac{Q}{r} = \frac{1}{4\pi\varepsilon_o}\frac{(q-Q)}{r}$. All potentials inside the outer shell are just shifted by an amount $V = -\frac{1}{4\pi\varepsilon_o}\frac{Q}{r_b}$. Therefore relative potentials within the shells are not affected. Thus (b) and (c) do not change. However, now that the potential does vary outside the spheres, there is an electric field there: $E = -\frac{\partial V}{\partial r} = -\frac{\partial}{\partial r}\left(\frac{kq}{r} + \frac{-kQ}{r}\right) = \frac{kq}{r^2}\left(1 - \frac{Q}{q}\right)$.

24-33: (a) $V_{ab} = kq\left(\frac{1}{r_a} - \frac{1}{r_b}\right) = 500\ \text{V} \Rightarrow q = \frac{500\ \text{V}}{k\left(\frac{1}{0.012\ \text{m}} - \frac{1}{0.096\text{m}}\right)} = 7.62 \times 10^{-10}\ \text{C}$.

(b)

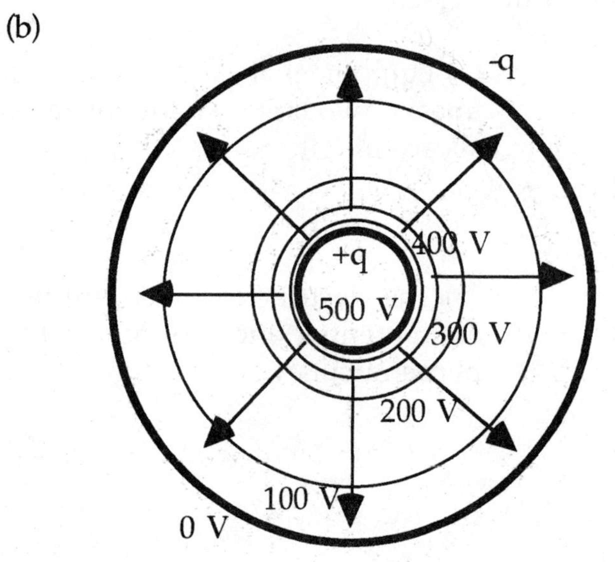

(c) The equipotentials are closest when the electric field is largest.

24-34: (a) $E_x = -\frac{\partial V}{\partial x} = -\frac{\partial}{\partial x}\left(\frac{kQ}{2a}\ln\left(\frac{\sqrt{a^2 + x^2} + a}{\sqrt{a^2 + x^2} - a}\right)\right)$

$$\Rightarrow E_x = -\frac{kQ}{2a}\left[\frac{\partial}{\partial x}\ln(\sqrt{a^2 + x^2} + a) - \frac{\partial}{\partial x}\ln(\sqrt{a^2 + x^2} - a)\right]$$

$$= -\frac{kQ}{2a}\left[\frac{x(a^2 + x^2)^{-1/2}}{\sqrt{a^2 + x^2} + a} - \frac{x(a^2 + x^2)^{-1/2}}{\sqrt{a^2 + x^2} - a}\right] = \frac{kQ}{x\sqrt{a^2 + x^2}}$$

$$\Rightarrow E_x = \frac{(2a\lambda)}{4\pi\varepsilon_0 xa\sqrt{1 + x^2 / a^2}} = \frac{1}{2\pi\varepsilon_0 x\sqrt{1 + x^2 / a^2}}\frac{\lambda}{}.$$

(b) The potential was evaluated at y and z equal to zero, and thus shows no dependence on them. However, the electric field depends upon the derivative of the potential, and the potential could still have a functional dependence on the variables y and z, and hence E_y and E_z may be non-zero.

24-35: (a) $E_x = -\frac{\partial V}{\partial x} = -\frac{\partial}{\partial x}(axy + by^2 + cy) = -ay = -3.00y\ \text{V/m}$.

$$E_y = -\frac{\partial V}{\partial y} = -\frac{\partial}{\partial y}(axy + by^2 + cy) = -ax - 2by - c = (-3.00x + 4.00y - 5.00)\ \text{V/m}.$$

$$E_z = -\frac{\partial V}{\partial z} = -\frac{\partial}{\partial z}(axy + by^2 + cy) = 0.$$

The electric field equals zero when $y = 0,\ 3x + 5 = 0 \Rightarrow x = -5/3$ and any z.

24-36: (a) $E_x = -\dfrac{\partial V}{\partial x} = -\dfrac{\partial}{\partial x}\left(\dfrac{kQ}{\sqrt{x^2 + y^2 + z^2}}\right) = \dfrac{kQx}{(x^2 + y^2 + z^2)^{3/2}} = \dfrac{kQx}{r^3}$.

Similarly, $E_y = \dfrac{kQy}{r^3}$ and $E_z = \dfrac{kQz}{r^3}$.

(b) So from (a), $\vec{E} = \dfrac{kQ}{r^2}\left(\dfrac{x\hat{i}}{r} + \dfrac{y\hat{j}}{r} + \dfrac{z\hat{k}}{r}\right) = \dfrac{kQ}{r^2}\hat{r}$, which agrees with Eq. (22-7).

24-37: (a) There is no dependence of the potential on x or z, and so it has no components in those directions. However, there is y dependence:

$$E_y = -\frac{\partial V}{\partial y} = -C \Rightarrow \vec{E} = -C\hat{j},\ \text{for } 0 < y < d.$$

and $\vec{E} = 0$, for $y > d$, since the potential is constant there.

(b) Infinite parallel plates of opposite charge could yield this electric field, where the surface charge is $\sigma = C\varepsilon_o$.

24-38: $W = q\Delta V = \dfrac{1}{2}m(v_f^2 - v_i^2) \Rightarrow v_f^2 = v_i^2 + \dfrac{2q\Delta V}{m}$

$\Rightarrow v_f^2 = (3.00 \times 10^6 \text{ m/s})^2 + \dfrac{2(1.60 \times 10^{-19} \text{ C})(108 \text{ V})}{9.11 \times 10^{-31} \text{ kg}}$

$\Rightarrow v_f = \sqrt{4.69 \times 10^{13}} \text{ m/s} = 6.85 \times 10^6 \text{ m/s}$.

24-39: (a) $t = \dfrac{0.060 \text{ m}}{8.00 \times 10^6 \text{ m/s}} = 7.5 \times 10^{-9}$ s, also $F = Eq \Rightarrow a = \dfrac{Eq}{m}$

$d_y = \dfrac{1}{2}at^2 = \dfrac{1}{2}\dfrac{Eq}{m}t^2 = \dfrac{(1400 \text{ N/C})(1.60 \times 10^{-19} \text{ C})(7.5 \times 10^{-9} \text{ s})^2}{2(9.11 \times 10^{-13} \text{ kg})} = 6.92 \times 10^{-3}$ m.

(b) Angle $\theta = \arctan(v_y / v_x) = \arctan((Eqm/t)/v_x) = \arctan(1.84/8.00) = 13.0°$.

(c) The distance below center of the screen is:

$D = d_y + v_y t = 6.92 \times 10^{-3} \text{ m} + (1.84 \times 10^6 \text{ m/s})\dfrac{0.120 \text{ m}}{8.00 \times 10^6 \text{ m/s}} = 0.0346$ m.

24-40: (a) $F = Eq = (1.80 \times 10^4 \text{ N/C})(1.60 \times 10^{-19} \text{ C}) = 2.88 \times 10^{-15}$ N.

(b) $a = F/m = (2.88 \times 10^{-15} \text{ N})/(9.11 \times 10^{-31} \text{ kg}) = 3.16 \times 10^{15} \text{ m/s}^2$.

24-41: The derivation is exactly the same as for Eq. (24-35) except for two extra terms $E_z(x, y, z + \Delta l)$ and $E_z(x, y, z - \Delta l)$, leading to a factor of $1/6$ in front of the expression when the average is taken.

Because the potential is calculated as the average of surrounding values, no point can ever be a local maximum or minimum, since that could not be obtained by average surrounding points. Therefore, the derivative of the potential never yields zero, (equivalent to finding maximum and minimum points), and so the electric field is never zero. Thus, any charged particle in that region will always feel a force and will never be in static equilibrium.

24-42: (a) $d((x,y),(x+cE_x/E, y+cE_y/E)) = \sqrt{\left(\left(\frac{cE_x}{E}+x\right)-x\right)^2 + \left(\left(\frac{cE_y}{E}+y\right)-y\right)^2}$

$$\Rightarrow d((x,y),(x+cE_x/E, y+cE_y/E)) = \sqrt{\left(\frac{cE_x}{E}\right)^2 + \left(\frac{cE_y}{E}\right)^2} = \frac{c}{E}\sqrt{E_x^{\ 2}+E_y^{\ 2}} = c.$$

(b) The direction of the line is: $\frac{\Delta y}{\Delta x} = \frac{cE_y/E}{cE_x/E} = \frac{E_y}{E_x}$ which is the direction of the electric field.

24-43 to 24-49:
These are spreadsheet or computer exercises using the relaxation method outlined in detail in section 24-8, and based upon the steps given on p756-7.

24-50: To remove the electrons to infinity, one charge is moved away from both the nucleus and the other electron (two potential energy terms) and the remaining electron just is removed to infinity from the nucleus (one potential energy term). So the total energy required is:

$$U = (U_{ee} + U_{en}) + U_{en} = \left(\frac{k(-e)}{d}(2e)\frac{d^3}{R^3} + \frac{k(-e)(-e)}{2d}\right) + \frac{k(-e)}{d}(2e)\frac{d^3}{R^3}$$

$$\Rightarrow \text{Energy required E} = -U = \frac{ke^2}{d}\left(\frac{4d^3}{R^3} - \frac{1}{2}\right)$$

24-51: (a) $V = \frac{kq}{r} = \frac{k(92)(1.60\times10^{-19}\text{ C})}{7.40\times10^{-19}\text{ m}} = 1.79\times10^7\text{ V}.$

(b) $U = \frac{kq^2}{r} = \frac{k(92\times1.60\times10^{-19}\text{ C})^2}{12.0\times10^{-15}\text{ m} + 2(7.40\times10^{-19}\text{ m})} = 7.28\times10^{-11}\text{ J} = 4.55\times10^8\text{ eV}.$

24-52: (a) $U = 2kq^2\left(-\frac{1}{d}+\frac{1}{2d}-\frac{1}{3d}+\ldots\right) = -\frac{2kq^2}{d}\sum_{i=1}^{\infty}\frac{(-1)^{i-1}}{i}.$

(b) $U = -\frac{2kq^2}{d}\ln(2).$

(c) The potential energy is the same for the negative ions - the equations are identical if we examine (a).

(d) If $d = 2.82\times10^{-10}$ m, then $U = -\frac{2k(1.60\times10^{-19}\text{ C})^2\ln(2)}{2.82\times10^{-10}\text{ m}} = -1.13\times10^{-18}\text{ J}.$

(e) The real energy (-0.80×10^{-19} J) is about 70% of that calculated above.

24-53: (a) $U_e = \frac{-2ke^2}{r} = \frac{-2k(1.60\times10^{-19}\text{ C})^2}{0.535\times10^{-10}\text{ m}} = -8.60\times10^{-18}\text{ J}.$

(b) If all the kinetic energy goes into potential energy:

$$U_t = U_e + K = -8.60 \times 10^{-18}\ \text{J} + 1.32 \times 10^{-20}\ \text{J} = \frac{2ke^2}{\sqrt{d^2 + x^2}}$$

$$\Rightarrow x^2 = \frac{4k^2e^4}{U_t^2} - d^2 = 2.87 \times 10^{-21}\ \text{m}^2 - (5.35 \times 10^{-11}\ \text{m})^2 = 8.82 \times 10^{-24}\ \text{m}^2$$

(Note that we must be careful to keep all significant figures along the way.)

$\Rightarrow x = 2.97 \times 10^{-12}$ m.

24-54: $F_e = mg\tan\theta = (3.20 \times 10^{-3}\ \text{kg})(9.80\ \text{m/s}^2)\tan(30°) = 0.0181\ \text{N}$.

But also: $F_e = Eq = \dfrac{Vq}{d} \Rightarrow V = \dfrac{Fd}{q} = \dfrac{(0.0181\ \text{N})(0.0500\ \text{m})}{5.80 \times 10^{-6}\ \text{C}} = 156\ \text{V}$.

24-55: (a) $U = kq^2\left(-\frac{3}{d} + \frac{3}{\sqrt{2}d} - \frac{1}{\sqrt{3}d}\right) + kq^2\left(-\frac{2}{d} + \frac{3}{\sqrt{2}d} - \frac{1}{\sqrt{3}d}\right) + kq^2\left(-\frac{2}{d} + \frac{2}{\sqrt{2}d} - \frac{1}{\sqrt{3}d}\right)$

$+ kq^2\left(-\frac{1}{d} + \frac{2}{\sqrt{2}d} - \frac{1}{\sqrt{3}d}\right) + kq^2\left(-\frac{2}{d} + \frac{1}{\sqrt{2}d}\right) + kq^2\left(-\frac{1}{d} + \frac{1}{\sqrt{2}d}\right) + kq^2\left(-\frac{1}{d}\right)$

$$\Rightarrow U = kq^2\left(-\frac{12}{d} + \frac{12}{\sqrt{2}d} - \frac{4}{\sqrt{3}d}\right) = -\frac{12kq^2}{d}\left(\frac{1}{d} - \frac{1}{\sqrt{2}d} - \frac{1}{3\sqrt{3}d}\right).$$

(b) The fact that the electric potential energy is less than zero means that is energetically favourable for the crystal ions to be together.

24-56: (a) $a_e = \dfrac{F}{m_e} = \dfrac{Ee}{m_e} = \dfrac{Ve}{dm_e} = \dfrac{(1400\ \text{V})(1.60 \times 10^{-19}\ \text{C})}{(0.0400\ \text{m})(9.11 \times 10^{-31}\ \text{kg})} = 6.15 \times 10^{15}\ \text{m/s}^2$.

$$a_p = \frac{F}{m_p} = \frac{Ee}{m_p} = \frac{Ve}{dm_p} = \frac{(1400\ \text{V})(1.60 \times 10^{-19}\ \text{C})}{(0.0400\ \text{m})(1.67 \times 10^{-27}\ \text{kg})} = 3.35 \times 10^{12}\ \text{m/s}^2$$

The two particles will meet when their positions are equal:

$$x_e = \frac{1}{2}a_e t^2 = d - \frac{1}{2}a_p t^2 = x_p \Rightarrow t = \sqrt{\frac{2d}{a_e + a_p}} = 3.61 \times 10^{-9}\ \text{s}.$$

So the distanc efrom the positive plate is just the proton's position:

$$\Rightarrow x_p = \frac{1}{2}a_p t^2 = \frac{1}{2}(3.35 \times 10^{12}\ \text{m/s}^2)(3.61 \times 10^{-9}\ \text{s})^2 = 2.18 \times 10^{-5}\ \text{m}.$$

(b) $W = qEd = \dfrac{1}{2}mv^2 \Rightarrow v \propto \dfrac{1}{\sqrt{m}} \Rightarrow \dfrac{v_e}{v_p} = \sqrt{\dfrac{m_p}{m_e}} = 42.8$.

(c) $\dfrac{K_e}{K_p} = 1$.

24-57: (a) $W_E = \Delta K - W_F = 6.20 \times 10^{-5}\ \text{J} - 8.45 \times 10^{-5}\ \text{J} = -2.25 \times 10^{-5}\ \text{J}$.

(b) $W_E = Vq \Rightarrow V = \dfrac{W_E}{q} = \dfrac{-2.25 \times 10^{-5}\ \text{J}}{4.30 \times 10^{-9}\ \text{C}} = -5230\ \text{V}$.

(c) $W = qEd \Rightarrow E = \frac{W}{qd} = \frac{2.25 \times 10^{-5}\text{ J}}{(4.30 \times 10^{-9}\text{ C})(0.0500\text{ m})} = 1.05 \times 10^{5}\text{ C}.$

24-58: (a) $\frac{mv^2}{r} = \frac{ke^2}{r^2} \Rightarrow v = \sqrt{\frac{ke^2}{mr}}.$

(b) $K = \frac{1}{2}mv^2 = \frac{1}{2}\frac{ke^2}{r} = -\frac{1}{2}U.$

(c) $E = K + U = \frac{1}{2}U = -\frac{1}{2}\frac{ke^2}{r} = -\frac{1}{2}\frac{k(1.60 \times 10^{-19}\text{ C})^2}{5.29 \times 10^{-11}\text{ m}} = -2.17 \times 10^{-18}\text{ J} = -13.6\text{ eV}.$

24-59: (a) $V = Cx^{4/3} \Rightarrow C = (160\text{ V})(0.0146\text{ m})^{-4/3} = 4.48 \times 10^{4}\text{ V/m}^{4/3}.$

(b) $E = -\frac{\partial V}{\partial x} = -\frac{4}{3}Cx^{1/3} = -\frac{4}{3}(4.48 \times 10^{4}\text{ V/m}^{4/3})x^{1/3} = (-5.98 \times 10^{4}\, x^{1/3})\text{ V/m}.$

(c) $F = Ee = (-5.98 \times 10^{4}(0.00730)^{1/3}\text{ V/m})(1.60 \times 10^{-19}\text{ C}) = 1.86 \times 10^{-15}\text{ N}.$

24-60: (a) $V(0.03,0) = \frac{k(-3.00 \times 10^{-9}\text{ C})}{0.0300\text{ m}} + \frac{k(2.00 \times 10^{-9}\text{ C})}{\sqrt{(0.0300^2 + 0.0500^2)}\text{ m}} = -591\text{ V}.$

$V(0.03,0.05) = \frac{k(-3.00 \times 10^{-9}\text{ C})}{\sqrt{(0.0300^2 + 0.0500^2)}\text{ m}} + \frac{k(2.00 \times 10^{-9}\text{ C})}{0.0300\text{ m}} = 137\text{ V}.$

(b) $W = q\Delta V = (-6.00 \times 10^{-9}\text{ C})(728\text{ V}) = -4.37 \times 10^{-6}\text{ J}.$
Note that the work done by the field is negative, since the charge is moved AGAINST the electric field.

24-61: (a) (*i*) $V = \frac{\lambda}{2\pi\varepsilon_0}(\ln(b/a) - \ln(b/b)) = \frac{\lambda}{2\pi\varepsilon_0}\ln(b/a).$

(*ii*) $V = \frac{\lambda}{2\pi\varepsilon_0}(\ln(b/r) - \ln(b/b)) = \frac{\lambda}{2\pi\varepsilon_0}\ln(b/r).$

(*iii*) $V = 0.$

(b) $V_{ab} = V(a) - V(b) = \frac{\lambda}{2\pi\varepsilon_0}\ln(b/a).$

(c) Between the cylinders: $V = \frac{\lambda}{2\pi\varepsilon_0}\ln(b/r) = \frac{V_{ab}}{\ln(b/a)}\ln(b/r)$

$\therefore E = -\frac{\partial V}{\partial r} = -\frac{V_{ab}}{\ln(b/a)}\frac{\partial}{\partial r}(\ln(b/r)) = \frac{V_{ab}}{\ln(b/a)}\frac{1}{r}.$

(d) The potential difference between the two cylinders is identical to that in part (b) even if the outer cylinder has no charge.

24-62: Using the results of **24-61**, we can calculate the potential difference:

$E = \frac{V_{ab}}{\ln(b/a)}\frac{1}{r} \Rightarrow V_{ab} = E\ln(b/a)r$

$\Rightarrow V_{ab} = (6.00 \times 10^{4}\text{ N/C})(\ln(0.02/50 \times 10^{-6}))50 \times 10^{-6}\text{ m} = 5390\text{ V}.$

24-63: (a) From **24-61**, $E = \frac{V_{ab}}{\ln(b/a)}\frac{1}{r} = \frac{60{,}000\ \text{V}}{\ln(0.120/8.00\times10^{-5})}\frac{1}{0.0600\ \text{m}}$

$\Rightarrow E = 1.37\times10^{5}\ \text{V/m}.$

(b) $F = Eq = 10mg \Rightarrow q = \frac{10(3.00\times10^{-5}\ \text{kg})(9.80\ \text{m/s}^2)}{1.37\times10^{5}\ \text{V/m}} = 2.15\times10^{-8}\ \text{C}.$

24-64: (a) From Example 24-12: $V(x) = \frac{kQ}{2a}\ln\left[\frac{\sqrt{a^2+x^2}+a}{\sqrt{a^2+x^2}-a}\right] = \frac{kQ}{2a}\ln\left[\frac{\sqrt{1+a^2/x^2}+a/x}{\sqrt{1+a^2/x^2}-a/x}\right]$

If $a << x$, $\sqrt{1+a^2/x^2} \pm a/x \approx 1+\frac{1}{2}\left(\frac{a}{x}\right)^2 \pm \frac{a}{x} \approx 1 \pm \frac{a}{x}$, and $\ln(1+\alpha) \approx \alpha + \frac{1}{2}\alpha^2 + \ldots$

$\Rightarrow V(x) \approx \frac{kQ}{2a}\left[\left(\frac{a}{x}+\frac{1}{2}\left(\frac{a}{x}\right)^2+\ldots\right)-\left(-\frac{a}{x}+\frac{1}{2}\left(\frac{a}{x}\right)^2+\ldots\right)\right] = \frac{kQ}{2a}\left[\frac{2a}{x}\right] = \frac{kQ}{x}.$

That is, the finite rod acts like a point charge when you are a long way from it.

(b) From Example 24-12: $V(x) = \frac{kQ}{2a}\ln\left[\frac{\sqrt{a^2+x^2}+a}{\sqrt{a^2+x^2}-a}\right] = \frac{kQ}{2a}\ln\left[\frac{\sqrt{1+x^2/a^2}+1}{\sqrt{1+x^2/a^2}-1}\right]$

If $x << a$, $\sqrt{1+x^2/a^2} \pm 1 \approx 1 \pm 1 + \frac{1}{2}\left(\frac{a}{x}\right)^2$, and $\ln(1+\alpha) \approx \alpha + \frac{1}{2}\alpha^2 + \ldots$

$\Rightarrow V(x) \approx \frac{kQ}{2a}\left[\ln\left(\frac{(2+x^2/2a^2)}{(x^2/2a^2)}\right)\right] = \frac{kQ}{2a}\left[\ln\left(\frac{4a^2}{x^2}+1\right)\right] \approx \frac{kQ}{a}\ln(2a/x) = \frac{Q}{4\pi\varepsilon_0 a}\ln(2a/x).$

Thus $\lambda \equiv \frac{Q}{2a}$, and $R = 2a$, which is the only natural length in the problem.

24-65: (a) $dV = \frac{kQ}{\sqrt{x^2+r^2}}\left[\frac{2\pi r dr}{\pi R^2}\right] = \frac{2kQ}{R^2}\frac{rdr}{\sqrt{x^2+r^2}}$

$V = \int_0^R dV = \frac{2kQ}{R^2}\int_0^R \frac{rdr}{\sqrt{x^2+r^2}} = \frac{2kQ}{R^2}\left(z^{1/2}\right)\Big|_{z=x^2}^{z=x^2+R^2} = \frac{2kQ}{R^2}\left[\sqrt{x^2+R^2}-x\right].$

$\Rightarrow E_x = -\frac{\partial V}{\partial x} = -\frac{2kQ}{R^2}\left[\frac{x}{\sqrt{x^2+R^2}}-1\right] = \frac{\sigma}{2\varepsilon_0}\left[1-\frac{1}{\sqrt{1+R^2/x^2}}\right].$

24-66: Recall from Example 24-12 for a line of charge of length a:

$$V = \frac{kQ}{a}\ln\left[\frac{\sqrt{a^2/4+x^2}+a/2}{\sqrt{a^2/4+x^2}-a/2}\right]$$

(a) For a square with two sets of oppositely charged sides, the potentials cancel and $V = 0$.

(b) If all sides have the same charge we have:

$V = \frac{4kQ}{a}\ln\left[\frac{\sqrt{a^2/4+x^2}+a/2}{\sqrt{a^2/4+x^2}-a/2}\right]$, but here x = a/2, so:

$$\Rightarrow V = \frac{4kQ}{a}\ln\left[\frac{\sqrt{a^2+4x^2}+a}{\sqrt{a^2+4x^2}-a}\right] = \frac{4kQ}{a}\ln\left[\frac{(\sqrt{2}+1)}{(\sqrt{2}-1)}\right].$$

24-67: (a) Recall: $r < R$: $E = \frac{\rho r}{2\varepsilon_o} \Rightarrow V = -\int_R^r E \cdot d\vec{r} = -\frac{\rho}{2\varepsilon_o}\int_R^r r\,dr = -\frac{\rho}{4\varepsilon_o}(r^2 - R^2)$

So with $\lambda = \pi R^2 \rho$, $V = -k\lambda(r^2 / R^2 - 1)$.

(b) For $r < R$: $E = \frac{\rho R^2}{2\varepsilon_o r} \Rightarrow V = -\int_R^r \vec{E}\cdot d\vec{r} = -\frac{\rho R^2}{2\varepsilon_o}\int_R^r \frac{dr}{r} = -\frac{\lambda}{2\pi\varepsilon_o}\ln\left[\frac{r}{R}\right] = -2k\lambda\ln\left[\frac{r}{R}\right].$

24-68: $dV = \frac{1}{4\pi\varepsilon_o}\frac{dq}{r} = \frac{1}{4\pi\varepsilon_o}\frac{\lambda dl}{a} = \frac{1}{4\pi\varepsilon_o}\frac{Q}{\pi a}\frac{dl}{a} = \frac{1}{4\pi\varepsilon_o}\frac{Qd\theta}{\pi a} \Rightarrow V = \frac{1}{4\pi\varepsilon_o}\int_0^\pi \frac{Qd\theta}{\pi a} = \frac{1}{4\pi\varepsilon_o}\frac{Q}{a}.$

24-69: From Example 22-10, we have: $E_x = \frac{1}{4\pi\varepsilon_o}\frac{Qx}{(x^2+a^2)^{3/2}}$

$$\Rightarrow V = -\frac{Q}{4\pi\varepsilon_o}\int_\infty^x \frac{x'}{(x'^2+a^2)^{3/2}}dx' = \frac{Q}{4\pi\varepsilon_o}u^{-1/2}\Big|_{u=\infty}^{u=x^2+a^2} = \frac{1}{4\pi\varepsilon_o}\frac{Q}{\sqrt{x^2+a^2}} = \text{Eq. 24-16.}$$

24-70: From Example 22-10, we have:

$r > R$: $E = \frac{kQ}{r^2} \Rightarrow V = -kQ\int_\infty^r \frac{dr'}{r'^2} = \frac{kQ}{r}$

$r < R$: $E = \frac{kQr}{R^3} \Rightarrow V = -\int_\infty^R \vec{E}\cdot d\vec{r} - \int_R^r \vec{E}\cdot d\vec{r}' = \frac{kQ}{R} - \frac{kQ}{R^3}\int_R^r r'dr'$

$$\Rightarrow V = \frac{kQ}{R} - \frac{kQ}{R^3}\frac{1}{2}r'^2\Big|_R^r = \frac{kQ}{R} - \frac{kQ}{2R} + \frac{kQr^2}{2R^3} \quad \therefore V = \frac{kQ}{2R}\left[1 + \frac{r^2}{R^2}\right]$$

(b)

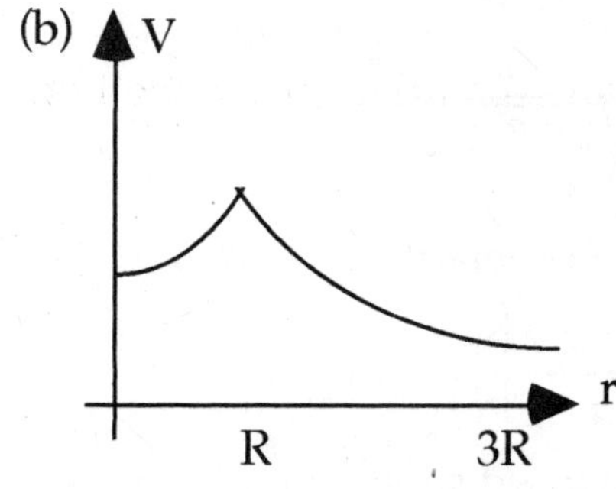

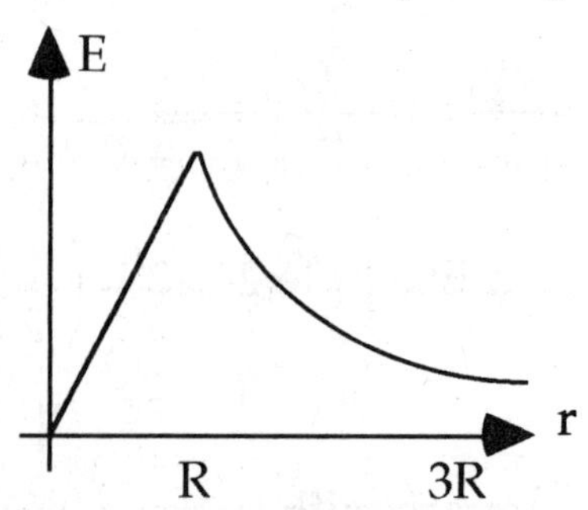

24-71: (a) S_1 and S_3: $V_{13} = -\int_0^{0.2}(3x\hat{i} + 4y\hat{j})\cdot\hat{j}dy = -4\,\frac{1}{2}y^2\Big|_0^{0.2} = -0.080\ \text{V}$, and so S_1 is at a higher potential.

(b) S_2 and S_4: $V_{24} = -\int_0^{0.2} (3x\hat{i} + 4y\hat{j}) \cdot \hat{k} dz = 0$, so both are at the same potential.

(c) S_5 and S_6: $V_{56} = -\int_0^{0.2} (3x\hat{i} + 4y\hat{j}) \cdot \hat{i} dx = -3 \frac{1}{2} x^2 \Big|_0^{0.2} = -0.060 \text{ V}$, and so S_6 is at a higher potential.

24-72: (a) At $r = c$: $V = -\int_\infty^c \frac{kq}{r^2} dr = \frac{kq}{c}$.

(b) At $r = b$: $V = -\int_\infty^c \vec{E} \cdot d\vec{r} - \int_c^b \vec{E} \cdot d\vec{r} = \frac{kq}{c} - 0 = \frac{kq}{c}$.

(c) At $r = a$: $V = -\int_\infty^c \vec{E} \cdot d\vec{r} - \int_c^b \vec{E} \cdot d\vec{r} - \int_b^a \vec{E} \cdot d\vec{r} = \frac{kq}{c} + kq \int_b^a \frac{dr}{r^2} = kq \left[\frac{1}{c} + \frac{1}{b} - \frac{1}{a} \right]$.

(d) At $r = 0$: $V = kq \left[\frac{1}{c} + \frac{1}{b} - \frac{1}{a} \right]$ since it is inside a metal sphere, and thus at the same potential as its surface.

24-73: Using the electric field from Problem **23-27**, the potential difference between the conducting sphere and insulating shell is:

$$V = -\int_{2R}^{R} \vec{E} \cdot d\vec{r} = -\int_{2R}^{R} \frac{kQ}{r^2} dr = kQ \left[\frac{1}{R} - \frac{1}{2R} \right] \Rightarrow V = \frac{kQ}{2R}.$$

24-74: Using the electric field from Problem **23-35**, the potential difference between the inner and outer shells is:

$$V_{ab} = -\int_b^a \vec{E} \cdot d\vec{r} = -\int_b^a \frac{2k\lambda}{r} dr = -2k\lambda \ln(a/b) \Rightarrow V = 2k\lambda \ln(b/a).$$

24-75: Using the electric field from Problem **23-44**, the potential difference between the two face of the uniformly charged slab is:

$$V = -\int_{-d}^{d} \vec{E} \cdot d\vec{r} = -\int_{-d}^{d} \frac{\rho x}{2\varepsilon_o} dx = \frac{\rho}{2\varepsilon_o} \left(\frac{x^2}{2} \right) \Big|_{-d}^{d} \Rightarrow V = 0.$$

24-76: (a) $V = \frac{kQ}{r} = \frac{k(-1.70 \times 10^{-12} \text{ C})}{4.50 \times 10^{-4} \text{ m}} = -34.0 \text{ V}$.

(b) The volume doubles, so the radius increases by the cube root of two: $R_{new} = \sqrt[3]{2} R = 5.67 \times 10^{-4} \text{ m}$ and the new charge is $Q_{new} = 2Q = -3.40 \times 10^{-12} \text{ C}$.

So the new potential is: $V_{new} = \frac{kQ_{new}}{R_{new}} = \frac{k(-3.40 \times 10^{-12} \text{ C})}{5.67 \times 10^{-4} \text{ m}} = -54.0 \text{ V}$.

24-77: (a) $dV_p = \frac{kdq}{z+x} = \frac{kQ}{a}\frac{dz}{z+x} \Rightarrow V = \frac{kQ}{a}\int_0^a \frac{dz}{z+x} = \frac{kQ}{a}\ln\left(\frac{x+a}{x}\right) = \frac{kQ}{a}\ln\left(1+\frac{a}{x}\right).$

(b) $dV_R = \frac{kQ}{a}\frac{dz}{r} = \frac{kQ}{a}\frac{dz}{\sqrt{z^2+y^2}} \Rightarrow V_R = \frac{kQ}{a}\int_0^a \frac{dz}{\sqrt{z^2+y^2}} = \frac{kQ}{a}\ln\left(\frac{\sqrt{a^2+y^2}+a}{y-a}\right).$

(c) $x >> a$: $V_p \approx \frac{kQ}{a}\frac{a}{x} = \frac{kQ}{x}$, since $\ln(1+\alpha) \approx \alpha$.

$y >> a$: $V_R \approx \frac{kQ}{a}\frac{a}{y} = \frac{kQ}{y}$, since $\ln\left(\frac{\sqrt{a^2+y^2}+a}{y-a}\right) \approx \ln\left(\frac{y+a}{y}\right) = \ln\left(1+\frac{a}{y}\right) \approx \frac{a}{y}.$

24-78: Set the alpha particle's kinetic energy equal to its potential energy:

$$K = U \Rightarrow 12\text{ MeV} = \frac{k(2e)(79e)}{r} \Rightarrow r = \frac{k(158)(1.60\times10^{-19}\text{ C})^2}{(12\times10^6\text{ eV})(1.60\times10^{-19}\text{ J/eV})} = 1.90\times10^{-14}\text{ m}.$$

24-79: (a) $V = \frac{kQ_B}{R_B} = \frac{kQ_A}{R_A} = \frac{kQ_B}{R_A/3} \Rightarrow Q_A = 3Q_B \Rightarrow \frac{Q_B}{Q_A} = \frac{1}{3}.$

(b) $E_B = -\left.\frac{\partial V}{\partial r}\right|_{r=R_B} = \frac{kQ_B}{R_B^2} = \frac{k(Q_A/3)}{(R_A/3)^2} = \frac{3kQ_A}{R_A^2} = 3E_A \Rightarrow \frac{E_B}{E_A} = 3.$

24-80: (a) We have $V(x,y,z) = ax^2 + ay^2 - 2az^2$. So:

$\vec{E} = -\frac{\partial V}{\partial x}\hat{i} - \frac{\partial V}{\partial y}\hat{j} - \frac{\partial V}{\partial z}\hat{k} = -2ax\hat{i} - 2ay\hat{j} + 4az\hat{k}.$

(b) A charge is moved in along the z-axis. So the work done is given by:

$W = q\int_{z_o}^{0} \vec{E}\cdot\hat{k}\,dz = q\int_{z_o}^{0} 4az\,dz = -2aqz_o^{\,2} \Rightarrow a = \frac{W}{-2qz_o^{\,2}}$

$\Rightarrow a = \frac{-5.00\times10^{-5}\text{ J}}{-2(2.0\times10^{-6}\text{ C})(0.01\text{ m}^2)} = 1250\text{ V/m}^2.$

(c) $\vec{E}(0,0,0.1) = -2a(0)\hat{i} - 2a(0)\hat{j} + 4a(0.100\text{ m})\hat{k} = 500\hat{k}\text{ V/m}.$

(d) In every plane parallel to the x-y plane, z is constant, so:

$V(x,y,z) = ax^2 + ay^2 + C \Rightarrow x^2 + y^2 = \frac{V-C}{a} \equiv R^2$, which is the equation for a circle since R is constant as long as we have constant potential on those planes.

(e) $V = 6250$ V, and $z = \sqrt{2}$ m: $x^2 + y^2 = \frac{6250\text{ V} + 2(1250\text{ V/m}^2)(\sqrt{2}\text{ m})^2}{1250\text{ V/m}^2} = 9\text{ m}^2.$

Thus the radius of the circle is 3.00 m.

24-81: (a) $E = \frac{kQ_1}{R_1^2} = \frac{k(3.50\times10^{-8}\text{ C})}{(0.240\text{ m})^2} = 5470\text{ N/C}.$

$V = \frac{kQ_1}{R_1} = \frac{k(3.50\times10^{-8}\text{ C})}{0.240\text{ m}} = 1310\text{ V}.$

(b) After electrostatic equilibrium is reached, we have:

$$Q_1 + Q_2 = 3.50 \times 10^{-8}\ \text{C and } V_1 = V_2 \Rightarrow \frac{Q_1}{R_1} = \frac{Q_2}{R_2} \Rightarrow Q_1 = Q_2 \frac{R_1}{R_2}$$

$$\Rightarrow Q_2 \frac{R_1}{R_2} + Q_2 = 3.50 \times 10^{-8}\ \text{C} \Rightarrow Q_2 = \frac{3.50 \times 10^{-8}\ \text{C}}{1 + \frac{0.240}{0.062}} = 7.2 \times 10^{-9}\ \text{C}$$

$$\Rightarrow Q_1 = 3.5 \times 10^{-8}\ \text{C} - 7.2 \times 10^{-9}\ \text{C} = 2.8 \times 10^{-8}\ \text{C}.$$

(c) The new potential is the same at each sphere's surface:

$$V_1 = \frac{kQ_1}{R_1} = \frac{k(2.8 \times 10^{-8}\ \text{C})}{0.240\ \text{m}} \Rightarrow V_1 = V_2 = 1040\ \text{V}.$$

(d) The new electric field is not the same at each sphere's surface:

$$E_1 = \frac{kQ_1}{R_1^2} = \frac{k(2.8 \times 10^{-8}\ \text{C})}{(0.240\ \text{m})^2} = 4400\ \text{N/C}$$

$$E_2 = \frac{kQ_2}{R_2^2} = \frac{k(7.2 \times 10^{-9}\ \text{C})}{(0.062\ \text{m})^2} = 1.7 \times 10^4\ \text{N/C}.$$

24-82: (a) From Problem **23-43** we have the electric field:

$$r \geq R:\ E = \frac{kQ}{r^2} \Rightarrow V = -\int_\infty^r \frac{kQ}{r'^2} dr' = \frac{kQ}{r}, \text{ which is the potential of a point charge.}$$

$$\text{(b) } r \leq R:\ E = \frac{kQ}{r^2}\left[4\frac{r^3}{R^3} - 3\frac{r^4}{R^4}\right] \Rightarrow V = -\int_\infty^R E dr' - \int_R^r E dr'$$

$$\Rightarrow V = \frac{kQ}{R}\left[1 - 2\frac{r^2}{R^2} + 2\frac{R^2}{R^2} + \frac{r^3}{R^3} - \frac{R^3}{R^3}\right] = \frac{kQ}{R}\left[\frac{r^3}{R^3} - 2\frac{r^2}{R^2} + 2\right].$$

24-83: (a) $E_i = E_f \Rightarrow 2\left[\frac{1}{2}m_p v^2\right] = \frac{ke^2}{2r_p} \Rightarrow v = \sqrt{\frac{k(1.60 \times 10^{-19}\ \text{C})^2}{2(1.2 \times 10^{-15}\ \text{m})(1.67 \times 10^{-27}\ \text{kg})}}$

$$\Rightarrow v = 7.58 \times 10^6\ \text{m/s}.$$

(b) For a helium-helium collision, the charges and masses change from (a):

$$v = \sqrt{\frac{k(2(1.60 \times 10^{-19}\ \text{C}))^2}{(3.5 \times 10^{-15}\ \text{m})(2.99)(1.67 \times 10^{-27}\ \text{kg})}} = 7.26 \times 10^6\ \text{m/s}.$$

$$\text{(c) } K = \frac{3kT}{2} = \frac{mv^2}{2} \Rightarrow T_p = \frac{m_p v^2}{3k} = \frac{(1.67 \times 10^{-27}\ \text{kg})(7.58 \times 10^6\ \text{m/s})^2}{3(1.38 \times 10^{-23}\ \text{J/K})} = 2.3 \times 10^9\ \text{K}$$

$$\Rightarrow T_{He} = \frac{m_{He} v^2}{3k} = \frac{(2.99)(1.67 \times 10^{-27}\ \text{kg})(7.26 \times 10^6\ \text{m/s})^2}{3(1.38 \times 10^{-23}\ \text{J/K})} = 6.4 \times 10^9\ \text{K}.$$

(d) These calculations were based on the particles' average speed. The distribution of speeds ensures that there are always a certain percentage with a speed greater than the average speed, and these particles can undergo the necessary reactions in the sun's core.

24-84: (a) The two daughter nuclei have half the volume of the original uranium nucleus, so their radii are smaller by a factor of the cube root of 2:

$$r = \frac{7.4 \times 10^{-15}\text{ m}}{\sqrt[3]{2}} = 5.9 \times 10^{-15}\text{ m}.$$

(b) $U = \frac{k(46e)^2}{2r} = \frac{k(46)^2(1.60 \times 10^{-19}\text{ C})^2}{1.17 \times 10^{-14}\text{ m}} = 4.14 \times 10^{-11}\text{ J}$

Each daughter has half of the potential energy turn into its kinetic energy when far from each other, so:

$K = U/2 = (4.15 \times 10^{-11}\text{ J})/2 = 2.07 \times 10^{-11}\text{ J}.$

(c) If we have 10.0 kg of uranium, then the number of nuclei is:

$n = \frac{10.0\text{ kg}}{236\text{ u}(1.66 \times 10^{-27}\text{ kg/u})} = 2.55 \times 10^{25}$ nuclei . And each releases energy U:

$E = nU = (2.55 \times 10^{25})(4.15 \times 10^{-11}\text{ J}) = 1.06 \times 10^{15}\text{ J} = 253$ kilotons of TNT.

(d) We could call an atomic bomb an "electric" bomb since the electric potential energy provides the kinetic energy of the particles.

24-85: Angular momentum and energy must be conserved, so:

$mv_1 b = mv_2 r_2$ and $E_1 = E_2 \Rightarrow E_1 = \frac{1}{2}mv_2{}^2 + \frac{kq_1q_2}{r_2}$ and $E_1 = 12\text{ MeV} = 1.92 \times 10^{-18}\text{ J}$

Substituting in for v_2 we find:

$E_1 = E_1\frac{b^2}{r_2{}^2} + \frac{kq_1q_2}{r_2} \Rightarrow (E_1)r_2{}^2 - (kq_1q_2)r_2 - E_1b^2 = 0$, and note $q_1 = 2e$ and $q_2 = 79e$.

(i) $b = 10^{-12}\text{ m} \Rightarrow r_2 = 1.01 \times 10^{-12}\text{ m}.$

(ii) $b = 10^{-13}\text{ m} \Rightarrow r_2 = 1.10 \times 10^{-13}\text{ m}.$

(iii) $b = 10^{-14}\text{ m} \Rightarrow r_2 = 2.32 \times 10^{-14}\text{ m}.$

24-86: (a) $r \le a$: $V = \frac{\rho a^2}{18\varepsilon_0}\left[1 - 3\frac{r^2}{a^2} + 2\frac{r^3}{a^3}\right]$ and $E = -\frac{\partial V}{\partial r}$

$$\Rightarrow E = -\frac{\rho_0 a^2}{18\varepsilon_0}\left[-6\frac{r}{a^2} + 6\frac{r^2}{a^3}\right] = \frac{\rho_0 a}{3\varepsilon_0}\left[\frac{r}{a} - \frac{r^2}{a^2}\right].$$

$r \ge a$: $V = 0$ and $E = -\frac{\partial V}{\partial r} = 0.$

(b) $r \le a$: $E_r 4\pi r^2 = \frac{Q_r}{\varepsilon_0} = \frac{\rho_0 a}{3\varepsilon_0}\left[\frac{r}{a} - \frac{r^2}{a^2}\right]4\pi r^2$

$$E_{r+dr}4\pi(r^2 + 2rdr) = \frac{Q_{r+dr}}{\varepsilon_0} = \frac{\rho_0 a}{3\varepsilon_0}\left[\frac{r+dr}{a} - \frac{(r^2 + 2rdr)}{a^2}\right]4\pi(r^2 + 2rdr)$$

$$\Rightarrow \frac{Q_{r+dr} - Q_r}{\varepsilon_0} = \frac{\rho(r)4\pi r^2 dr}{\varepsilon_0} \approx \frac{\rho_0 a 4\pi r^2 dr}{3\varepsilon_0}\left[-\frac{2r}{a^2} + \frac{2}{a} - \frac{2r}{a^2} + \frac{1}{a}\right]$$

$$\Rightarrow \rho(r) = \frac{\rho_0}{3}\left[3 - \frac{4r}{a}\right] = \rho_0\left[1 - \frac{4r}{3a}\right].$$

(c) $r \geq a$: $\rho(r) = 0$, so the total charge enclosed will be given by:

$$Q = 4\pi\int_0^a \rho(r) r^2 dr = 4\pi\rho_o \int_o^a \left[r^2 - \frac{4r^3}{3a}\right]dr = 4\pi\rho_o \left[\frac{1}{3}r^3 - \frac{r^4}{3a}\right]_o^a = 0.$$ Therefore,

by Gauss's Law, the electric field must equal zero for any position $r \geq a$.

24-87: (a) $v_{cm} = \dfrac{m_1 v_1 + m_2 v_2}{m_1 + m_2} = \dfrac{(6 \times 10^{-5}\ \text{kg})(400\ \frac{\text{m}}{\text{s}}) + (3 \times 10^{-5}\ \text{kg})(1300\ \frac{\text{m}}{\text{s}})}{6.0 \times 10^{-5}\ \text{kg} + 3.0 \times 10^{-5}\ \text{kg}} = 700\ \text{m/s}.$

(b) $E_{rel} = \dfrac{1}{2}m_1 v_1^2 + \dfrac{1}{2}m_2 v_2^2 + \dfrac{kq_1q_2}{r} - \dfrac{1}{2}(m_1 + m_2)v_{cm}^2.$

After expanding the center of mass velocity and collecting like terms:

$$\Rightarrow E_{rel} = \frac{1}{2}\frac{m_1 m_2}{m_1 + m_2}\left[v_1^2 + v_2^2 - 2v_1 v_2\right] + \frac{kq_1q_2}{r} = \frac{1}{2}\mu(v_1 - v_2)^2 + \frac{kq_1q_2}{r}.$$

(c) $E_{rel} = \dfrac{1}{2}(2.0 \times 10^{-5}\ \text{kg})(900\ \text{m/s})^2 + \dfrac{k(2.0 \times 10^{-6}\ \text{C})(-5.0 \times 10^{-6}\ \text{C})}{0.0090\ \text{m}} = -1.9\ \text{J}.$

(d) Since the energy is less than zero, the system is "bound".

(e) The maximum separation is when the velocity is zero:

$$1.9\ \text{J} = \frac{kq_1q_2}{r} \Rightarrow r = \frac{k(2.0 \times 10^{-6}\ \text{C})(-5.0 \times 10^{-6}\ \text{C})}{1.9\ \text{J}} = 0.047\ \text{m}.$$

(f) Now using $v_1 = 400\ \text{m/s}$ and $v_2 = 1800\ \text{m/s}$, we find:

$E_{rel} = +9.6\ \text{J}$ so the particles do escape, and the final relative velocity is:

$$|v_1 - v_2| = \sqrt{\frac{2E_{rel}}{\mu}} = \sqrt{\frac{2(9.6\ \text{J})}{2.0 \times 10^{-5}\ \text{kg}}} = 980\ \text{m/s}.$$

24-88: For an infiitesimal slice of a finite cylinder, we have the potential:

$$dV = \frac{kdQ}{\sqrt{(x-z)^2 + R^2}} = \frac{kQ}{L}\frac{dz}{\sqrt{(x-z)^2 + R^2}}$$

$$\Rightarrow V = \frac{kQ}{L}\int_{-L/2}^{L/2}\frac{dz}{\sqrt{(x-z)^2 + R^2}} = \frac{kQ}{L}\int_{-L/2-x}^{L/2-x}\frac{du}{\sqrt{u^2 + R^2}}$$ where $u = x - z$.

$$\Rightarrow V = \frac{kQ}{L}\ln\left[\frac{\sqrt{(L/2-x)^2 + R^2} + (L/2-x)}{\sqrt{(L/2+x)^2 + R^2} - L/2 - x}\right]$$ on the cylinder's axis.

(b) For $L \ll R$:

$$V \approx \frac{kQ}{L}\ln\left[\frac{\sqrt{(L/2-x)^2 + R^2} + L/2 - x}{\sqrt{(L/2+x)^2 + R^2} - L/2 - x}\right] \approx \frac{kQ}{L}\ln\left[\frac{\sqrt{x^2 - xL + R^2} + L/2 - x}{\sqrt{x^2 + xL + R^2} - L/2 - x}\right]$$

$$\Rightarrow V \approx \frac{kQ}{L}\ln\left[\frac{\sqrt{1 - xL/(R^2 + x^2)} + (L/2 - x)/\sqrt{R^2 + x^2}}{\sqrt{1 + xL/(R^2 + x^2)} + (-L/2 - x)/\sqrt{R^2 + x^2}}\right]$$

$$\Rightarrow V \approx \frac{kQ}{L}\ln\left[\frac{1 - xL/2(R^2 + x^2) + (L/2 - x)/\sqrt{R^2 + x^2}}{1 + xL/2(R^2 + x^2) + (-L/2 - x)/\sqrt{R^2 + x^2}}\right]$$

$$\Rightarrow V \approx \frac{kQ}{L}\ln\left[\frac{1+L/2\sqrt{R^2+x^2}}{1-L/2\sqrt{R^2+x^2}}\right] = \frac{kQ}{L}\left(\ln\left[1+\frac{L}{2\sqrt{R^2+x^2}}\right]-\ln\left[1-\frac{L}{2\sqrt{R^2+x^2}}\right]\right)$$

$$\Rightarrow V \approx \frac{kQ}{L}\frac{2L}{2\sqrt{x^2+R^2}} = \frac{kQ}{\sqrt{x^2+R^2}},$$ which is the same as for a ring.

(c) $$E = -\frac{\partial V}{\partial x} = \frac{2kQ\left(\sqrt{(L-2x)^2+4R^2}-\sqrt{(L+2x)^2+4R^2}\right)}{\sqrt{(L-2x)^2+4R^2}\cdot\sqrt{(L+2x)^2+4R^2}}.$$

24-89: (a) $F_g = mg = \frac{4\pi r^3}{3}\rho g = qV_{ab}/d = qE = F_e \Rightarrow q = \frac{4\pi}{3}\frac{\rho r^3 gd}{V_{ab}}$.

(b) $F_g = mg = \frac{4\pi r^3}{3}\rho g = 6\pi\eta r v_t = F_v \Rightarrow r = \sqrt{\frac{9\eta v_t}{2\rho g}}$

$$\Rightarrow q = \frac{4\pi}{3}\frac{\rho g d}{V_{ab}}\left[\sqrt{\frac{9\eta v_t}{2\rho g}}\right]^3 = 18\pi\frac{d}{V_{ab}}\sqrt{\frac{\eta^3 v_t^3}{2\rho g}}.$$

(c) $$q = 18\pi\frac{10^{-3}\text{ m}}{9.16\text{ V}}\sqrt{\frac{(1.81\times10^{-5}\text{ Ns/m}^2)^3(10^{-3}\text{ m}/32.3\text{ s})^3}{2(824\text{ kg/m}^3)(9.80\text{ m/s}^2)}} = 6.44\times10^{-19}\text{ C} = 4e.$$

The drop's radius is 5.59×10^{-7} m.

Chapter 25: Capacitance and Dielectrics

25-1: (a) $V = \dfrac{Q}{C} = \dfrac{0.346 \times 10^{-6}\ \text{C}}{5.00 \times 10^{-10}\ \text{F}} = 692\ \text{V}.$

(b) $A = \dfrac{Cd}{\varepsilon_o} = \dfrac{(5.00 \times 10^{-10}\ \text{F})(0.453 \times 10^{-3}\ \text{m})}{\varepsilon_o} = 0.0256\ \text{m}^2.$

(c) $E = \dfrac{V}{d} = \dfrac{692\ \text{V}}{0.453 \times 10^{-3}\ \text{m}} = 1.53 \times 10^6\ \text{V/m}.$

(d) $E = \dfrac{\sigma}{\varepsilon_o} \Rightarrow \sigma = \varepsilon_o E = \varepsilon_o (1.53 \times 10^6\ \text{V/m}) = 1.35 \times 10^{-5}\ \text{C/m}^2.$

25-2: (a) $V = Ed = (4.77 \times 10^6\ \text{V/m})(4.79 \times 10^{-3}\ \text{m}) = 2.28 \times 10^4\ \text{V}.$

(b) $A = \dfrac{Cd}{\varepsilon_o} = \dfrac{Qd}{V\varepsilon_o} = \dfrac{(5.16 \times 10^{-8}\ \text{C})(4.79 \times 10^{-3}\ \text{m})}{(2.28 \times 10^4\ \text{V})\varepsilon_o} = 1.22 \times 10^{-3}\ \text{m}^2.$

(c) $C = \dfrac{Q}{V} = \dfrac{5.16 \times 10^{-8}\ \text{C}}{2.28 \times 10^4\ \text{V}} = 2.26 \times 10^{-12}\ \text{F}.$

25-3: $V_o = \dfrac{Q_o}{C},\ V_1 = \dfrac{Q_1}{C}$, and $V_o - V_1 = 50\ \text{V} = \dfrac{\Delta Q}{C}$

$\Rightarrow \Delta Q = (6.17 \times 10^{-6}\ \text{F})(50\ \text{V}) = 3.09 \times 10^{-4}\ \text{C}.$

25-4: (a) For two concentric spherical shells, the capacitance is:

$C = \dfrac{1}{k}\left(\dfrac{r_a r_b}{r_b - r_a}\right) \Rightarrow kCr_b - kCr_a = r_a r_b \Rightarrow r_b = \dfrac{kCr_a}{kC - r_a}$

$\Rightarrow r_b = \dfrac{k(150 \times 10^{-12}\ \text{F})(0.200\ \text{m})}{k(150 \times 10^{-12}\ \text{F}) - 0.200\ \text{m}} = 0.235\ \text{m}$, so the distance between the spheres is 0.035 m.

(b) $V = 220\ \text{V}$, and $Q = CV = (150 \times 10^{-12}\ \text{F})(220\ \text{V}) = 3.30 \times 10^{-8}\ \text{C}.$

25-5: (a) $C = \dfrac{1}{k}\left(\dfrac{r_b r_a}{r_b - r_a}\right) = \dfrac{1}{k}\left(\dfrac{(0.15\ \text{m})(0.12\ \text{m})}{0.15\ \text{m} - 0.12\ \text{m}}\right) = 6.68 \times 10^{-11}\ \text{F}.$

(b) The electric field at a distance of 12.1 cm:

$E = \dfrac{kQ}{r^2} = \dfrac{kCV}{r^2} = \dfrac{k(6.68 \times 10^{-11}\ \text{F})(140\ \text{V})}{(0.121\ \text{m})^2} = 5740\ \text{N/C}.$

(c) The electric field at a distance of 14.9 cm:

$E = \dfrac{kQ}{r^2} = \dfrac{kCV}{r^2} = \dfrac{k(6.68 \times 10^{-11}\ \text{F})(140\ \text{V})}{(0.149\ \text{m})^2} = 3780\ \text{N/C}.$

(d) For a spherical capacitor, the electric field is not constant between the surfaces.

25-6: (a) $\dfrac{C}{L} = \dfrac{2\pi\varepsilon_o}{\ln(r_b / r_a)} \Rightarrow \ln(r_b / r_a) = \dfrac{2\pi\varepsilon_o}{C/L} = \dfrac{2\pi\varepsilon_o}{69 \times 10^{-12}\ \text{F/m}} = 0.81 \Rightarrow \dfrac{r_b}{r_a} = 2.2.$

(b) $\frac{Q}{L} = V\frac{C}{L} = (2.0 \text{ V})(69 \times 10^{-12} \text{ F/m}) = 1.4 \times 10^{-10} \text{ C/m}.$

25-7: (a) Since the outer conductor is at a highter potential it is positively charged, and the magnitude is:

$$Q = CV = \frac{2\pi\varepsilon_0 LV}{\ln(r_b / r_a)} = \frac{2\pi\varepsilon_0(3.5 \text{ m})(0.35 \text{ V})}{\ln(0.0040 \text{ m} / 0.0025 \text{ m})} = 1.4 \times 10^{-10} \text{ C}.$$

(b) $C/L = \frac{2\pi\varepsilon_0}{\ln(r_b / r_a)} = \frac{2\pi\varepsilon_0}{\ln(4.0 \text{ mm} / 2.5 \text{ mm})} = 1.2 \times 10^{-10} \text{ F/m}.$

25-8: (a) $\frac{1}{C_{eq}} = \frac{1}{C_1} + \frac{1}{C_2} = \frac{1}{(4.0 \times 10^{-6} \text{ F})} + \frac{1}{(6.0 \times 10^{-6} \text{ F})} = 4.17 \times 10^5 \text{ F}^{-1}$

$\Rightarrow Q = VC_{eq} = 67.5 \text{ V} / 4.17 \times 10^5 \text{ F}^{-1} = 1.62 \times 10^{-4} \text{ C}.$

(b) $V_1 = Q / C_1 = 1.62 \times 10^{-4} \text{ C} / 4.0 \times 10^{-6} \text{ F} = 40.5 \text{ V}.$

$V_2 = Q / C_2 = 1.62 \times 10^{-4} \text{ C} / 6.0 \times 10^{-6} \text{ F} = 27.0 \text{ V}.$

25-9: (a) $Q_1 = VC_1 = (67.5 \text{ V})(4.0 \times 10^{-6} \text{ F}) = 2.7 \times 10^{-4} \text{ C}.$

$Q_2 = VC_2 = (67.5 \text{ V})(6.0 \times 10^{-6} \text{ F}) = 4.05 \times 10^{-4} \text{ C}.$

(b) For parallel capacitors, the voltage over each is the same, and equals the voltage source: 67.5 V.

25-10: (a) $\frac{1}{C_{eq}} = \frac{1}{C_1 + C_2} + \frac{1}{C_3} = \frac{1}{((2.0 + 4.0) \times 10^{-6} \text{ F})} + \frac{1}{(9.0 \times 10^{-6} \text{ F})} \Rightarrow C_{eq} = 3.60 \times 10^{-6} \text{ F}.$

$Q_3 = Q_1 + Q_2 = VC_{eq} = (61.5 \text{ V})(3.60 \times 10^{-6} \text{ F}) = 2.21 \times 10^{-4} \text{ C}.$

The charge on C_1 is half that of C_2 since they are at the same potential, but the capacitance of C_2 is twice that of C_1. Therefore:

$Q_3 = 3Q_1 = 2.21 \times 10^{-4} \text{ C} \Rightarrow Q_1 = 7.38 \times 10^{-5} \text{ C}$, and $Q_2 = 1.48 \times 10^{-4} \text{ C}$.

(b) $V_2 = V_1 = Q_1 / C_1 = (7.38 \times 10^{-5} \text{ C}) / (2.00 \times 10^{-6} \text{ F}) = 36.9 \text{ V}.$

And $V_3 = 61.5 \text{ V} - 36.9 \text{ V} = 24.6 \text{ V}.$

(c) The potential difference between a and d: $V_{ad} = V_1 = V_2 = 36.9 \text{ V}.$

25-11: (a) $\frac{1}{C_{eq}} = \frac{1}{C_{12} + C_3} + \frac{1}{C_4} = \frac{1}{(1.00 \,\mu\text{F} + 2.0 \,\mu\text{F})} + \frac{1}{(2.0 \,\mu\text{F})} \Rightarrow C_{eq} = 1.20 \,\mu\text{F}.$

Then, $Q_{123} = Q_4 = Q_{total} = C_{eq}V = (1.20 \times 10^{-6} \text{ F})(40.4 \text{ V}) = 4.85 \times 10^{-5} \text{ C}.$

and $2Q_{12} = Q_3 \Rightarrow Q_{12} = \frac{Q_{total}}{3} = \frac{4.85 \times 10^{-5} \text{ C}}{3} = 1.62 \times 10^{-5} \text{ C}$, and $Q_3 = 3.23 \times 10^{-5} \text{ C}.$

But also, $Q_1 = Q_2 = Q_{12} = 1.62 \times 10^{-5} \text{ C}.$

(b) $V_1 = Q_1 / C_1 = (1.62 \times 10^{-5} \text{ C}) / (2.00 \times 10^{-6} \text{ F}) = 8.08 \text{ V} = V_2.$

$V_3 = Q_3 / C_3 = (3.23 \times 10^{-5} \text{ C}) / (2.00 \times 10^{-6} \text{ F}) = 16.2 \text{ V}.$

$V_4 = Q_4 / C_4 = (4.85 \times 10^{-5} \text{ C}) / (2.00 \times 10^{-6} \text{ F}) = 24.2 \text{ V}.$

(c) $V_{ad} = V_{ab} - V_4 = 40.4\text{ V} - 24.3\text{ V} = 16.2\text{ V}$.

25-12: (a) and (b) The equivalent resistance of the combination is 6.0 μF, therefore the total charge on the network is:
$Q = C_{eq}V_{ab} = (6.0\ \mu\text{F})(25\text{ V}) = 1.5 \times 10^{-4}\text{ C}$. This is also the charge on the 9.0 μF capacitor because it is connected in series with the point b. So:

$$V_9 = \frac{Q_9}{C_9} = \frac{1.5 \times 10^{-4}\text{ C}}{9.0 \times 10^{-6}\text{ F}} = 16.7\text{ V}.$$

Then $V_3 = V_{11} = V_{12} + V_6 = V - V_9 = 25\text{ V} - 16.7\text{ V} = 8.3\text{ V}$.

$\Rightarrow Q_3 = C_3V_3 = (3.0\ \mu\text{F})(8.3\text{ V}) = 2.5 \times 10^{-5}\text{ C}$.

$\Rightarrow Q_{11} = C_{11}V_{11} = (11\ \mu\text{F})(8.3\text{ V}) = 9.2 \times 10^{-5}\text{ C}$.

$\Rightarrow Q_6 = Q_{12} = Q - Q_3 - Q_{11} = 1.5 \times 10^{-4}\text{ C} - 2.5 \times 10^{-5}\text{ C} - 9.2 \times 10^{-5}\text{ C} = 3.3 \times 10^{-5}\text{ C}$.

So now the final voltages can be calculated:

$$V_6 = \frac{Q_6}{C_6} = \frac{3.3 \times 10^{-5}\text{ C}}{6.0 \times 10^{-6}\text{ F}} = 5.5\text{ V}.$$

$$V_{12} = \frac{Q_{12}}{C_{12}} = \frac{3.3 \times 10^{-5}\text{ C}}{12 \times 10^{-6}\text{ F}} = 2.8\text{ V}.$$

(c) Since the 3 μF, 11 μF and 6 μF capacitors are connected in parallel and are in series with the 9 μF capacitor, their charges must add up to that of the 9 μF capacitor. Similarly, the charge on the 3 μF, 11 μF and 12 μF capacitors must add up to the same as that of the 9 μF capacitor, which is the same as the whole network.

25-13: Capacitances in parallel simply add, so:

$$\frac{1}{C_{eq}} = \frac{1}{8.0\ \mu\text{F}} = \left(\frac{1}{(11 + 4.0 + x)\ \mu\text{F}} + \frac{1}{9.0\ \mu\text{F}}\right) \Rightarrow (15 + x)\ \mu\text{F} = 72\ \mu\text{F} \Rightarrow x = 57\ \mu\text{F}.$$

25-14: $C_{eq} = \left(\frac{1}{C_1} + \frac{1}{C_2}\right)^{-1} = \left(\frac{d_1}{\varepsilon_o A} + \frac{d_2}{\varepsilon_o A}\right)^{-1} = \frac{\varepsilon_o A}{d_1 + d_2}$. So the combined capacitance for two capacitors in series is the same as that for a capacitor of area A and separation $(d_1 + d_2)$.

25-15: $C_{eq} = C_1 + C_2 = \frac{\varepsilon_o A_1}{d} + \frac{\varepsilon_o A_2}{d} = \frac{\varepsilon_o (A_1 + A_2)}{d}$. So the combined capacitance for two capacitors in parallel is that of a single capacitor of their combined area $(A_1 + A_2)$ and common plate separation d.

26-16: (a) $C = Q/V = (0.0240\ \mu\text{C})/(220\text{ V}) = 1.20 \times 10^{-10}\text{ F}$.

(b) $C = \frac{\varepsilon_o A}{d} \Rightarrow A = \frac{Cd}{\varepsilon_o} = \frac{(1.20 \times 10^{-10}\text{ F})(0.0012\text{ m})}{\varepsilon_o} = 0.0163\text{ m}^2$.

(c) $E_{max} = V_{max} / d \Rightarrow V_{max} = E_{max} d = (3.00 \times 10^6 \text{ V/m})(0.0012 \text{ m}) = 3600 \text{ V}$.

(d) $U = \dfrac{Q^2}{2C} = \dfrac{(2.40 \times 10^{-8} \text{ C})^2}{2(1.20 \times 10^{-10} \text{ F})} = 2.40 \times 10^{-6} \text{ J}$.

26-17 $U = \dfrac{1}{2} CV^2 = \dfrac{1}{2}(3.00 \times 10^{-4} \text{ F})(276 \text{ V}) = 11.4 \text{ J}$.

26-18 (a) $V = Q / C = (4.36 \ \mu\text{C}) / (1.00 \times 10^{-9} \text{ F}) = 4360 \text{ V}$.

(b) Since the charge is kept constant while the separation doubles, that means that the capacitance halves and the voltage doubles to 8720 V.

(c) $U = \dfrac{1}{2} CV^2 = \dfrac{1}{2}(1.00 \times 10^{-9} \text{ F})(4360 \text{ V})^2 = 9.50 \times 10^{-3} \text{ J}$. Now if the separation is doubled, the capacitance halves, and the energy stored doubles. So the amount of work done to move the plates equals the difference in energy stored in the capacitor, which is 9.50×10^{-3} J.

25-19: $E = V / d = (500 \text{ V}) / (0.00400 \text{ m}) = 1.25 \times 10^5 \text{ V/m}$.

And $u = \dfrac{1}{2}\varepsilon_o E^2 = \dfrac{1}{2}\varepsilon_o (1.25 \times 10^5 \text{ V/m})^2 = 0.0692 \text{ J/m}^3$.

25-20: (a) $u = \dfrac{1}{2}\varepsilon_o E^2 = \dfrac{1}{2}\varepsilon_o \left(\dfrac{1}{4\pi\varepsilon_o} \dfrac{q}{r^2}\right)^2 = \dfrac{1}{32\pi^2 \varepsilon_o} \dfrac{(4.00 \times 10^{-9} \text{ C})^2}{(0.250 \text{ m})^4} = 1.47 \times 10^{-6} \text{ J/m}^3$.

(b) If the charge was -4.00 nC, the electric field energy would remain the same.

25-21: (a) $U = \dfrac{1}{2} QV \Rightarrow Q = \dfrac{2U}{V} = \dfrac{2(5.40 \times 10^{-9} \text{ J})}{3.00 \text{ V}} = 3.60 \times 10^{-9} \text{ C}$.

(b) $\dfrac{C}{L} = \dfrac{2\pi\varepsilon_o}{\ln(r_a / r_b)} \Rightarrow \dfrac{r_a}{r_b} = \exp(2\pi\varepsilon_o L / C) = \exp(2\pi\varepsilon_o LV / Q)$

$\Rightarrow \dfrac{r_a}{r_b} = \exp(2\pi\varepsilon_o (25.0 \text{ m})(3.00 \text{ V}) / (3.60 \times 10^{-9} \text{ C})) = 3.19$.

25-22: (a) For a spherical capacitor:

$C = \dfrac{1}{k} \dfrac{r_a r_b}{r_b - r_a} = \dfrac{1}{k} \dfrac{(0.120 \text{ m})(0.140 \text{ m})}{(0.140 \text{ m} - 0.120 \text{ m})} = 9.35 \times 10^{-11} \text{ F}$

$\Rightarrow V = Q / C = (5.30 \times 10^{-9} \text{ C}) / (9.35 \times 10^{-11} \text{ F}) = 56.7 \text{ V}$.

(b) $U = \dfrac{1}{2} CV^2 = \dfrac{(9.35 \times 10^{-11} \text{ F})(56.7 \text{ V})^2}{2} = 1.50 \times 10^{-7} \text{ J}$.

25-23: (a) $u = \dfrac{1}{2}\varepsilon_o E^2 = \dfrac{\varepsilon_o}{2}\left(\dfrac{kq}{r^2}\right)^2 = \dfrac{\varepsilon_o}{2}\left(\dfrac{kVC}{r^2}\right)^2 = \dfrac{\varepsilon_o}{2} \dfrac{k^2 (140 \text{ V})^2 (6.67 \times 10^{-11} \text{ F})^2}{(0.121 \text{ m})^4}$

$\Rightarrow u = 1.46 \times 10^{-4} \text{ J/m}^3$.

(b) The same calculation for $r = 14.9$ cm $\Rightarrow u = 6.33 \times 10^{-5} \text{ J/m}^3$.

(c) No, the electric energy density is NOT constant within the spheres.

25-24: (a) If the separation distance is halved while the charge is kept fixed, then the capacitance increases and the stored energy, which was 6.45 J, decreases since $U = Q^2/2C$. Therefore the new energy is 3.23 J.
(b) If the voltage is kept fixed while the separation is decreased by one half, then the doubling of the capacitance leads to a doubling of the stored energy to 12.9 J, using $U = CV^2/2$, when V is held constant throughout.

25-25: (a) $U_o = \dfrac{q^2}{2C} = \dfrac{xq^2}{2\varepsilon_o A}$.

(b) Increase the separation by $dx \Rightarrow U = \dfrac{(x+dx)q^2}{2\varepsilon_o A} = U_o(1 + dx/x)$.

(c) the work done in increasing the separation is given by:

$dW = U - U_o = \dfrac{dxq^2}{2\varepsilon_o A} = Fdx \Rightarrow F = \dfrac{q^2}{2\varepsilon_o A}$.

(d) The force is not simply equal to qE since the energy can be thought of as being stored either in the charges on the plates, or in the electric field between the plates, but not a combination.

25-26: (a) $Q = CV = (2.00 \times 10^{-5}\ \text{F})(900\ \text{V}) = 0.0180\ \text{C}$.
(b) They must have equal potential difference, and their combined charge must add up to the original charge. Therefore:

$\dfrac{Q_1}{C_1} = \dfrac{Q_2}{C_2} \Rightarrow \dfrac{Q_1}{20\mu\text{F}} = \dfrac{Q_2}{10\mu\text{F}} \Rightarrow Q_1 = 2Q_2$ and also $Q_1 + Q_2 = Q$.

$\Rightarrow Q_2 = \dfrac{Q}{3} = 0.006\ \text{C}$ and $Q_1 = 0.012\ \text{C}$.

(c) $U = \dfrac{1}{2}\left(\dfrac{Q_1^2}{C_1} + \dfrac{Q_2^2}{C_2}\right) = \dfrac{1}{2}\left(\dfrac{(0.012\ \text{C})^2}{20\ \mu\text{F}} + \dfrac{(0.006\ \text{C})^2}{10\ \mu\text{F}}\right) = 5.4\ \text{J}$.

(d) The original U was $U = \dfrac{1}{2}QV = \dfrac{1}{2}(0.0180\ \text{C})(900\ \text{V}) = 8.1\ \text{J} \Rightarrow \Delta U = -2.7\ \text{J}$.

25-27: $C = \dfrac{\varepsilon_o A}{d} = \dfrac{\varepsilon_o AE}{V} \Rightarrow A = \dfrac{CV}{\varepsilon_o E} = \dfrac{(1.37 \times 10^{-9}\ \text{F})(6000\ \text{V})}{\varepsilon_o(2.00 \times 10^7\ \text{V/m})} = 0.0464\ \text{m}^2$.

25-28: Placing a dielectric between the plates just results in the replacement of ε for ε_o in the derivation of Eq.(25-19). One can follow exactly the procedure as shown for Eq. (25-11).

25-29: (a) $\sigma_i = \varepsilon_o E = \varepsilon_o(1.80 \times 10^5\ \text{V/m}) = 1.59 \times 10^{-6}\ \text{C/m}^2$.

(b) $K = \dfrac{E_o}{E} = \dfrac{3.60 \times 10^5\ \text{V/m}}{1.80 \times 10^5\ \text{V/m}} = 2.00$.

25-30: (a) $E_o = KE = (4.50)(1.40 \times 10^6 \text{ V/m}) = 6.30 \times 10^6 \text{ V/m}$.

(b) $\sigma_i = \sigma\left(1 - \frac{1}{K}\right) = (5.58 \times 10^{-5} \text{ C/m}^2)(1 - 1/4.50) = 4.34 \times 10^{-5} \text{ C/m}^2$.

(c) $U = \frac{1}{2}CV^2 = uAd = \frac{1}{2}K\varepsilon_o E^2 Ad \Rightarrow U/A = \frac{1}{2}K\varepsilon_o E^2 d$

$\Rightarrow U/A = \frac{1}{2}(4.50)\varepsilon_o(1.40 \times 10^6 \text{ V/m})^2(0.0016 \text{ m}) = 0.0624 \text{ J/m}^2$.

25-31: (a) $U_o = \frac{1}{2}C_o V^2 \Rightarrow V = \sqrt{\frac{2U_o}{C_o}} = \sqrt{\frac{2(1.99 \times 10^{-5} \text{ J})}{(2.55 \times 10^{-7} \text{ F})}} = 12.5 \text{ V}$.

(b) $U = \frac{1}{2}KC_o V^2 \Rightarrow K = \frac{2U_o}{C_o V^2} = \frac{2(4.68 \times 10^{-5} \text{ J})}{(2.55 \times 10^{-7} \text{ F})(12.5 \text{ V})^2} = 2.35$.

25-32: (a) $E = \frac{\sigma}{K\varepsilon_o} \Rightarrow K = \frac{\sigma}{E\varepsilon_o} = \frac{Q}{EA\varepsilon_o} = \frac{1.80 \times 10^{-7} \text{ C}}{(3.40 \times 10^5 \text{ V/m})(0.0040 \text{ m}^2)\varepsilon_o} = 15.0$.

(b) $\sigma_i = \frac{Q}{A}\left(1 - \frac{1}{K}\right) = \left(\frac{1.80 \times 10^{-7} \text{ C}}{0.0040 \text{ m}^2}\right)(1 - 1/15.0) = 4.20 \times 10^{-5} \text{ C/m}^2$

$\Rightarrow Q_i = \sigma_i A = (4.20 \times 10^{-5} \text{ C/m}^2)(0.0040 \text{ m}^2) = 1.68 \times 10^{-7} \text{ C}$.

25-33: (a) $\varepsilon = K\varepsilon_o = (2.2)\varepsilon_o = 1.9 \times 10^{-11} \text{ C}^2/\text{Nm}^2$.

(b) $V_{max} = E_{max}d = (7.0 \times 10^7 \text{ V/m})(2.0 \times 10^{-3} \text{ m}) = 1.4 \times 10^5 \text{ V}$.

(c) $E = \frac{\sigma}{K\varepsilon_o} \Rightarrow \sigma = \varepsilon E = (1.9 \times 10^{-11} \text{ C}^2/\text{Nm}^2)(7.0 \times 10^7 \text{ V/m}) = 1.3 \times 10^{-3} \text{ C/m}^2$.

And $\sigma_i = \sigma\left(1 - \frac{1}{K}\right) = (1.3 \times 10^{-3} \text{ C/m}^2)(1 - 1/2.2) = 0.73 \times 10^{-3} \text{ C/m}^2$.

25-34: (a) $\Delta Q = Q - Q_o = (K-1)Q_o = (K-1)C_o V_o = (2.1)(2.5 \times 10^{-7} \text{ F})(24 \text{ V}) = 1.3 \times 10^{-5} \text{ C}$.

(b) $Q_i = Q\left(1 - \frac{1}{K}\right) = (1.86 \times 10^{-5} \text{ C})(1 - 1/3.1) = 1.26 \times 10^{-5} \text{ C}$.

(c) The addition of the mylar doesn't affect the electric field since the induced charge cancels the additional charge drawn to the plates.

25-35: (a) Eq. (25-22): $\oint K\vec{E} \cdot d\vec{A} = \frac{Q_{free}}{\varepsilon_o} \Rightarrow KEA = \frac{Q_{free}}{\varepsilon_o} \Rightarrow E = \frac{Q_{free}}{K\varepsilon_o A} = \frac{Q_{free}}{\varepsilon A}$.

(b) $V = Ed = \frac{Q_{free}d}{\varepsilon A}$.

(c) $C = \frac{Q}{V} = \frac{\varepsilon A}{d} = K\frac{\varepsilon_o A}{d} = KC_o$.

25-36: (a) $\oint K\vec{E} \cdot d\vec{A} = \frac{Q_{free}}{\varepsilon_o} \Rightarrow KE4\pi d^2 = \frac{q}{\varepsilon_o} \Rightarrow E = \frac{q}{4\pi\varepsilon d^2}$.

(b) $\oint \vec{E}\cdot d\vec{A} = \frac{q_{total}}{\varepsilon_o} = \frac{q_f + q_b}{\varepsilon_o} \Rightarrow KE4\pi d^2 = \frac{q + q_b}{\varepsilon_o} \Rightarrow E = \frac{q + q_b}{4\pi\varepsilon_o d^2} \Rightarrow q_{total} = q + q_b = q / K.$

(c) The total bound charge is $q_b = q\left(\frac{1}{K} - 1\right)$.

25-37: (a) The power output is 600 W, and 95 % of the original energy is converted.
$\Rightarrow E = Pt = (600\text{ W})(0.01\text{ s}) = 6.0\text{ J} \therefore E_o = \frac{6.0\text{ J}}{0.95} = 6.32\text{ J}.$

(b) $U = \frac{1}{2}CV^2 \Rightarrow V = \sqrt{\frac{2U}{C}} = \sqrt{\frac{2(6.32\text{ J})}{7.54 \times 10^{-3}\text{ F}}} = 129\text{ V}.$

25-38: $C_o = \frac{A\varepsilon_o}{d} = \frac{(4.90 \times 10^{-5}\text{ m}^2)\varepsilon_o}{6.00 \times 10^{-4}\text{ m}} = 7.23 \times 10^{-13}\text{ F} \Rightarrow C = C_o + 0.3\text{ pF} = 1.02 \times 10^{-12}\text{ F}.$

But $C = \frac{A\varepsilon_o}{d'} \Rightarrow d' = \frac{A\varepsilon_o}{C} = \frac{(4.90 \times 10^{-5}\text{ m}^2)\varepsilon_o}{1.02 \times 10^{-12}\text{ F}} = 4.24 \times 10^{-4}\text{ m}.$
Therefore the key must be depressed by a distance of:
$600 \times 10^{-4}\text{ m} - 4.24 \times 10^{-4}\text{ m} = 0.18\text{ mm}.$

25-39: (a) $d << r_a$: $C = \frac{2\pi\varepsilon_o L}{\ln(r_b / r_a)} = \frac{2\pi\varepsilon_o L}{\ln((d + r_a)/r_a)} = \frac{2\pi\varepsilon_o L}{\ln(1 + d/r_a)} \approx \frac{2\pi r_a L\varepsilon_o}{d} = \frac{\varepsilon_o A}{d}.$

(b) At the scale of part (a) the cylinders appear to be flat, and so the capacitance should appear like that of flat plates.

25-40: (a) $C = \frac{\varepsilon_o A}{d} = \frac{\varepsilon_o (0.18\text{ m})^2}{5.8 \times 10^{-3}\text{ m}} = 4.9 \times 10^{-11}\text{ F}.$

(b) $Q = CV = (4.9 \times 10^{-11}\text{ F})(50\text{ V}) = 2.5 \times 10^{-9}\text{ C}.$

(c) $E = V/d = (50\text{ V})/(5.8 \times 10^{-3}\text{ m}) = 8600\text{ V/m}.$

(d) $U = \frac{1}{2}CV^2 = \frac{1}{2}(4.9 \times 10^{-11}\text{ F})(50\text{ V})^2 = 6.2 \times 10^{-8}\text{ J}.$

(e) If the battery is disconnected, so the charge remains constant, and the plates are pulled further apart to 0.0116 m, then the calculations above can be carried out just as before, and we find:

(a) $C = 2.5 \times 10^{-11}\text{ F}$ (b) $Q = 2.5 \times 10^{-9}\text{ C}$

(c) $E = 4300\text{ V/m}$ (d) $U = \frac{Q^2}{2C} = \frac{(2.5 \times 10^{-9}\text{ C})^2}{2(2.5 \times 10^{-11}\text{ F})} = 1.3 \times 10^{7}\text{ J}.$

25-41: If the plates are pulled out as in **25-40** but now the battery is connected, ensuring that the voltage remains constant. This time we find:

(a) $C = 2.5 \times 10^{-11}\text{ F}$ (b) $Q = 1.24 \times 10^{-9}\text{ C}$

(c) $E = 4310\text{ V/m}$ (d) $U = \frac{CV^2}{2} = \frac{(2.5 \times 10^{-11}\text{ F})(50\text{ V})^2}{2} = 3.09 \times 10^{-8}\text{ J}.$

25-42:

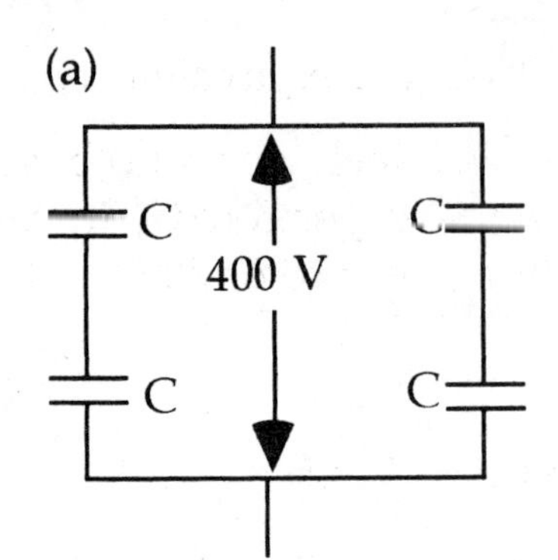

$C_{eq} = \frac{C}{2} + \frac{C}{2} = C$. So the total capacitance is the same as each individual capacitor, and the voltage is split over each so that $V = 200$ V.

(b) If one capacitor is a moderately good conductor, the total capacitance will increase, and one capacitor will have greater than 250 V over it.

25-43: (a) $\frac{1}{C_{eq}} = \frac{1}{C_1} + \frac{1}{C_2 + \left(\frac{1}{C_3} + \frac{1}{C_4}\right)^{-1}} + \frac{1}{C_5}$

$\Rightarrow \frac{1}{C_{eq}} = \frac{1}{4.6\ \mu\text{F}} + \frac{1}{2.3\ \mu\text{F} + \left(\frac{1}{2.3\ \mu\text{F}} + \frac{1}{2.3\ \mu\text{F}}\right)^{-1}} + \frac{1}{4.6\ \mu\text{F}} = 7.25 \times 10^{5}\ \text{F}^{-1} \Rightarrow C_{eq} = 1.38\ \mu\text{F}$.

(b) $Q = CV = (1.38\ \mu\text{F})(540\ \text{V}) = 7.45 \times 10^{-4}\ \text{C} = Q_1 = Q_5$.

$\Rightarrow V_1 = V_5 = (7.45 \times 10^{-4}\ \text{C}) / (4.6 \times 10^{-6}\ \text{F}) = 162\ \text{V}$.

So $V_2 = 540 - 2(162) = 216\ \text{V} \Rightarrow Q_2 = (216\ \text{V})(2.3\ \mu\text{F}) = 4.97 \times 10^{-4}\ \text{C}$.

Also $V_3 = V_4 = \frac{1}{2}(216\ \text{V}) = 108\ \text{V} \Rightarrow Q_3 = Q_4 = (108\ \text{V})(2.3\ \mu\text{F}) = 2.48 \times 10^{-4}\ \text{C}$.

25-44: (a) $C_{eq} = \left(\frac{1}{2.0\ \mu\text{F}} + \frac{1}{3.0\ \mu\text{F}}\right)^{-1} = 1.2 \times 10^{-6}\ \text{F}$

$\Rightarrow Q = C_{eq}V = (1.2 \times 10^{-6}\ \text{F})(600\ \text{V}) = 7.2 \times 10^{-4}\ \text{C}$

and $V_2 = Q / C_2 = (7.2 \times 10^{-4}\ \text{C}) / (2.0\ \mu\text{F}) = 360\ \text{V} \Rightarrow V_3 = 600\ \text{V} - 360\ \text{V} = 240\ \text{V}$.

(b) Disconnecting them from the voltage source and reconnecting them to themselves we must have equal potential difference, and the sum of their charges must be the sum of the original charges:

$Q_1 = C_1V$ and $Q_2 = C_2V \Rightarrow 2Q = Q_1 + Q_2 = (C_1 + C_2)V$

$\Rightarrow V = \frac{2Q}{C_1 + C_2} = \frac{2(7.20 \times 10^{-4}\ \text{C})}{5.00 \times 10^{-6}\ \text{F}} = 144\ \text{V}$.

$\Rightarrow Q_1 = (2.00 \times 10^{-6}\ \text{F})(144\ \text{V}) = 2.88 \times 10^{-4}\ \text{C}$.

$\Rightarrow Q_2 = (3.00 \times 10^{-6}\ \text{F})(144\ \text{V}) = 4.32 \times 10^{-4}\ \text{C}$.

25-45: (a) Reducing the furthest right leg yields $C = \left(\frac{1}{9.3\ \mu\text{F}} + \frac{1}{9.3\ \mu\text{F}} + \frac{1}{9.3\ \mu\text{F}}\right)^{-1} = 3.1\ \mu\text{F} = C_1 / 3$.

It combines in parallel with a $C_2 \Rightarrow C = 6.2\ \mu\text{F} + 3.1\ \mu\text{F} = 9.3\ \mu\text{F} = C_1$.

So the next reduction is the same as the first: $C = 3.1\ \mu\text{F} = C_1 / 3$. And the next is the same as the second, leaving 3 C_1's in series so $C_{eq} = 3.1\ \mu\text{F} = C_1 / 3$.

(b) For the three capacitors nearest points a and b:

$Q_{C_1} = C_{eq}V = (3.1 \times 10^{-6}\ \text{F})(840\ \text{V}) = 2.6 \times 10^{-3}\ \text{C}$

and $Q_{C_2} = C_2V_2 = (6.2 \times 10^{-6}\ \text{F})(840\ \text{V})/3 = 1.7 \times 10^{-3}\ \text{C}$.

(c) $V_{cd} = \frac{1}{3}(280\ \text{V}) = 93\ \text{V}$, since the total voltage drop over the equivalent capacitance of the part of the circuit from the junctions between a,c and d,b is 280 V, and the equivalent capacitance is that of three equal capacitors C_1 in series. V_{cd} is the voltage over just one of those capacitors, i.e. 1/3 of 280 V.

25-46: Originally: $Q_1 = C_1V_1 = (6.0\ \mu\text{F})(24\ \text{V}) = 1.4 \times 10^{-4}\ \text{C}$

$Q_2 = C_2V_2 = (3.0\ \mu\text{F})(24\ \text{V}) = 7.2 \times 10^{-5}\ \text{C}$, and $C_{eq} = C_1 + C_2 = 9.0\ \mu\text{F}$.

So the original energy stored is $U = \frac{1}{2}C_{eq}V^2 = \frac{1}{2}(9.0 \times 10^{-6}\ \text{F})(24\ \text{V})^2 = 2.6 \times 10^{-3}\ \text{J}$

Disconnect and flip the capacitors, so now the total charge is

$Q = Q_2 - Q_1 = 7.2 \times 10^{-5}\ \text{C}$, and the equivalent capacitance is the still the same,

$C_{eq} = 9.0\ \mu\text{F}$. So the new energy stored is: $U = \frac{Q^2}{2C_{eq}} = \frac{(7.2 \times 10^{-5}\ \text{C})^2}{2(9.0 \times 10^{-6}\ \text{F})} = 2.9 \times 10^{-4}\ \text{J}$

$\Rightarrow \Delta U = 2.9 \times 10^{-4}\ \text{J} - 2.6 \times 10^{-3}\ \text{J} = -2.3 \times 10^{-3}\ \text{J}$.

25-47: (a) $C_{eq} = 1.00\ \mu\text{F} + 2.00\ \mu\text{F} = 3.00\ \mu\text{F}$, and

$Q_{total} = C_{eq}V = (3.00\ \mu\text{F})(1200\ \text{V}) = 3.60 \times 10^{-3}\ \text{C}$.

The voltage over each is 1200 V since they are in parallel.

So: $Q_1 = C_1V_1 = (1.00\ \mu\text{F})(1200\ \text{V}) = 1.20 \times 10^{-3}\ \text{C}$.

$Q_2 = C_2V_2 = (2.00\ \mu\text{F})(1200\ \text{V}) = 2.40 \times 10^{-3}\ \text{C}$.

(b) $Q_{total} = 2.4 \times 10^{-3}\ \text{C} - 1.2 \times 10^{-3}\ \text{C} = 1.2 \times 10^{-3}\ \text{C}$, and still $C_{eq} = 3.00\ \mu\text{F}$, so the voltage is $V = Q/C = (1.2 \times 10^{-3}\ \text{C})/(3.00\ \mu\text{F}) = 400\ \text{V}$, and the new charges:

$Q_1 = C_1V_1 = (1.00\ \mu\text{F})(400\ \text{V}) = 4.00 \times 10^{-4}\ \text{C}$.

$Q_2 = C_2V_2 = (2.00\ \mu\text{F})(400\ \text{V}) = 8.00 \times 10^{-4}\ \text{C}$.

25-48: (a) With the switch open: $C_{eq} = \left(\left(\frac{1}{3\ \mu\text{F}} + \frac{1}{6\ \mu\text{F}}\right)^{-1} + \left(\frac{1}{3\ \mu\text{F}} + \frac{1}{6\ \mu\text{F}}\right)^{-1}\right) = 4.00\ \mu\text{F}$

$\Rightarrow Q_{total} = C_{eq}V = (4.00\ \mu\text{F})(360\ \text{V}) = 1.44 \times 10^{-3}\ \text{C}$. By symmetry, each capacitor carries $7.20 \times 10^{-4}\ \text{C}$. The voltages are then just calculated via $V = Q/C$. So: $V_{ad} = Q/C_3 = 240\ \text{V}$, and $V_{ac} = Q/C_6 = 120\ \text{V} \Rightarrow V_{cd} = V_{ad} - V_{ac} = 120\ \text{V}$.

(b) When the switch is closed, the points c and d must be at the same potential, so the equivalent capacitance is:

$$C_{eq} = \left(\frac{1}{(3+6)\ \mu\text{F}} + \frac{1}{(3+6)\ \mu\text{F}}\right)^{-1} = 4.5\ \mu\text{F}.$$

$\Rightarrow Q_{total} = C_{eq}V = (4.50\ \mu\text{F})(360\ \text{V}) = 1.62 \times 10^{-3}\ \text{C}$, and each capacitor has the same potential difference of 180 V.

(c) $\Delta Q = 1.62 \times 10^{-3}\ \text{C} - 1.44 \times 10^{-3}\ \text{C} = 1.80 \times 10^{-4}\ \text{C}$.

25-49: (a) $C_{eq} = \left(\frac{1}{7.2\ \mu\text{F}} + \frac{1}{7.2\ \mu\text{F}} + \frac{1}{3.6\ \mu\text{F}}\right) = 1.80\ \mu\text{F}$

$\Rightarrow Q = C_{eq}V = (1.8\ \mu\text{F})(24\ \text{V}) = 4.3 \times 10^{-5}\ \text{C}.$

(b) $U = \frac{1}{2}CV^2 = \frac{1}{2}(1.8\ \mu\text{F})(24\ \text{V})^2 = 5.2 \times 10^{-4}\ \text{J}.$

(c) If the capacitors are all in parallel, then:

$C_{eq} = (7.2\ \mu\text{F} + 7.2\ \mu\text{F} + 3.6\ \mu\text{F}) = 18\ \mu\text{F}$ and $Q = 3(4.3 \times 10^{-5}\ \text{C}) = 1.3 \times 10^{-4}\ \text{C}$,

and $V = Q/C = (1.3 \times 10^{-6}\ \text{C})/(18\ \mu\text{F}) = 7.2\ \text{V}$.

(d) $U = \frac{1}{2}CV^2 = \frac{1}{2}(18\ \mu\text{F})(7.2\ \text{V})^2 = 4.7 \times 10^{-4}\ \text{J}.$

25-50: (a) $u = \frac{1}{2}\varepsilon_o E^2 = \frac{1}{2}\varepsilon_o\left(\frac{\lambda}{2\pi\varepsilon_o r}\right)^2 = \frac{\lambda^2}{8\pi^2\varepsilon_o r^2}.$

(b) $U = \int u\,dV = 2\pi L\int u r\,dr = \frac{L\lambda^2}{4\pi\varepsilon_o}\int_{r_a}^{r_b}\frac{dr}{r} \Rightarrow \frac{U}{L} = \frac{\lambda^2}{4\pi\varepsilon_o}\ln(r_b/r_a).$

(c) Using Eq. (25-9): $U = \frac{Q^2}{2C} = \frac{Q^2}{4\pi\varepsilon_o L}\ln(r_b/r_a) = \frac{\lambda^2 L}{4\pi\varepsilon_o}\ln(r_b/r_a) = U$ of part (b).

25-51: (b) $C = \frac{Q}{V} = \frac{Q}{(Q/4\pi\varepsilon_o R)} = 4\pi\varepsilon_o R$, where we've chosen V = 0 at infinity.

(c) $C_{earth} = 4\pi\varepsilon_o R_{earth} = 4\pi\varepsilon_o(6.4 \times 10^6\ \text{m}) = 7.1 \times 10^{-4}\ \text{F}.$

25-52: (a) $r < R$: $u = \frac{1}{2}\varepsilon_o E^2 = 0.$

(b) $r > R$: $u = \frac{1}{2}\varepsilon_o E^2 = \frac{1}{2}\varepsilon_o\left(\frac{Q}{4\pi\varepsilon_o r^2}\right)^2 = \frac{Q^2}{32\pi^2\varepsilon_o r^4}.$

(c) $U = \int u\,dV = 4\pi\int_R^\infty r^2 u\,dr = \frac{Q^2}{8\pi\varepsilon_o}\int_R^\infty\frac{dr}{r^2} = \frac{Q^2}{8\pi\varepsilon_o R}.$

(d) This energy is equal to $\frac{1}{2}\frac{Q^2}{4\pi\varepsilon_o R}$ which is just the energy required to assemble all the charge into a spherical distribution. (Note, being aware of double counting gives the factor of 1/2 in front of the familiar potential energy formula for a charge Q a distance R from another charge Q.)

(e) From Eq. (25-9): $U = \frac{Q^2}{2C} = \frac{Q^2}{8\pi\varepsilon_o R}$ from part (c) $\Rightarrow C = 4\pi\varepsilon_o R$, as in **Pr. (25-51)**.

25-53: (a) $r < R$: $u = \frac{1}{2}\varepsilon_o E^2 = \frac{1}{2}\varepsilon_o\left(\frac{kQr}{R^3}\right)^2 = \frac{kQ^2r^2}{8\pi R^6}.$

(b) $r > R$: $u = \frac{1}{2}\varepsilon_o E^2 = \frac{1}{2}\varepsilon_o\left(\frac{kQ}{r^2}\right)^2 = \frac{kQ^2}{8\pi r^4}$.

(c) $r < R$: $U = \int u dV = 4\pi\int_0^R r^2 u dr = \frac{kQ^2}{2R^6}\int_0^R r^4 dr = \frac{kQ^2}{10R}$.

$r > R$: $U = \int u dV = 4\pi\int_R^\infty r^2 u dr = \frac{kQ^2}{2}\int_R^\infty \frac{dr}{r^2} = \frac{kQ^2}{2R} \Rightarrow U = \frac{3kQ^2}{5R}$.

25-54: (a) $Q = CV = \frac{K\varepsilon_o A}{d}V = \frac{(3.00)\varepsilon_o(0.200\ \text{m}^2)(3000\ \text{V})}{1.00\times 10^{-2}\ \text{m}} = 1.59\times 10^{-6}\ \text{C}$.

(b) $Q_i = Q(1-1/K) = (1.59\times 10^{-6}\ \text{C})(1-1/3.00) = 1.06\times 10^{-6}\ \text{C}$.

(c) $E = \frac{\sigma}{\varepsilon} = \frac{Q}{K\varepsilon_o A} = \frac{1.59\times 10^{-6}\ \text{C}}{(3.00)\varepsilon_o(0.200\ \text{m}^2)} = 2.99\times 10^5\ \text{V/m}$.

(d) $U = \frac{1}{2}QV = \frac{1}{2}(1.59\times 10^{-6}\ \text{C})(3000\ \text{V}) = 2.39\times 10^{-3}\ \text{J}$.

(e) $u = \frac{U}{Ad} = \frac{2.39\times 10^{-3}\ \text{J}}{(0.200\ \text{m}^2)(0.0100\ \text{m})} = 1.19\ \text{J/m}^3$.

(f) In this case, one does work by pushing the slab into the capacitor since the constant potential requires more charges to be brought onto the plates. When the charge is kept constant, the field pulls the dielectric into the gap, with the field (or charges) doing the work.

25-55: (a) $E = \frac{\sigma}{K\varepsilon_o} = \frac{0.50\times 10^{-3}\ \text{C/m}^2}{(5.4)\varepsilon_o} = 1.0\times 10^7\ \text{V/m}$.

(b) $V = Ed = (1.0\times 10^7\ \text{V/m})(5.0\times 10^{-9}\ \text{m}) = 0.052\ \text{V}$.

(c) volume $= 10^{-16}\ \text{m}^3 \Rightarrow R \approx 2.88\times 10^{-6}\ \text{m}$

$\Rightarrow$ shell volume $= 4\pi R^2 d = 4\pi(2.88\times 10^{-6}\ \text{m})^2(5.0\times 10^{-9}\ \text{m}) = 5.2\times 10^{-19}\ \text{m}^3$

$\Rightarrow U = uV = (\frac{1}{2}K\varepsilon_o E^2)V = \frac{1}{2}(5.4)\varepsilon_o(1.0\times 10^7\ \text{V/m})(5.2\times 10^{-19}\ \text{m}^3) = 1.25\times 10^{-15}\ \text{J}$.

25-56: (a) This situation is analagous to having two capacitors C_1 in series, each with separation $\frac{1}{2}(d-a)$. Therefore $C = \left(\frac{1}{C_1}+\frac{1}{C_1}\right)^{-1} = \frac{1}{2}C_1 = \frac{1}{2}\frac{\varepsilon_o A}{(d-a)/2} = \frac{\varepsilon_o A}{d-a}$.

(b) $C = \frac{\varepsilon_o A}{d-a} = \frac{\varepsilon_o A}{d}\frac{d}{d-a} = C_o\frac{d}{d-a}$.

(c) As $a\to 0$, $C\to C_o$. And as $a\to d$, $C\to\infty$.

25-57: $C_{eq} = \left(\left(\frac{\varepsilon_1 A}{d/2}\right)^{-1} + \left(\frac{\varepsilon_2 A}{d/2}\right)^{-1}\right)^{-1} = \left(\left(\frac{d}{2\varepsilon_1 A}\right)+\left(\frac{d}{2\varepsilon_2 A}\right)\right)^{-1} = \left(\frac{d}{2\varepsilon_o A}\left(\frac{1}{K_1}+\frac{1}{K_2}\right)\right)^{-1}$

$$\Rightarrow C_{eq} = \frac{2\varepsilon_o A}{d}\left(\frac{K_1 K_2}{K_1 + K_2}\right).$$

25-58: This situation is analagous to having two capacitors in parallel, each with an area $\frac{A}{2}$. So: $C_{eq} = C_1 + C_2 = \frac{\varepsilon_1 A/2}{d} + \frac{\varepsilon_2 A/2}{d} = \frac{\varepsilon_o A}{2d}(K_1 + K_2)$.

25-59:

(a)

a A B C b + −

(b) $C = 2\left(\frac{\varepsilon A}{d}\right) = \frac{2(4.2)\varepsilon_o(0.075\text{ m})^2}{4.5\times10^{-4}\text{ m}}$

$\Rightarrow C = 9.3\times10^{-10}\text{ F}.$

25-60: (a) The capacitors are in parallel so:

$$C = \frac{\varepsilon_{eff} WL}{d} = \frac{\varepsilon_o W(L-h)}{d} + \frac{K\varepsilon_o Wh}{d} = \frac{\varepsilon_o WL}{d}\left(1 + \frac{Kh}{L} - \frac{h}{L}\right) \Rightarrow K_{eff} = \left(1 + \frac{Kh}{L} - \frac{h}{L}\right).$$

(b) For gasolene, with K = 1.95:

$\frac{1}{4}$ full: $K_{eff}(h = \frac{L}{4}) = 1.24$; $\frac{1}{2}$ full: $K_{eff}(h = \frac{L}{2}) = 1.48$; $\frac{3}{4}$ full: $K_{eff}(h = \frac{3L}{4}) = 1.71$.

(c) For methanol, with K = 33:

$\frac{1}{4}$ full: $K_{eff}(h = \frac{L}{4}) = 9$; $\frac{1}{2}$ full: $K_{eff}(h = \frac{L}{2}) = 17$; $\frac{3}{4}$ full: $K_{eff}(h = \frac{3L}{4}) = 25$.

(d) This kind of fuel tank sensor will work best for methanol since it has the greater range of K_{eff} values.

25-61: (a) We are to show the transformation from one circuit to the other:

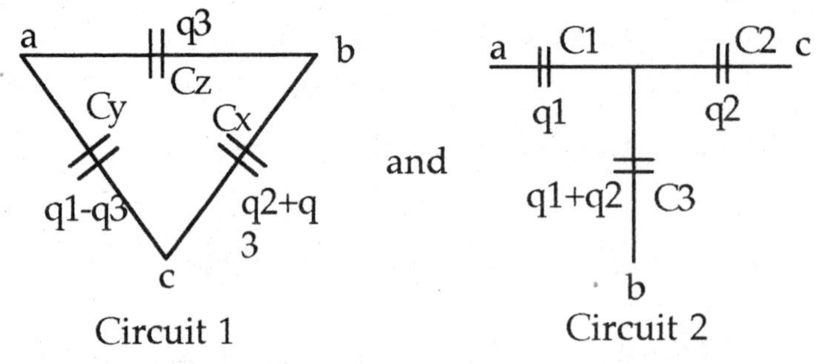

From Circuit 1: $V_{ac} = \frac{q_1 - q_3}{C_y}$ and $V_{bc} = \frac{q_2 + q_3}{C_x}$, where q_3 is derived from V_{ab}:

$$V_{ab} = \frac{q_3}{C_z} = \frac{q_1 - q_3}{C_y} - \frac{q_2 + q_3}{C_x} \Rightarrow q_3 = \frac{C_x C_y C_z}{C_x + C_y + C_z}\left(\frac{q_1}{C_y} - \frac{q_2}{C_x}\right) \equiv K\left(\frac{q_1}{C_y} - \frac{q_2}{C_x}\right)$$

From Circuit 2: $V_{ac} = \frac{q_1}{C_1} + \frac{q_1 + q_2}{C_3} = q_1\left(\frac{1}{C_1} + \frac{1}{C_3}\right) + q_2\frac{1}{C_3}$ and

$$V_{bc} = \frac{q_2}{C_2} + \frac{q_1 + q_2}{C_3} = q_1\frac{1}{C_3} + q_2\left(\frac{1}{C_2} + \frac{1}{C_3}\right).$$

Setting the coefficients of the charges equal to each other in matching potential equations from the two circuits results in three independent equations relating the two sets of capacitances. The set of equations are:

$$\frac{1}{C_1}=\frac{1}{C_y}\left(1-\frac{1}{KC_y}-\frac{1}{KC_x}\right),\ \frac{1}{C_2}=\frac{1}{C_x}\left(1-\frac{1}{KC_y}-\frac{1}{KC_x}\right)\text{ and }\frac{1}{C_3}=\frac{1}{KC_yC_x}.$$

From these, subbing in the expression for K, we get:

$$C_1=\left(C_xC_y+C_yC_z+C_zC_x\right)/\,C_x.$$
$$C_2=\left(C_xC_y+C_yC_z+C_zC_x\right)/\,C_y.$$
$$C_3=\left(C_xC_y+C_yC_z+C_zC_x\right)/\,C_z.$$

(b) Using the transformation of part (a) we have:

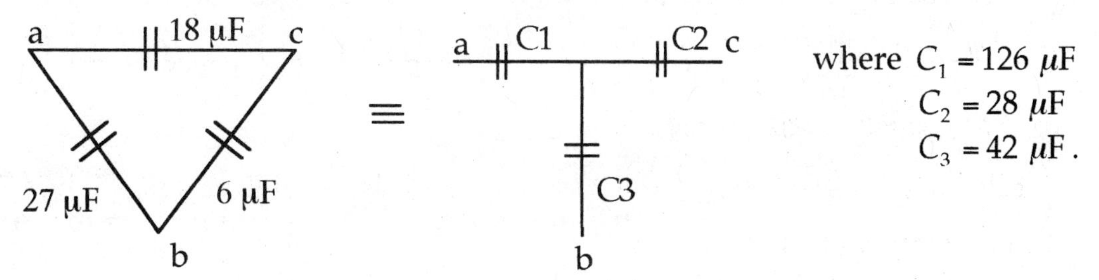

where $C_1=126\ \mu\text{F}$
$C_2=28\ \mu\text{F}$
$C_3=42\ \mu\text{F}$.

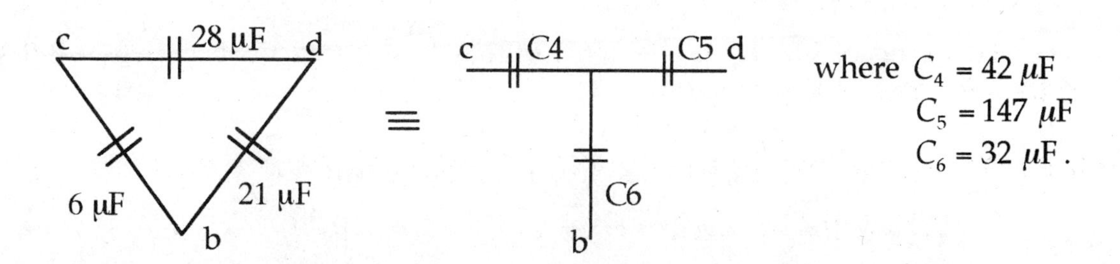

where $C_4=42\ \mu\text{F}$
$C_5=147\ \mu\text{F}$
$C_6=32\ \mu\text{F}$.

Now the total equivalent capacitance is:

$$C_{eq}=\left(\frac{1}{72\ \mu\text{F}}+\frac{1}{126\ \mu\text{F}}+\frac{1}{34.8\ \mu\text{F}}+\frac{1}{147\ \mu\text{F}}+\frac{1}{72\ \mu\text{F}}\right)^{-1}=14.0\ \mu\text{F},$$ where the 34.8 μF

comes from: $34.8\ \mu\text{F}=\left(\left(\frac{1}{42\ \mu\text{F}}+\frac{1}{32\ \mu\text{F}}\right)^{-1}+\left(\frac{1}{28\ \mu\text{F}}+\frac{1}{42\ \mu\text{F}}\right)^{-1}\right).$

(c) The circuit diagram can be re-drawn as shown on the next page. The overall charge is given by:

$Q=C_{eq}V=(14.0\ \mu\text{F})(36\ \text{V})\Rightarrow Q=5.04\times10^{-4}\ \text{C}.$

And this is also the charge over the 72 μF capacitors.

$$\Rightarrow V_{72}=\frac{5.04\times10^{-4}\ \text{C}}{72\times10^{-6}\ \text{F}}=7.0\ \text{V}.$$

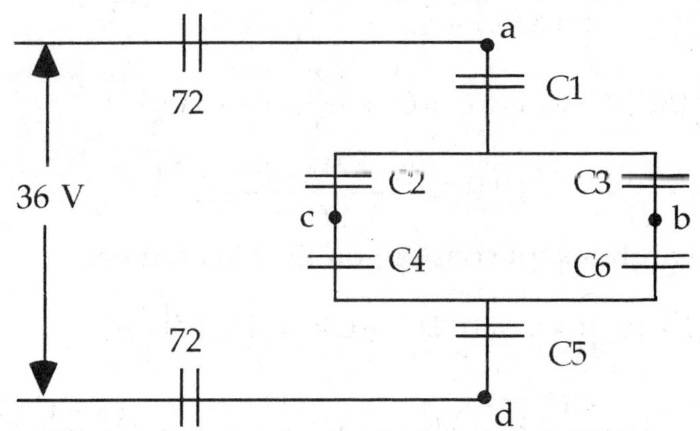

Next we will find the voltage over the numbered capacitors, and their associated voltages. Then those voltages will be changed back into the voltage of the original capacitors, and then their charges.

$Q_{C_1} = Q_{C_5} = Q_{72} = 5.04 \times 10^{-4}$ C

$\Rightarrow V_{C_5} = \dfrac{5.04 \times 10^{-4} \text{ C}}{147 \times 10^{-6} \text{ F}} = 3.43 \text{ V}$ and $V_{C_1} = \dfrac{5.04 \times 10^{-4} \text{ C}}{126 \times 10^{-6} \text{ F}} = 4.00 \text{ V}$.

$\Rightarrow V_{C_2C_4} = V_{C_3C_6} = (36.0 - 7.00 - 7.00 - 4.00 - 3.43) \text{ V} = 14.6 \text{ V}$.

But $C_{eq}(C_2C_4) = \left(\dfrac{1}{C_2} + \dfrac{1}{C_4}\right)^{-1} = 16.8\ \mu\text{F}$ and $C_{eq}(C_3C_6) = \left(\dfrac{1}{C_3} + \dfrac{1}{C_6}\right)^{-1} = 18.2\ \mu\text{F}$, so:

$Q_{C_2} = Q_{C_4} = V_{C_2C_4} C_{eq(C_2C_4)} = 2.45 \times 10^{-4}$ C, $Q_{C_3} = Q_{C_6} = V_{C_3C_6} C_{eq(C_3C_6)} = 2.64 \times 10^{-4}$ C.

$\Rightarrow V_{C_2} = \dfrac{Q_{C_2}}{C_2} = 8.8 \text{ V},\ V_{C_3} = \dfrac{Q_{C_3}}{C_3} = 6.3 \text{ V},\ V_{C_4} = \dfrac{Q_{C_4}}{C_4} = 5.8 \text{ V},\ V_{C_6} = \dfrac{Q_{C_6}}{C_6} = 8.3 \text{ V}$.

$\Rightarrow V_{ac} = V_{C_1} + V_{C_2} = V_{18} = 13 \text{ V} \Rightarrow Q_{18} = C_{18}V_{18} = 2.3 \times 10^{-4}$ C.

$V_{ab} = V_{C_1} + V_{C_3} = V_{27} = 10 \text{ V} \Rightarrow Q_{27} = C_{27}V_{27} = 2.8 \times 10^{-4}$ C.

$V_{cd} = V_{C_4} + V_{C_5} = V_{28} = 9 \text{ V} \Rightarrow Q_{28} = C_{28}V_{28} = 2.6 \times 10^{-4}$ C.

$V_{bd} = V_{C_5} + V_{C_6} = V_{21} = 12 \text{ V} \Rightarrow Q_{21} = C_{21}V_{21} = 2.5 \times 10^{-4}$ C.

$V_{bc} = V_{C_3} - V_{C_2} = V_6 = 2.5 \text{ V} \Rightarrow Q_6 = C_6V_6 = 1.5 \times 10^{-5}$ C.

25-62: (a) The force between the two parallel plates is:

$$F = qE = \frac{q\sigma}{2\varepsilon_o} = \frac{q^2}{2\varepsilon_o A} = \frac{(CV)^2}{2\varepsilon_o A} = \frac{\varepsilon_o^2 A^2}{z^2}\frac{V^2}{2\varepsilon_o A} = \frac{\varepsilon_o A V^2}{2z^2}.$$

(b) When V = 0, the separation is just z_o. So:

$$F_{4\ springs} = 4k(z_o - z) = \frac{\varepsilon_o A V^2}{2z^2} \Rightarrow 2z^3 - 2z^2 z_o + \frac{\varepsilon_o A V^2}{4k} = 0.$$

(c) For $A = 0.300 \text{ m}^2$, $z_o = 1.2 \times 10^{-3}$ m, $k = 25$ N / m, and $V = 120$ V,

$2z^3 - (2.4 \times 10^{-3} \text{ m})z^2 + 3.82 \times 10^{-10} \text{ m}^3 = 0 \Rightarrow z = 0.537$ mm, 1.014 mm.

(d) Stable equilibrium occurs if a slight displacement from equilibrium yields a force back toward the equilibrium point. If one evaluates the forces at small displacements from the equilibrium positions above, the 1.014 mm separation is seen to be stable, but not the 0.537 mm separation.

25-63: (a) $C_o = \frac{\varepsilon_o}{D}((L-x)L + xKL) = \frac{\varepsilon_o L}{D}(L+(K-1)x)$.

(b) $\Delta U = \frac{1}{2}(\Delta C)V^2$ where $C = C_o + \frac{\varepsilon_o L}{D}(-dx + dxK)$

$\Rightarrow \Delta U = \frac{1}{2}\left(\frac{\varepsilon_o L dx}{D}(K-1)\right)V^2 = \frac{(K-1)\varepsilon_o V^2 L}{2D}dx$.

(c) If the charge is kept constant on the plates, then:

$Q = \frac{\varepsilon_o LV}{D}(L+(k-1)x)$, and $U = \frac{1}{2}CV^2 = \frac{1}{2}C_o V^2\left(\frac{C}{C_o}\right)$

$\Rightarrow U \approx \frac{C_o V^2}{2}\left(1 - \frac{\varepsilon_o L}{DC_o}(K-1)dx\right) \Rightarrow \Delta U = U - U_o = -\frac{(K-1)\varepsilon_o V^2 L}{2D}dx$.

(d) Since $dU = -Fdx = -\frac{(K-1)\varepsilon_o V^2 L}{2D}dx$, then the force is in the opposite direction to the motion dx, meaning that the slab feels a force pushing it out.

25-64: (a) For a normal spherical capacitor: $C_o = 4\pi\varepsilon_0\left(\frac{r_a r_b}{r_b - r_a}\right)$. Here we have, in effect, two parallel capacitors, C_L and C_U.

$C_L = \frac{KC_o}{2} = 2\pi K\varepsilon_0\left(\frac{r_a r_b}{r_b - r_a}\right)$ and $C_U = \frac{C_o}{2} = 2\pi\varepsilon_0\left(\frac{r_a r_b}{r_b - r_a}\right)$.

(b) Using a hemispherical Gaussian surface for each respective half:

$E_L\frac{4\pi r^2}{2} = \frac{Q_L}{K\varepsilon_o} \Rightarrow E_L = \frac{Q_L}{2\pi K\varepsilon_o r^2}$ and $E_U\frac{4\pi r^2}{2} = \frac{Q_U}{\varepsilon_o} \Rightarrow E_U = \frac{Q_U}{2\pi\varepsilon_o r^2}$.

But $Q_L = VC_L$ and $Q_U = VC_U$, $Q_L + Q_U = Q$.

So: $Q_L = \frac{VC_o K}{2} = KQ_U \Rightarrow Q_U(1+K) = Q \Rightarrow Q_U = \frac{Q}{1+K}$ and $Q_L = \frac{KQ}{1+K}$.

$\Rightarrow E_L = \frac{KQ}{1+K}\frac{1}{2\pi K\varepsilon_o r^2} = \frac{2}{1+K}\frac{Q}{4\pi\varepsilon_o r^2}$ and $E_U = \frac{Q}{1+K}\frac{1}{2\pi K\varepsilon_o r^2} = \frac{2}{1+K}\frac{Q}{4\pi\varepsilon_o r^2}$.

(c) The free charge density on upper and lower hemispheres are:

$\left(\sigma_{f_{r_a}}\right)_U = \frac{Q_U}{4\pi r_a^2} = \frac{Q}{4\pi r_a^2(1+K)}$ and $\left(\sigma_{f_{r_b}}\right)_U = \frac{Q_U}{4\pi r_b^2} = \frac{Q}{4\pi r_b^2(1+K)}$.

$\left(\sigma_{f_{r_a}}\right)_L = \frac{Q_L}{4\pi r_a^2} = \frac{KQ}{4\pi r_a^2(1+K)}$ and $\left(\sigma_{f_{r_b}}\right)_L = \frac{Q_L}{4\pi r_b^2} = \frac{KQ}{4\pi r_b^2(1+K)}$.

(d) $\sigma_{i_{r_a}} = \sigma_{f_{r_a}}(1 - 1/K) = \frac{(K-1)}{K}\frac{Q}{4\pi r_a^2}\frac{K}{K+1} = \frac{K-1}{K+1}\frac{Q}{4\pi r_a^2}$.

$\sigma_{i_{r_b}} = \sigma_{f_{r_b}}(1 - 1/K) = \frac{(K-1)}{K}\frac{Q}{4\pi r_b^2}\frac{K}{K+1} = \frac{K-1}{K+1}\frac{Q}{4\pi r_b^2}$.

(e) There is zero bound charge on the flat surface of the dielectric-air interface, or else that would imply a circumferential electric field, or that the electric field changed as we went around the sphere.

Chapter 26: Current, Resistance and Electromotive Force

26-1: $Q = It = (4.8\ \text{A})(2)(3600\ \text{s}) = 3.5\times10^{4}\ \text{C}.$

26-2: (a) $I = \dfrac{Q_{total}}{t} = \dfrac{(5.04\times10^{18} + 1.61\times10^{18})(1.60\times10^{-19}\ \text{C})}{1\ \text{s}} = 1.06\ \text{A}.$
(b) The current is in the direction of the proton flow.

26-3: (a) $Q = \int_0^{10} I dt = \int_0^{10} (3 + 0.73t^2)dt = 3t\Big|_0^{10} + \dfrac{0.73}{3}t^3\Big|_0^{10} = 273\ \text{C}.$
(b) The same charge would flow in 10 seconds if there was a constant current of: $I = Q/t = (273\ \text{C})/(10\ \text{s}) = 27.3\ \text{A}.$

26-4: (a) Current is given by $I = \dfrac{Q}{t} = \dfrac{72\ \text{C}}{70(60\ \text{s})} = 1.7\times10^{-2}\ \text{A}.$
(b) $I = nqv_dA$

$$\Rightarrow v_d = \frac{I}{nqA} = \frac{1.7\times10^{-2}\ \text{A}}{(5.8\times10^{28})(1.6\times10^{-19}\ \text{C})(\pi(1.3\times10^{-3}\ \text{m})^2)} = 3.5\times10^{-7}\ \text{m/s}.$$

26-5: (a) $v_d = \dfrac{I}{nqA} = \dfrac{3.55\ \text{A}}{(8.5\times10^{28})(1.6\times10^{-19}\ \text{C})(\pi/4)(0.255\times10^{-3}\ \text{m})^2)} = 5.12\times10^{-3}\ \text{m/s}$

$\Rightarrow$ travel time $= \dfrac{d}{v_d} = 110\ \text{s}.$
(b) If the diameter is now 1.45 mm, the time can be calculated using the formula above or comparing the ratio of the areas, and yields a time of 3500 s.
(c) The drift velocity depends on the diameter of the wire as an inverse square relationship.

26-6: $Q_{total} = (n_e + n_p)e = (6.45\times10^{16} + 4.18\times10^{16})(1.60\times10^{-19}\ \text{C}) = 0.0170\ \text{C}$

$$\Rightarrow I = \frac{Q_{total}}{t} = \frac{0.0170\ \text{C}}{1\ \text{s}} = 0.0170\ \text{A} = 17.0\ \text{mA}.$$

26-7: (a) silver: $E = \rho J = \dfrac{\rho I}{A} = \dfrac{(1.47\times10^{-8}\ \Omega/\text{m}^3)(0.470\ \text{A})}{(\pi/4)(2.59\times10^{-3}\ \text{m})^2} = 1.31\times10^{-3}\ \text{V/m}.$

(b) nichrome: $E = \rho J = \dfrac{\rho I}{A} = \dfrac{(100\times10^{-8}\ \Omega/\text{m}^3)(0.470\ \text{A})}{(\pi/4)(2.59\times10^{-3}\ \text{m})^2} = 8.92\times10^{-2}\ \text{V/m}.$

26-8: (a) $J = \dfrac{I}{A} = \dfrac{4.6\ \text{A}}{(1.8\times10^{-3}\ \text{m})^2} = 1.4\times10^{6}\ \text{A/m}^2.$
(b) $E = \rho J = (1.72\times10^{-8}\ \Omega\cdot\text{m})(1.4\times10^{6}\ \text{A/m}^2) = 0.024\ \text{V/m}.$
(c) Time to travel the wire's length:

$$t = \frac{l}{v_d} = \frac{l\,nqA}{I} = \frac{(5.0\ \text{m})(8.5\times10^{28}/\text{m}^3)(1.6\times10^{-19}\ \text{C})(1.8\times10^{-3}\ \text{m})^2}{4.6\ \text{A}} = 4.8\times10^{4}\ \text{s}$$

26-9: $R = \dfrac{\rho L}{A} = \dfrac{(1.72\times10^{-8}\ \Omega\text{m})(35.0\ \text{m})}{(\pi/4)(2.05\times10^{-3}\ \text{m})^2} = 0.182\ \Omega.$

26-10: $R = \dfrac{\rho L}{A} \Rightarrow L = \dfrac{RA}{\rho} = \dfrac{(1.00\ \Omega)(\pi/4)(0.75\times10^{-3}\ \text{m})^2}{1.72\times10^{-8}\ \Omega\cdot\text{m}} = 25.7\ \text{m}.$

26-11: (a) $I = \dfrac{J}{A} = \dfrac{E}{\rho A} = \dfrac{0.49\ \text{V/m}}{(2.75\times10^{-8}\ \Omega\text{m})(\pi/4)(0.84\times10^{-3}\ \text{m})^2} = 9.9\ \text{A}.$

(b) $V = IR = \dfrac{I\rho L}{A} = \dfrac{(9.9\ \text{A})(2.75\times10^{-8}\ \Omega\text{m})(12.0\ \text{m})}{(\pi/4)(0.84\times10^{-3}\ \text{m})^2} = 5.9\ \text{V}.$

(c) $R = \dfrac{V}{I} = \dfrac{5.9\ \text{V}}{9.9\ \text{A}} = 0.60\ \Omega.$

26-12: $R_{Al} = R_{Cu} \Rightarrow \dfrac{\rho_{Al}L}{A_{Al}} = \dfrac{\rho_{Cu}L}{A_{Cu}} \Rightarrow \dfrac{\pi d_{Al}^{\ 2}}{4\rho_{Al}} = \dfrac{\pi d_{Cu}^{\ 2}}{4\rho_{Cu}} \Rightarrow d_{Al} = d_{Cu}\sqrt{\dfrac{\rho_{Al}}{\rho_{Cu}}}$

$\Rightarrow d_{Al} = (2.2\ \text{mm})\sqrt{\dfrac{2.75\times10^{-8}\ \Omega\cdot\text{m}}{1.72\times10^{-8}\ \Omega\cdot\text{m}}} = 2.8\ \text{mm}.$

26-13: (a) $E = \rho J = \dfrac{RAJ}{L} = \dfrac{RI}{L} = \dfrac{V}{L} = \dfrac{8.52\ \text{V}}{8.00\ \text{m}} = 1.07\ \text{V/m}.$

(b) $\rho = \dfrac{RA}{L} = \dfrac{V}{JL} = \dfrac{8.52\ \text{V}}{(3.4\times10^{7}\ \text{A/m}^2)(8.00\ \text{m})} = 3.1\times10^{-8}\ \Omega\text{m}.$

26-14: $\rho = \dfrac{RA}{L} = \dfrac{VA}{IL} = \dfrac{(3.80\ \text{V})\pi(6.54\times10^{-4}\ \text{m})^2}{(17.6\ \text{A})(2.50\ \text{m})} = 11.6\times10^{-8}\ \Omega\cdot\text{m}.$

26-15: $R = \dfrac{V}{I} = \dfrac{\rho L}{A} = \dfrac{\rho L}{\pi r^2} \Rightarrow r = \sqrt{\dfrac{I\rho L}{\pi V}} = \sqrt{\dfrac{(6.00\ \text{A})(2.00\times10^{-8}\ \Omega\text{m})(1.70\ \text{m})}{\pi(1.50\ \text{V})}} = 6.58\times10^{-4}\ \text{m}.$

26-16: Because the density does not change, volume stays the same, so $LA = (2L)(A/2)$ and the area is halved. So the resistance becomes: $R = \dfrac{\rho(2L)}{A/2} = 4\dfrac{\rho L}{A} = 4R_o$. That is, four times the original resistance.

26-17: (a) If 250 strands of wire are placed side by side, we are effectively increasing the area of the current carrier by 250. so the resistance is smaller by that factor: $R = 4.35\times10^{-6}\ \Omega/250 = 1.74\times10^{-8}\ \Omega.$
(b) If 250 strands of wire are placed end to end, we are effectively increasing the length of the wire by 250, and so $R = (4.35\times10^{-6}\ \Omega)250 = 1.09\times10^{-3}\ \Omega.$

26-18: $\dfrac{R - R_o}{R_o} = \alpha(T_f - T_i)$

$\Rightarrow \alpha = \dfrac{R - R_o}{(T_f - T_i)R_o} = \dfrac{1.654\ \Omega - 1.504\ \Omega}{(36.0°\text{C} - 20.0°\text{C})(1.504\ \Omega)} = 6.23\times10^{-3}\ °\text{C}^{-1}.$

26-19: (a) $R_f - R_i = R_i\alpha(T_f - T_i) \Rightarrow R_f = 100\Omega + 100\Omega(0.0004°\text{C}^{-1})(17.3°\text{C}) = 99\Omega$.
(b) $R_f - R_i = R_i\alpha(T_f - T_i) \Rightarrow R_f = 0.019\Omega + 0.019\Omega(-0.0005°\text{C}^{-1})(34°\text{C}) = 0.0187\Omega$.

26-20: $T_f - T_i = \dfrac{R_f - R_i}{\alpha R_i}; \quad T_f = T_i + \dfrac{R_f - R_i}{\alpha R_i} = \dfrac{214.0\Omega - 217.3\Omega}{(-0.0005°\text{C}^{-1})(217.3\Omega)} + 4°\text{C} = 34.4°\text{C}.$

26-21: (a) When there is no current flowing, the voltmeter reading is simply the emf of the battery: $\mathcal{E} = 1.56\ \text{V}$.
(b) The voltage over the internal resistance is:
$V_r = 1.56\,\text{V} - 1.45\ \text{V} = 0.11\ \text{V} \Rightarrow r = \dfrac{V}{I} = \dfrac{0.11\ \text{V}}{1.3\ \text{A}} = 0.085\ \Omega.$

26-22: (a) A voltmeter placed over the battery terminals reads the emf: $\mathcal{E} = 12.0\ \text{V}$.
(b) There is no current flowing, so $V_r = 0$.
(c) The voltage reading over the switch is that over the battery: $V_s = 12.0\ \text{V}$
(d) Having closed the switch:
$I = 12.0\ \text{V}/4.48\ \Omega = 2.68\ \text{A} \Rightarrow V_{ab} = 12.0\ \text{V} - (2.68\ \text{A})(0.28\ \Omega) = 11.3\ \text{V}.$
$V_r = IR = (2.68\ \text{A})(4.20\ \Omega) = 11.3\ \text{V}.$
$V_s = 0$, since all the voltage has been "used up" in the circuit.

26-23: (a) $r = \mathcal{E}/I = 1.50\ \text{V}/14.8\ \text{A} = 0.101\ \Omega$.
(b) $r = \mathcal{E}/I = 1.50\ \text{V}/5.2\ \text{A} = 0.29\ \Omega$.
(c) $r = \mathcal{E}/I = 12.0\ \text{V}/1000\ \text{A} = 0.0120\ \Omega$.

26-24: (a) $V_r = \mathcal{E} - V_{ab} = 24.0\ \text{V} - 20.2\ \text{V} = 3.8\ \text{V} \Rightarrow r = 3.8\ \text{V}/4.00\ \text{A} = 0.950\ \Omega$.
(b) $V_R = 20.2\ \text{V} \Rightarrow R = 20.2\ \text{V}/4.00\ \text{A} = 5.05\ \Omega$.

26-25: (a) An ideal voltmeter has infinite resistance, so there would be NO current through the $2.0\ \Omega$ resistor.
(b) $V_{ab} = \mathcal{E} = 5.0\ \text{V}$, since there is no current there is no voltage lost over the internal resistance.
(c) The voltmeter reading is therefore 5.0 V since with no current flowing, it measures the terminal voltage of the battery.

26-26: (a) The current is counter-clockwise, because the 16 V battery determines the direction of current flow. Its magnitude is given by:
$$I = \frac{\sum \varepsilon}{\sum R} = \frac{16.0\ \text{V} - 8.0\ \text{V}}{1.6\ \Omega + 5.0\ \Omega + 1.4\ \Omega + 9.0\ \Omega} = 0.47\ \text{A}.$$
(b) $V_{ab} = 16.0\ \text{V} - (1.6\ \Omega)(0.47\ \text{A}) = 15.2\ \text{V}$.

(c) $V_{ac} = (5.0\ \Omega)(0.47\ \text{A}) + (1.4\ \Omega)(0.47\ \text{A}) + 8.0\ \text{V} = 11.0\ \text{V}$.

(d)

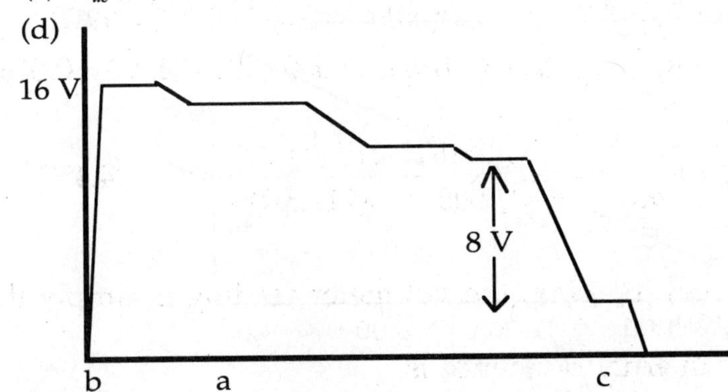

26-27: (a) Now the current flows clockwise since both batteries point in that direction: $I = \dfrac{\sum \varepsilon}{\sum R} = \dfrac{16.0\ \text{V} + 8.0\ \text{V}}{1.6\ \Omega + 5.0\ \Omega + 1.4\ \Omega + 9.0\ \Omega} = 1.41\ \text{A}$.

(b) $V_{ab} = 16.0\ \text{V} - (1.6\ \Omega)(1.41\ \text{A}) = 13.7\ \text{V}$.

(c) $V_{ac} = -(5.0\ \Omega)(1.41\ \text{A}) - (1.4\ \Omega)(1.41\ \text{A}) - 8.0\ \text{V} = -17.0\ \text{V}$.

(d)

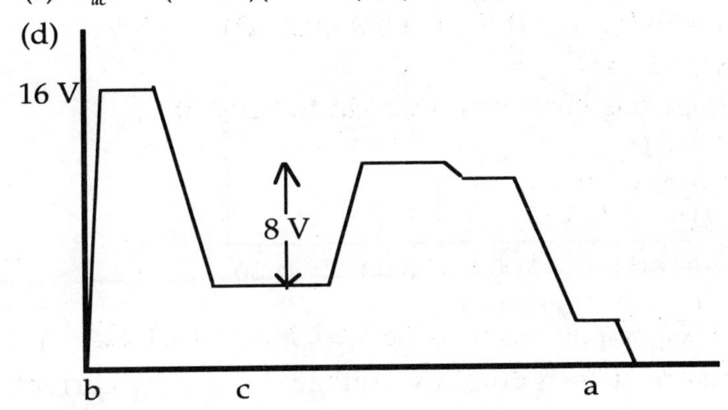

26-28: (a) $V_{bc} = 2.1\ \text{V} \Rightarrow I = V_{bc} / R_{bc} = 2.1\ \text{V}/9.0\ \text{V} = 0.23\ \text{A}$.

(b) $\sum \varepsilon = \sum IR \Rightarrow 8.0\ \text{V} = ((1.6 + 9.0 + 1.4 + R)\Omega)(0.23\ \text{A}) \Rightarrow R = \dfrac{5.24}{0.23} = 22.7\ \Omega$.

(c)

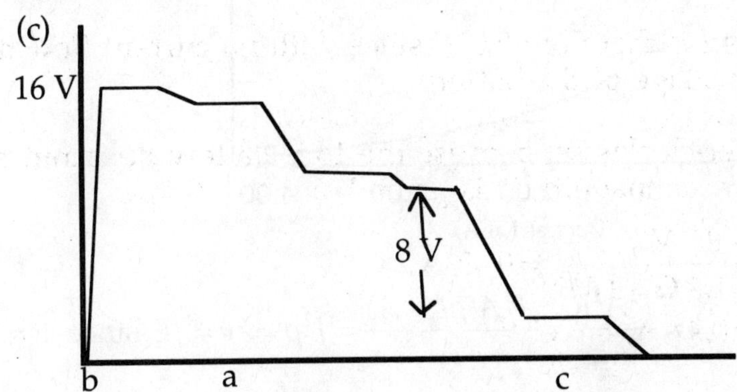

26-29: (a) Nichrome wire:

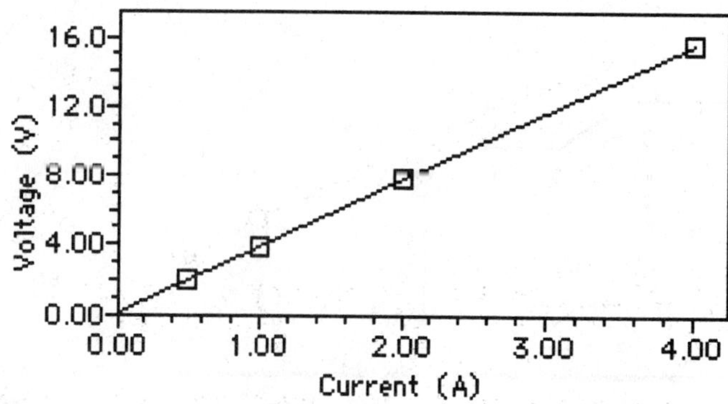

(b) The Nichrome wire does obey Ohm's Law since it is a straight line.
(c) The resistance is the voltage divided by current which is 3.88 Ω.

26-30: (a) Thyrite resistor:

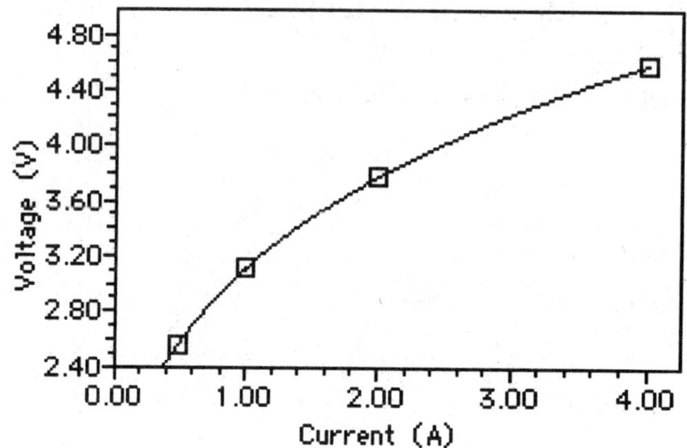

(b) The Thyrite is non-Ohmic since the plot is curved.
(c) Calculating the resistance at each point by voltage divided by current:

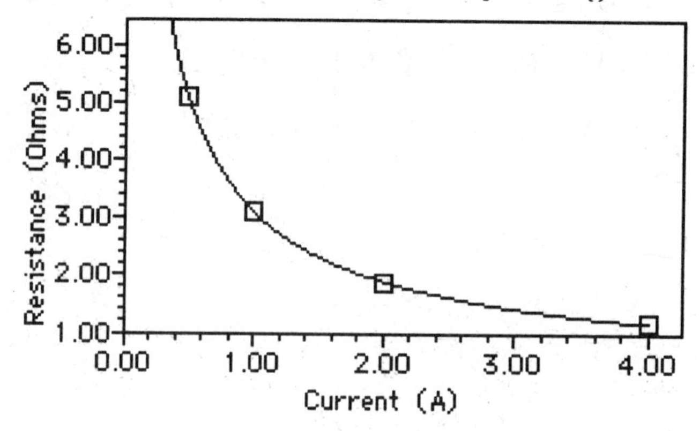

26-31: (a) $P = I^2R \Rightarrow p = \dfrac{P}{Vol} = \dfrac{I^2R}{AL} = \dfrac{J^2A^2R}{AL} = \dfrac{J^2A(\rho L / A)}{L} = J^2\rho \Rightarrow p = JE$ since $E = \rho J$.
(b) From (a) $p = J^2\rho$.
(c) Since $J = E / \rho$, (a) becomes $p = E^2 / \rho$.

26-32: $P = V^2 / R \Rightarrow R = V^2 / P = (18 \text{ V})^2 / 369 \text{ W} = 0.878 \ \Omega$.

26-33: $P = VI = (650 \text{ V})(0.80 \text{ A}) = 520 \text{ W}$.

26-34: $W = Pt = IVt = (0.29\text{ A})(12\text{ V})(4.5)(3600\text{ s}) = 5.6 \times 10^4\text{ J}$.

26-35: (a) $P = V^2 / R \Rightarrow R = V^2 / P = (120\text{ V})^2 / 540\text{ W} = 26.7\ \Omega$.
(b) $I = V / R = 120\text{ V} / 26.7\ \Omega = 4.5\text{ A}$.
(c) If the voltage is just 110 V, then $I = 4.13\text{ A} \Rightarrow P = VI = 454\text{ W}$.

26-36: (a) $I = \sum \varepsilon / R = 3.0\text{ V}/17\ \Omega = 0.18\text{ A}$.
(b) $W = Pt = IVt = (0.18\text{ A})(3.0\text{ V})(5)(3600\text{ s}) = 9720\text{ J}$.
(c) Now if the power to the bulb is 0.27 W,

$$P = I^2R \Rightarrow 0.27\text{ W} = \left(\frac{3.0\text{ V}}{17\ \Omega + R}\right)^2 (17\ \Omega) \Rightarrow (17\ \Omega + R)^2 = 567\ \Omega^2 \Rightarrow R = 6.8\ \Omega.$$

26-37: (a) $W = Pt = IVt = (50\text{ A})(12\text{ V})(3600\text{ s}) = 2.16 \times 10^6\text{ J}$.
(b) To release this much energy we need a volume of gasoline given by:

$$m = \frac{2.16 \times 10^6\text{ J}}{46{,}000\text{ J/g}} = 47.0\text{ g} \Rightarrow \text{Vol} = \frac{m}{\rho} = \frac{0.047\text{ kg}}{900\text{ kg/m}^3} = 0.0522\text{ m}^3 = 52.2\text{ liters}.$$

(c) To recharge the battery: $t = W / P = (600\text{ Wh}) / (500\text{ W}) = 1.2\text{ h}$.

26-38: (a) $I = \varepsilon/(R + r) = 12\text{ V}/10\ \Omega = 1.2\text{ A} \Rightarrow P = \mathcal{E}I = (12\text{ V})(1.2\text{ A}) = 14.4\text{ W}$. This is less than the previous value of 24 W.
(b) The work dissipated in the battery is just: $P = I^2r = (1.2\text{ A})^2(2.0\ \Omega) = 2.9\text{ W}$. This is less than 8 W, the amount found in Example (26-9).
(c) The net power output of the battery is $14.4\text{ W} - 2.9\text{ W} = 11.5\text{ W}$. This is less than 16 W, the amount found in Example (26-9).

26-39: (a) $I = V / R = 12\text{ V} / 6\ \Omega = 2.0\text{ A} \Rightarrow P = \mathcal{E}I = (12\text{ V})(2.0\text{ A}) = 24\text{ W}$.
(b) The power dissipated in the battery is $P = I^2r = (2.0\text{ A})^2(1.0\ \Omega) = 4.0\text{ W}$.
(c) The power delivered is then $24\text{ W} - 4\text{ W} = 20\text{ W}$.

26-40: (a) $I = \sum \varepsilon / R_{total} = 8.0\text{ V}/17\ \Omega = 0.47\text{ A} \Rightarrow P_{5\Omega} = I^2R = (0.47\text{ A})^2(5.0\ \Omega) = 1.1\text{ W}$ and $P_{9\Omega} = I^2R = (0.47\text{ A})^2(9.0\ \Omega) = 2.0\text{ W}$.
(b) $P_{16V} = \mathcal{E}I - I^2r = (16\text{ V})(0.47\text{ A}) - (0.47\text{ A})^2(1.6\ \Omega) = 7.2\text{ W}$.
(c) $P_{8V} = \varepsilon I = (8.0\text{ V})(0.47\text{ A}) = 3.8\text{ W}$.
(d) If we add up the other power losses we find $P = 7.2\text{ W} - 1.1\text{ W} - 2.0\text{ W} - 0.31\text{ W} = 3.8\text{ W}$, which is the same as that dissipated over the 8.0 V battery.

26-41: From Eq. (26-24), $\rho = \dfrac{m}{ne^2\tau}$.

$$\Rightarrow \tau = \frac{m}{ne^2\rho} = \frac{9.11 \times 10^{-31}\text{ kg}}{(5.80 \times 10^{28}\text{ m}^{-3})(1.60 \times 10^{-19}\text{ C})^2(1.47 \times 10^{-8}\ \Omega\text{m})} = 4.16 \times 10^{-14}\text{ s}.$$

26-42: (a) $R = \frac{\rho L}{A} = \frac{(5.0\ \Omega\cdot\text{m})(1.6\ \text{m})}{(\pi/4)(0.10\ \text{m})^2} = 1020\ \Omega.$

(b) $V = IR = (0.10\ \text{A})(1020\ \Omega) = 102\ \text{V}.$

(c) $P = I^2R = (0.10\ \text{A})^2(1020\ \Omega) = 10.2\ \text{W}.$

26-43: (a) $I = V / R = 20\ \text{kV} / 12\ \text{k}\Omega = 1.7\ \text{A}.$

(b) $P = I^2R = (1.7\ \text{A})^2(10{,}000\ \Omega) = 28\ \text{kW}.$

(c) If we want the current to be 1.0 mA, then the internal resistance must be:

$R + r = \frac{20{,}000\ \text{V}}{0.001\ \text{A}} = 2.0\times10^7\ \Omega \Rightarrow R = 20\ \text{M}\Omega - 10\ \text{k}\Omega \approx 20\ \text{M}\Omega.$

26-44: (a) $I = \frac{V}{R} \Rightarrow J = \frac{I}{A} = \frac{V}{RA} = \frac{V}{(\rho L/A)A} = \frac{V}{\rho L}$ So to make the current density a maximum, we need the length between faces to be as small as possible, which means $L = d$. So the potential difference should be applied to those faces which are a distance d apart. This maximum current density is $J_{MAX} = \frac{V}{\rho d}$.

(b) For a maximum current $I = \frac{V}{R} = \frac{VA}{\rho L} = JA$ must be a maximum. The maximum area is presented by the faces that are a distance d apart, and these two faces also have the greatest current density, so again, the potential should be placed over the faces a distance d apart. This maximum current is $I_{MAX} = 6\frac{Vd}{\rho}$.

26-45: (a) $\rho = \frac{RA}{L} = \frac{(0.0625\ \Omega)(2.00\times10^{-3}\ \text{m})^2}{12.0\ \text{m}} = 2.08\times10^{-8}\ \Omega\text{m}.$

(b) $I = JA = \frac{EA}{\rho} = \frac{(1.28\ \text{V/m})(2.00\times10^{-3}\ \text{m})^2}{2.08\times10^{-8}\ \Omega\text{m}} = 246\ \text{A}.$

(c) $v_d = \frac{J}{nq} = \frac{E}{\rho nq} = \frac{1.28\ \text{V/m}}{(2.08\times10^{-8}\ \Omega\text{m})(8.5\times10^{28}\ \text{m}^{-3})(1.6\times10^{-19}\ \text{C})} = 4.52\times10^{-3}\ \text{m/s}.$

26-46: (a) $I = \frac{V}{R} = \frac{V}{R_{Cu} + R_{Ag}}$ and $R_{Cu} = \frac{\rho_{Cu}L_{Cu}}{A_{Cu}} = \frac{(1.72\times10^{-8}\ \Omega\cdot\text{m})(1.8\ \text{m})}{(\pi/4)(8.0\times10^{-4}\ \text{m})^2} = 0.062\ \Omega$, and

$R_{Ag} = \frac{\rho_{Ag}L_{Ag}}{A_{Ag}} = \frac{(1.47\times10^{-8}\ \Omega\cdot\text{m})(1.2\ \text{m})}{(\pi/4)(8.0\times10^{-4}\ \text{m})^2} = 0.035\ \Omega \Rightarrow I = \frac{5.0\ \text{V}}{0.062\ \Omega + 0.035\ \Omega} = 52\ \text{A}.$

So the current in the copper wire is 52 A.

(b) The current in the silver wire is 52 A, the same as that in the copper wire or else charge would build up at their interface.

(c) $E_{Cu} = J\rho_{Cu} = \frac{IR_{Cu}}{L_{Cu}} = \frac{(52\ \text{A})(0.062\ \Omega)}{1.8\ \text{m}} = 1.77\ \text{V/m}.$

(d) $E_{Ag} = J\rho_{Ag} = \frac{IR_{Ag}}{L_{Ag}} = \frac{(52\ \text{A})(0.035\ \Omega)}{1.2\ \text{m}} = 1.50\ \text{V/m}.$

(e) $V_{Ag} = IR_{Ag} = (52\ \text{A})(0.035\ \Omega) = 1.8\ \text{V}.$

26-47: (a) The current must be the same in both sections of the wire, so the current in the thin end is 3.0 mA.

(b) $E_{1.8\text{mm}} = \rho J = \dfrac{\rho I}{A} = \dfrac{(1.72\times10^{-8}\ \Omega\text{m})(3.00\times10^{-3}\ \text{A})}{(\pi/4)(1.80\times10^{-3}\ \text{A})^2} = 2.03\times10^{-5}\ \text{V/m}.$

(c) $E_{0.9\text{mm}} = \rho J = \dfrac{\rho I}{A} = \dfrac{(1.72\times10^{-8}\ \Omega\text{m})(3.00\times10^{-3}\ \text{A})}{(\pi/4)(0.90\times10^{-3}\ \text{A})^2} = 8.11\times10^{-5}\ \text{V/m}\ (= 4E_{1.8\text{mm}}).$

(d) $V = E_{1.8\text{mm}}L_{1.8\text{mm}} + E_{0.9\text{mm}}L_{0.9\text{mm}}$

$\Rightarrow V = (2.03\times10^{-5}\ \text{V/m})(1.50\ \text{m}) + (8.11\times10^{-5}\ \text{V/m})(2.50\ \text{m}) = 2.33\times10^{-4}\ \text{V}.$

26-48: (a) $\dfrac{K}{\text{volume}} = n\left(\dfrac{1}{2}mv_d^{\,2}\right)$

$\Rightarrow \dfrac{K}{\text{volume}} = \dfrac{1}{2}(8.5\times10^{28}\ \text{m}^{-3})(9.11\times10^{-31}\ \text{kg})(1.5\times10^{-4}\ \text{m/s})^2 = 8.7\times10^{-10}\ \text{J/m}^3.$

(b) $U = qV = ne(\text{volume})V = (8.5\times10^{28}\ \text{m}^{-3})(1.6\times10^{-19}\ \text{C})(10^{-6}\ \text{m}^3)(1.0\ \text{V}) = 13600\ \text{J}$

And the kinetic energy in 1.0 cm^3 is $K = (8.7\times10^{-10}\ \text{J/m}^3)(10^{-6}\ \text{m}) = 8.7\times10^{-16}\ \text{J}$

So $\dfrac{U}{K} = \dfrac{13600\ \text{J}}{8.7\times10^{-16}\ \text{J}} = 1.6\times10^{19}.$

26-49:

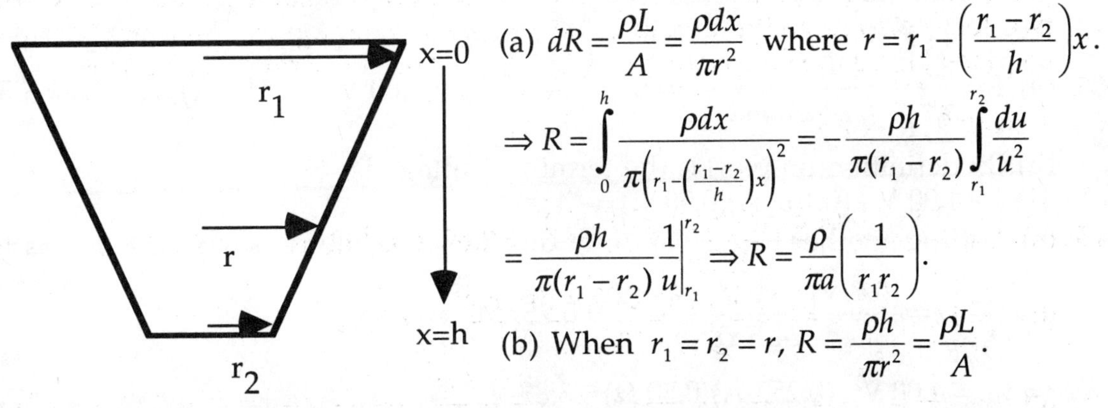

(a) $dR = \dfrac{\rho L}{A} = \dfrac{\rho dx}{\pi r^2}$ where $r = r_1 - \left(\dfrac{r_1 - r_2}{h}\right)x.$

$\Rightarrow R = \displaystyle\int_0^h \frac{\rho dx}{\pi\left(r_1 - \left(\frac{r_1-r_2}{h}\right)x\right)^2} = -\frac{\rho h}{\pi(r_1 - r_2)}\int_{r_1}^{r_2}\frac{du}{u^2}$

$= \dfrac{\rho h}{\pi(r_1 - r_2)}\dfrac{1}{u}\Big|_{r_1}^{r_2} \Rightarrow R = \dfrac{\rho}{\pi a}\left(\dfrac{1}{r_1 r_2}\right).$

(b) When $r_1 = r_2 = r$, $R = \dfrac{\rho h}{\pi r^2} = \dfrac{\rho L}{A}.$

26-50: (a) $dR = \dfrac{\rho dr}{4\pi r^2} \Rightarrow R = \dfrac{\rho}{4\pi}\displaystyle\int_a^b \frac{dr}{r^2} = -\frac{\rho}{4\pi}\frac{1}{r}\Big|_a^b = \frac{\rho}{4\pi}\left(\frac{1}{a} - \frac{1}{b}\right).$

(b) $I = \dfrac{V_{ab}}{R} = \dfrac{V_{ab}4\pi ab}{\rho(b-a)} \Rightarrow J = \dfrac{I}{A} = \dfrac{V_{ab}4\pi ab}{\rho(b-a)4\pi r^2} = \dfrac{V_{ab}ab}{\rho(b-a)r^2}.$

(c) If the thickness of the shells is small, we have the resistance given by:

$R = \dfrac{\rho}{4\pi}\left(\dfrac{1}{a} - \dfrac{1}{b}\right) = \dfrac{\rho(b-a)}{4\pi ab} \approx \dfrac{\rho L}{4\pi a^2} = \dfrac{\rho L}{A}$, where $L = b - a$.

26-51: $E = \rho J$ and $E = \dfrac{\sigma}{K\varepsilon} = \dfrac{Q}{AK\varepsilon} \Rightarrow \rho J = \dfrac{Q}{AK\varepsilon} \Rightarrow AJ = I = \dfrac{Q}{K\varepsilon\rho} =$ leakage current.

26-52: Initially: $R_o = V/I_o = (120\ \text{V})/(1.44\ \text{A}) = 83.3\ \Omega.$

Finally: $R_f = V/I_f = (120\ \text{V})/(1.33\ \text{A}) = 90.2\ \Omega.$

And $\frac{R_f}{R_o} = 1 + \alpha(T_f - T_o) \Rightarrow (T_f - T_o) = \frac{1}{\alpha}\left(\frac{R_f}{R_o} - 1\right) = \frac{1}{4.5 \times 10^{-4}\ °C^{-1}}\left(\frac{90.2\ \Omega}{83.3\ \Omega} - 1\right)$
$\Rightarrow T_f - T_o = 184°C \Rightarrow T_f = 184°C + 20°C = 204°C.$

26-53: (a) $R = \frac{\rho L}{A} = \frac{(9.5 \times 10^{-7}\ \Omega \cdot m)(0.15\ m)}{(\pi/4)(0.0012\ m)^2} = 0.126\ \Omega.$

(b) $\rho(T) = \rho_o(1 + \alpha\Delta T) \Rightarrow \rho(60°C) = (9.5 \times 10^{-7}\Omega \cdot m)(1 + (0.00088(C°)^{-1})(60°C)$
$\Rightarrow \rho(60°C) = 1.00 \times 10^{-6}\ \Omega \cdot m \Rightarrow \Delta\rho = 5.0 \times 10^{-8}\ \Omega \cdot m.$

(c) $\Delta V = \beta V_o \Delta T \Rightarrow A\Delta L = A(\beta L_o \Delta T) \Rightarrow \Delta L = \beta L_o \Delta T = (18 \times 10^{-5}(C°)^{-1})(0.15\ m)(60°C)$
$\Rightarrow \Delta L = 1.6 \times 10^{-3}\ m = 1.6\ mm.$

(d) $R = \frac{\rho L}{A} \Rightarrow \Delta R = \frac{\Delta\rho L}{A} + \frac{\rho \Delta L}{A}$

$\Rightarrow \Delta R = \frac{(5.0 \times 10^{-8}\ \Omega \cdot m)(0.15\ m)}{(\pi/4)(0.0012\ m)^2} + \frac{(95 \times 10^{-8}\ \Omega \cdot m)(1.6 \times 10^{-3}\ m)}{(\pi/4)(0.0012\ m)^2} = 8.0 \times 10^{-3}\ \Omega.$

(e) From Eq. (26-12), $\alpha = \frac{1}{\Delta T}\left(\frac{R}{R_o} - 1\right) = \frac{1}{60°C}\left(\frac{0.134\ \Omega}{0.126\ \Omega} - 1\right) = 1.06 \times 10^{-3}(C°)^{-1}.$

This value is 20% greater than the temperature coefficient of resistivity and therefore is an important change caused by the length increase.

26-54: (a) $I = \frac{\sum \mathcal{E}}{\sum R} = \frac{8.0\ V - 4.0\ V}{24.0\ \Omega} = 0.167\ A \Rightarrow V_{ad} = 8.00\ V - (0.167\ A)(8.50\ \Omega) = 6.58\ V.$

(b) By definition on p. 811, the terminal voltage is
$V_{bc} = -4.00\ V + (0.167\ A)(0.50\ \Omega) = -4.08\ V.$

(c) Adding another battery at point d in the opposite sense to the 8.0 V battery:

$I = \frac{\sum \mathcal{E}}{\sum R} = \frac{10.3\ V - 8.0\ V + 4.0\ V}{24.5\ \Omega} = 0.257\ A$, and so

$\Rightarrow V_{bc} = 4.00\ V - (0.257\ A)(0.50\ \Omega) = 3.87\ V.$

26-55: (a) $V_{ab} = \varepsilon - Ir \Rightarrow 8.6\ V = \varepsilon - (3.00\ A)r$ and $10.1\ V = \mathcal{E} + (2.00\ A)r$

$\Rightarrow 10.1\ V = (8.6\ V + (3.00\ A)r) + (2.00\ A)r \Rightarrow r = \frac{10.1\ V - 8.6\ V}{5.00\ A} = 0.30\ \Omega.$

(b) $\mathcal{E} = 8.6\ V + (3.00\ A)(0.30\ \Omega) = 9.5\ V.$

26-56: (a) $V = 2.00I + 0.48I^2 \Rightarrow 3.0\ V = (2.00\ V/A)I + (0.48\ V/A^2)I^2.$
Solving the quadratic equation yields $I = 1.17$ A or -5.34 A, so the appropriate current through the semiconductor is $I = 1.17$ A.
(b) If the current $I = 2.34$ A,
$\Rightarrow V = (2.00\ V/A)(2.34\ A) + (0.48\ V/A^2)(2.34\ A)^2 = 7.3\ V.$

26-57: $I = \frac{V}{R_{total}} = \frac{V}{R + (\alpha I + \beta I^2)/I} = \frac{V}{R + \alpha + \beta I} \Rightarrow \beta I^2 + (R + \alpha)I - V = 0$
$\Rightarrow 1.3I^2 + 9.7I - 12 = 0 \Rightarrow I = 1.08\ A.$

26-58: (a) $r = \frac{\mathcal{E}}{I} = \frac{6.35\ \text{V}}{3.92\ \text{A}} = 1.62\ \Omega \Rightarrow I = \frac{\mathcal{E}}{R+r} = \frac{6.35\ \text{V}}{1.62\ \Omega + 2.4\ \Omega} = 1.6\ \text{A}.$

(b) If the resistance is that of **Ex. (26-30)**, we use the graph, and trial and error, for different current values to find when the calculated current through the thyrite resistor $I = \frac{\mathcal{E} - Ir}{R_{thyrite}}$ equals the original current value.

This works for I = 1.7 A.

(c) The terminal voltage at this current is
$V_{ab} = \mathcal{E} - Ir = 6.35\ \text{V} - (1.7\ \text{A})(1.62\ \Omega) = 3.6\ \text{V}.$

26-59: (a) With an ammeter in the circuit: $I = \frac{\mathcal{E}}{r + R + R_A} \Rightarrow \mathcal{E} = I_A(r + R + R_A).$

So with no ammeter: $I = \frac{\mathcal{E}}{r+R} = I_A\left(\frac{r + R + R_A}{r+R}\right) = I_A\left(1 + \frac{R_A}{r+R}\right).$

(b) We want: $\frac{I}{I_A} = \left(1 + \frac{R_A}{r+R}\right) \approx 1.01 \Rightarrow \frac{R_A}{r+R} \approx 0.01$

$\Rightarrow R_A \approx (0.01)(0.75\ \Omega + 6.5\ \Omega) = 0.073\ \Omega.$

(c) This is a maximum value, since any larger resistance makes the current even less than it would be with out it. That is, since the ammeter is in series, ANY resistance it has increases the circuit resistance and makes the reading less accurate.

26-60: (a) With a voltmeter in the circuit: $I = \frac{\mathcal{E}}{r + R_V} \Rightarrow V_{ab} = \mathcal{E} - Ir = \mathcal{E}\left(1 - \frac{r}{r + R_V}\right).$

(b) We want: $\frac{V_{ab}}{\mathcal{E}} = \left(1 - \frac{r}{r + R_V}\right) \approx 0.99 \Rightarrow \frac{r}{r + R_V} \approx 0.01$

$\Rightarrow R_V \approx \frac{r - 0.01r}{0.01} = 99r = 99 \cdot 0.75\Omega = 74\ \Omega.$

(c) This is the minimum resistance necessary - any greater resistance leads to less current flow and hence less potential loss over the battery's internal resistance.

26-61: (a) $V_{ab} = \mathcal{E} - Ir = 12.0\ \text{V} - (-15.0\ \text{A})(0.31\ \Omega) = 17\ \text{V}.$

(b) $E = Pt = IVt = (15\ \text{A})(17\ \text{V})(4)(3600\ \text{s}) = 3.6 \times 10^6\ \text{J}.$

(c) $E_{diss} = P_{diss}t = I^2Rt = (15\ \text{A})^2(0.31\ \Omega)(4)(3600\ \text{s}) = 1.0 \times 10^6\ \text{J}.$

(d) Discharged at 15 A: $I = \frac{\mathcal{E}}{r+R} \Rightarrow R = \frac{\mathcal{E} - Ir}{I} = \frac{12.0\ \text{V} - (15\ \text{A})(0.31\ \Omega)}{15\ \text{A}} = 0.49\ \Omega.$

(e) $E = Pt = IVt = (15\ \text{A})(17\ \text{V})(4)(3600\ \text{s}) = 3.6 \times 10^6\ \text{J}.$

(f) Since the current through the internal resistance is the same as before, there is the same energy dissipated as in (c): $E_{diss} = 1.0 \times 10^6\ \text{J}.$

(g) The energy originally supplied went into the battery and some was also lost over the internal resistance. So the stored energy was less than was needed to charge it. Then when discharging, even more energy is lost over the internal resistance, and what is left is dissipated over the external resistor.

26-62: (a) $V_{ab} = \varepsilon - Ir = 12.0\ \text{V} - (-30\ \text{A})(0.31\ \Omega) = 21\ \text{V}$.

(b) $E = Pt = IVt = (30\ \text{A})(21\ \text{V})(2)(3600\ \text{s}) = 4.6 \times 10^6\ \text{J}$.

(c) $E_{diss} = P_{diss}t = I^2Rt = (30\ \text{A})^2(0.31\ \Omega)(2)(3600\ \text{s}) = 2.0 \times 10^6\ \text{J}$.

(d) Discharged at 30 A: $I = \dfrac{\mathcal{E}}{r+R} \Rightarrow R = \dfrac{\mathcal{E} - Ir}{I} = \dfrac{12.0\ \text{V} - (30\ \text{A})(0.31\ \Omega)}{30\ \text{A}} = 0.09\ \Omega$.

(e) $E = Pt = I^2Rt = (30\ \text{A})^2(0.09\ \Omega)(2)(3600) = 0.58 \times 10^6\ \text{J}$

(f) Since the current through the internal resistance is the same as before, there is the same energy dissipated as in (c): $E_{diss} = 2.0 \times 10^6\ \text{J}$.

(g) Again, the energy originally supplied went into the battery and some was also lost over the internal resistance. So the stored energy was less than was needed to charge it. Then when discharging, even more energy is lost over the internal resistance, and what is left is dissipated over the external resistor. This time, at a higher current, much more energy is lost over the internal resistance.

26-63: (a) $I = \dfrac{\sum \varepsilon}{\sum R} = \dfrac{12.0\ \text{V} - 8.0\text{V}}{10.0\ \Omega} = 0.40\ \text{A}$.

(b) $P_{total} = I^2R_{total} = (0.40\ \text{A})^2(10\ \Omega) = 1.6\ \text{W}$.

(c) Power generated in $\mathcal{E}_1$, $P = \mathcal{E}_1 I = (12.0\ \text{V})(0.40\ \text{A}) = 4.8\ \text{W}$.

(d) Rate of electrical energy transferred to chemical energy in $\mathcal{E}_2$, $P = \mathcal{E}_2 I = (8.0\ \text{V})(0.40\ \text{A}) = 3.2\ \text{W}$.

(e) Note (c) = (b) + (d), and so the rate of creation of electrical energy equals its rate of dissipation.

26-64: (a) $R_{steel} = \dfrac{\rho L}{A} = \dfrac{(2.0 \times 10^{-7}\ \Omega\text{m})(1.8\ \text{m})}{(\pi/4)(0.020\ \text{m})^2} = 1.15 \times 10^{-3}\ \Omega$

$R_{Cu} = \dfrac{\rho L}{A} = \dfrac{(1.72 \times 10^{-8}\ \Omega\text{m})(35\ \text{m})}{(\pi/4)(0.0090\ \text{m})^2} = 9.46 \times 10^{-3}\ \Omega$

$\Rightarrow V = IR = I(R_{steel} + R_{Cu}) = (15000\ \text{A})(1.15 \times 10^{-3}\ \Omega + 9.46 \times 10^{-3}\ \Omega) = 160\ \text{V}$.

(b) $E = Pt = I^2Rt = (15000\ \text{A})^2(0.0106\ \Omega)(80 \times 10^{-6}\ \text{s}) = 191\ \text{J}$.

26-65: (a) The line voltage and wire diameter are what must be considered in household wiring, along with the current to be drawn.

(b) $P = VI \Rightarrow I = \dfrac{P}{V} = \dfrac{3500\ \text{W}}{120\ \text{V}} = 29.2\ \text{A}$, so the 10 gauge wire is necessary, since it can carry up to 30 A.

(c) $P = I^2R = \dfrac{I^2\rho L}{A} = \dfrac{(29.2\ \text{A})^2(1.72 \times 10^{-8}\ \Omega\text{m})(35\ \text{m})}{(\pi/4)(0.0259\ \text{m})^2} = 97\ \text{W}$.

(d) If 8-gauge wire is used, $P = \dfrac{I^2\rho L}{A} = \dfrac{(29.2\ \text{A})^2(1.72 \times 10^{-8}\ \Omega\text{m})(35\ \text{m})}{(\pi/4)(0.0326\ \text{m})^2} = 61\ \text{W}$

$\Rightarrow \Delta E = \Delta Pt = (35.9\ \text{W})(365)(12\ \text{h}) = 157\ \text{kWh}$

$\Rightarrow \text{Savings} = (157\ \text{kWh})(\$0.11/\text{kWh}) = \$17.30$.

26-66: (a) We need to heat the water in 8 minutes, so the heat and power required are: $Q = mc_v\Delta T = (0.250\text{ kg})(4190\text{ J/kg°C})(80°\text{C}) = 83800\text{ J}$

$$\Rightarrow P = \frac{Q}{t} = \frac{83800\text{ J}}{8(60\text{ s})} = 175\text{ W}.$$

But $P = \dfrac{V^2}{R} \Rightarrow R = \dfrac{V^2}{P} = \dfrac{(120\text{ V})^2}{175\text{ W}} = 82.5\ \Omega.$

(b) The resistance changes with temperature, so we must integrate in order to get the amount of energy deposited in the water.

$$dE = Pdt = \frac{V^2}{R}dt \Rightarrow E = \int_0^{480} \frac{V^2 dt}{R_o(1+\alpha\Delta T)} \text{ but } \Delta T = \frac{80t}{480} \text{ so } E = \int_0^{480} \frac{V^2 dt}{R_o(1+\frac{\alpha t}{6})}$$

$$\Rightarrow E = \frac{6V^2}{\alpha R_o}\ln(1+\tfrac{\alpha t}{6})\Big|_0^{480} = \frac{6(120\text{ V})^2}{(0.0004°\text{C}^{-1})R_o}\ln(1+(80°\text{C})(0.0004°\text{C}^{-1})) = \frac{6.80\times10^6\ \Omega\cdot\text{J}}{R_o}$$

$$\Rightarrow R_o = \frac{6.80\times10^6\ \Omega\cdot\text{J}}{E} = \frac{6.80\times10^6\ \Omega\cdot\text{J}}{83800\text{ J}} = 81.2\ \Omega.$$

But $R = \dfrac{\rho L}{A} = \dfrac{\rho L^2}{\text{vol}} \Rightarrow L = \sqrt{\dfrac{R\cdot\text{vol}}{\rho}} = \sqrt{\dfrac{(81.2\ \Omega)(2.5\times10^{-5}\text{ m}^3)}{1.00\times10^{-6}\ \Omega/\text{m}}} = 45\text{ m}.$

Now the radius of the wire can be simply calculated from the volume:

$$\text{vol} = L(\pi r^2) \Rightarrow r = \sqrt{\frac{\text{vol}}{\pi L}} = \sqrt{\frac{2.5\times10^{-5}\text{ m}^3}{\pi(45\text{ m})}} = 4.2\times10^{-4}\text{ m}.$$

26-67: (a) $\sum F = ma = |q|E \Rightarrow \dfrac{|q|}{m} = \dfrac{a}{E}.$

(b) If the electric field is constant, $V_{bc} = EL \Rightarrow \dfrac{|q|}{m} = \dfrac{aL}{V_{bc}}.$

(c) The free charges are "left behind" so the left end of the rod is negatively charged, while the right end is positively charged. Thus the right end is at the higher potential.

(d) $a = \dfrac{V_{bc}|q|}{mL} = \dfrac{(1.0\times10^{-3}\text{ V})(1.6\times10^{-19}\text{ C})}{(9.11\times10^{-31}\text{ kg})(0.50\text{ m})} = 3.5\times10^8\text{ m/s}^2.$

(e) Performing the experiment in a rotational way enables one to keep the experimental apparatus in a localized area - whereas an acceleration like that obtained in (d) if linear, would quickly have the apparatus moving a high speeds, and large distances.

26-68: (a) $I = \dfrac{\mathcal{E}}{r+R} \Rightarrow P = \mathcal{E}I - I^2 r \Rightarrow \dfrac{dP}{dI} = \mathcal{E} - 2Ir = 0$ for maximum power output.

$$\Rightarrow I_{P_{max}} = \frac{1}{2}\frac{\mathcal{E}}{r} = \frac{1}{2}I_{\text{short circuit}}.$$

(b) For the maximum power output of (a), $I = \dfrac{\mathcal{E}}{r+R} = \dfrac{1}{2}\dfrac{\mathcal{E}}{r} \Rightarrow r + R = 2r \Rightarrow R = r.$

Then, $P = I^2 R = \left(\dfrac{\mathcal{E}}{2r}\right)^2 r = \dfrac{\mathcal{E}^2}{4r}.$

26-69: (a) $\alpha = \frac{1}{\rho}\left(\frac{d\rho}{dT}\right) = -\frac{n}{T} \Rightarrow -\frac{ndT}{T} = \frac{d\rho}{\rho} \Rightarrow \ln(T^{-n}) = \ln(\rho) \Rightarrow \rho = \frac{a}{T^n}.$

(b) $n = -\alpha T = -(-5\times10^{-4}(\text{K})^{-1})(293\ \text{K}) = 0.15.$

$\rho = \frac{a}{T^n} \Rightarrow a = \rho T^n = (3.5\times10^{-5}\ \Omega\cdot\text{m})(293\ \text{K})^{0.15} = 8.0\times10^{-5}.$

(c) $T = -196°\text{C} = 77\ \text{K}: \rho = \frac{8.0\times10^{-5}}{(77\ \text{K})^{0.15}} = 4.3\times10^{-5}\ \Omega\cdot\text{m}.$

$T = 300°\text{C} = 573\ \text{K}: \rho = \frac{8.0\times10^{-5}}{(573\ \text{K})^{0.15}} = 3.2\times10^{-5}\ \Omega\cdot\text{m}.$

26-70: (a) $\varepsilon = IR + IR_d \Rightarrow 2.00\ \text{V} = I(1.0\ \Omega) + V \Rightarrow 2 = I_o[\exp(eV/kT)-1] + V.$

(b) $I_o = 1.50\times10^{-3}\ \text{A},\ T = 293\ \text{K} \Rightarrow 1333 = \exp[39.6V - 667] + 667V.$

Trial and error shows that the right hand side (rhs) above, for specific V values, equals 1333 V, when $V = 0.179$ V. The current then is just $I = I_o\exp[39.6\ V - 1] = (1.5\times10^{-3}\ \text{A})[\exp 39.6(0.179) - 1] = 1.80\ \text{A}.$

26-71: (a) $R = \frac{\rho L}{A} \Rightarrow dR = \frac{\rho dx}{A} = \frac{\rho_o \exp[-x/L]dx}{A} \Rightarrow R = \frac{\rho_o}{A}\int_0^L \exp[-x/L]dx$

$\Rightarrow R = \frac{\rho_o}{A}[-L\exp[-x/L]]_o^L = \frac{\rho_o L}{A}(1-e^{-1}) \Rightarrow I = \frac{V_o}{R} = \frac{V_o A}{\rho_o L(1-e^{-1})}.$

(b) $E(x) = -\frac{\partial V}{\partial x} = -\frac{\partial(IR)}{\partial x} = -\frac{\partial}{\partial x}\left(\frac{I\rho_o L e^{-x/L}}{A}\right) = \frac{I\rho_o e^{-x/L}}{A} = \frac{V_o e^{-x/L}}{L(1-e^{-1})}.$

(c) $V(x) = V_o\frac{e^{-x/L}}{(1-e^{-1})} + C \Rightarrow V(0) = V_o = \frac{V_o}{(1-e^{-1})} + C \Rightarrow C = \frac{-V_o e^{-1}}{(1-e^{-1})}$

$\Rightarrow V(x) = V_o\frac{(e^{-x/L} - e^{-1})}{(1-e^{-1})}.$

(d) Graphs of resistivity, electric field and potential from x = 0 to L.

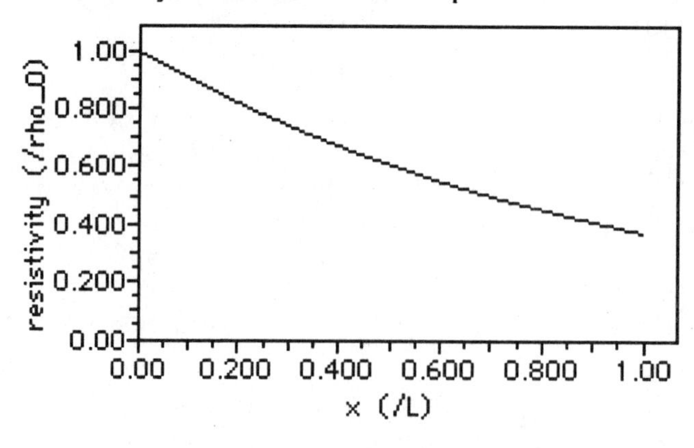

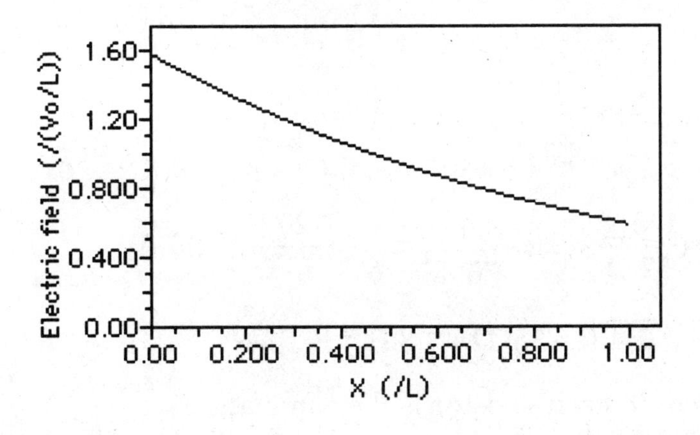
Electric field (/(Vo/L))
1.60
1.20
0.800
0.400
0.00
0.00
0.200
0.400
0.600
0.800
1.00
X (/L)

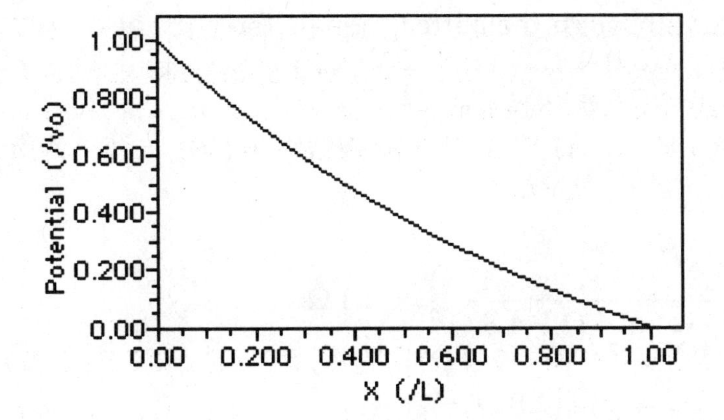
Potential (/Vo)
1.00
0.800
0.600
0.400
0.200
0.00
0.00
0.200
0.400
0.600
0.800
1.00
X (/L)

Chapter 27: Direct-Current Circuits

27-1: (a) $R_{eq} = \left(\frac{1}{52} + \frac{1}{48}\right)^{-1} = 25\ \Omega.$

(b) $I = \frac{V}{R_{eq}} = \frac{120\ \text{V}}{25\ \Omega} = 4.8\ \text{A}.$

(c) $I_{52\Omega} = \frac{V}{R} = \frac{120\ \text{V}}{52\ \Omega} = 2.3\ \text{A};\ I_{48\Omega} = \frac{V}{R} = \frac{120\ \text{V}}{48\ \Omega} = 2.5\ \text{A}.$

27-2: (a) $R_{eq} = \sum R_i = 2.4\ \Omega + 3.6\ \Omega + 4.8\ \Omega = 10.8\ \Omega.$

(b) The current in each resistor is the same and is $I = \frac{\mathcal{E}}{R_{eq}} = \frac{54\ \text{V}}{10.8\ \Omega} = 5.0\ \text{A}.$

(c) The current through the battery equals the current of (b), 5.0 A.

(d) $V_{2.4} = IR_{2.4} = (5.0\ \text{A})(2.4\ \Omega) = 12\ \text{V};\ V_{3.6} = IR_{3.6} = (5.0\ \text{A})(3.6\ \Omega) = 18\ \text{V};$
$V_{4.8} = IR_{4.8} = (5.0\ \text{A})(4.8\ \Omega) = 24\ \text{V}.$

(e) $P_{2.4} = I^2R_{2.4} = (5.0\ \text{A})^2(2.4\ \Omega) = 60\ \text{W};\ P_{3.6} = I^2R_{3.6} = (5.0\ \text{A})^2(3.6\ \Omega) = 90\ \text{W};$
$P_{4.8} = I^2R_{4.8} = (5.0\ \text{A})^2(4.8\ \Omega) = 120\ \text{W}.$

27-3: (a) $R_{eq} = \left(\frac{1}{2.4\ \Omega} + \frac{1}{3.6\ \Omega} + \frac{1}{4.8\ \Omega}\right)^{-1} = 1.1\ \Omega.$

(b) $I_{2.4} = \mathcal{E} / R_{2.4} = (54\ \text{V}) / (2.4\ \Omega) = 23\ \text{A};\ I_{3.6} = \mathcal{E} / R_{3.6} = (54\ \text{V}) / (3.6\ \Omega) = 15\ \text{A};$
$I_{4.8} = \mathcal{E} / R_{4.8} = (54\ \text{V}) / (4.8\ \Omega) = 11\ \text{A}.$

(c) $I_{total} = \mathcal{E} / R_{total} = (54\ \text{V}) / (1.1\ \Omega) = 49\ \text{A}.$

(d) When in parallel, all resistors have the same potential difference over them, so here all have V = 54 V.

(e) $P_{2.4} = I^2R_{2.4} = (23\ \text{A})^2(2.4\ \Omega) = 1200\ \text{W};\ P_{3.6} = I^2R_{3.6} = (15\ \text{A})^2(3.6\ \Omega) = 810\ \text{W};$
$P_{4.8} = I^2R_{4.8} = (11\ \text{A})^2(4.8\ \Omega) = 610\ \text{W}.$

27-4: $R_{eq} = \left(\left(\frac{1}{3.00\ \Omega} + \frac{1}{6.00\ \Omega}\right)^{-1} + \left(\frac{1}{12.0\ \Omega} + \frac{1}{4.00\ \Omega}\right)^{-1}\right) = 5.00\ \Omega.$

$I_{total} = \mathcal{E} / R_{total} = (60.0\ \text{V}) / (5.00\ \Omega) = 12.0\ \text{A}$

$I_{12} = \frac{4}{12+4}(12.0) = 3.00\ \text{A}; I_4 = \frac{12}{12+4}(12.0) = 9.00\ \text{A};$

$I_3 = \frac{6}{3+6}(12.0) = 8.00\ \text{A}; I_6 = \frac{3}{3+6}(12.0) = 4.00\ \text{A}.$

27-5: $R_{eq} = \left(\frac{1}{3.00\ \Omega + 1.00\ \Omega} + \frac{1}{5.00\ \Omega + 7.00\ \Omega}\right)^{-1} = 3.00\ \Omega.$

$I_{total} = \mathcal{E} / R_{total} = (48.0\ \text{V}) / (3.00\ \Omega) = 16.0\ \text{A}.$

$I_5 = I_7 = \frac{4}{4+12}(16.0) = 4.00\ \text{A}; I_1 = I_3 = \frac{12}{4+12}(16.0) = 12.0\ \text{A}.$

27-6: $R_{eq} = \left(\frac{1}{R_1} + \frac{1}{R_2}\right)^{-1} = \left(\frac{R_1 + R_2}{R_1 R_2}\right)^{-1} \Rightarrow R_{eq} = \frac{R_1 R_2}{R_1 + R_{2\,1}}.$

$\Rightarrow R_{eq} = R_1 \frac{R_2}{R_1 + R_2} < R_1$ and $R_{eq} = R_2 \frac{R_1}{R_1 + R_2} < R_2.$

27-7: (a) $P = \frac{V^2}{R} \Rightarrow V = \sqrt{PR} = \sqrt{(5.0\text{ W})(10{,}000\ \Omega)} = 220\text{ V}.$

(b) $P = \frac{V^2}{R} = \frac{(220\text{ V})^2}{15{,}000\ \Omega} = 3.2\text{ W}.$

27-8: (a) $R_{40W} = \frac{V^2}{P} = \frac{(120\text{ V})^2}{40\ W} = 360\ \Omega;\ R_{150W} = \frac{V^2}{P} = \frac{(120\text{ V})^2}{150\ W} = 96\ \Omega.$

$\Rightarrow I_{40W} = I_{150W} = \frac{\mathcal{E}}{R} = \frac{240\text{ V}}{456\ \Omega} = 0.526\text{ A}.$

(b) $P_{40W} = I^2R = (0.526\text{ A})^2(360\ \Omega) = 100\text{ W};\ P_{150W} = I^2R = (0.526\text{ A})^2(96\ \Omega) = 27\text{ W}.$

(c) The 40 W bulb burns out quickly because the power it delivers (100 W) is 2.5 times its rated value.

27-9: (a) $I = \frac{\mathcal{E}}{R} = \frac{120\text{ V}}{1500\ \Omega} = 0.080\text{ A}.$

(b) $P_{500} = I^2R = (0.080\text{ A})^2(500\ \Omega) = 3.2\text{ W};\ P_{1000} = I^2R = (0.080\text{ A})^2(1000\ \Omega) = 6.4\text{ W}$

$\Rightarrow P_{total} = 3.2\text{ W} + 6.4\text{ W} = 9.6\text{ W}.$

(c) When in parallel, the equivalent resistance becomes:

$R_{eq} = \left(\frac{1}{500\ \Omega} + \frac{1}{1000\ \Omega}\right)^{-1} = 333\ \Omega \Rightarrow I_{total} = \frac{\mathcal{E}}{R_{eq}} = \frac{120\text{ V}}{333\ \Omega} = 0.360\text{ A}.$

$I_{500} = \frac{1000}{500 + 1000}(0.36\text{ A}) = 0.24\text{ A};\ I_{1000} = \frac{500}{500 + 1000}(0.36\text{ A}) = 0.12\text{ A}.$

(d) $P_{500} = I^2R = (0.24\text{ A})^2(500\ \Omega) = 29\text{ W};\ P_{1000} = I^2R = (0.12\text{ A})^2(1000\ \Omega) = 14\text{ W}$

$\Rightarrow P_{total} = 29\text{ W} + 14\text{ W} = 43\text{ W}.$

(e) The 1000 Ωresistor is brighter when the resistors are in series, and the 500 Ω is brighter when in parallel. The greatest total light output is when they are in parallel.

27-10: (a) The three resistors R_2, R_3 and R_4 are in parallel, so:

$R_{234} = \left(\frac{1}{R_2} + \frac{1}{R_3} + \frac{1}{R_4}\right)^{-1} = \left(\frac{1}{3.75\ \Omega} + \frac{1}{9.50\ \Omega} + \frac{1}{6.25\ \Omega}\right)^{-1} = 1.88\ \Omega$

$\Rightarrow R_{eq} = R_1 + R_{234} = 2.50\ \Omega + 1.88\ \Omega = 4.38\ \Omega.$

(b) $I_1 = \frac{\mathcal{E}}{R_{eq}} = \frac{4.50\text{ V}}{4.38\ \Omega} = 1.03\text{ A} \Rightarrow V_1 = I_1R_1 = (1.03\text{ A})(2.50\ \Omega) = 2.57\text{ V}.$

$\Rightarrow V_{R_{234}} = I_1R_{eq} = (1.03\text{ A})(1.88\ \Omega) = 1.93\text{ V} \Rightarrow I_2 = \frac{V_{R_{234}}}{R_2} = \frac{1.93\text{ V}}{3.75\ \Omega} = 0.515\text{ A},$

$I_3 = \frac{V_{R_{234}}}{R_3} = \frac{1.93\text{ V}}{9.50\ \Omega} = 0.203\text{ A}$ and $I_4 = \frac{V_{R_{234}}}{R_4} = \frac{1.93\text{ V}}{6.25\ \Omega} = 0.309\text{ A}.$

27-11: Using the same circuit as in **27-10,** with all resistances the same:

$$R_{eq} = R_1 + R_{234} = R_1 + \left(\frac{1}{R_2} + \frac{1}{R_3} + \frac{1}{R_4}\right)^{-1} = 6.00\ \Omega + \left(\frac{3}{6.00\ \Omega}\right)^{-1} = 8.00\ \Omega.$$

(a) $I_1 = \frac{\mathcal{E}}{R_{eq}} = \frac{3.00\text{ V}}{8.00\ \Omega} = 0.375\text{ A},\ I_2 = I_3 = I_4 = \frac{1}{3}I_1 = 0.125\text{ A}.$

(b) $P_1 = I_1^2R_1 = (0.375\text{ A})^2(6.00\ \Omega) = 0.844\text{ W},\ P_2 = P_3 = P_4 = \frac{1}{9}P_1 = 0.0938\text{ W}.$

(c) If there is a break at R_4, then the equivalent resistance increases:

$R_{eq} = R_1 + R_{23} = R_1 + \left(\frac{1}{R_2} + \frac{1}{R_3}\right)^{-1} = 6.00\ \Omega + \left(\frac{2}{6.00\ \Omega}\right)^{-1} = 9.00\ \Omega.$ And so:

$$I_1 = \frac{\mathcal{E}}{R_{eq}} = \frac{3.00\text{ V}}{9.00\ \Omega} = 0.333\text{ A},\ I_2 = I_3 = \frac{1}{2}I_1 = 0.167\text{ A}.$$

(d) $P_1 = I_1^2R_1 = (0.333\text{ A})^2(6.00\ \Omega) = 0.667\text{ W},\ P_2 = P_3 = \frac{1}{4}P_1 = 0.167\text{ W}.$

(e) So R_2 and R_3 are brighter than before, while R_1 is fainter. The amount of current flow is all that determines the power output of these bulbs since their resistances are equal.

27-12: (a) The filaments must be connected such that the current can flow through each separately, and also through both in parallel, yielding three possibile current flows. The parallel situation always has less resistance than any of the individual members, so it will give the highest power output of 150 W, while the other two must give power outputs of 50 W and 100 W.

$50\text{ W} = \frac{V^2}{R_1} \Rightarrow R_1 = \frac{(120\text{ V})^2}{50\text{ W}} = 289\ \Omega,$ and $100\text{ W} = \frac{V^2}{R_2} \Rightarrow R_2 = \frac{(120\text{ V})^2}{100\text{ W}} = 144\ \Omega.$

Check for parallel: $P = \frac{V^2}{\left(\frac{1}{R_1} + \frac{1}{R_2}\right)^{-1}} = \frac{(120\text{ V})^2}{\left(\frac{1}{289\ \Omega} + \frac{1}{144\ \Omega}\right)^{-1}} = \frac{(120\text{ V})^2}{96\ \Omega} = 150\text{ W}.$

(b) If R_1 burns out, the 100 W setting stays the same, the 50 W setting does not work and the 150 W setting goes to 100 W: brightnesses of zero, medium and medium.

(c) If R_2 burns out, the 50 W setting stays the same, the 100 W setting does not work, and the 150 W setting is now 50 W: brightnesses of low, zero and low.

27-13: The total power dissipated in the four resistors of Fig. 27-8a is given by the sum of:

$P_2 = I^2R_2 = (0.5\text{ A})^2(2\ \Omega) = 0.5\text{ W},\ P_3 = I^2R_3 = (0.5\text{ A})^2(3\ \Omega) = 0.75\text{ W},$

$P_4 = I^2R_4 = (0.5\text{ A})^2(4\ \Omega) = 1\text{ W},\ P_7 = I^2R_7 = (0.5\text{ A})^2(7\ \Omega) = 1.8\text{ W}.$

$\Rightarrow P_{total} = P_2 + P_3 + P_4 + P_7 = 4\text{ W}.$

27-14: (a) If the 12 V battery is removed and then replaced with the opposite polarity, the current will flow in the clockwise direction, with magnitude:

$$I = \frac{\sum \mathcal{E}}{\sum R} = \frac{12\text{ V} + 4\text{ V}}{16\ \Omega} = 1\text{ A}.$$

(b) $V_{ab} = (R_4 + R_7)I - \mathcal{E}_4 = (4\ \Omega + 7\ \Omega)(1\text{ A}) - 4\text{ V} = 7\text{ V}.$

27-15: (a)

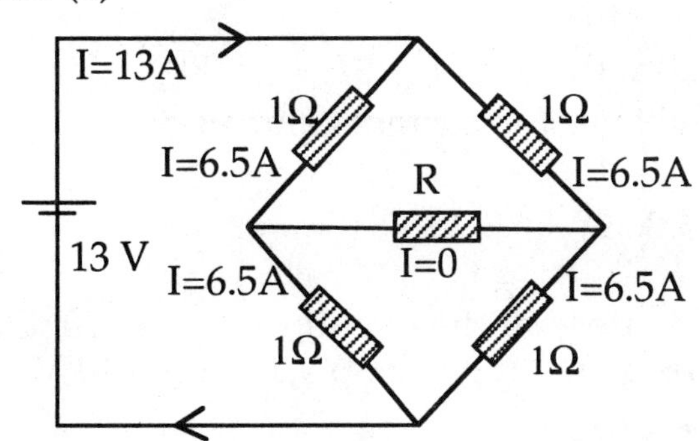

Since all the external resistors are equal, the current must be symmetrical through them. That is, there can be no current through the resistor R for that would imply an imbalance in currents through the other resistors. With no current going through R, the circuit is like that shown below at left.

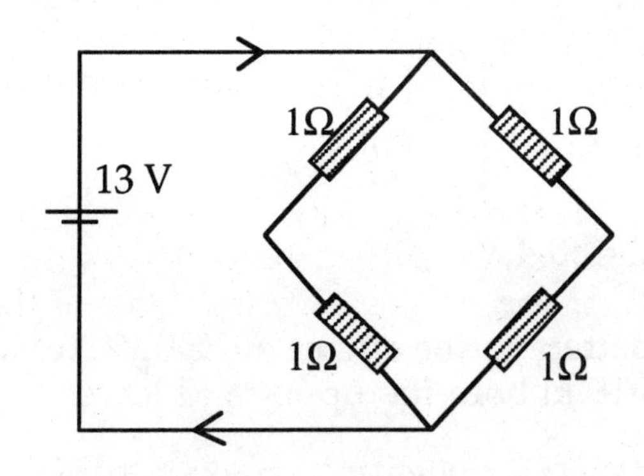

So the equivalent resistance of the circuit is

$$R_{eq} = \left(\frac{1}{2\Omega} + \frac{1}{2\Omega}\right)^{-1} = 1\ \Omega \Rightarrow I_{total} = \frac{13\text{ V}}{1\ \Omega} = 13\text{ A}.$$

$\Rightarrow I_{\text{each leg}} = \frac{1}{2} I_{total} = 6.5\text{ A}$, and no current passes through R.

(b) As worked out above, $R_{eq} = 1\ \Omega$.

(c) $V_{ab} = 0$, since no current flows.

(d) R does not show up since no current flows through it.

27-16: From the given currents in the diagram, the current through the middle branch of the circuit must be 1.00 A (the difference between 2.00 A and 1.00 A). We now use Kirchoff's Rules, passing counter-clockwise around the top loop:

$20.0\text{ V} - (1.00\text{ A})(6.00\ \Omega + 1.00\ \Omega) + (1.00\text{ A})(4.00\ \Omega + 1.00\ \Omega) - \mathcal{E}_1 = 0$

$\Rightarrow \mathcal{E}_1 = 18.0\text{ V}.$

Now travelling around the external loop of the circuit:

$20.0\text{ V} - (1.00\text{ A})(6.00\ \Omega + 1.00\ \Omega) - (2.00\text{ A})(1.00\ \Omega + 2.00\ \Omega) - \mathcal{E}_2 = 0$

$\Rightarrow \mathcal{E}_2 = 7.0\text{ V}.$

And $V_{ab} = -(1.00\text{ A})(4.00\ \Omega + 1.00\ \Omega) + 18.0\text{ V} = +13.0\text{ V}.$

27-17: (a) $I_R = 6.00\text{ A} - 4.00\text{ A} = 2.00\text{ A}$.
(b) Using a Kirchoff loop around the outside of the circuit:
$28.0\text{ V} - (6.00\text{ A})(3.00\ \Omega) - (2.00\text{ A})R = 0 \Rightarrow R = 5.00\ \Omega$.
(c) Using a counter-clockwise loop in the bottom half of the circuit:
$\mathcal{E} - (6.00\text{ A})(3.00\ \Omega) - (4.00\text{ A})(6.00\ \Omega) = 0 \Rightarrow \mathcal{E} = 42.0\text{ V}$.
(d) If the circuit is broken at point x, then the current in the 28 V battery is:

$$I = \frac{\sum \mathcal{E}}{\sum R} = \frac{28.0\text{ V}}{3.00\ \Omega + 5.00\ \Omega} = 3.50\text{ A}.$$

27-18: From the circuit in Fig. 27-30, we use Kirchoff's Rules to find the currents, I_1 to the left through the 20 V battery, I_2 to the right through 5 V battery, and I_3 to the right through the 10 Ω resistor:
Upper loop: $10.0\text{ V} - (2.00\ \Omega + 3.00\ \Omega)I_1 - (1.00\ \Omega + 4.00\ \Omega)I_2 - 5.00\text{ V} = 0$
$\Rightarrow 5.0\text{ V} - (5.00\ \Omega)I_1 - (5.00\ \Omega)I_2 = 0 \Rightarrow I_1 + I_2 = 1.00\text{ A}$.
Lower loop: $5.00\text{ V} + (1.00\ \Omega + 4.00\ \Omega)I_2 - (10.0\ \Omega)I_3 = 0$
$\Rightarrow 5.00\text{ V} + (5.00\ \Omega)I_2 - (10.0\ \Omega)I_3 = 0 \Rightarrow I_2 - 2I_3 = -1.00\text{ A}$.
Along with $I_1 = I_2 + I_3$, we can solve for the three currents and find:
$I_1 = 0.800\text{ A},\ I_2 = 0.200\text{ A},\ I_3 = 0.600\text{ A}$.
(b) $V_{ab} = -(0.200\text{ A})(4.00\ \Omega) - (0.800\text{ A})(3.00\ \Omega) = -3.20\text{ V}$.

27-19: After reversing the polarity of the 10 V battery in the circuit of Fig. 27-30, the only change in the equations from **27-18** is the upper loop where the 10 V battery is:
Upper loop: $-10.0\text{ V} - (2.00\ \Omega + 3.00\ \Omega)I_1 - (1.00\ \Omega + 4.00\ \Omega)I_2 - 5.00\text{ V} = 0$
$\Rightarrow -15.0\text{ V} - (5.00\ \Omega)I_1 - (5.00\ \Omega)I_2 = 0 \Rightarrow I_1 + I_2 = -3.00\text{ A}$.
Lower loop: $5.00\text{ V} + (1.00\ \Omega + 4.00\ \Omega)I_2 - (10.0\ \Omega)I_3 = 0$
$\Rightarrow 5.00\text{ V} + (5.00\ \Omega)I_2 - (10.0\ \Omega)I_3 = 0 \Rightarrow I_2 - 2I_3 = -1.00\text{ A}$.
Along with $I_1 = I_2 + I_3$, we can solve for the three currents and find:
$I_1 = -1.60\text{ A},\ I_2 = -1.40\text{ A},\ I_3 = -0.200\text{ A}$.
(b) $V_{ab} = +(1.40\text{ A})(4.00\ \Omega) + (1.60\text{ A})(3.00\ \Omega) = 10.4\text{ V}$.

27-20: After switching the 5 V battery for a 15 V battery in the circuit of Fig. 27-30, there is a change in the equations from **27-18** in both the upper and lower loops:
Upper loop: $10.0\text{ V} - (2.00\ \Omega + 3.00\ \Omega)I_1 - (1.00\ \Omega + 4.00\ \Omega)I_2 - 15.00\text{ V} = 0$
$\Rightarrow -5.0\text{ V} - (5.00\ \Omega)I_1 - (5.00\ \Omega)I_2 = 0 \Rightarrow I_1 + I_2 = -1.00\text{ A}$.
Lower loop: $15.00\text{ V} + (1.00\ \Omega + 4.00\ \Omega)I_2 - (10.0\ \Omega)I_3 = 0$
$\Rightarrow 15.00\text{ V} + (5.00\ \Omega)I_2 - (10.0\ \Omega)I_3 = 0 \Rightarrow I_2 - 2I_3 = -3.00\text{ A}$.
Along with $I_1 = I_2 + I_3$, we can solve for the three currents and find:
$I_1 = 0,\ I_2 = -1.00\text{ A},\ I_3 = 1.00\text{ A}$.
(b) $V_{ab} = +(1.00\text{ A})(4.00\ \Omega) = 4.00\text{ V}$.

27-21: (a) The sum of the currents that enter the junction below the 3 Ω resistor equals 3.00 Ω + 5.00 Ω = 8.00 Ω.
(b) Using the lower left loop: $\mathcal{E}_1 - (4.00\ \Omega)(3.00\text{ A}) - (3.00\ \Omega)(8.00\text{ A}) = 0$
$\Rightarrow \mathcal{E}_1 = 36.0\text{ V}$.

Using the lower right loop: $\mathcal{E}_2 - (6.00\ \Omega)(5.00\ \text{A}) - (3.00\ \Omega)(8.00\ \text{A}) = 0$
$\Rightarrow \mathcal{E}_2 = 54.0\ \text{V}$.

(c) Using the top loop: $54.0\ \text{V} - R(2.00\ \text{A}) - 36.0\ \text{V} = 0 \Rightarrow R = \dfrac{18.0\ \text{V}}{2.00\ \text{A}} = 9.00\ \Omega.$

27-22: Given that the full-scale deflection current is 300 μA and the coil resistance is 50.0 Ω:

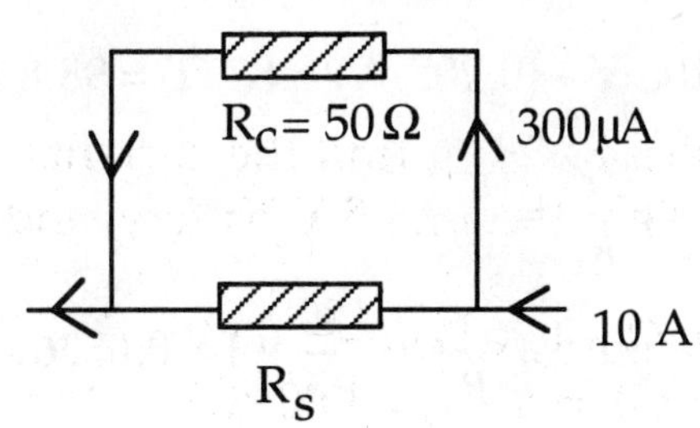

(a) For a 10 A ammeter, the two resistances are in parallel:
$V_c = V_s \Rightarrow I_c R_c = I_s R_s$
$\Rightarrow (300 \times 10^{-6}\ \text{A})(50.0\ \Omega) = (10.0\ \text{A} - 300 \times 10^{-6}\ \text{A})R_s$
$\Rightarrow R_s = 1.50\ \Omega$

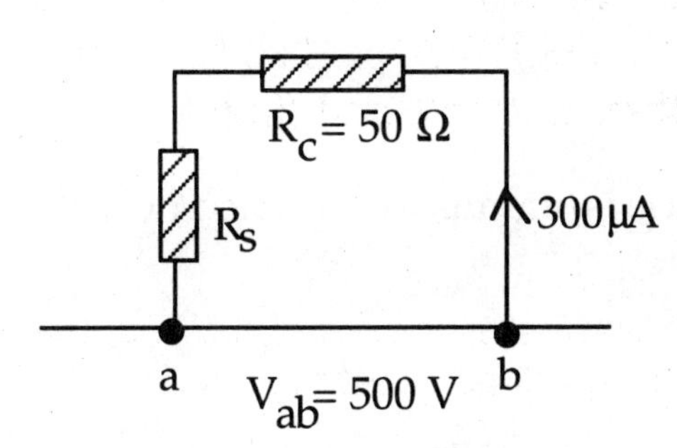

(b) For a 300 V voltmeter, the resistances are in series:

$$V_{ab} = I(R_c + R_s) \Rightarrow R_s = \frac{V_{ab}}{I} - R_c$$

$$\Rightarrow R_s = \frac{500\ \text{V}}{300 \times 10^{-6}\ \text{A}} - 50.0\ \Omega = 1.67 \times 10^{6}\ \Omega.$$

27-23: The full-scale deflection current is 0.0192 A, and we wish a full-scale reading for 10.0 A.

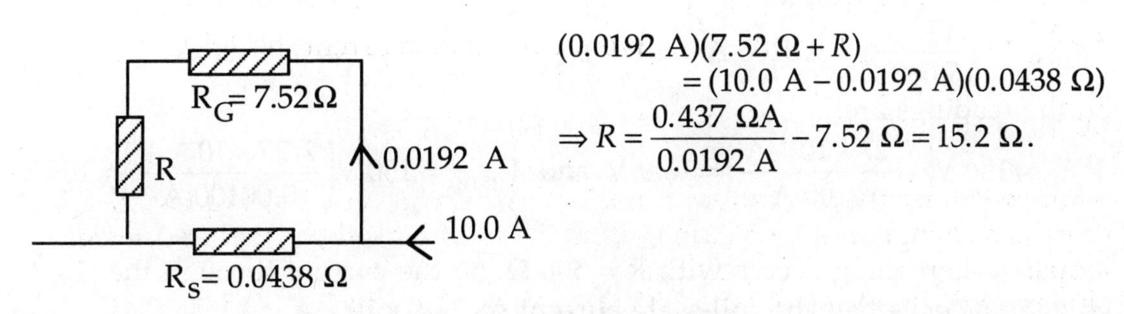

$$(0.0192\ \text{A})(7.52\ \Omega + R) = (10.0\ \text{A} - 0.0192\ \text{A})(0.0438\ \Omega)$$

$$\Rightarrow R = \frac{0.437\ \Omega\text{A}}{0.0192\ \text{A}} - 7.52\ \Omega = 15.2\ \Omega.$$

27-24: The current for full scale deflection is 0.01 A. From the circuit we can derive three equations:
(i) $(R_1 + R_2 + R_3)(0.100\ \text{A} - 0.0100\ \text{A}) = 36.0\ \Omega(0.0100\ \text{A})$
$\Rightarrow R_1 + R_2 + R_3 = 4.00\ \Omega$.
(ii) $(R_1 + R_2)(1.00\ \text{A} - 0.0100\ \text{A}) = (36.0\ \Omega + R_3)(0.0100\ \text{A})$
$\Rightarrow R_1 + R_2 - 0.0101 R_3 = 0.364\ \Omega$.
(iii) $R_1(10.0\ \text{A} - 0.0100\ \text{A}) = (36.0\ \Omega + R_2 + R_3)(0.0100\ \text{A})$
$\Rightarrow R_1 - 0.001 R_2 - 0.001 R_3 = 0.0360\ \Omega$.
From (i) and (ii) $\Rightarrow R_3 = 3.60\ \Omega$.
From (ii) and (iii) $\Rightarrow R_2 = 0.360\ \Omega$. And so $\Rightarrow R_1 = 0.040\ \Omega$.

27-25: From the 3 V range:
$(1.50\times10^{-3}\ \text{A})(35.0\ \Omega+R_1)=3.00\ \text{V}\Rightarrow R_1=1970\ \Omega\Rightarrow R_{overall}=2000\ \Omega.$
From the 15 V range:
$(1.50\times10^{-3}\ \text{A})(35.0\ \Omega+R_1+R_2)=15.0\ \text{V}\Rightarrow R_2=8000\ \Omega\Rightarrow R_{overall}=10{,}000\ \Omega.$
From the 150 V range:
$(1.50\times10^{-3}\ \text{A})(35.0\ \Omega+R_1+R_2+R_3)=15.0\ \text{V}\Rightarrow R_2=90{,}000\ \Omega\rightarrow R_{overall}=100\ \text{k}\Omega.$

27-26: (a) $I=\dfrac{\mathcal{E}}{R_{total}}=\dfrac{100\ \text{V}}{484\ \Omega}=0.207\ \text{A}\Rightarrow V=\mathcal{E}-Ir=100\ \text{V}-(0.207\ \text{A})(5.83\ \Omega)=98.8\ \Omega.$

(b) $V=\varepsilon-Ir=\varepsilon-\dfrac{\varepsilon r}{r+R_V}=\dfrac{\varepsilon R_V}{r+R_V}=\dfrac{\varepsilon}{(r/R_V)+1}\Rightarrow\dfrac{r}{R_V}=\dfrac{\varepsilon}{V}-1.$

Now if V is to be off by no more than 5% it requires: $\dfrac{r}{R_V}=\dfrac{100}{95}-1=0.0526.$

27-27: (a) When the galvanometer reading is zero:
$\mathcal{E}_2=IR_{cb}$ and $\mathcal{E}_1=IR_{ab}\Rightarrow\mathcal{E}_2=\mathcal{E}_1\dfrac{R_{cb}}{R_{ab}}=\mathcal{E}_1\dfrac{x}{l}.$
(b) The value of the galvanometer's resistance is unimportant since no current flows through it.
(c) $\mathcal{E}_2=\mathcal{E}_1\dfrac{x}{l}=(14.18\ \text{V})\dfrac{0.676\ \text{m}}{1.000\ \text{m}}=9.59\ \text{V}.$

27-28: Two voltmeters with different resistances are connected in series across a 120 V line. So the current flowing is $I=\dfrac{V}{R_{total}}=\dfrac{120\ \text{V}}{165\times10^3\ \Omega}=7.27\times10^{-4}\ \text{A}.$
But the current required for full-scale deflection for each voltmeter is:
$I_{fsd(15k\Omega)}=\dfrac{150\ \text{V}}{15{,}000\ \Omega}=0.0100\ \text{A}$ and $I_{fsd(150k\Omega)}=\dfrac{150\ \text{V}}{150{,}000\ \Omega}=0.00100\ \text{A}.$
So the readings are:
$V_{15k\Omega}=150\ \text{V}\left(\dfrac{7.27\times10^{-4}\ \text{A}}{0.0100\ \text{A}}\right)=10.9\ \text{V}$ and $V_{150k\Omega}=150\ \text{V}\left(\dfrac{7.27\times10^{-4}\ \text{A}}{0.00100\ \text{A}}\right)=109\ \text{V}.$

27-29: A half-scale reading occurs with R = 500 Ω. So the current through the galvanometer is half the full-scale current of 3.40×10^{-3} A.
$\Rightarrow\mathcal{E}=IR_{total}\Rightarrow1.50\ \text{V}=(1.70\times10^{-3}\ \text{A})(20.0\ \Omega+500\ \Omega+R_s)\Rightarrow R_s=362\ \Omega.$

27-30: (a) When the wires are shorted, the full-scale deflection current is obtained:
$\mathcal{E}=IR_{total}\Rightarrow3.14\ \text{V}=(1.50\times10^{-3}\ \text{A})(83.0\ \Omega+R)\Rightarrow R=2010\ \Omega.$
(b) If the resistance $R_x=600\ \Omega$: $I=\dfrac{V}{R_{total}}=\dfrac{3.14\ \text{V}}{83.0\ \Omega+2010\ \Omega+R}=1.17\ \text{mA}.$

(c) $I_x=\dfrac{\mathcal{E}}{R_{total}}=\dfrac{3.14\ \text{V}}{83.0\ \Omega+2010\ \Omega+R_x}\Rightarrow R_x=\dfrac{3.14\ \text{V}}{I_x}-2093\ \Omega.$

So: $I_x = \frac{1}{4} I_{fsd} = 3.75 \times 10^{-4}\ \text{A} \Rightarrow R_x = \frac{3.14\ \text{V}}{3.75 \times 10^{-4}\ \text{A}} - 2093\ \Omega = 6280\ \Omega.$

$I_x = \frac{1}{2} I_{fsd} = 7.50 \times 10^{-4}\ \text{A} \Rightarrow R_x = \frac{3.14\ \text{V}}{7.50 \times 10^{-4}\ \text{A}} - 2093\ \Omega = 2090\ \Omega.$

$I_x = \frac{3}{4} I_{fsd} = 1.13 \times 10^{-3}\ \text{A} \Rightarrow R_x = \frac{3.14\ \text{V}}{1.13 \times 10^{-3}\ \text{A}} - 2093\ \Omega = 698\ \Omega.$

27-31: $[RC] = \left[\frac{V}{I}\frac{Q}{V}\right] = \left[\frac{Q}{I}\right] = \left[\frac{Q}{Q/t}\right] = [t]$

27-32: An uncharged capacitor is placed into a circuit.

(a) At the instant the circuit is completed, there is no voltage over the capacitor, since it has no charge stored.

(b) All the voltage of the battery is lost over the resistor, so $V_R = \mathcal{E} = 273\ \text{V}$.

(c) There is no charge on the capacitor.

(d) The current through the resistor is $i = \frac{\mathcal{E}}{R_{total}} = \frac{273\ \text{V}}{6030\ \Omega} = 0.0453\ \text{A}$.

(e) After a long time has passed:

The voltage over the capacitor balances the emf: $V_c = 273\ \text{V}$.

The voltage over the resistor is zero.

The capacitor's charge is $q = Cv_c = (6.74 \times 10^{-6}\ \text{F})(273\ \text{V}) = 1.84 \times 10^{-3}\ \text{C}$.

The current in the circuit is zero.

27-33: (a) $i = \frac{q}{RC} = \frac{7.83 \times 10^{-8}\ \text{C}}{(5.30 \times 10^{5}\ \Omega)(3.43 \times 10^{-10}\ \text{F})} = 4.31 \times 10^{-4}\ \text{A}.$

(b) $\tau = RC = (5.30 \times 10^{5}\ \Omega)(3.43 \times 10^{-10}\ \text{F}) = 1.82 \times 10^{-4}\ \text{s}.$

27-34: $v = v_o e^{-\tau/RC} \Rightarrow C = \frac{\tau}{R\ln(v_o / v)} = \frac{5.00\ \text{s}}{(2.25 \times 10^{-6}\ \Omega)(\ln(15/5))} = 2.02 \times 10^{-6}\ \text{F}.$

27-35: (a) The time constant $RC = (0.895 \times 10^{6}\ \Omega)(12.4 \times 10^{-6}\ \text{F}) = 11.1\ \text{s}$. So at:

t = 0 s: $q = C\mathcal{E}(1 - e^{-t/RC}) = 0.$

t = 5 s: $q = C\mathcal{E}(1 - e^{-t/RC}) = (12.4 \times 10^{-6}\ \text{F})(60.0\ \text{V})(1 - e^{-(5.0\ \text{s})/(11.1\ \text{s})}) = 2.70 \times 10^{-4}\ \text{C}.$

t = 10 s: $q = C\mathcal{E}(1 - e^{-t/RC}) = (12.4 \times 10^{-6}\ \text{F})(60.0\ \text{V})(1 - e^{-(10.0\ \text{s})/(11.1\ \text{s})}) = 4.42 \times 10^{-4}\ \text{C}.$

t = 20 s: $q = C\mathcal{E}(1 - e^{-t/RC}) = (12.4 \times 10^{-6}\ \text{F})(60.0\ \text{V})(1 - e^{-(20.0\ \text{s})/(11.1\ \text{s})}) = 6.21 \times 10^{-4}\ \text{C}.$

t = 100 s: $q = C\mathcal{E}(1 - e^{-t/RC}) = (12.4 \times 10^{-6}\ \text{F})(60.0\ \text{V})(1 - e^{-(100\ \text{s})/(11.1\ \text{s})}) = 7.44 \times 10^{-4}\ \text{C}.$

(b) The current at time t is given by: $i = \frac{\mathcal{E}}{R} e^{-t/RC}$. So at:

t = 0 s: $i = \frac{60.0\text{V}}{8.95 \times 10^{5}\ \Omega} e^{-0/11.1} = 6.70 \times 10^{-5}\ \text{A}.$

t = 5 s: $i = \frac{60.0\text{V}}{8.95 \times 10^{5}\ \Omega} e^{-5/11.1} = 4.27 \times 10^{-5}\ \text{A}.$

t = 10 s: $i = \frac{60.0\text{V}}{8.95 \times 10^{5}\ \Omega} e^{-10/11.1} = 2.72 \times 10^{-5}\ \text{A}.$

$$t = 20\text{ s: } i = \frac{60.0\text{V}}{8.95\times10^5\ \Omega}e^{-20/11.1} = 1.11\times10^{-5}\text{ A}.$$

$$t = 100\text{ s: } i = \frac{60.0\text{V}}{8.95\times10^5\ \Omega}e^{-100/11.1} = 8.20\times10^{-9}\text{ A}.$$

(c) Charge against time: Current against time:

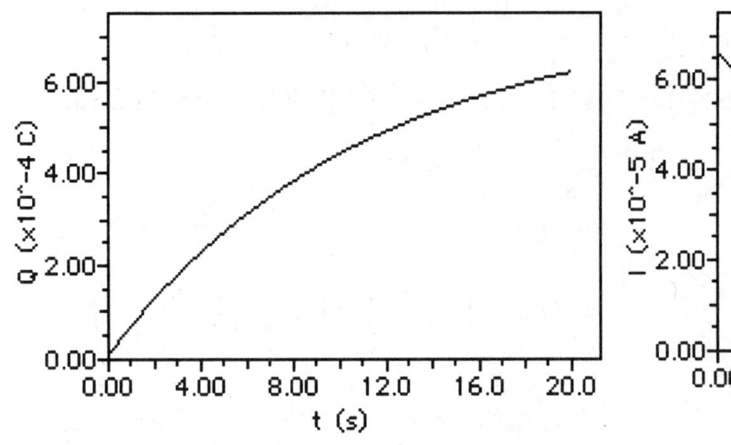

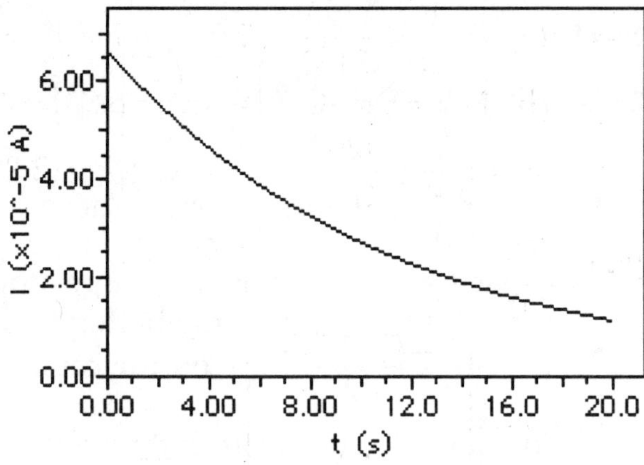

27-36: (a) $Q = CV = (7.50\times10^{-6}\text{ F})(36.0\text{ V}) = 2.70\times10^{-4}\text{ C}.$

(b) $q = Q(1-e^{-t/RC}) \Rightarrow e^{-t/RC} = 1-\dfrac{q}{Q} \Rightarrow R = \dfrac{-t}{C\ln(1-q/Q)}.$

After $t = 3\times10^{-3}$ s: $R = \dfrac{-3\times10^{-3}\text{ s}}{(7.50\times10^{-6}\text{ F})(\ln(1-225/270))} = 223\ \Omega.$

(c) If the charge is to be 99% of final value:

$\dfrac{q}{Q} = (1-e^{-t/RC}) \Rightarrow t = -RC\ln(1-q/Q) = -(223\ \Omega)(7.5\times10^{-6}\text{ F})\ln(0.01) = 7.71\times10^{-3}\text{ s}.$

27-37: (a) The time constant $RC = (950\ \Omega)(4.00\times10^{-5}\text{ F}) = 0.0380\text{ s}.$

$t = 0.05$ s: $q = C\mathcal{E}(1-e^{-t/RC}) = (4.00\times10^{-5}\text{ F})(24.0\text{ V})(1-e^{-0.05/0.0380}) = 7.02\times10^{-4}\text{ C}.$

(b) $i = \dfrac{\mathcal{E}}{R}e^{-t/RC} = \dfrac{24.0\text{ V}}{950\ \Omega}e^{-0.05/0.0380} = 6.78\times10^{-3}\text{ A}.$

$\Rightarrow V_R = IR = (6.78\times10^{-3}\text{ A})(950\ \Omega) = 6.44\text{ V}$ and $V_c = 24.0\text{ V} - 6.44\text{ V} = 17.6\text{ V}.$

(c) Once the switch is thrown, $V_R = V_C = 17.6\text{ V}.$

(d) After $t = 0.05$ s: $q = Q_o e^{-t/RC} = (7.02\times10^{-4}\text{ C})e^{-0.05/0.0380} = 1.88\times10^{-4}\text{ C}.$

27-38: (a) $I = \dfrac{P}{V} = \dfrac{3500\text{ W}}{240\text{ V}} = 15\text{ A}.$ So we need at least 12 gauge wire (good up to 20 A).

(b) $P = \dfrac{V^2}{R} \Rightarrow R = \dfrac{V^2}{P} = \dfrac{(240\text{ V})^2}{3500\text{ W}} = 16\ \Omega.$

(c) At 10 c / kWhr $\Rightarrow$ in 1 hour, cost = (10 c / kWhr)(1 hr)(3.5 kW) = 35 c.

27-39: We want to trip a 20 A circuit breaker:

$I = \dfrac{1350\text{ W}}{120\text{ V}} + \dfrac{P}{120\text{ V}} \Rightarrow$ With $P = 1200$ W: $I = \dfrac{1350\text{ W}}{120\text{ V}} + \dfrac{1200\text{ W}}{120\text{ V}} = 21.3\text{ A}.$ (The next lowest setting caused 18.8 A of current to be drawn.)

27-40: The current gets split evenly between all the parallel bulbs. A single bulb will draw $I = \frac{P}{V} = \frac{45\text{ W}}{120\text{ V}} = 0.38\text{ A} \Rightarrow$ Number of bulbs $\le \frac{20\text{ A}}{0.38\text{ A}} = 53.3$. So you can attach 53 bulbs safely.

27-41: (a) $I = \frac{V}{R} = \frac{120\text{ V}}{24\ \Omega} = 5.0\text{ A} \Rightarrow P = IV = (5.0\text{ A})(120\text{ V}) = 600\text{ W}.$

(b) At T = 280°C, $R = R_0(1 + \alpha\Delta T) = 24\ \Omega(1 + (2.8 \times 10^{-3}(\text{C}°)^{-1}(257°\text{C})) = 41.3\ \Omega.$

$\Rightarrow I = \frac{V}{R} = \frac{120\text{ V}}{41.3\text{ A}} = 2.9\text{ A} \Rightarrow P = (2.9\text{ A})(120\text{ V}) = 350\text{ W}.$

27-42:

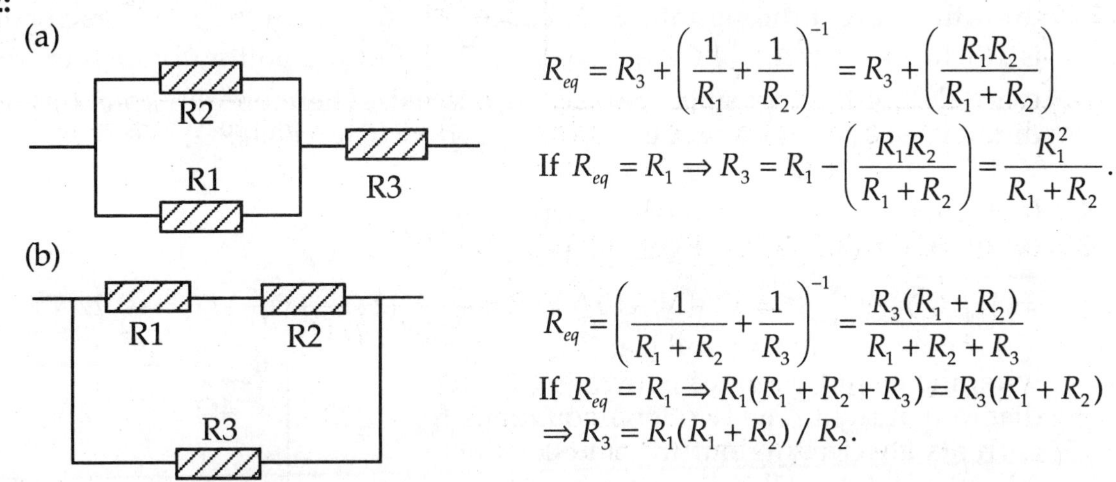

(a) $R_{eq} = R_3 + \left(\frac{1}{R_1} + \frac{1}{R_2}\right)^{-1} = R_3 + \left(\frac{R_1R_2}{R_1 + R_2}\right)$

If $R_{eq} = R_1 \Rightarrow R_3 = R_1 - \left(\frac{R_1R_2}{R_1 + R_2}\right) = \frac{R_1^2}{R_1 + R_2}.$

(b) $R_{eq} = \left(\frac{1}{R_1 + R_2} + \frac{1}{R_3}\right)^{-1} = \frac{R_3(R_1 + R_2)}{R_1 + R_2 + R_3}$

If $R_{eq} = R_1 \Rightarrow R_1(R_1 + R_2 + R_3) = R_3(R_1 + R_2)$

$\Rightarrow R_3 = R_1(R_1 + R_2)/R_2.$

27-43: (a) We wanted a total resistance of 800 Ω and power of 3.2 W from a combination of individual resistors of 800 Ω and 1.6 W power-rating.

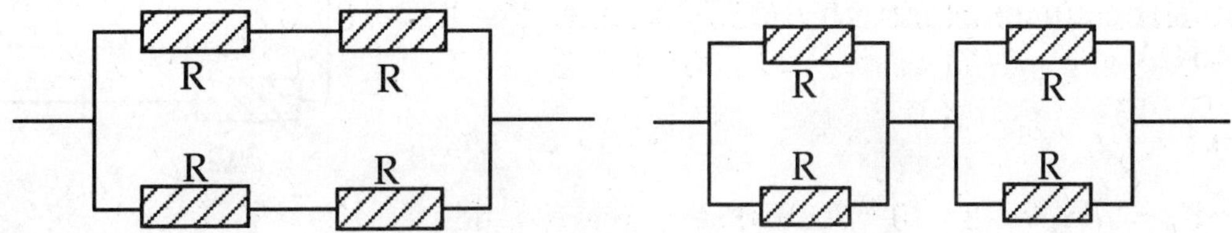

(b) The current is given by: $I = \sqrt{P/R} = \sqrt{3.2\text{ W}/800\ \Omega} = 0.63\text{ A}$. In each leg half the current flows, so the power in each resistor in each combination is the same: $P = (I/2)^2R = (.032\text{ A})^2(800\ \Omega) = 0.80\text{ W}$.

27-44: (a) The equivalent resistance of the two bulbs is 1.0 Ω. So the current is:

$I = \frac{V}{R_{total}} = \frac{8.0\text{ V}}{1.0\ \Omega + 0.80\ \Omega} = 4.4\text{ A} \Rightarrow$ the current through each bulb is 2.2 A.

$V_{bulb} = \mathcal{E} - Ir = 8.0\text{ V} - (4.4\text{ A})(0.80\Omega) = 4.4\text{ V} \Rightarrow P_{bulb} = IV = (2.2\text{ A})(4.4\text{ V}) = 9.9\text{ W}$

(b) If one bulb burns out, then

$I = \frac{V}{R_{total}} = \frac{8.0\text{ V}}{2.0\ \Omega + 0.80\ \Omega} = 2.9\text{ A} \Rightarrow P = I^2R = (2.9\text{ A})^2(2.0\ \Omega) = 16.3\text{ W}$, so the remaining bulb is brighter than before.

27-45: The maximum allowed power is when the total current is the maximum allowed value of $I = \sqrt{P/R} = \sqrt{23\text{ W}/1.8\ \Omega} = 3.6\text{ A}$. Then half the current flows through the parallel resistors and the maximum power is:
$P_{\text{max}} = (I/2)^2 R + (I/2)^2 R + I^2 R = \frac{3}{2} I^2 R = \frac{3}{2}(3.6\text{ A})^2 (1.8\ \Omega) = 35\text{ W}$.

27-46: (a) $R_{eq}(8,16,16) = \left(\frac{1}{8\ \Omega} + \frac{1}{16\ \Omega} + \frac{1}{16\ \Omega}\right)^{-1} = 4.0\ \Omega;\ R_{eq}(9,18) = \left(\frac{1}{9\ \Omega} + \frac{1}{18\ \Omega}\right)^{-1} = 6.0\ \Omega$.

So the circuit is equivalent to the one shown at right. Thus:
$R_{eq} = \left(\frac{1}{6\ \Omega + 6\ \Omega} + \frac{1}{20\ \Omega + 4\ \Omega}\right)^{-1} = 8.0\ \Omega$

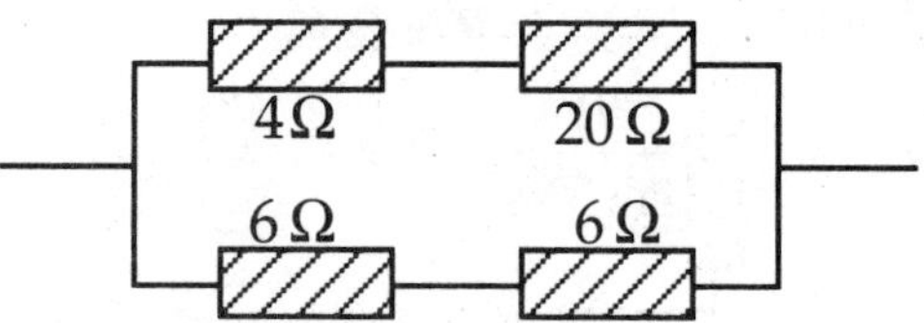

(b) If the current through the 8 Ω resistor is 1.6 A, then the top branch current is $I(8,16,16) = 1.6\text{ A} + \frac{1}{2}1.6\text{ A} + \frac{1}{2}1.6\text{ A} = 3.2\text{ A}$. But the bottom branch current is twice that of the top, since its resistance is half. Therefore the potential difference between x and a is $V_{xa} = IR_{eq}(9,18) = (6.4\text{ A})(6.00\ \Omega) = 38\text{ V}$.

26-47: (a) Going around the complete loop, we have:
$\sum \mathcal{E} - \sum IR = 12.0\text{ V} - 8.0\text{ V} - I(9.0\ \Omega) = 0 \Rightarrow I = 0.44\text{ A}$.
$\Rightarrow V_{ab} = \sum \varepsilon - \sum IR = 12.0\text{ V} - 10.0\text{ V} - (0.44\text{ A})(2\ \Omega + 1\ \Omega + 1\ \Omega) = +0.22\text{ V}$.

(b) If now the points a and b are connected by a wire, the circuit becomes equivalent to the diagram at right. The two loop equations for currents are (leaving out the units):
$12 - 10 - 4I_1 + 4I_2 = 0 \Rightarrow I_2 = I_1 - 0.5$ and
$10 - 8 - 4I_2 - 5I_3 = 2 - 4I_2 - 5(I_1 + I_2) = 0$
$\Rightarrow 2 - (4I_1 - 2) - 5I_1 - 5I_1 + 2.5 = 0 \Rightarrow I_1 = 0.464\text{ A}$.
Thus the current through the 12 V battery is 0.464 A.

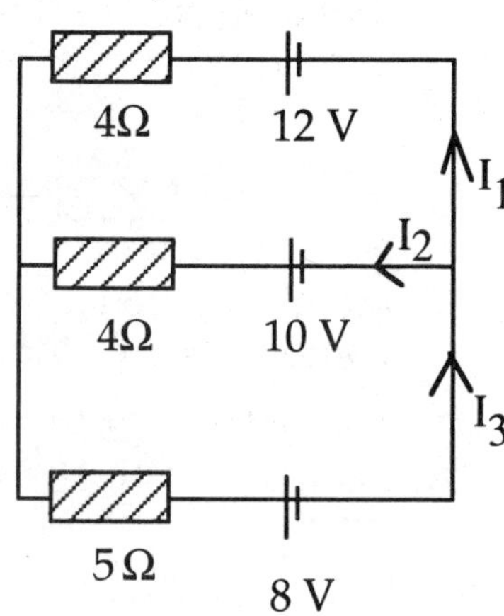

27-48: For three identical resistors in series, $P_s = \dfrac{V^2}{3R}$. If they are now in parallel over the same voltage, $P_p = \dfrac{V^2}{R_{eq}} = \dfrac{V^2}{R/3} = \dfrac{9V^2}{3R} = 9P_s = 9(76\text{ W}) = 684\text{ W}$.

27-49: Top left loop: $12 - 5(I_2 - I_3) - 1I_2 = 0 \Rightarrow 12 - 6I_2 + 5I_3 = 0$.
Top right loop: $9 - 8(I_1 + I_3) - 1I_1 = 0 \Rightarrow 9 - 9I_1 - 8I_3 = 0$.
Bottom loop: $12 - 10I_3 - 9 + 1I_1 - 1I_2 = 0 \Rightarrow 3 + I_1 - I_2 - 10I_3 = 0$.
Solving these three equations for the currents yields:
$I_1 = 0.848\text{ A},\ I_2 = 2.14\text{ A}$, and $I_3 = 0.171\text{ A}$.

27-50: Outside loop: $24 - 7(2.5) - 3(2.5 - I_\varepsilon) = 0 \Rightarrow I_\varepsilon = 0.333\text{ A}$.
Right loop: $\varepsilon - 7(2.5) - 2(0.333) = 0 \Rightarrow \varepsilon = 18.2\text{ V}$.

27-51: Left loop: $20 - 14 - 2I_1 + 4(I_2 - I_1) = 0 \Rightarrow 6 - 6I_1 + 4I_2 = 0$.
Right loop: $36 - 5I_2 - 4(I_2 - I_1) = 0 \Rightarrow 36 + 4I_1 - 9I_2 = 0$.
Solving these two equations for the currents yields:
$I_1 = 5.21\ \text{A} = I_{2\Omega}$, $I_2 = 6.32\ \text{A} = I_{5\Omega}$, and $I_{4\Omega} = I_2 - I_1 = 1.11\ \text{A}$.

27-52: (a) Using the currents as defined on the circuit diagram below we obtain three equations to solve for the currents:

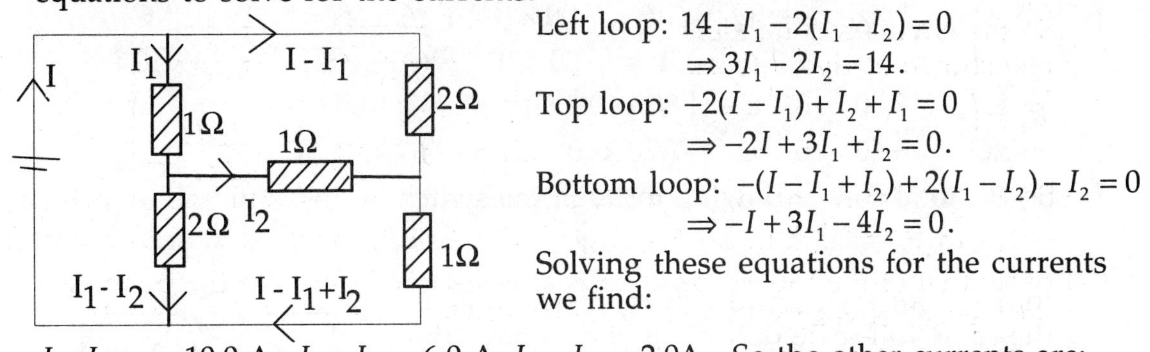

Left loop: $14 - I_1 - 2(I_1 - I_2) = 0$
$\Rightarrow 3I_1 - 2I_2 = 14$.
Top loop: $-2(I - I_1) + I_2 + I_1 = 0$
$\Rightarrow -2I + 3I_1 + I_2 = 0$.
Bottom loop: $-(I - I_1 + I_2) + 2(I_1 - I_2) - I_2 = 0$
$\Rightarrow -I + 3I_1 - 4I_2 = 0$.
Solving these equations for the currents we find:

$I = I_{battery} = 10.0\ \text{A}$; $I_1 = I_{R_1} = 6.0\ \text{A}$; $I_2 = I_{R_3} = 2.0\text{A}$. So the other currents are:
$I_{R_2} = I - I_1 = 4.0\ \text{A}$; $I_{R_4} = I_1 - I_2 = 4.0\ \text{A}$; $I_{R_5} = I - I_1 + I_2 = 6.0\ \text{A}$.

27-53: (a) When the switch is open, only the outer resistances have current through them. So the equivalent resistance of them is:

$$R_{eq} = \left(\tfrac{1}{6\ \Omega + 3\ \Omega} + \tfrac{1}{3\ \Omega + 6\ \Omega}\right)^{-1} = 4.50\ \Omega \Rightarrow I = \frac{V}{R_{eq}} = \frac{36.0\ \text{V}}{4.50\ \Omega} = 8.00\ \text{A}$$

$\Rightarrow V_{ab} = (\tfrac{1}{2}8.00\ \text{A})(3.00\ \Omega) - (\tfrac{1}{2}8.00\ \text{A})(6.00\ \Omega) = -12.0\ \text{V}$.
(b) If the switch is closed, the circuit geometry and resistance ratios become identical to that of **27-52** and the same analysis can be carried out. However, we can also use symmetry to infer the following:
$I_{6\Omega} = \tfrac{2}{3}I_{3\Omega}$, and $I_{switch} = \tfrac{1}{3}I_{3\Omega}$. From the left loop as in **27-52**:
$36\ \text{V} - (\tfrac{2}{3}I_{3\Omega})(6\ \Omega) - I_{3\Omega}(3\ \Omega) = 0 \Rightarrow I_{3\Omega} = 5.14\ \text{A} \Rightarrow I_{switch} = \tfrac{1}{3}I_{3\Omega} = 1.71\ \text{A}$.

(c) $I_{battery} = \tfrac{2}{3}I_{3\Omega} + I_{3\Omega} = \tfrac{5}{3}I_{3\Omega} = 8.57\ \text{A} \Rightarrow R_{eq} = \dfrac{\mathcal{E}}{I_{battery}} = \dfrac{36.0\ \text{V}}{8.57\ \text{A}} = 4.20\ \Omega$.

27-54: (a) With an open switch: $V_{ab} = \mathcal{E} = 18.0\ \text{V}$, since equilibrium has been reached.
(b) Point "a" is at a higher potential since it is directly connected to the positive terminal of the battery.
(c) When the switch is closed:
$18.0\ \text{V} = I(6.00\ \Omega + 3.00\ \Omega) \Rightarrow I = 2.00\ \text{A} \Rightarrow V_b = (2.00\ \text{A})(3.00\ \Omega) = 6.00\ \text{V}$.
(d) Initially the capacitor's charges were:
$Q_3 = CV = (3.00 \times 10^{-6}\ \text{F})(18.0\ \text{V}) = 5.40 \times 10^{-5}\ \text{C}$.
$Q_6 = CV = (6.00 \times 10^{-6}\ \text{F})(18.0\ \text{V}) = 1.08 \times 10^{-4}\ \text{C}$.
After the switch is closed:
$Q_3 = CV = (3.00 \times 10^{-6}\ \text{F})(18.0\ \text{V} - 12.0\ \text{V}) = 1.80 \times 10^{-5}\ \text{C}$.
$Q_6 = CV = (6.00 \times 10^{-6}\ \text{F})(18.0\ \text{V} - 6.0\ \text{V}) = 7.20 \times 10^{-5}\ \text{C}$.
So both capacitors lose $3.60 \times 10^{-5}\ \text{C}$.

27-55: (a) With an open switch: $Q = C_{eq}V = (2.00\times10^{-6}\text{ F})(18.0\text{ V}) = 3.60\times10^{-5}\text{ C}.$

Also, there is a current in the left branch: $I = \dfrac{18.0\text{ V}}{6.00\ \Omega + 3.00\ \Omega} = 2.00\text{ A}.$

So $V_{ab} = V_{6\mu F} - V_{6\Omega} = \dfrac{Q_{6\mu F}}{C} - IR_{6\Omega} = \dfrac{3.6\times10^{-5}\text{ C}}{6.0\times10^{-6}\text{ F}} - (2.0\text{ A})(6.0\ \Omega) = -6.00\text{ V}.$

(b) Point "b" is at the higher potential.

(c) If the switch is closed: $V_b = V_a = (2.00\text{ A})(3.00\ \Omega) = 6.00\text{ V}.$

(d) New charges are: $Q_3 = CV = (3.00\times10^{-6}\text{ F})(6.0\text{ V}) = 1.80\times10^{-5}\text{ C}.$

$Q_6 = CV = (6.00\times10^{-6}\text{ F})(-12.0\text{ V}) = -7.20\times10^{-5}\text{ C}.$

$\Rightarrow \Delta Q_3 = +3.60\times10^{-5}\text{ C} - (1.80\times10^{-5}\text{ C}) = +1.80\times10^{-5}\text{ C}.$

$\Rightarrow \Delta Q_6 = -3.60\times10^{-5}\text{ C} - (-7.20\times10^{-5}\text{ C}) = +3.60\times10^{-5}\text{ C}.$

So the total charge flowing through the switch is $5.40\times10^{-5}\text{ C}.$

27-56:

In order for the second galvanometer to give the same full-scale deflection and to have the same resistance as the first, we need two additional resistances as shown at right.

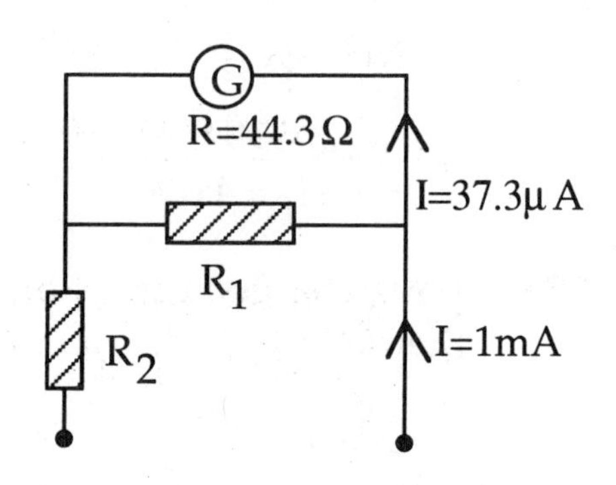

So:

$(37.3\ \mu\text{A})(44.3\ \Omega) = (962.7\ \mu\text{A})R_1 \Rightarrow R_1 = 1.72\ \Omega.$

And for the total resistance to be 200 Ω:

$200 = R_2 + \left(\frac{1}{44.3\ \Omega} + \frac{1}{1.72\ \Omega}\right)^{-1} \Rightarrow R_2 = 198\ \Omega.$

27-57: (a) $I = \dfrac{90\text{ V}}{429\ \Omega + \left(\frac{1}{R_V} + \frac{1}{582\ \Omega}\right)^{-1}}$ and $V = \mathcal{E} - IR \Rightarrow 44.6\text{ V} = 90\text{ V} - \dfrac{(90\text{ V})(429\ \Omega)}{429\ \Omega + \left(\frac{1}{R_V} + \frac{1}{582\ \Omega}\right)^{-1}}$

$\Rightarrow \left(\frac{1}{R_V} + \frac{1}{582\ \Omega}\right)^{-1} = 421\ \Omega \Rightarrow R_V = 1530\ \Omega.$

(b) If the voltmeter is connected over the 429 Ω resistor, then:

$R_{eq} = 582\ \Omega + \left(\frac{1}{1530} + \frac{1}{429\ \Omega}\right)^{-1} = 335\ \Omega \Rightarrow I = \dfrac{90\text{ V}}{335\ \Omega} = 0.0982\text{ A}.$

$\Rightarrow V_{429\Omega} = 90\text{ V} - (0.0982\text{ A})(582\ \Omega) = 32.9\text{ V}.$

27-58: (a) $R_{eq} = 100\text{ k}\Omega + \left(\frac{1}{200\text{ k}\Omega} + \frac{1}{40\text{ k}\Omega}\right)^{-1} = 126\text{ k}\Omega \Rightarrow I = \dfrac{0.250\text{ kV}}{126\text{ k}\Omega} = 1.98\times10^{-3}\text{ A}.$

$\Rightarrow V_{200k\Omega} = IR = (1.98\times10^{-3}\text{ A})\left(\frac{1}{2.00\times10^{5}\ \Omega} + \frac{1}{3.00\times10^{4}\ \Omega}\right)^{-1} = 51.7\text{ V}.$

(b) If $V_R = 3.00\times10^{6}\ \Omega$, then we carry out the same calculations as above to find $R_{eq} = 287\text{ k}\Omega \Rightarrow I = 8.70\times10^{-4}\text{ A} \Rightarrow V_{200k\Omega} = 163\text{ V}.$

(c) If $V_R = \infty$, then we find $R_{eq} = 300\text{ k}\Omega \Rightarrow I = 8.33\times10^{-4}\text{ A} \Rightarrow V_{200k\Omega} = 167\text{ V}.$

27-59: $I = \dfrac{110\text{ V}}{(20{,}000\ \Omega + R)} \Rightarrow V = 110\text{ V} - \dfrac{(110\text{ V})R}{(20{,}000\ \Omega + R)} = 56\text{ V}$

$\Rightarrow (54\text{ V})(20{,}000\ \Omega + R) = (110\text{ V})R \Rightarrow R = 19{,}300\ \Omega.$

27-60: (a) $V = IR + IR_A \Rightarrow R = \frac{V}{I} - R_A$. The true resistance R is always less than the reading because in the circuit the ammeter's resistance causes the current to be less than it should. Thus the smaller current requires the resistance R to be calculated larger than it should be.

(b) $I = \frac{V}{R} + \frac{V}{R_V} \Rightarrow R = \frac{VR_V}{IR_V - V} = \frac{V}{I - V/R_V}$. Now the current measured is greater than that through the resistor, so $R = V/I_R$ is always greater than V/I.

(c) (a): $P = I^2R = I^2(V/I - R_A) = IV - I^2R_A$.

(b): $P = V^2/R = V(I - V/R_V) = IV - V^2/R_V$.

27-61: (a) When the bridge is balanced, no current flows through the galvanometer:

$$I_G = 0 \Rightarrow V_N = V_P \Rightarrow NI_{NM} = PI_{PX} \Rightarrow N\frac{(P+X)}{(P+X+N+M)} = P\frac{(N+M)}{(P+X+N+M)}$$

$$\Rightarrow N(P+X) = P(N+M) \Rightarrow NX = PM \Rightarrow X = \frac{PM}{N}.$$

(b) $X = \frac{(1000\ \Omega)(21.46\ \Omega)}{10.00\ \Omega} = 2146\ \Omega$.

27-62: For a charged capacitor, connected into a circuit:

$$I_o = \frac{Q_o}{RC} \Rightarrow Q_o = I_oRC = (0.380\ \text{A})(4840\ \Omega)(9.46\times10^{-10}\ \text{F}) = 1.74\times10^{-6}\ \text{C}.$$

27-63: $\mathcal{E} = I_oR \Rightarrow R = \frac{\mathcal{E}}{I_o} = \frac{200\ \text{V}}{8.6\times10^{-4}\ \text{A}} = 2.3\times10^5\ \Omega \Rightarrow C = \frac{\tau}{R} = \frac{5.7\ \text{s}}{2.3\times10^5\ \Omega} = 2.5\times10^{-5}\ \text{F}$.

27-64: (a) $U_o = \frac{Q_o^2}{2C} = \frac{(0.0636\ \text{C})^2}{2(4.5\times10^{-6}\ \text{F})} = 450\ \text{J}$.

(b) $P_o = I_o^2R = \left(\frac{Q_o}{RC}\right)^2 R = \frac{(0.0636\ \text{C})^2}{(8.60\times10^4\ \Omega)(4.5\times10^{-6}\ \text{F})^2} = 2320\ \text{W}$.

(c) When $U = \frac{1}{2}U_o = \frac{1}{2}\frac{Q_o^2}{2C}$

$$\Rightarrow Q = \frac{Q_o}{\sqrt{2}} \Rightarrow P = \left(\frac{Q}{RC}\right)^2 R = \frac{1}{2}\left(\frac{Q_o}{RC}\right)^2 R = \frac{1}{2}P_o = 1160\ \text{W}.$$

27-65: If $i = 0.0185\ \text{A} \Rightarrow V_C = 180\ \text{V} - (0.0185\ \text{A})(7250\ \Omega) = 45.9\ \text{V}$.

$\Rightarrow Q = CV = (3.40\times10^{-6}\ \text{F})(45.9\ \text{V}) = 1.56\times10^{-4}\ \text{C}$.

27-66: (a) (i) $P_R = \frac{V^2}{R} = \frac{(150\ \text{V})^2}{4.80\ \Omega} = 4690\ \text{W}$ (ii) $P_C = \frac{dU}{dt} = \frac{1}{2C}\frac{d(q^2)}{dt} = \frac{iq}{C} = 0$.

(iii) $P_\varepsilon = \mathcal{E}I = (150\ \text{V})\frac{150\ \text{V}}{4.8\ \Omega} = 4690\ \text{W}$.

(b) After a long time, $i = 0 \Rightarrow P_R = 0,\ P_C = 0,\ P_\varepsilon = 0$.

(c) When $q = \frac{1}{2}Q_f \Rightarrow e^{-t/RC} = 0.5 \Rightarrow i = \frac{1}{2}I_o = \frac{150\text{ V}}{2(4.8\text{ W})} = 15.6\text{ A}.$

$\Rightarrow P_R = I^2R = (15.6\text{ A})^2(4.8\ \Omega) = 1170\text{ W}.$

$P_C = \frac{iq}{C} = \frac{iQ_f}{2C} = \frac{i\mathcal{E}}{2} = \frac{(15.6\text{ A})(150\text{ V})}{2} = 1170\text{ W}.$

And $P_\varepsilon = \mathcal{E}I = (150\text{ V})(15.6\text{ A}) = 2340\text{ W}.$

27-67: (a) $E_{total} = \int_0^\infty P_\varepsilon dt = \int_0^\infty \mathcal{E}Idt = \frac{\mathcal{E}^2}{R}\int_0^\infty e^{-t/RC}dt = \mathcal{E}^2C(1) = \mathcal{E}^2C.$

(b) $E_R = \int_0^\infty P_R dt = \int_0^\infty i^2R\,dt = \frac{\mathcal{E}^2}{R}\int_0^\infty e^{-2t/RC}dt = \frac{1}{2}\mathcal{E}^2C.$

(c) $U = \frac{Q_o^{\ 2}}{2C} = \frac{V^2C}{2} = \frac{1}{2}\mathcal{E}^2C = E_{total} - E_R.$

(d) One half of the energy is stored in the capacitor, regardless of the size of the resistor.

27-68: $i = -\frac{Q_o}{RC}e^{-t/RC} \Rightarrow P = i^2R = \frac{Q_o^{\ 2}}{RC^2}e^{-2t/RC} \Rightarrow E = \frac{Q_o^{\ 2}}{RC^2}\int_0^\infty e^{-2t/RC}dt = \frac{Q_o^{\ 2}}{RC^2}\frac{RC}{2} = \frac{Q_o^{\ 2}}{2C} = U_o.$

27-69: (a) We will say that a capacitor is discharged if its charge is less than that of one electron. The time this takes is then given by:

$q = Q_oe^{-t/RC} \Rightarrow t = RC\ln(Q_o / e)$

$\Rightarrow t = (5.7\times10^5\ \Omega)(8.5\times10^{-7}\text{ F})\ln(4.0\times10^{-6}\text{ C}/1.6\times10^{-19}\text{ C}) = 14.9\text{ s}.$

(b) As shown in (a), $t = \tau\ln(Q_o / q)$, and so the number of time constants required to discharge the capacitor is independent of R and C, and depends only on the initial charge.

27-70: (a) The equivalent capacitance and time constant are:

$C_{eq} = \left(\frac{1}{3\ \mu\text{F}} + \frac{1}{6\ \mu\text{F}}\right)^{-1} = 2.00\ \mu\text{F} \Rightarrow \tau = R_{total}C_{eq} = (6.00\ \Omega)(2.00\ \mu\text{F}) = 1.20\times10^{-5}\text{ s}.$

(b) After $t = 1.20\times10^{-5}$ s, $q = Q_f(1-e^{-t/RC_{eq}}) = C_{eq}\mathcal{E}(1-e^{-t/RC_{eq}})$

$\Rightarrow V_{3\mu\text{F}} = \frac{q}{C_{3\mu\text{F}}} = \frac{C_{eq}\mathcal{E}}{C_{3\mu\text{F}}}(1-e^{-t/RC_{eq}}) = \frac{(2.0\ \mu\text{F})(12\text{ V})}{3.0\ \mu\text{F}}(1-e^{-1}) = 5.06\text{ V}.$

27-71: We can re-draw the circuit as shown below:

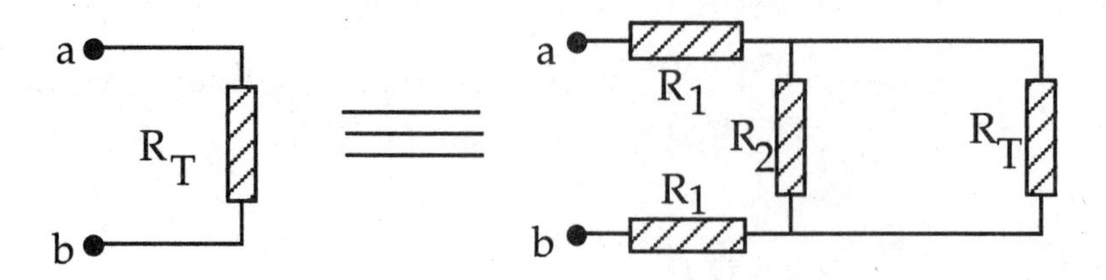

$$\Rightarrow R_T = 2R_1 + \left(\frac{1}{R_2} + \frac{1}{R_T}\right)^{-1} = 2R_1 + \frac{R_2 R_T}{R_2 + R_T} \Rightarrow R_T^{\,2} - 2R_1 R_T - 2R_1 R_2 = 0.$$

$$\Rightarrow R_T = R_1 \pm \sqrt{R_1^{\,2} + 2R_1 R_2} \text{ but } R_T > 0 \Rightarrow R_T = R_1 + \sqrt{R_1^{\,2} + 2R_1 R_2}.$$

27-72:

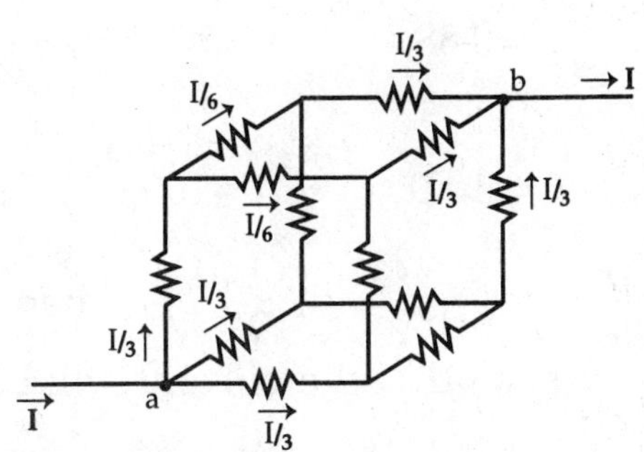

Let current I enter at a and exit at b. At a there are three equivalent branches, so current is $I/3$ in each. At the next junction point there are two equivalent branches so each gets current $I/6$. Then at b there are three equivalent branches with current $I/3$ in each. The voltage drop from a to b then is $V = \left(\frac{I}{3}\right)R + \left(\frac{I}{6}\right)R + \left(\frac{I}{3}\right)R = \frac{5}{6}IR.$ This must be the same as $V = IR_{eq}$, so $R_{eq} = \frac{5}{6}R.$

27-73: (a) The circuit can be re-drawn as follows:

a R1 c R2 RT b R1 d ≡ a R1 c Req b R1 d

Then $V_{cd} = V_{ab}\dfrac{R_{eq}}{2R_1 + R_{eq}} = V_{ab}\dfrac{1}{2R_1/R_{eq} + 1}$ and $R_{eq} = \dfrac{R_2 R_T}{R_2 + R_T}.$

But $\beta = \dfrac{2R_1(R_T + R_2)}{R_T R_2} = \dfrac{2R_1}{R_{eq}} \Rightarrow V_{cd} = V_{ab}\dfrac{1}{1+\beta}.$

(b) Recall $V_1 = \dfrac{V_0}{(1+\beta)} \Rightarrow V_2 = \dfrac{V_1}{(1+\beta)} = \dfrac{V_0}{(1+\beta)^2} \Rightarrow V_n = \dfrac{V_{n-1}}{(1+\beta)} = \dfrac{V_0}{(1+\beta)^n}.$

If $R_1 = R_2 \Rightarrow R_T = R_1 + \sqrt{R_1^{\,2} + 2R_1R_1} = R_1(1+\sqrt{3})$ and $\beta = \dfrac{2(2+\sqrt{3})}{1+\sqrt{3}} = 2.73.$

So, for the n-th segment to have 1% of the original voltage, we need:

$$\frac{1}{(1+\beta)^n} = \frac{1}{(1+2.73)^n} \le 0.01 \Rightarrow n = 4:\ V_4 = 0.005V_0.$$

(c) $R_T = R_1 + \sqrt{R_1^{\,2} + 2R_1R_2}$

$$\Rightarrow R_T = 6400\ \Omega + \sqrt{(6400\ \Omega)^2 + 2(6400\ \Omega)(8.0\times10^8\ \Omega)} = 3.2\times10^6\ \Omega.$$

$$\Rightarrow \beta = \frac{2(6400\ \Omega)(3.2\times10^6\ \Omega + 8.0\times10^8\ \Omega)}{(3.2\times10^6\ \Omega)(8.0\times10^8\ \Omega)} = 4.0\times10^{-3}.$$

(d) Along a length of 2.0 mm of axon, there are 2000 segments each 1.0 μm long. The voltage therefore attenuates by:

$$V_{2000} = \frac{V_0}{(1+\beta)^{2000}} \Rightarrow \frac{V_{2000}}{V_0} = \frac{1}{(1+4.0\times10^{-3})^{2000}} = 3.4\times10^{-4}.$$

(e) If $R_2 = 3.3\times10^{12}\ \Omega \Rightarrow R_T = 2.1\times10^{8}\ \Omega$ and $\beta = 6.2\times10^{-5}$.

$$\Rightarrow \frac{V_{2000}}{V_0} = \frac{1}{(1+6.2\times10^{-5})^{2000}} = 0.88.$$

27-74: (a) Fully charged: $Q = CV = (10.0\times10^{-12}\ \text{F})(1000\ \text{V}) = 1.00\times10^{-8}\ \text{C}$.

(b) $i_o = \dfrac{\mathcal{E} - V_{C'}}{R} = \dfrac{\mathcal{E}}{R} - \dfrac{q}{RC'} \Rightarrow i(t) = \left(\frac{\mathcal{E}}{R} - \frac{q}{RC'}\right)e^{-t/RC'}$, where $C' = 1.1C$.

(c) We need a resistance such that the current will be greater than 1 μA for longer than 200 μs.

$$\Rightarrow i(200\ \mu s) = 1.0\times10^{-6}\ \text{A} = \frac{1}{R}\left(1000\ \text{V} - \frac{1.0\times10^{-8}\ \text{C}}{1.1(1.0\times10^{-11}\ \text{F})}\right)e^{-\frac{2.0\times10^{-4}\,\text{s}}{R(11\times10^{-12}\,\text{F})}}$$

$$\Rightarrow 1.0\times10^{-6}\ \text{A} = \frac{1}{R}(90.9)e^{-(1.8\times10^{7}\,\Omega)/R} \Rightarrow 18.3R - R\ln R - 1.8\times10^{7} = 0.$$

Solving for R numerically we find $7.15\times10^{6}\ \Omega \le R \le 7.01\times10^{7}\ \Omega$.
If the resistance is too small, then the capacitor discharges too quickly, and if the resistance is too large, the current is not large enough.

27-75: (a) Using Kirchoff's Rules on the circuit we find:
Left loop: $92 - 140I_1 - 210I_2 + 55 = 0 \Rightarrow 147 - 140I_1 - 210I_2 = 0.$
Right loop: $57 - 35I_3 - 210I_2 + 55 = 0 \Rightarrow 112 - 210I_2 - 35I_3 = 0.$
Currents: $\Rightarrow I_1 - I_2 + I_3 = 0.$
Solving for the three currents we have:
$I_1 = 0.300\ \text{A},\ I_2 = 0.500\ \text{A},\ I_3 = 0.200\ \text{A}.$

(b) Leaving only the 92 V battery in the circuit:
Left loop: $92 - 140I_1 - 210I_2 = 0.$
Right loop: $-35I_3 - 210I_2 = 0.$
Currents: $I_1 - I_2 + I_3 = 0.$
Solving for the three currents: $I_1 = 0.541\ \text{A},\ I_2 = 0.077\ \text{A},\ I_3 = -0.464\ \text{A}.$

(c) Leaving only the 57 V battery in the circuit:
Left loop: $140I_1 + 210I_2 = 0.$
Right loop: $57 - 35I_3 - 210I_2 = 0.$
Currents: $I_1 - I_2 + I_3 = 0.$
Solving for the three currents: $I_1 = -0.287\ \text{A},\ I_2 = 0.192\ \text{A},\ I_3 = 0.480\ \text{A}.$

(d) Leaving only the 55 V battery in the circuit:
Left loop: $55 - 140I_1 - 210I_2 = 0.$
Right loop: $55 - 35I_3 - 210I_2 = 0.$
Currents: $I_1 - I_2 + I_3 = 0.$
Solving for the three currents: $I_1 = 0.046\ \text{A},\ I_2 = 0.231\ \text{A},\ I_3 = 0.185\ \text{A}.$

(e) If we sum the currents from the previous three parts we find:
$I_1 = 0.300\ \text{A},\ I_2 = 0.500\ \text{A},\ I_3 = 0.200\ \text{A}$, just as in part (a).

(f) Changing the 57 V battery for an 80 V battery just affects the calculation in part (c). It changes to:

Left loop: $140I_1 + 210I_2 = 0$.

Right loop: $80 - 35I_3 - 210I_2 = 0$.

Currents: $I_1 - I_2 + I_3 = 0$.

Solving for the three currents: $I_1 = -0.403\ \text{A},\ I_2 = 0.269\ \text{A},\ I_3 = 0.672\ \text{A}$.

So the total current for the full circuit is the sum of (b), (d) and (f) above:

$I_1 = 0.184\ \text{A},\ I_2 = 0.576\ \text{A},\ I_3 = 0.392\ \text{A}$.

Chapter 28: Magnetic Field and Magnetic Forces

28-1:

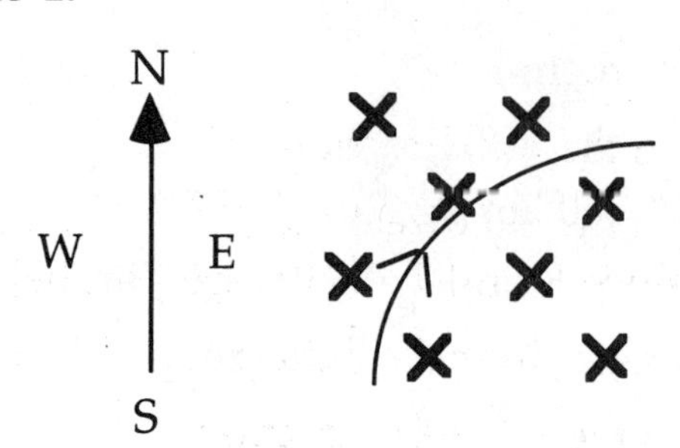

By the right hand rule, the charge is negative.

28-2: $\vec{F} = m\vec{a} = q\vec{v}\times\vec{B} \Rightarrow \vec{a} = \dfrac{q\vec{v}\times\vec{B}}{m}$

$$\Rightarrow \vec{a} = \frac{(1.22\times10^{-8}\ \text{C})(3.0\times10^{5}\ \text{m/s})(\text{-}0.815\ \text{T})(\hat{j}\times\hat{i})}{1.81\times10^{-3}\ \text{kg}} = +(1.65\ \text{m/s}^2)\hat{k}.$$

28-3: (a) $\vec{F} = m\vec{a} = q\vec{v}\times\vec{B} = (-2.46\times10^{-6}\ \text{C})(4.19\times10^{4}\ \text{m/s})(1.40\ \text{T})(\hat{j}\times\hat{i})$

$\Rightarrow \vec{F} = (1.45\times10^{-3}\ \text{N})\hat{k}.$

(b) $\vec{F} = m\vec{a} = q\vec{v}\times\vec{B}$

$\Rightarrow \vec{F} = (-2.46\times10^{-6}\ \text{C})(1.40\ \text{T})\left[(4.19\times10^{4}\ \text{m/s})(\hat{j}\times\hat{k}) + (\text{-}3.85\times10^{4}\ \text{m/s})(\hat{i}\times\hat{k})\right]$

$\Rightarrow \vec{F} = -(1.46\times10^{-3}\ \text{N})\hat{i} - (1.34\times10^{-3}\ \text{N})\hat{j}.$

28-4: Need a force from the magnetic field to balance the downward gravitational force. It's magnitude is:

$$qvB = mg \Rightarrow B = \frac{mg}{qv} = \frac{(3.90\times10^{-4}\ \text{kg})(9.80\ \text{m/s}^2)}{(2.50\times10^{-8}\ \text{C})(4.00\times10^{4}\ \text{m/s})} = 3.82\ \text{T}.$$

The right hand rule requires the magnetic field to be to the south, since the velocity is westward, and the force is upwards.

28-5: $F = |q|vB\sin\phi \Rightarrow v = \dfrac{F}{|q|B\sin\phi} = \dfrac{4.60\times10^{-15}\ \text{N}}{(1.6\times10^{-19}\ \text{C})(3.5\times10^{-3}\ \text{T})\sin 40°} = 1.28\times10^{7}\ \text{m/s}.$

28-6: (a) The smallest possible acceleration is zero, when the motion is parallel to the magnetic field. The greatest acceleration is when the velocity and magnetic field are at right angles:

$$a = \frac{qvB}{m} = \frac{(1.60\times10^{-8}\ \text{C})(2.50\times10^{6}\ \text{m/s})(7.4\times10^{-2}\ \text{T})}{(1.67\times10^{-27}\ \text{kg})} = 1.77\times10^{13}\ \text{m/s}^2.$$

(b) If $a = \dfrac{1}{2}(1.77\times10^{13}\ \text{m/s}^2) = \dfrac{qvB\sin\phi}{m} \Rightarrow \sin\phi = 0.5 \Rightarrow \phi = 30.0°.$

28-7:

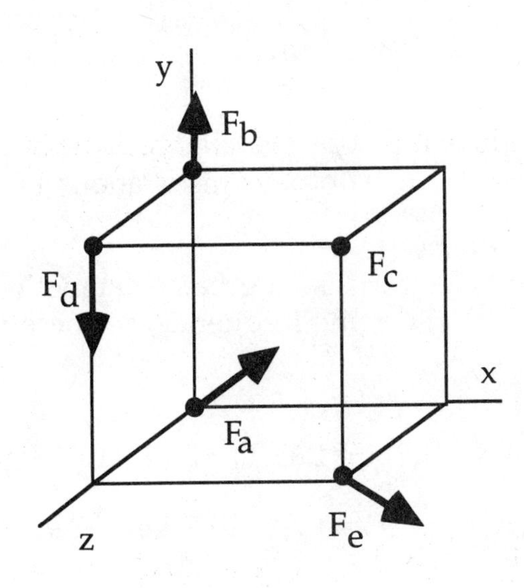

Let $F_0 = qvB$, then:

$F_a = F_0$ in the $-\hat{k}$ direction

$F_b = F_0$ in the $+\hat{j}$ direction

$F_c = 0$, since B and velocity are parallel

$F_d = F_0 \sin 45°$ in the $-\hat{j}$ direction

$F_e = F_0$ in the $-(\hat{j}+\hat{k})$ direction

28-8: (a) $\Phi_B(abcd) = \vec{B}\cdot\vec{A} = -(0.385\text{ T})(0.400\text{ m})(0.300\text{ m}) = -0.0462\text{ Wb}$.

(b) $\Phi_B(befc) = \vec{B}\cdot\vec{A} = 0$.

(c) $\Phi_B(aefd) = \vec{B}\cdot\vec{A} = BA\cos\phi = \frac{4}{5}(0.385\text{ T})(0.500\text{ m})(0.300\text{ m}) = 0.0462\text{ Wb}$.

(d) The net flux through the rest of the surfaces is zero since they are parallel to the x-axis, so the total flux is the sum of all parts above, which is zero.

28-9: (a) $\vec{B} = \left[(5.00\text{ T/m}^2)y^2\right]\hat{j}$ and we can calculate the flux through each surface. Note that there is no flux for the surface at y=0, nor through any surfaces parallel to the y-axis. Thus, the total flux through the closed surface is:

$\Phi_B(abe) = \vec{B}\cdot\vec{A} = (5.00\text{ T/m}^2)(0.300\text{ m})^2\,\frac{1}{2}(0.400\text{ m})(0.300\text{ m}) = -0.0270\text{ Wb}$.

(b) The student's claim is implausible since it would require the existence of a magnetic monopole to result in a net non-zero flux through the closed surface.

28-10: (a) The total flux must be zero, so the flux through the remaining surfaces must be -0.250 Wb.
(b) The shape of the surface is unimportant, just that it is closed.

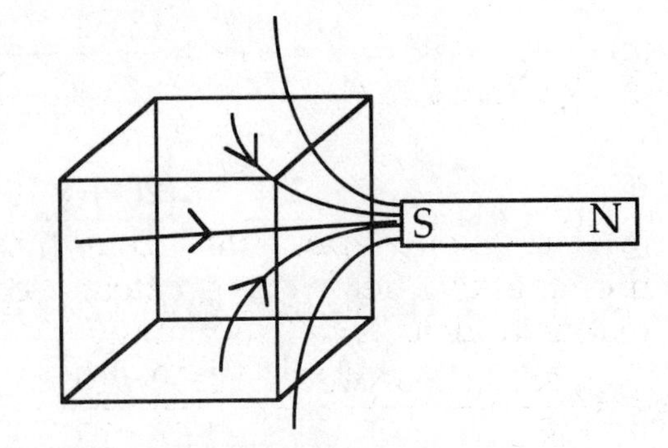

28-11: (a) $\Phi_B = \vec{B}\cdot\vec{A} = (1.16\text{ T})\pi(0.374\text{ m})^2 = 0.510\text{ Wb}$.

(b) $\Phi_B = \vec{B}\cdot\vec{A} = (1.16\text{ T})\pi(0.374\text{ m})^2\cos 73.5° = 0.145\text{ Wb}$.

(c) $\Phi_B = 0$ since $\vec{B}\perp\vec{A}$.

28-12: $R = \dfrac{mv}{qB} = \dfrac{(9.11\times 10^{-31}\text{ kg})(3.7\times 10^{7}\text{ m/s})}{(1.60\times 10^{-19}\text{ C})(0.0877\text{ T})} = 2.40\times 10^{-3}\text{ m}.$

28-13: (a) $B = \dfrac{m2\pi f}{|q|} = \dfrac{(9.11\times 10^{-31}\text{ kg})2\pi(2.00\times 10^{12}\text{ Hz})}{(1.60\times 10^{-19}\text{ C})} = 71.5\text{ T}$. This is about 1.6 times the greatest magnitude yet obtained on earth.
(b) Protons have a greater mass than the electrons, so a greater magnetic field would be required to accelerate them with the same frequency, so there would be no advantage in using them.

28-14: (a) $p = mv = m\left(\dfrac{RqB}{m}\right) = RqB = (0.468\text{ m})(4.8\times 10^{-19}\text{ C})(1.65\text{ T}) = 3.71\times 10^{-19}\text{ kgm/s}.$
(b) $L = Rp = R^2qB = (0.468\text{ m})^2(4.8\times 10^{-19}\text{ C})(1.65\text{ T}) = 1.73\times 10^{-19}\text{ kgm}^2/\text{s}.$

28-15: (a) $B = \dfrac{mv}{|q|R} = \dfrac{(9.11\times 10^{-31}\text{ kg})(2.94\times 10^{6}\text{ m/s})}{(1.60\times 10^{-19}\text{ C})(0.0500\text{ m})} = 3.35\times 10^{-4}\text{ T}.$
The direction of the magnetic field is into the page (the charge is negative).
(b) The time to complete half a circle is just the distance traveled divided by the velocity: $t = \dfrac{D}{v} = \dfrac{\pi R}{v} = \dfrac{\pi(0.0500\text{ m})}{2.94\times 10^{6}\text{ m/s}} = 5.34\times 10^{-8}\text{ s}.$

28-16: (a) $B = \dfrac{mv}{qR} = \dfrac{(1.67\times 10^{-27}\text{ kg})(2.94\times 10^{6}\text{ m/s})}{(1.60\times 10^{-19}\text{ C})(0.0500\text{ m})} = 0.614\text{ T}$
The direction of the magnetic field is out of the page (the charge is positive).
(b) The time to complete half a circle is unchanged: $t = 5.34\times 10^{-8}\text{ s}.$

28-17: (a) $v = \dfrac{qRB}{m} = \dfrac{(1.60\times 10^{-19}\text{ C})(0.0348\text{ m})(1.50\text{ T})}{(3.34\times 10^{-27}\text{ kg})} = 2.50\times 10^{6}\text{ m/s}.$
(b) $t = \dfrac{D}{v} = \dfrac{\pi R}{v} = \dfrac{\pi(0.0348\text{ m})}{2.50\times 10^{6}\text{ m/s}} = 4.37\times 10^{-8}\text{ s}.$
(c) $\dfrac{1}{2}mv^2 = qV \Rightarrow V = \dfrac{mv^2}{2q} = \dfrac{(3.34\times 10^{-27}\text{ kg})(2.50\times 10^{6}\text{ m/s})^2}{2(1.60\times 10^{-19}\text{ C})} = 65{,}300\text{ V}.$

28-18: $\dfrac{1}{2}mv^2 = qV \Rightarrow v = \sqrt{\dfrac{2qV}{m}} = \sqrt{\dfrac{2(1.6\times 10^{-19}\text{ C})(450\text{ V})}{(1.16\times 10^{-26}\text{ kg})}} = 1.11\times 10^{5}\text{ m/s}.$
$\Rightarrow R = \dfrac{mv}{qB} = \dfrac{(1.16\times 10^{-26}\text{ kg})(1.11\times 10^{5}\text{ m/s})}{(1.60\times 10^{-19}\text{ C})(0.723\text{ T})} = 0.0112\text{ m}.$

28-19: $\dfrac{1}{2}mv^2 = |q|\Delta V \Rightarrow v = \sqrt{\dfrac{2|q|\Delta V}{m}} = \sqrt{\dfrac{2(1.6\times 10^{-19}\text{ C})(2.0\times 10^{4}\text{ V})}{(9.11\times 10^{-31}\text{ kg})}} = 8.38\times 10^{7}\text{ m/s}.$
$\Rightarrow B = \dfrac{mv}{|q|R} = \dfrac{(9.11\times 10^{-31}\text{ kg})(8.38\times 10^{7}\text{ m/s})}{(1.60\times 10^{-19}\text{ C})(0.130\text{ m})} = 3.67\times 10^{-3}\text{ T}.$

28-20: The initial velocity is all in the y-direction, and we want the pitch to equal the radius of curvature $\Rightarrow d_x = v_x T = \dfrac{mv_y}{qB} = R$. But $T = \dfrac{2\pi}{\omega} = \dfrac{2\pi m}{qB}$.

$$\Rightarrow \frac{2\pi m v_x}{qB} = \frac{mv_y}{qB} \Rightarrow \frac{v_y}{v_x} = 2\pi = \tan\theta \Rightarrow \theta = 81.0°.$$

28-21: (a) The radius of the path is unaffected, but the pitch of the helix varies with time as the proton is accelerated in the x-direction.

(b) $T = \dfrac{2\pi}{\omega} = \dfrac{2\pi m}{qB} = \dfrac{2\pi(1.67\times10^{-27}\text{ kg})}{(1.60\times10^{-19}\text{ C})(0.500\text{ T})} = 1.31\times10^{-7}\text{ s}$, $t = T/2$, and

$$a_x = \frac{F}{m} = \frac{qE}{m} = \frac{(1.6\times10^{-19}\text{ C})(-3.00\times10^4\text{ V/m})}{1.67\times10^{-27}\text{ kg}} = -2.87\times10^{12}\text{ m/s}.$$

$$d_x = v_{ox}t + \tfrac{1}{2}a_x t^2 = (1.5\times10^5\text{ m/s})(6.56\times10^{-8}\text{ s}) + \frac{(-2.87\times10^{12}\text{ m/s}^2)(6.56\times10^{-8}\text{ s})^2}{2}$$

$$\Rightarrow d_x = 3.67\times10^{-3}\text{ m}.$$

28-22: (a) $v = E/B = (3.94\times10^5\text{ V/m})/(0.0462\text{ T}) = 8.53\times10^6\text{ m/s}$.

(b)

v

B_{out}

E

(c) $R = \dfrac{mv}{|q|B} = \dfrac{(9.11\times10^{-31}\text{ kg})(8.53\times10^6\text{ m/s})}{(1.60\times10^{-19}\text{ C})(0.0462\text{ T})}$

$\Rightarrow R = 1.05\times10^{-3}\text{ m}.$

28-23: $R = \dfrac{mv}{qB} = \dfrac{mE}{qB^2} \Rightarrow m = \dfrac{RqB^2}{E} = \dfrac{(0.899\text{ m})(1.60\times10^{-19}\text{ C})(0.600\text{ T})}{(1.20\times10^6\text{ V/m})} = 4.32\times10^{-26}\text{ kg}$

$$\Rightarrow m(\text{amu}) = \frac{4.32\times10^{-26}\text{ kg}}{1.66\times10^{-27}\text{ kg}} = 26\text{ u}.$$

28-24: (a) $E = vB = (3.48\times10^6\text{ m/s})(1.20\text{ T}) = 4.18\times10^6\text{ V/m}$.
(b) $E = V/d \Rightarrow V = Ed = (4.18\times10^6\text{ V/m})(4.96\times10^{-3}\text{ m}) = 20{,}700\text{ V}$.

28-25: $I = \dfrac{F}{lB} = \dfrac{0.22\text{ N}}{(0.200\text{ m})(0.087\text{ T})} = 13\text{ A}.$

28-26: $F = IlB = (10.8\text{ A})(0.100\text{ m})(1.01\text{ T}) = 1.09\text{ N}$.

28-27: The wire lies on the x-axis, and the force on 1 cm of it is

(a) $\vec{F} = I\vec{l}\times\vec{B} = (7.00\text{ A})(0.010\text{ m})(-0.65\text{ T})(\hat{\imath}\times\hat{\jmath}) = -(0.046\text{ N})\hat{k}$.

(b) $\vec{F} = I\vec{l}\times\vec{B} = (7.00\text{ A})(0.010\text{ m})(+0.56\text{ T})(\hat{\imath}\times\hat{k}) = -(0.039\text{ N})\hat{\jmath}$.

(c) $\vec{F} = I\vec{l}\times\vec{B} = (7.00\text{ A})(0.010\text{ m})(-0.31\text{ T})(\hat{\imath}\times\hat{\imath}) = 0$.

(d) $\vec{F} = I\vec{l} \times \vec{B} = (7.00 \text{ A})(0.010 \text{ m})(-0.28 \text{ T})(\hat{i} \times \hat{k}) = -(0.020 \text{ N})\hat{j}$.

(e) $\vec{F} = I\vec{l} \times \vec{B} = (7.00 \text{ A})(0.010 \text{ m})\left[(0.74 \text{ T})(\hat{i} \times \hat{j}) + (-0.36 \text{ T})(\hat{i} \times \hat{k})\right]$

$\Rightarrow \vec{F} = (0.025 \text{ N})\hat{j} + (0.052 \text{ N})\hat{k}$.

28-28: (a) $F = IlB = (8.00 \text{ A})(0.0100 \text{ m})(6.72 \text{ T}) = 0.538 \text{ N}$, and by the right hand rule, the easterly magnetic field results in a northerly force.
(b) If the field is southerly, then the force is to the east, and of the same magnitude as part (a), $F = 0.538 \text{ N}$.
(c) If the field is 30° south of west, the force is 30° east of south (90° counter-clockwise from the field) and still of the same magnitude, $F = 0.538 \text{ N}$.

28-29: (a) $\tau = NIBA\sin\phi \Rightarrow \tau_{\text{max}} = (12)(2.7 \text{ A})(0.56 \text{ T})\pi(0.065\text{m}/2)^2 \sin 90° = 0.0602 \text{ Nm}$.
(b) The torque on the loop is one-half of the maximum value when $\sin\phi = 0.5 \Rightarrow \phi = 30°$.

28-30: $\tau_{\text{max}} = NIBA\sin 90° = (600)(0.0613 \text{ A})(0.267 \text{ T})(0.050 \text{ m})(0.12 \text{ m}) = 0.0589 \text{ Nm}$.

29-31: $\Delta U = U_f - U_i = -\mu B\cos 180° + \mu B\cos 0° = 2\mu B = 2(1.20 \text{ Am}^2)(0.728 \text{ T}) = 1.75 \text{ J}$.

28-32: (a) $\tau = IBA = (4.1 \text{ A})(0.19 \text{ T})(0.050 \text{ m})(0.080 \text{ m}) = 3.1 \times 10^{-3} \text{ Nm}$.
(b) $\mu = IA = (4.1 \text{ A})(0.050 \text{ m})(0.080 \text{ m}) = 0.016 \text{ Am}^2$.
(c) Maximum torque will occur when the area is largest, which means a circle: $2\pi R = 2(0.050 \text{ m} + 0.080 \text{ m}) \Rightarrow R = 0.041 \text{ m}$.
$\Rightarrow \tau_{\text{max}} = IBA = (4.1 \text{ A})(0.19 \text{ T})\pi(0.041 \text{ m})^2 = 4.2 \times 10^{-3} \text{ Nm}$.

28-33: (a) $\phi = 90°$: $\tau = NIAB\sin(90°) = NIAB$, direction: $\hat{k} \times \hat{j} = -\hat{i}$, $U = -N\mu B\cos\phi = 0$.
(b) $\phi = 0$: $\tau = NIAB\sin(0) = 0$, no direction, $U = -N\mu B\cos\phi = -NIAB$.
(c) $\phi = 90°$: $\tau = NIAB\sin(90°) = NIAB$, direction: $-\hat{k} \times \hat{j} = \hat{i}$, $U = -N\mu B\cos\phi = 0$.
(d) $\phi = 180°$: $\tau = NIAB\sin(180°) = 0$, no direction, $U = -N\mu B\cos(180°) = NIAB$.

28-34: (a) $V_{ab} = \mathcal{E} + Ir \Rightarrow I = \dfrac{V_{ab} - \mathcal{E}}{r} = \dfrac{120 \text{ V} - 105 \text{ V}}{4.3\ \Omega} = 3.5 \text{ A}$.
(b) $P_{supplied} = IV_{ab} = (3.5 \text{ A})(120 \text{ V}) = 420 \text{ W}$.
(c) $P_{mech} = IV_{ab} - I^2 r = 420 \text{ W} - (3.5 \text{ A})^2(4.3\ \Omega) = 370 \text{ W}$.

28-35: (a) $I_f = \dfrac{120 \text{ V}}{128\ \Omega} = 0.94 \text{ A}$.
(b) $I_r = I_{total} - I_f = 4.87A - 0.94A = 3.93A$.
(c) $V = \mathcal{E} + I_r R_r \Rightarrow \mathcal{E} = V - I_r R_r = 120 \text{ V} - (3.93 \text{ A})(5.9\ \Omega) = 97 \text{ V}$.
(d) $P_{mech} = \mathcal{E} I_r = (97 \text{ V})(3.93 \text{ A}) = 380 \text{ W}$.

28-36: (a) Field current $I_f = \dfrac{120 \text{ V}}{243\ \Omega} = 0.494 \text{ A}$.
(b) Rotor current $I_r = I_{total} - I_f = 4.56 \text{ A} - 0.494 \text{ A} = 4.07 \text{ A}$.

(c) $V = \mathcal{E} + I_r R_r \Rightarrow \mathcal{E} = V - I_r R_r = 120\ \text{V} - (4.07\ \text{A})(3.8\ \Omega) = 105\ \text{V}.$

(d) $P_f = I_f^2 R_f = (0.494\ \text{A})^2 (243\ \Omega) = 59\ \text{W}.$

(e) $P_r = I_r^2 R_r = (4.07\ \text{A})^2 (3.8\ \Omega) = 63\ \text{W}.$

(f) *Power input* $= (120\ \text{V})(4.56\ \text{A}) = 547\ \text{W}.$

(g) $\textit{Efficiency} = \dfrac{\text{P}_{output}}{\text{P}_{input}} = \dfrac{((120\ \text{V})(4.56\ \text{A}) - 59\ \text{W} - 63\ \text{W} - 50\ \text{W})}{547\ \text{W}} = \dfrac{375\ \text{W}}{547\ \text{W}} = 0.685.$

28-37: (a) $v_d = \dfrac{J}{n|q|} = \dfrac{I}{An|q|} = \dfrac{150\ \text{A}}{(0.0145\ \text{m})(2.9\times10^{-4}\ \text{m})(5.85\times10^{28}\ \text{m}^{-3})(1.6\times10^{-19}\ \text{C})}$

$\Rightarrow v_d = 3.81\times10^{-3}\ \text{m/s}.$

(b) $E_z = v_d B_y = (3.8\times10^{-3}\ \text{m/s})(1.5\ \text{T}) = 5.7\times10^{-3}\ \text{N/C}$, in the +z-direction (negative charge).

(c) $V_{Hall} = zE_z = (0.0145\ \text{m})(5.7\times10^{-3}\ \text{N/C}) = 8.3\times10^{-5}\ \text{V}.$

28-38: $n = \dfrac{J_x B_y}{|q|E_z} = \dfrac{IB_y}{A|q|E_z} = \dfrac{IB_y z_1}{A|q|\varepsilon_z} = \dfrac{IB_y}{y_1|q|\varepsilon} = \dfrac{(100\ \text{A})(4.30\ \text{T})}{(2.9\times10^{-4}\ \text{m})(1.6\times10^{-19}\ \text{C})(2.10\times10^{-4}\ \text{V})}$

$\Rightarrow n = 4.4\times10^{28}$ electrons per cubic meter.

28-39: To pass undeflected, $E = vB = (3.29\times10^3\ \text{m/s})(0.515\ \text{T}) = 1690\ \text{N/C}.$

(a) If $q = +2.92\times10^{-9}\ \text{C}$, the electric field direction is given by $-\left(\hat{i}\times(-\hat{j})\right) = \hat{k}$, since it must point in the opposite direction to the magnetic force.

(b) If $q = -2.92\times10^{-9}\ \text{C}$, the electric field direction is given by $\left(-\hat{i}\times(-\hat{j})\right) = \hat{k}$, since it must point in the same direction as the magnetic force, which has swapped from part (a).

28-40: The electron's initial velocity comes from the work done by the voltage:

$$\frac{1}{2}mv^2 = |q|V_{ab} \Rightarrow v = \sqrt{\frac{2|q|V_{ab}}{m}} = \sqrt{\frac{2(1.6\times10^{-19}\ \text{C})(7500\ \text{V})}{9.11\times10^{-31}\ \text{kg}}} = 5.13\times10^7\ \text{m/s}.$$

The electron's acceleration perpendicular to the screen can now be calculated:

$$a = \frac{qvB}{m} = \frac{(1.6\times10^{-19}\ \text{C})(5.13\times10^7\ \text{m/s})(5.0\times10^{-5}\ \text{T})}{9.11\times10^{-31}\ \text{kg}} = 4.5\times10^{14}\ \text{m/s}^2.$$

The time of flight $t = d/v = (0.40\ \text{m})/(5.13\times10^7\ \text{m/s}) = 7.8\times10^{-9}\ \text{s}.$

So the deflection is $x = \frac{1}{2}at^2 = \frac{1}{2}(4.5\times10^{14}\ \text{m/s}^2)(7.8\times10^{-9}\ \text{s})^2 = 0.014\ \text{m}.$

28-41: $\vec{F}_2 = q\vec{v}_2\times\vec{B} = (4.97\times10^{-9}\ \text{C})\begin{vmatrix}\hat{i} & \hat{j} & \hat{k}\\ 0 & 0 & 1.62\times10^4\ \text{m/s}\\ B_x & B_y & B_z\end{vmatrix} = (4.00\times10^{-5}\ \text{N})\hat{i}$

$\Rightarrow B_y = \dfrac{(4.00\times10^{-5}\ \text{N})}{-(1.62\times10^4\ \text{m/s})(4.97\times10^{-9}\ \text{C})} = -0.497\ \text{T}$, and B_x must be zero since the particle felt no force in the y-direction.

$$\vec{F}_1 = q\vec{v}_1 \times \vec{B} = q\begin{vmatrix} \hat{i} & \hat{j} & \hat{k} \\ \frac{1}{\sqrt{2}}v_1 & \frac{1}{\sqrt{2}}v_1 & 0 \\ 0 & B_y & B_z \end{vmatrix} = \left(\frac{B_z v_1}{\sqrt{2}}\right)\hat{i} - \left(\frac{B_z v_1}{\sqrt{2}}\right)\hat{j} + \left(\frac{B_y v_1}{\sqrt{2}}\right)\hat{k} = F_1\hat{k} \Rightarrow B_z = 0.$$

$$\rightarrow \vec{B} = (\ 0.497\ \text{T})\hat{j}.$$

28-42: $\vec{F}_1 = (qv_0B_y)\hat{k} + (qv_0B_z)(-\hat{j}) = qv_0(B_y\hat{k} - B_z\hat{j})$

$$\Rightarrow \vec{F} = (3.5\times10^{-8}\ \text{C})(\text{-}5.89\times10^{5}\ \text{m/s})\left[(-0.522\ \text{T})\hat{k} - (0.322\ \text{T})\hat{j})\right]$$

$$\Rightarrow \vec{F} = (6.64\times10^{-3}\ \text{N})\hat{j} + (1.08\times10^{-2}\ \text{N})\hat{k}$$

28-43: $f = \dfrac{\omega}{2\pi} = \dfrac{qB}{2\pi m} \Rightarrow \dfrac{f_e}{f_\alpha} = \dfrac{q_eB/2\pi m_e}{q_\alpha B/2\pi m_\alpha} = \dfrac{em_\alpha}{2em_e} = \dfrac{1}{2}\dfrac{6.65\times10^{-27}\ \text{kg}}{9.11\times10^{-31}\ \text{kg}} = 3650.$

28-44: (a) $K = 1.6\ \text{MeV} = (1.6\times10^{6}\ \text{eV})(1.6\times10^{-19}\ \text{J/eV}) = 2.56\times10^{-13}\ \text{J}.$

$$\Rightarrow v = \sqrt{\frac{2K}{m}} = \sqrt{\frac{2(2.56\times10^{-13}\ \text{J})}{1.67\times10^{-27}\ \text{kg}}} = 1.75\times10^{7}\ \text{m/s}.$$

$$\Rightarrow R = \frac{mv}{qB} = \frac{(1.67\times10^{-27}\ \text{kg})(1.75\times10^{7}\ \text{m/s})}{(1.6\times10^{-19}\ \text{C})(3.5\ \text{T})} = 0.0523\ \text{m}.$$

Also, $\omega = \dfrac{v}{R} = \dfrac{1.75\times10^{7}\ \text{m/s}}{0.0523\ \text{m}} = 3.35\times10^{8}\ \text{rad/s}.$

(b) If the energy reaches the final value of 3.2 MeV, The velocity increases by $\sqrt{2}$, as does the radius, to 0.0740 m. The angular frequency is unchanged from part (a) at $3.35\times10^{8}\ \text{rad/s}$.

28-45: (a) $v_{\text{max}} = \dfrac{qBR}{m} = \dfrac{(1.6\times10^{-19}\ \text{C})(1.5\ \text{T})(0.50\ \text{m})}{1.67\times10^{-27}\ \text{kg}} = 7.2\times10^{7}\ \text{m/s}.$

$$\Rightarrow E_{\text{max}} = \frac{1}{2}mv_{\text{max}}^{2} = \frac{(1.67\times10^{-27}\ \text{kg})(7.2\times10^{7}\ \text{m/s})^2}{2} = 4.3\times10^{-12}\ \text{J} = 27\ \text{MeV}.$$

(b) $T = \dfrac{2\pi R}{v} = \dfrac{2\pi(0.50\ \text{m})}{7.2\times10^{7}\ \text{m/s}} = 4.4\times10^{-8}\ \text{s}.$

(c) If the energy was to be doubled, then the speed would have to be increased by $\sqrt{2}$, as would the magnetic field. Therefore the new magnetic field would be $B_{new} = \sqrt{2}B_o = 2.1\ \text{T}$.

(d) For alpha particles, $E_{\text{max}}(\alpha) = E_{\text{max}}(p)\dfrac{m_p}{m_\alpha}\dfrac{q_\alpha^{\ 2}}{q_p^{\ 2}} = E_{\text{max}}(p)\dfrac{m_p}{(4m_p)}\dfrac{(2q_p)^2}{q_p^{\ 2}} = E_{\text{max}}(p).$

28-46: (a) $F = \sqrt{F_x^{\ 2} + F_y^{\ 2}} = (1.53 \times 10^{-4}\ \text{N})\sqrt{3^2 + 4^2} = 7.65 \times 10^{-4}\ \text{N}.$

(b) $\vec{F} = q\vec{v} \times \vec{B} = q\begin{vmatrix} \hat{i} & \hat{j} & \hat{k} \\ 0 & v & 0 \\ B_x & B_y & B_z \end{vmatrix}$

$\Rightarrow (1.53 \times 10^{-4}\text{N})\left[3\hat{i} - 4\hat{k}\right] = (1.45 \times 10^{-6}\ \text{C})(1.98 \times 10^3\ \text{m/s})\left[B_z\hat{i} - B_x\hat{k}\right]$

$\Rightarrow B_x = 0.213\ \text{T},\ \ B_z = 0.156\ \text{T}.$

(c)

$B = \sqrt{B_x^{\ 2} + B_y^{\ 2} + B_z^{\ 2}} \Rightarrow B_y = \pm\sqrt{B^2 - B_x^{\ 2} - B_z^{\ 2}} = \pm\sqrt{(0.50\ \text{T})^2 - (0.21\ \text{T})^2 - (0.16\ \text{T})^2}$

$\Rightarrow B_y = \pm 0.42\ \text{T}.$

28-47: $v = \dfrac{E}{B} = \dfrac{1.88 \times 10^4\ \text{N/C}}{0.701\ \text{T}} = 2.68 \times 10^4\ \text{m/s},$ and $R = \dfrac{mv}{qB}$, so:

$$R_{24} = \frac{24(1.66 \times 10^{-27}\ \text{kg})(2.68 \times 10^4\ \text{m/s})}{(1.60 \times 10^{-19}\ \text{C})(0.701\ \text{T})} = 9.53 \times 10^{-3}\ \text{m}.$$

$$R_{25} = \frac{25(1.66 \times 10^{-27}\ \text{kg})(2.68 \times 10^4\ \text{m/s})}{(1.60 \times 10^{-19}\ \text{C})(0.701\ \text{T})} = 9.92 \times 10^{-3}\ \text{m}.$$

$$R_{26} = \frac{26(1.66 \times 10^{-27}\ \text{kg})(2.68 \times 10^4\ \text{m/s})}{(1.60 \times 10^{-19}\ \text{C})(0.701\ \text{T})} = 10.32 \times 10^{-3}\ \text{m}.$$

So the total spread of the particles is 0.79 mm.

28-48: $F_x = q(v_yB_z - v_zB_y) = 0.$

$F_y = q(v_zB_x - v_xB_z) = (7.82 \times 10^{-8}\ \text{C})(\text{-}4.95 \times 10^4\ \text{m/s})(0.300\ \text{T}) = -1.16 \times 10^{-3}\ \text{N}.$

$F_z = q(v_xB_y - v_yB_x) = -(7.82 \times 10^{-8}\ \text{C})(8.82 \times 10^4\ \text{m/s})(0.300\ \text{T}) = -2.07 \times 10^{-3}\ \text{N}.$

28-49: (a) $\vec{F} = q\vec{v} \times \vec{B} = q\left[(v_yB_z)\hat{i} - (v_xB_z)\hat{j}\right] \Rightarrow F^2 = q^2\left[(v_yB_z)^2 + (v_xB_z)^2\right]$

$$\Rightarrow q^2 = \frac{F^2}{B_z^{\ 2}}\frac{1}{(v_y)^2 + (v_x)^2}$$

$$\Rightarrow q = \frac{1.75\ \text{N}}{0.220\ \text{T}}\sqrt{\frac{1}{\left[-3(1.22 \times 10^6\ \text{m/s})\right]^2 + \left[4(1.22 \times 10^6\ \text{m/s})\right]^2}} = 1.30 \times 10^{-6}\ \text{C}.$$

(b) $\vec{a} = \dfrac{\vec{F}}{m} = \dfrac{q\vec{v} \times \vec{B}}{m} = \dfrac{q}{m}\left[(v_yB_z)\hat{i} - (v_xB_z)\hat{j}\right]$

$$\Rightarrow \vec{a} = \frac{1.30 \times 10^{-6}\ \text{C}}{1.53 \times 10^{-15}\ \text{kg}}(1.22 \times 10^6\ \text{m/s})(-0.220\ \text{T})\left[-3\hat{i} - 4\hat{j}\right]$$

$\Rightarrow \vec{a} = (2.28 \times 10^{14}\ \text{m/s}^2)\left[3\hat{i} + 4\hat{j}\right].$

(c) The motion is helical since the force is in the xy-plane but the velocity has a z-component. The radius of the circular part of the motion is:

$$R = \frac{mv}{qB} = \frac{(1.53 \times 10^{-15}\ \text{kg})(5)(1.22 \times 10^6\ \text{m/s})}{(1.30 \times 10^{-6}\ \text{C})(0.220\ \text{T})} = 0.0326\ \text{m}.$$

(d) $f = \frac{\omega}{2\pi} = \frac{qB}{2\pi m} = \frac{(1.30\times10^{-6}\text{ C})(0.220\text{ T})}{2\pi(1.53\times10^{-15}\text{ kg})} = 29.8\text{ MHz}.$

(e) After two complete cycles, the x and y values are back to their original values, x=R and y=0, but z has changed.

$$z = 2Tv_z = \frac{2v_z}{f} = \frac{2(-12)(1.22\times10^6\text{ m/s})}{2.98\times10^7\text{ Hz}} = -0.983\text{ m}$$

28-50: (a) $\frac{mv^2}{R} = qE \Rightarrow v = \sqrt{\frac{qER}{m}} = \sqrt{\frac{qV_{ab}}{m\ln(b/a)}} = \sqrt{\frac{(1.6\times10^{-19}\text{ C})(120\text{ V})}{(9.11\times10^{-31}\text{ kg})\ln(5.00/1.00)}}$

$\Rightarrow v = 2.32\times10^6\text{ m/s}.$

(b) $\frac{mv^2}{R} = q(E+vB) \Rightarrow \left(\frac{m}{R}\right)v^2 - (qB)v - qE = 0$

$\Rightarrow (2.28\times10^{-29})v^2 - (2.08\times10^{-23})v - (1.23\times10^{-16}) = 0$

$\Rightarrow v = 2.82\times10^6\text{ m/s}$ or $-1.91\times10^6\text{ m/s}$, but we need the positive velocity to get the correct force, so $v = 2.82\times10^6\text{ m/s}.$

(c) If the direction of the magnetic field is reversed, then there is a smaller net force and a smaller velocity, and the value is the second root found in part (b), $\Rightarrow v = 1.91\times10^6\text{ m/s}.$

28-51: (a) $\frac{1}{2}mv_x{}^2 = qV \Rightarrow v_x = \sqrt{\frac{2qV}{m}}$. Also $a = \frac{qv_xB}{m}$, and $t = \frac{x}{v_x}$.

$$\Rightarrow y = \frac{1}{2}at^2 = \frac{1}{2}a\left(\frac{x}{v_x}\right)^2 = \frac{1}{2}\left(\frac{qv_xB}{m}\right)\left(\frac{x}{v_x}\right)^2 = \frac{1}{2}\left(\frac{qBx^2}{m}\right)\left(\frac{m}{2qV}\right)^{1/2} \Rightarrow y = Bx^2\left(\frac{q}{8mV}\right)^{1/2}.$$

(b) This can be used for isotope separation since the mass in the denominator leads to different locations for different isotopes.

28-52: (a) $\vec{F} = I\vec{l}\times\vec{B} = I(l\hat{j})\times\vec{B} = Il\left[B_z\hat{i} - (0)\hat{j} + (-B_x)\hat{k}\right]$

$\Rightarrow F_x = IlB_z = (8.00\text{ A})(0.150\text{ m})(0.538\text{ T}) = 0.646\text{ N}.$

$\Rightarrow F_y = 0$, since the wire is in the y-direction.

$\Rightarrow F_z = -IlB_x = (8.00\text{ A})(0.150\text{ m})(0.107\text{ T}) = -0.128\text{ N}.$

(b) $F = \sqrt{F_x{}^2 + F_z{}^2} = \sqrt{(0.646\text{ N})^2 + (0.128\text{ N})^2} = 0.659\text{ N}.$

28-53: (a) l_{ab}: $\vec{F} = I\vec{l}_{ab}\times\vec{B} = I(l_{ab}B)\hat{j}\times\hat{i} = -(4.92\text{ A})(0.500\text{ m})(0.435\text{ T})\hat{k} = (-1.07\text{ N})\hat{k}.$

l_{bc}: $\vec{F} = I\vec{l}_{bc}\times\vec{B} = I(l_{bc}B)\left[\frac{(\hat{i}-\hat{k})}{\sqrt{2}}\times\hat{i}\right] = -(4.92\text{ A})(0.500\text{ m})(0.435\text{ T})\hat{j} = (-1.07\text{ N})\hat{j}.$

l_{cd}: $\vec{F} = I\vec{l}_{cd}\times\vec{B} = I(l_{cd}B)\left[\frac{(\hat{k}-\hat{j})}{\sqrt{2}}\times\hat{i}\right] = -(4.92\text{ A})(0.500\text{ m})(0.435\text{ T})\left[\hat{j}+\hat{k}\right]$

$\Rightarrow \vec{F} = (1.07\text{ N})\left[\hat{j}+\hat{k}\right].$

l_{de}: $\vec{F} = I\vec{l}_{de} \times \vec{B} = Il_{de}B\left[-\hat{k} \times \hat{i}\right] = -(4.92\ \text{A})(0.500\ \text{m})(0.435\ \text{T})\hat{j} = (-1.07\ \text{N})\hat{j}$

l_{ef} : $\vec{F} = I\vec{l}_{ef} \times \vec{B} = I(l_{ef}B)(-\hat{i}) \times \hat{i} = 0$.

(b) Summing all the forces in part (a) we have $\vec{F}_{total} = (-1.07\ \text{N})\hat{j}$.

28-54: (a) $F = ILB$, to the right.

(b) We want $v = 1.12 \times 10^4\ \text{m/s} \Rightarrow v^2 = 2ad \Rightarrow d = \dfrac{v^2}{2a} = \dfrac{v^2 m}{2ILB}$.

$$\Rightarrow d = \frac{(1.12 \times 10^4\ \text{m/s})^2 (50\ \text{kg})}{2(1000\ \text{A})(1.0\ \text{m})(0.10\ \text{T})} = 3.14 \times 10^7\ \text{m} = 31{,}400\ \text{km!}$$

28-55: The current is to the left, so the force is into the plane.

$\sum F_y = N\cos\theta - mg = 0$ and $\sum F_x = N\sin\theta - F_B = 0$.

$$\Rightarrow F_B = mg\tan\theta = ILB \Rightarrow I = \frac{mg\tan\theta}{LB}.$$

28-56: (a) By examining a small piece of the wire (shown below) we find:

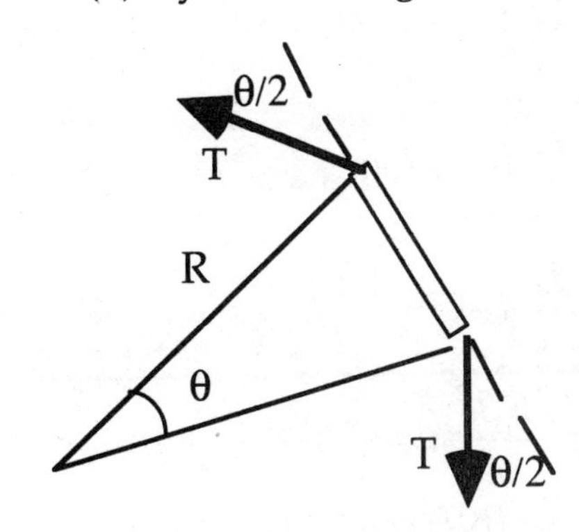

$$F_B = ILB = 2T\sin(\theta/2)$$

$$\Rightarrow ILB \approx \frac{2T\theta}{2} = \frac{2TL/R}{2} \Rightarrow R = \frac{T}{IB}.$$

(b) For a particle:

$$qvB = \frac{mv^2}{R} \Rightarrow B = \frac{mv}{Rq} = \frac{mvIB}{Tq} \Rightarrow v = \frac{Tq}{mI}.$$

28-57: Summing the torques on the wire from gravity and the magnetic field will enable us to find the magnetic field value.

$\tau_B = IAB\sin 60° = B(6.8\ \text{A})(0.060\ \text{m})(0.080\ \text{m})\sin 60° = (0.0283\ \text{N}\cdot\text{m/T})B$.

There are three sides to consider for the gravitational torque, leading to:

$\tau_g = m_6 g l_6 \sin\phi + 2m_8 g l_8 \sin\phi$

$\Rightarrow \tau_g = (9.8\ \text{m/s}^2)\sin 30°\left[(1.14 \times 10^{-3}\ \text{kg})(0.080\ \text{m}) + 2(1.52 \times 10^{-3}\ \text{kg})(0.040\ \text{m})\right]$

$$\Rightarrow \tau_g = 1.04 \times 10^{-3}\ \text{N}\cdot\text{m} \Rightarrow B = \frac{1.04 \times 10^{-3}\ \text{N}\cdot\text{m}}{0.0283\ \text{N}\cdot\text{m/T}} = 0.037\ \text{T}.$$

28-58: (a) $\tau = IAB\sin 60° = (15.0\ \text{A})(0.060\ \text{m})(0.080\ \text{m})(0.25\ \text{T})\sin 60° = 0.016\ \text{N}\cdot\text{m}$.

(b) $\tau = IAB\sin 30° = (15.0\ \text{A})(0.060\ \text{m})(0.080\ \text{m})(0.25\ \text{T})\sin 30° = 9.0 \times 10^{-3}\ \text{N}\cdot\text{m}$.

(c) If the loop was pivoted through its center, then there would be a torque on both sides of the loop parallel to the rotation axis. However the lever arm is only half as large, so the total torque in each case is identical to the values found in parts (a) and (b).

28-59:

(a)

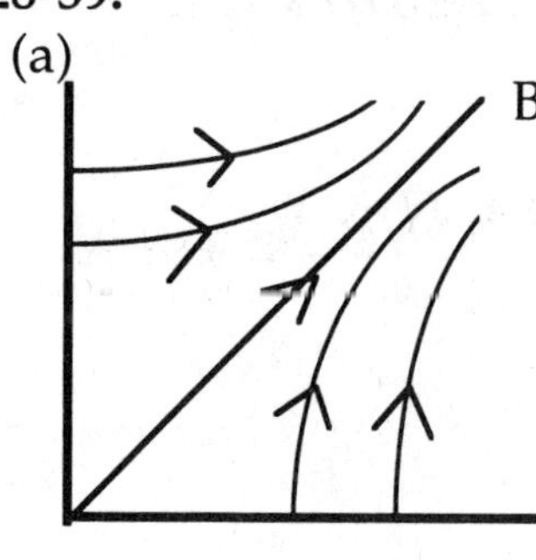

(b) Side 1: $\vec{F}=\int_0^L Id\vec{l}\times\vec{B}=I\int_0^L \frac{B_o y\,dy}{L}\hat{k}=\frac{1}{2}B_oLI\hat{i}.$

Side 2: $\vec{F}=\int_{0,y=L}^{L} Id\vec{l}\times\vec{B}=I\int_{0,y=L}^{L}\frac{B_o y\,dx}{L}\hat{j}=-IB_oL\hat{j}.$

Side 3: $\vec{F}=\int_{L,x=L}^{0} Id\vec{l}\times\vec{B}=I\int_{L,x=L}^{0}\frac{B_o y\,dy}{L}(-\hat{i})=-\frac{1}{2}IB_oL\hat{i}.$

Side 4: $\vec{F}=\int_{L,y=0}^{0} Id\vec{l}\times\vec{B}=I\int_{L,y=0}^{0}\frac{B_o y\,dx}{L}\hat{j}=0.$

(c) The sum of all forces is $\vec{F}_{total}=-IB_oL\hat{j}.$

28-60:

(a)

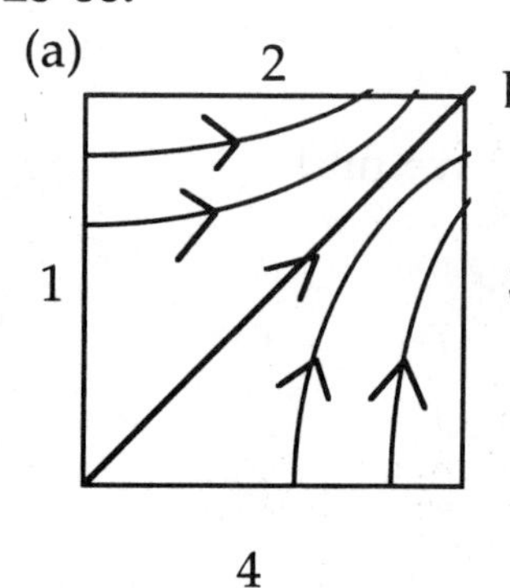

(b) Side 1: $\vec{F}=\int_0^L Id\vec{l}\times\vec{B}=I\int_0^L\frac{B_o y\,dy}{L}(-\hat{k})=-\frac{1}{2}B_oLI\hat{k}.$

Side 2: $\vec{F}=\int_0^L Id\vec{l}\times\vec{B}=I\int_0^L\frac{B_o x\,dx}{L}\hat{k}=\frac{1}{2}IB_oL\hat{k}.$

Side 3: $\vec{F}=\int_0^L Id\vec{l}\times\vec{B}=I\int_0^L\frac{B_o y\,dy}{L}\hat{k}=+\frac{1}{2}IB_oL\hat{k}.$

Side 4: $\vec{F}=\int_0^L Id\vec{l}\times\vec{B}=I\int_0^L\frac{B_o x\,dx}{L}(-\hat{k})=-\frac{1}{2}IB_oL\hat{k}.$

(c) If free to rotate about the x-axis $\Rightarrow \vec{\tau}=\vec{L}\times\vec{F}=\frac{IB_oL^2}{2}\hat{i}=\frac{1}{2}IAB_o\hat{i}.$

(d) If free to rotate about the y-axis $\Rightarrow \vec{\tau}=\vec{L}\times\vec{F}=-\frac{IB_oL^2}{2}\hat{j}=-\frac{1}{2}IAB_o\hat{j}.$

(e) The form of the torque $\vec{\tau}=\vec{\mu}\times\vec{B}$ is not appropriate, since the magnetic field is not constant.

28-61: The y-components of the magnetic field provide forces which cancel as you go around the loop. The x-components of the magnetic field, however, provide a net force in the -y-direction.

$$F=\int NIB\,dl\sin 60^\circ=NIB\sin 60^\circ\int_0^{2\pi R}dl=2\pi RNIB\sin 60^\circ$$

$$\Rightarrow F=2\pi(0.0065\text{ m})(40)(0.810\text{ A})(0.170\text{ T})\sin 60^\circ=0.195\text{ N}.$$

28-62: $\sum\vec{\tau}_i=\sum\vec{r}_i\times\vec{F}_i=\sum\vec{r}_i\times\vec{F}_i-\vec{r}_p\times\sum\vec{F}_i=\sum(\vec{r}_i-\vec{r}_p)\times\vec{F}_i=\sum\vec{\tau}_i(p).$

Note that we added a term after the second equals sign that was zero because the body is in translational equilibrium.

28-63: (a) $\Delta y=0.700\text{ m}-0.050\text{ m}=0.650\text{ m},$ and $v^2=0=v_o^{\,2}-2g\Delta y\Rightarrow v_o=\sqrt{2g\Delta y}.$

$\Rightarrow v_o=\sqrt{2(9.8\text{ m/s}^2)(0.650\text{ m})}=3.57\text{ m/s}.$

(b) In a distance of 0.05 m the wire's speed increases from zero to 3.57 m/s.

$$\Rightarrow a = \frac{v^2}{2\Delta y} = \frac{(3.57\ \text{m/s})^2}{2(0.050\ \text{m})} = 127\ \text{m/s}^2.$$

But $F = ILB = ma_B \Rightarrow I = \frac{m(g+a)}{LB} = \frac{(9.09\times10^{-5}\ \text{kg})((127+9.8)\ \text{m/s}^2)}{(0.25\ \text{m})(0.0127\ \text{T})} = 3.93\ \text{A}.$

(c) $V = IR \Rightarrow R = \frac{V}{I} = \frac{1.50\ \text{V}}{3.93\ \text{A}} = 0.382\ \Omega.$

28-64: (a) $I_u = \frac{dq}{dt} = \frac{\Delta q}{\Delta t} = \frac{q_u v}{2\pi r} \Rightarrow I_u = \frac{ev}{3\pi r}.$

(b) $\mu_u = I_u A = \frac{ev}{3\pi r}\pi r^2 = \frac{evr}{3}.$

(c) Since there are two down quarks, each of half the charge of the up quark,

$\mu_d = \mu_u = \frac{evr}{3} \Rightarrow \mu_{total} = \frac{2evr}{3}.$

(d) $v = \frac{3\mu}{2er} = \frac{3(9.66\times10^{-27}\ \text{A}\cdot\text{m}^2)}{2(1.60\times10^{-19}\ \text{C})(1.20\times10^{-15}\ \text{m})} = 7.55\times10^7\ \text{m/s}.$

28-65: (a) $\vec{\mu} = \vec{I}\times\vec{A} = (15.0\ \text{A})(5.28\times10^{-4}\ \text{m}^2)\,\hat{k} = (7.92\times10^{-3}\ \text{A}\cdot\text{m}^2)\,\hat{k}.$

(b) $\vec{\tau} = (1.00\times10^{-3}\ \text{N}\cdot\text{m})\left[-6\hat{i}+8\hat{j}\right] = \vec{\mu}\times\vec{B} = \mu\left[(-B_y)\hat{i} - (B_x)\hat{j}\right]$

$\Rightarrow B_y = \frac{6(1.00\times10^{-3}\ \text{N}\cdot\text{m})}{7.92\times10^{-3}\ \text{A}\cdot\text{m}^2} = 0.76\ \text{T}$ and $B_x = \frac{8(1.00\times10^{-3}\ \text{N}\cdot\text{m})}{7.92\times10^{-3}\ \text{A}\cdot\text{m}^2} = 1.01\ \text{T}.$

Also $B_z = \sqrt{B^2 - B_x{}^2 - B_y{}^2} = \sqrt{(2.6\ \text{T})^2 - (1.01\ \text{T})^2 - (0.76\ \text{T})^2} = \pm2.27\ \text{T}.$

But we also know that $U = -\mu B_z < 0 \Rightarrow B_z = +2.27\ \text{T}.$

28-66: (a) $d\vec{l} = dl d\hat{l} = Rd\theta\left[-\sin\theta\,\hat{i} + \cos\theta\,\hat{j}\right]$. Note that this implies that when θ=0, the line element points in the +y-direction, and when the angle is 90°, the line element points in the -x-direction. This is in agreement with the diagram.

$d\vec{F} = Id\vec{l}\times\vec{B} = IRd\theta\left[-\sin\theta\,\hat{i} + \cos\theta\,\hat{j}\right]\times(B_x\hat{i}) \Rightarrow d\vec{F} = IB_xRd\theta\left[-\cos\theta\,\hat{k}\right].$

(b) $\vec{F} = \int_0^{2\pi} -\cos\theta IB_xRd\theta\hat{k} = -IB_xR\int_0^{2\pi}\cos\theta d\theta\hat{k} = 0.$

(c) $d\vec{\tau} = \vec{r}\times d\vec{F} = R(\cos\theta\,\hat{i} + \sin\theta\,\hat{j})\times\left(IB_xRd\theta\left[-\cos\theta\,\hat{k}\right]\right)$

$\Rightarrow d\vec{\tau} = -R^2IB_x\,d\theta(\sin\theta\cos\theta\,\hat{i} - \cos^2\theta\,\hat{j}).$

(d) $\vec{\tau} = \int d\vec{\tau} = -R^2IB_x(\int_0^{2\pi}\sin\theta\cos\theta\,d\theta\,\hat{i} - \int_0^{2\pi}\cos^2\theta\,d\theta\,\hat{j}) = IR^2B_x\left(\frac{\theta}{2} + \frac{\sin2\theta}{4}\right)_0^{2\pi}\hat{j}$

$\Rightarrow \vec{\tau} = IR^2B_x\pi\hat{j} = I\pi R^2B_x\hat{j} = IA\hat{k}\times B_x\hat{i} \Rightarrow \vec{\tau} = \vec{\mu}\times\vec{B}.$

28-67: (a) $\oint\vec{B}\cdot d\vec{A} = \int_{top} B_z\,dA + \int_{barrel} B_r\,dA = \int_{top}(\beta L)dA + \int_{barrel} B_r\,dA = 0.$

$\Rightarrow 0 = \beta L\pi r^2 + B_r 2\pi rL \Rightarrow B_r(r) = -\frac{\beta r}{2}.$

(b) The two diagrams show views of the field lines from the top and side:

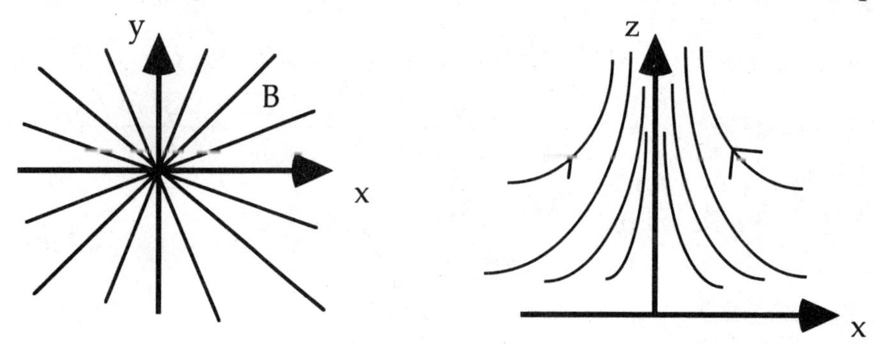

28-68: (a) $\Delta U = -(\vec{\mu}_f \cdot \vec{B} - \vec{\mu}_i \cdot \vec{B}) = -(\vec{\mu}_f - \vec{\mu}_i)\cdot \vec{B} = \left[-\mu(-\hat{k} - (-0.6\hat{i} + 0.8\hat{j}))\right]\cdot\left[B_o(3\hat{i} - 4\hat{j} - 12\hat{k})\right]$

$\Rightarrow \Delta U = IAB_o[(-0.6)(3) + (0.8)(-4) + (+1)(-12)]$

$\Rightarrow \Delta U = (50\text{ A})(6.46\times10^{-4}\text{ m}^2)(0.0133\text{ T})(\text{-}17) = -7.30\times10^{-3}\text{ J}.$

(b) $\Delta K = \frac{1}{2}I\omega^2 \Rightarrow \omega = \sqrt{\frac{2\Delta K}{I}} = \sqrt{\frac{2(7.30\times10^{-3}\text{ J})}{1.7\times10^{-6}\text{ kg}\cdot\text{m}^2}} = 92.7\text{ rad/s}.$

28-69: (a) $R = \frac{mv}{qB} = \frac{(1.51\times10^{-11}\text{ kg})(3.19\times10^{5}\text{ m/s})}{(4.64\times10^{-6}\text{ C})(0.500\text{ T})} = 2.08\text{ m}.$

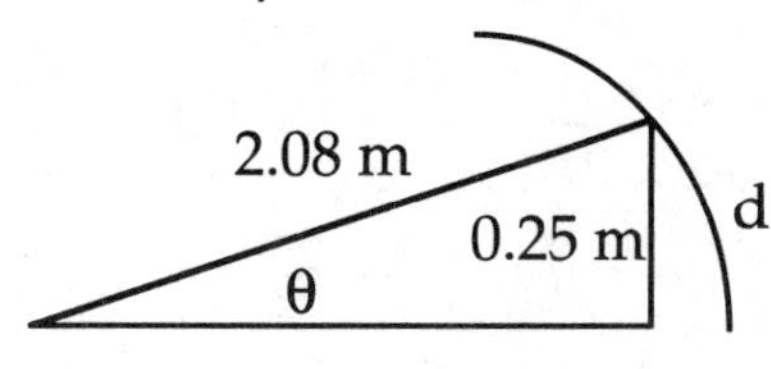

(b) The distance along the curve, d, is given by $d = R\theta = (2.08\text{ m})\sin^{-1}(0.25/2.08) = 0.251\text{ m}.$

And $t = \frac{d}{v} = \frac{0.251\text{ m}}{3.19\times10^5\text{ m/s}} = 7.86\times10^{-7}\text{ s}.$

(c) $\Delta x_1 = d\tan(\theta/2) = (0.25\text{ m})\tan(0.121/2) = 0.0151\text{ m}.$

(d) $\Delta x = \Delta x_1 + \Delta x_2 = 0.0151\text{ m} + (0.50\text{ m})\tan(6.92°) = 0.0758\text{ m}.$

28-70: (a) $\Delta p = FA = IlBA = JlB.$

(b) $J = \frac{\Delta p}{lB} = \frac{(1.00\text{ atm})(1.013\times10^5\text{ Pa/atm})}{(0.0200\text{ m})(1.86\text{ T})} = 2.72\times10^6\text{ A}\cdot\text{m}^2.$

28-71: (a) The maximum speed occurs at the top of the cycloidal path, and hence the radius of curvature is greatest there. Once the motion is beyond the top, the particle is being slowed by the electric field. As it returns to y = 0, the speed decreases, leading to a smaller magnetic force, until the particle stops completely. Then the electric field again provides the acceleration in the y-direction of the particle, leading to the repeated motion.

(b) $W = Fd = qEd = qEy = \frac{1}{2}mv^2 \Rightarrow v = \sqrt{\frac{2qEy}{m}}.$

(c) At the top, $F_y = qE - qvB = -\frac{mv^2}{R} = -\frac{m}{2y}\frac{2qEy}{m} = -qE \Rightarrow 2qE = qvB \Rightarrow v = \frac{2E}{B}.$

Chapter 29: Sources of Magnetic Field

29-1: For a charge with velocity $\vec{v} = (4.00 \times 10^6 \text{ m/s})\hat{i}$, the magnetic field produced at a position r away from the particle is $\vec{B} = \frac{\mu_o}{4\pi}\frac{q\vec{v}\times\hat{r}}{r^2}$. So for the cases below:

(a) $\vec{r} = (0.500 \text{ m})\hat{i} \Rightarrow \hat{v}\times\hat{r} = 0 \Rightarrow \vec{B} = 0.$

(b) $\vec{r} = (-0.500 \text{ m})\hat{j} \Rightarrow \hat{v}\times\hat{r} = -\hat{k}$

$$\Rightarrow \vec{B} = -\frac{\mu_o}{4\pi}\frac{qv}{r^2}\hat{k} = -\frac{\mu_o}{4\pi}\frac{(3.0\times10^{-6}\text{ C})(4.0\times10^6\text{ m/s})}{(0.50\text{ m})^2}\hat{k} = -(4.80\times10^{-6}\text{ T})\hat{k} \equiv -B_o\hat{k}.$$

(c) $\vec{r} = (0.500 \text{ m})\hat{k} \Rightarrow \hat{v}\times\hat{r} = -\hat{j}$

$$\Rightarrow \vec{B} = -\frac{\mu_o}{4\pi}\frac{qv}{r^2}\hat{j} = -B_o\hat{j} - (4.80\times10^{-6}\text{ T})\hat{j}.$$

(d) $\vec{r} = -(0.500 \text{ m})\hat{j} + (0.500 \text{ m})\hat{k} \Rightarrow \hat{v}\times\hat{r} = -\frac{1}{\sqrt{2}}(\hat{j}+\hat{k})$

$$\Rightarrow \vec{B} = -\frac{\mu_o}{4\pi}\frac{qv}{r^2}\frac{1}{\sqrt{2}}(\hat{j}+\hat{k}) = -B_o\frac{1}{2\sqrt{2}}(\hat{j}+\hat{k}) = -(1.70\times10^{-6}\text{ T})(\hat{j}+\hat{k}).$$

29-2: (a) From the right hand rule, the force on q from q' is up, since $B_{q'}$ is into the page.
(b) From the right hand rule, the force on q' from q is down, since B_q is into the page.

(c) $F_B = qvB = qv\frac{\mu_o}{4\pi}\frac{qv}{r^2} = \frac{\mu_o}{4\pi}\frac{q^2v^2}{r^2}$ and $F_E = \frac{1}{4\pi\varepsilon_o}\frac{q^2}{r^2}$

$$\Rightarrow \frac{F_B}{F_E} = \varepsilon_o\mu_o v^2 = \frac{v^2}{c^2} = \frac{(3.0\times10^6\text{ m/s})^2}{(3.0\times10^8\text{ m/s})^2} = 1.0\times10^{-4}.$$

29-3: The magnetic field is into the page at the origin, and the magnitude is

$$B = B_+ + B_- = \frac{\mu_o}{4\pi}\left(\frac{q_+v_+}{r_+^2} + \frac{q_-v_-}{r_-^2}\right)$$

$$\Rightarrow B = \frac{\mu_o}{4\pi}\left(\frac{(5.0\times10^{-6}\text{ C})(6.0\times10^5\text{ m/s})}{(0.300\text{ m})^2} + \frac{(3.0\times10^{-6}\text{ C})(8.0\times10^5\text{ m/s})}{(0.400\text{ m})^2}\right)$$

$$\Rightarrow B = 4.83\times10^{-6}\text{ T} \Rightarrow \vec{B} = -(4.83\times10^{-6}\text{ T})\hat{k}$$

29-4: $B_{total} = B + B' = \frac{\mu_o}{4\pi}\left(\frac{qv}{d^2} + \frac{q'v'}{d^2}\right)$

$$\Rightarrow B = \frac{\mu_o}{4\pi}\left(\frac{(4.0\times10^{-6}\text{ C})(7.5\times10^5\text{ m/s})}{(0.150\text{ m})^2} + \frac{(6.0\times10^{-6}\text{ C})(2.5\times10^5\text{ m/s})}{(0.150\text{ m})^2}\right)$$

$\Rightarrow B = 2.00\times10^{-5}$ T, into the page.

29-5: The wire carries current in the x-direction. The magnetic field of a small piece of wire $d\vec{B} = \frac{\mu_o}{4\pi}\frac{I\,d\vec{l} \times \hat{r}}{r^2}$ at different locations is therefore:

(a) $\vec{r} = (3.00\text{ m})\hat{i} \Rightarrow \hat{l} \times \hat{r} = 0 \Rightarrow \vec{B} = 0.$

(b) $\vec{r} = (3.00\text{ m})\hat{j} \rightarrow \hat{l} \times \hat{r} = \hat{k}$

$$\Rightarrow d\vec{B} = \frac{\mu_o}{4\pi}\frac{I\,dl\sin\theta}{r^2}\hat{k} = \frac{\mu_o}{4\pi}\frac{(8.00\text{ A})(2.0\times10^{-3}\text{ m})\sin(90°)}{(3.00\text{ m})^2}\hat{k} = (1.78\times10^{-10}\text{ T})\hat{k}.$$

(c) $\vec{r} = (3.00\text{ m})\hat{i} + (3.00\text{ m})\hat{j} \Rightarrow \hat{l} \times \hat{r} = \hat{k}$

$$\Rightarrow d\vec{B} = \frac{\mu_o}{4\pi}\frac{Idl\sin\theta}{r^2}\hat{k} = \frac{\mu_o}{4\pi}\frac{(8.00\text{ A})(2.0\times10^{-3}\text{ m})\sin(45°)}{(3.00\text{ m})^2 + (3.00\text{ m})^2}\hat{k} = (6.62\times10^{-11}\text{ T})\hat{k}.$$

(d) $\vec{r} = (3.00\text{ m})\hat{k} \Rightarrow \hat{l} \times \hat{r} = -\hat{j}$

$$\Rightarrow d\vec{B} = -\frac{\mu_o}{4\pi}\frac{I\,dl\sin\theta}{r^2}\hat{j} = -\frac{\mu_o}{4\pi}\frac{(8.00\text{ A})(2.0\times10^{-3}\text{ m})\sin(90°)}{(3.00\text{ m})^2}\hat{j} = -(1.78\times10^{-10}\text{ T})\hat{j}.$$

29-6: The magnetic field at the given points is:

$$dB_a = \frac{\mu_o}{4\pi}\frac{I\,dl\sin\theta}{r^2} = \frac{\mu_o}{4\pi}\frac{(200\text{ A})(0.00200\text{ m})}{(0.100\text{ m})^2} = 4.00\times10^{-6}\text{ T}.$$

$$dB_b = \frac{\mu_o}{4\pi}\frac{I\,dl\sin\theta}{r^2} = \frac{\mu_o}{4\pi}\frac{(200\text{ A})(0.00200\text{ m})\sin45°}{2(0.100\text{ m})^2} = 1.41\times10^{-6}\text{ T}.$$

$$dB_c = \frac{\mu_o}{4\pi}\frac{I\,dl\sin\theta}{r^2} = \frac{\mu_o}{4\pi}\frac{(200\text{ A})(0.00200\text{ m})}{(0.100\text{ m})^2} = 4.00\times10^{-6}\text{ T}.$$

$$dB_d = \frac{\mu_o}{4\pi}\frac{I\,dl\sin\theta}{r^2} = \frac{\mu_o}{4\pi}\frac{I\,dl\sin(0°)}{r^2} = 0.$$

$$dB_e = \frac{\mu_o}{4\pi}\frac{I\,dl\sin\theta}{r^2}$$

$$\Rightarrow dB_e = \frac{\mu_o}{4\pi}\frac{(200\text{ A})(0.00200\text{ m})}{3(0.100\text{ m})^2}\frac{\sqrt{2}}{\sqrt{3}}$$

$$\Rightarrow dB_e = 1.09\times10^{-6}\text{ T}.$$

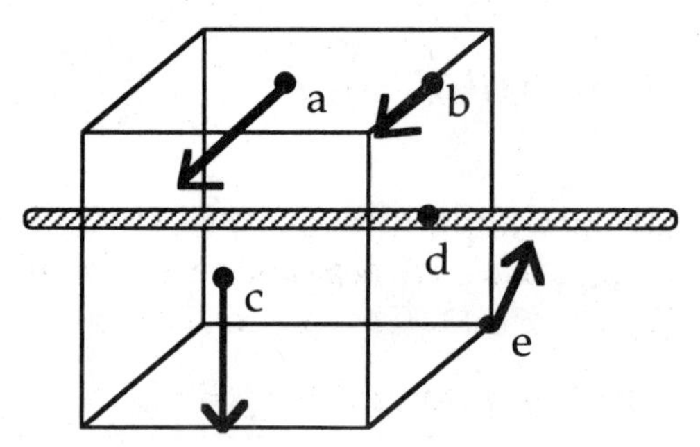

29-7: (a) From **Ex. (29-2(b))**:

$\hat{r} = \cos\theta\hat{i} + \sin\theta\hat{j} = \cos(150°)\hat{i} + \sin(150°)\hat{j} = -(0.866)\hat{i} + (0.500)\hat{j}.$

(b) $d\vec{l} \times \hat{r} = (-dl\,\hat{i}) \times (-(0.866)\hat{i} + (0.500)\hat{j}) = -dl(0.500)\hat{k}.$

(c) $d\vec{B} = \frac{\mu_o}{4\pi}\frac{I\,d\vec{l} \times \hat{r}}{r^2} = -\frac{\mu_o}{4\pi}\frac{I\,dl(0.500\text{ m})}{r^2}\hat{k} = -\frac{\mu_o}{4\pi}\frac{(125\text{ A})(0.010\text{ m})(0.500\text{ m})}{(1.20\text{ m})^2}\hat{k}$

$\Rightarrow d\vec{B} = -(4.3\times10^{-8}\text{ T})\hat{k}.$

29-8: (a) $B_o = \frac{\mu_o I}{2\pi r} \Rightarrow I = \frac{2\pi r B_o}{\mu_o} = \frac{2\pi(0.050\text{ m})(7.50\times10^{-4}\text{ T})}{\mu_o} = 190\text{ A}.$

(b) $B = \frac{\mu_o I}{2\pi r}$, so $B(r = 0.1\text{ m}) = \frac{B_o}{2} = 3.75\times10^{-4}\text{ T}$, $B(r = 0.2\text{ m}) = \frac{B_o}{4} = 1.88\times10^{-4}\text{ T}.$

29-9: $B = \frac{\mu_o I}{4\pi}\int_{-a}^{a}\frac{x\,dy}{(x^2+y^2)^{3/2}} = \frac{\mu_o I x}{4\pi}\frac{y}{x^2(x^2+y^2)^{1/2}}\Bigg|_{-a}^{a} = \frac{\mu_o I}{4\pi}\frac{2a}{x(x^2+a^2)^{1/2}}.$

29-10: (a) At $x = \frac{d}{2}$: $B = \frac{\mu_o I}{2\pi}\left(\frac{1}{d/2}+\frac{1}{3d/2}\right) = \frac{\mu_o I}{2\pi}\left(\frac{8}{3d}\right) = \frac{4\mu_o I}{3\pi d}$, in the $\hat{j}$ direction.
(b) The position $x = -\frac{d}{2}$ is symmetrical with that of part (a), so the magnetic field there is $B = \frac{4\mu_o I}{3\pi d}$, in the $\hat{j}$ direction.

29-11: (a) $B = \frac{\mu_o I}{2\pi r} = \frac{\mu_o(900\text{ A})}{2\pi(5.00\text{ m})} = 3.60\times10^{-5}$ T, to the south.
(b) Since the magnitude of the earth's magnetic field is 5.00×10^{-5} T, to the north, the total magnetic field is still to the north, although weaker.

29-12: The total magnetic field is the vector sum of the constant magnetic field and the wire's magnetic field. So:
(a) At (0,0,2 m): $\vec{B} = \vec{B}_o - \frac{\mu_o I}{2\pi r}\hat{i} = (1.75\times10^{-6}\text{ T})\hat{i} - \frac{\mu_o(5.00\text{ A})}{2\pi(2.00\text{ m})}\hat{i} = (1.25\times10^{-6}\text{ T})\hat{i}$.
(b) At (2 m,0,0): $\vec{B} = \vec{B}_o + \frac{\mu_o I}{2\pi r}\hat{k} = (1.75\times10^{-6}\text{ T})\hat{i} + \frac{\mu_o(5.00\text{ A})}{2\pi(2.00\text{ m})}\hat{k}$
$\Rightarrow \vec{B} = (1.75\times10^{-6}\text{ T})\hat{i} + (5.00\times10^{-7}\text{ T})\hat{k} = 1.82\times10^{-6}$ T, at $\theta = 15.9°$ from x to z.
(c) At (0,0,-0.5 m): $\vec{B} = \vec{B}_o + \frac{\mu_o I}{2\pi r}\hat{i} = (1.75\times10^{-6}\text{ T})\hat{i} + \frac{\mu_o(5.00\text{ A})}{2\pi(0.50\text{ m})}\hat{i} = (3.75\times10^{-6}\text{ T})\hat{i}$.

29-13: (a) At the point exactly midway between the wires, the two magnetic fields are in opposite directions and cancel.
(b) At a distance "a" above the top wire, the magnetic fields are in the same direction and add up: $\vec{B} = \frac{\mu_o I}{2\pi r_1}\hat{k} + \frac{\mu_o I}{2\pi r_2}\hat{k} = \frac{\mu_o I}{2\pi a}\hat{k} + \frac{\mu_o I}{2\pi(3a)}\hat{k} = \frac{2\mu_o I}{3\pi a}\hat{k}$.
(c) At the same distance as part (b), but below the lower wire, yields the same magnitude magnetic field but in the opposite direction: $\vec{B} = -\frac{2\mu_o I}{3\pi a}\hat{k}$.

29-14: (a) $F = \frac{\mu_o I_1 I_2 L}{2\pi r} = \frac{\mu_o(5.00\text{ A})(2.00\text{ A})(0.200\text{ m})}{2\pi(0.400\text{ m})} = 1.00\times10^{-6}$ N, and the force is repulsive since the currents are in opposite directions.
(b) Tripling the currents makes the force increase by a factor of nine to $F = 9.00\times10^{-6}$ N.

29-15: $\frac{F}{L} = \frac{\mu_o I_1 I_2}{2\pi r} \Rightarrow I_2 = \frac{F}{L}\frac{2\pi r}{\mu_o I_1} = (6.0\times10^{-5}\text{ N/m})\frac{2\pi(0.0500\text{ m})}{\mu_o(2.00\text{ A})} = 7.50$ A.
(b) The two wires repel since the currents are in opposite directions

29-16: On the top wire: $\frac{F}{L}=\frac{\mu_o I^2}{2\pi}\left(\frac{1}{d}-\frac{1}{2d}\right)=\frac{\mu_o I^2}{4\pi d}$, upward.

On the middle wire, the magnetic fields cancel so the force is zero.

On the bottom wire: $\frac{F}{L}=\frac{\mu_o I^2}{2\pi}\left(-\frac{1}{d}+\frac{1}{2d}\right)=\frac{\mu_o I^2}{4\pi d}$, downward.

29-17: We need the magnetic and gravitational forces to cancel:

$$\Rightarrow \left(\frac{m}{l}\right)Lg=\frac{\mu_o I^2 L}{2\pi r}\Rightarrow r=\frac{\mu_o I^2}{2\pi(m/l)g}=\frac{\mu_o(40.0\text{ A})^2}{2\pi(5.00\times10^{-3}\text{ kg/m})(9.80\text{ m/s}^2)}$$

$$\Rightarrow r=6.53\times10^{-3}\text{ m}.$$

29-18: (a) From Eq. (29-17), $B_{center}=\frac{\mu_o NI}{2a}=\frac{\mu_o(400)(0.300\text{ A})}{2(0.050\text{ m})}=1.51\times10^{-3}\text{ T}.$

(b) From Eq. (29-16), $B(x)=\frac{\mu_o NIa^2}{2(x^2+a^2)^{3/2}}$

$$\Rightarrow B(0.10\text{ m})=\frac{\mu_o(400)(0.300\text{ A})(0.050\text{ m})^2}{2((0.100\text{ m})^2+(0.050\text{ m})^2)^{3/2}}=1.35\times10^{-4}\text{ T}.$$

29-19: $B_{center}=\frac{\mu_o NI}{2a}\Rightarrow N=\frac{2aB_{center}}{\mu_o I}=\frac{B_{center}d}{\mu_o I}=\frac{(8.38\times10^{-4}\text{ T})(0.180\text{ m})}{\mu_o(2.50\text{ A})}=48.$

29-20: There is no magnetic field at the center of the loop from the straight sections. The magnetic field from the semi-circle is just half that of a complete loop:

$B=\frac{1}{2}B_{loop}=\frac{1}{2}\left(\frac{\mu_o I}{2R}\right)=\frac{\mu_o I}{4R}$, into the page.

29-21: As in **Ex. (29-20)**, there is no contribution from the straight wires, and now we have two oppositely oriented contributions from the two semi-circles:

$B=(B_1-B_2)=\frac{1}{2}\left(\frac{\mu_o}{2R}\right)(I_1-I_2)$, into the page. Note that if the two currents are equal, the magnetic field goes to zero at the center of the loop.

29-22: $\oint\vec{B}\cdot d\vec{l}=\mu_o I_{encl}=2.15\times10^{-5}\text{ T}\cdot\text{m}\Rightarrow I_{encl}=17.1\text{ A}.$

29-23: We will travel around the loops in the counter-clockwise direction.

(a) $I_{encl}=0\Rightarrow\oint\vec{B}\cdot d\vec{l}=0.$

(b) $I_{encl}=-I_1=-3.0\text{ A}\Rightarrow\oint\vec{B}\cdot d\vec{l}=-\mu_o(3.0\text{ A})=-3.8\times10^{-6}\text{ T}\cdot\text{m}.$

(c) $I_{encl}=-I_1+I_2=0\Rightarrow\oint\vec{B}\cdot d\vec{l}=0.$

(d) $I_{encl}=-I_1+I_2+I_3=5.0\text{ A}\Rightarrow\oint\vec{B}\cdot d\vec{l}=-\mu_o(5.0\text{ A})=+6.3\times10^{-6}\text{ T}\cdot\text{m}$

29-24: Using the formula for the magnetic field of a solenoid:

$B=\mu_o nI=\frac{\mu_o NI}{L}=\frac{\mu_o(500)(6.00\text{ A})}{(0.200\text{ m})}=0.0188\text{ T}.$

29-25: (a) $B=\dfrac{\mu_o NI}{L}\Rightarrow N=\dfrac{BL}{\mu_o I}=\dfrac{(0.190\text{ T})(0.800\text{ m})}{\mu_o(10.0\text{ A})}=12{,}100\text{ turns}$

$\Rightarrow n=\dfrac{N}{L}=\dfrac{12{,}100\text{ turns}}{0.800\text{ m}}=1.51\times10^4\text{ turns / m}.$

(b) The length of wire required is $2\pi rN=2\pi(0.0300\text{ m})(12{,}100)=2280\text{ m}.$

29-26: Outside a toroidal solenoid there is no magnetic field and inside it the magnetic field is given by $B=\dfrac{\mu_o NI}{2\pi r}$.

(a) $r=0.15\text{ m}$, which is outside the toroid, so $B=0$.

(b) $r=0.24\text{ m}\Rightarrow B=\dfrac{\mu_o NI}{2\pi r}=\dfrac{\mu_o(300)(6.20\text{ A})}{2\pi(0.240\text{ m})}=1.55\times10^{-3}\text{ T}.$

(c) $r=0.35\text{ m}$, which is outside the toroid, so $B=0$.

29-27: $B=\dfrac{\mu_o NI}{2\pi r}=\dfrac{\mu_o(500)(0.130\text{ A})}{2\pi(0.090\text{ m})}=1.44\times10^{-4}\text{ T}.$

29-28: Consider a coaxial cable where the currents run in OPPOSITE directions.

(a) For $a<r<b,\ I_{encl}=I\Rightarrow\oint\vec{B}\cdot d\vec{l}=\mu_o I\Rightarrow B2\pi r=\mu_o I\Rightarrow B=\dfrac{\mu_o I}{2\pi r}$.

(b) For $r>c$, the enclosed current is zero, so the magnetic field is also zero.

29-29: Consider a coaxial cable where the currents run in the SAME direction.

(a) For $a<r<b,\ I_{encl}=I_1\Rightarrow\oint\vec{B}\cdot d\vec{l}=\mu_o I_1\Rightarrow B2\pi r=\mu_o I_1\Rightarrow B=\dfrac{\mu_o I_1}{2\pi r}$.

(b) For $r>c,\ I_{encl}=I_1+I_2\Rightarrow\oint\vec{B}\cdot d\vec{l}=\mu_o(I_1+I_2)\Rightarrow B2\pi r=\mu_o(I_1+I_2)$

$\Rightarrow B=\dfrac{\mu_o(I_1+I_2)}{2\pi r}$.

29-30: $\left[\dfrac{\text{J}}{\text{T}}\right]=\left[\dfrac{\text{N}\cdot\text{m}}{\text{N}\cdot\text{s/C}\cdot\text{m}}\right]=\left[\dfrac{\text{C}\cdot\text{m}^2}{\text{s}}\right]=\left[\text{A}\cdot\text{m}^2\right].$

29-31:

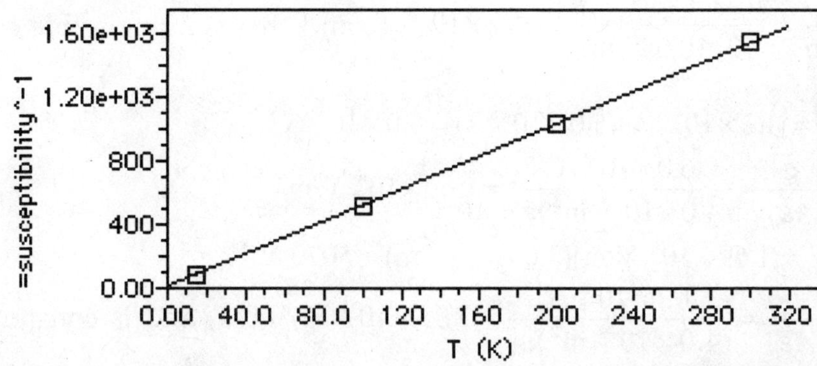

The material does obey Curie's Law because we have a straight line for temperature against one over the magnetic susceptibility. The Curie constant from the graph is $C=\dfrac{1}{\mu_o(\text{slope})}=\dfrac{1}{\mu_o(5.13)}=1.55\times10^5\text{ KA/Tm}.$

29-32: (a) $B = \frac{\mu NI}{2\pi r} = \frac{K_m \mu_o NI}{2\pi r} = \frac{\mu_o (90)(500)(0.40\text{ A})}{2\pi(0.080\text{ m})} = 0.045\text{ T}.$

(b) The fraction due to atomic currents is $B' = \frac{89}{90} B = \frac{89}{90}(0.045\text{ T}) = 0.0445\text{ T}.$

29-33: (a) If $K_m = 1400 \Rightarrow B = \frac{K_m \mu_o NI}{2\pi r} \Rightarrow I = \frac{2\pi r B}{K_m \mu_o N} = \frac{2\pi(0.0380\text{ m})(0.250\text{ T})}{\mu_o (1400)(400)} = 0.0848\text{ A}.$

(b) If $K_m = 5200 \Rightarrow I = \frac{1400}{5200} I_{part(a)} = 0.0228\text{ A}.$

29-34: (a) $B = \frac{K_m \mu_o NI}{2\pi r} \Rightarrow K_m = \frac{2\pi r B}{\mu_o NI} = \frac{2\pi(0.400\text{ m})(1.56\text{ T})}{\mu_o (600)(1.75\text{ A})} = 2971.$

(b) $\chi_m = K_m - 1 = 2970.$

29-35: (a) The magnetic field from the solenoid alone is:

$B_o = \mu_o nI = \mu_o (5000\text{ m}^{-1})(0.20\text{ A})$

$\Rightarrow B_o = 1.26 \times 10^{-3}\text{ T}.$

But $M = \frac{K-1}{\mu_o} B_o = \frac{1399}{\mu_o}(1.26 \times 10^{-3}\text{ T})$

$\Rightarrow M = 1.40 \times 10^6\text{ A/m}.$ And also $\frac{B}{B_o} = K_m = 1400.$

(b)

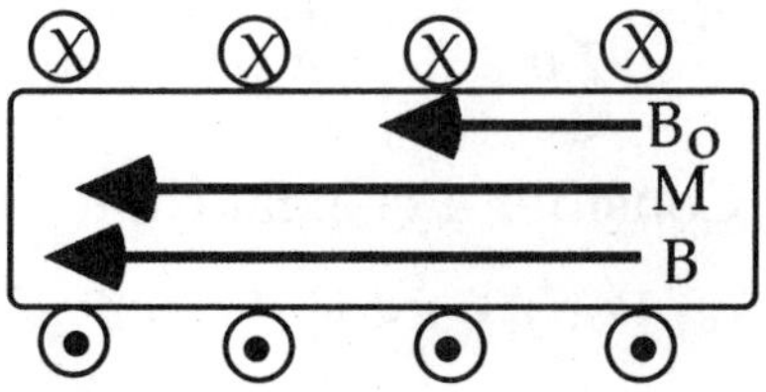

29-36: (a) $E = \rho J = \frac{\rho I}{A} = \frac{(2.0 \times 10^{-8}\ \Omega\text{m})(30\text{ A})}{4.0 \times 10^{-6}\text{ m}^2} = 0.15\text{ V/m}.$

(b) $\frac{dE}{dt} = \frac{d}{dt}\left(\frac{\rho I}{A}\right) = \frac{\rho}{A}\frac{dI}{dt} = \frac{2.0 \times 10^{-8}\ \Omega\text{m}}{4.0 \times 10^{-6}\text{ m}^2}(6000\text{ A/s}) = 30\text{ V/m}\cdot\text{s}.$

(c) $j_D = \varepsilon_o \frac{dE}{dt} = \varepsilon_o (30\text{ V/m}\cdot\text{s}) = 2.7 \times 10^{-10}\text{ A/m}^2.$

(d) $i_D = j_D A = (2.7 \times 10^{-10}\text{ A/m}^2)(4.0 \times 10^{-6}\text{ m}^2) = 1.1 \times 10^{-15}\text{ A}$

$\Rightarrow B = \frac{\mu_o I}{2\pi r} = \frac{\mu_o (1.1 \times 10^{-15}\text{ A})}{2\pi(0.050\text{ m})} = 1.2 \times 10^{-4}\text{ T},$ and this is a negligible contribution.

29-37: (a) $q = i_c t = (1.2 \times 10^{-3}\text{ A})(5.0 \times 10^{-6}\text{ s}) = 6.0 \times 10^{-9}\text{ C}.$

$E = \frac{\sigma}{\varepsilon_o} = \frac{q}{A\varepsilon_o} = \frac{6.0 \times 10^{-9}\text{ C}}{(4.0 \times 10^{-4}\text{ m}^2)\varepsilon_o} = 1.69 \times 10^6\text{ V/m}$

$\Rightarrow V = Ed = (1.69 \times 10^6\text{ V/m})(3.00 \times 10^{-3}\text{ m}) = 5070\text{ V}.$

(b) $\frac{dE}{dt} = \frac{i_c}{A\varepsilon_o} = \frac{1.2 \times 10^{-3}\text{ A}}{(4.0 \times 10^{-4}\text{ m}^2)\varepsilon_o} = 3.39 \times 10^{11}\text{ V/m}\cdot\text{s},$ and is constant in time.

(c) $j_D = \varepsilon_o \frac{dE}{dt} = \varepsilon_o (3.39 \times 10^{11}\text{ V/m}\cdot\text{s}) = 3.00\text{ A/m}^2$

$\Rightarrow i_D = j_D A = (3.00\text{ A/m}^2)(4.0 \times 10^{-4}\text{ m}^2) = 1.2 \times 10^{-3}\text{ A},$ which is the same as i_c.

29-38: (a) $Q = CV = \left(\frac{\varepsilon A}{d}\right)V = \frac{(3.00)\varepsilon_o(4.00\times10^{-4}\ \text{m}^2)(160\ \text{V})}{2.00\times10^{-3}\ \text{m}} = 8.50\times10^{-10}\ \text{C}.$

(b) $\frac{dQ}{dt} = i_c = 5.00\times10^{-3}\ \text{A}.$

(c) $j_D = \varepsilon\frac{dE}{dt} = K\varepsilon_o\frac{i_c}{K\varepsilon_o A} = \frac{i_c}{A} = j_c \Rightarrow i_D = i_c = 5.00\times10^{-3}\ \text{A}.$

29-39: (a) $j_D = \varepsilon_o\frac{dE}{dt} = \varepsilon_o\frac{i_c}{\varepsilon_o A} = \frac{i_c}{A} = \frac{0.450\ \text{A}}{7.85\times10^{-3}\ \text{m}^2} = 57.3\ \text{A/m}^2.$

(b) $\frac{dE}{dt} = \frac{j_D}{\varepsilon_o} = \frac{57.3\ \text{A}/\text{m}^2}{\varepsilon_o} = 6.47\times10^{12}\ \text{V}/\text{m}\cdot\text{s}.$

(c) $r < R: B = \frac{\mu_o}{2\pi}\frac{r}{R^2}i_c = \frac{\mu_o}{2\pi}\frac{0.025\ \text{m}}{(0.050\ \text{m})^2}(0.450\ \text{A}) = 9.0\times10^{-7}\ \text{T}.$

(d) $r > R: B = \frac{\mu_o}{2\pi}\frac{i_c}{r} = \frac{\mu_o}{2\pi}\frac{(0.450\ \text{A})}{(0.10\ \text{m})} = 9.0\times10^{-7}\ \text{T}.$

29-40: (a) $\vec{B} = \frac{\mu_o}{4\pi}\frac{q\vec{v}\times\hat{r}}{r^2} = \frac{\mu_o}{4\pi}\frac{q}{r^2}\begin{vmatrix}\hat{i} & \hat{j} & \hat{k}\\ v_x & v_y & v_z\\ 1 & 0 & 0\end{vmatrix} = \frac{\mu_o}{4\pi}\frac{q}{r^2}\left(v_z\hat{j} - v_y\hat{k}\right) = (8.0\times10^{-6}\ \text{T})\hat{k}$

$\Rightarrow \frac{\mu_o}{4\pi}\frac{q}{r^2}v_z = 0 \Rightarrow v_z = 0$ and $-\frac{\mu_o}{4\pi}\frac{q}{r^2}v_y = 8.0\times10^{-6}\ \text{T}$

$\Rightarrow v_y = -\frac{4\pi(8.0\times10^{-6}\ \text{T})(0.20\ \text{m})^2}{\mu_o(-6.0\times10^{-3}\ \text{C})} = 533\ \text{m/s}.$

And $v_x = \pm\sqrt{v^2 - v_y^{\,2} - v_z^{\,2}} = \pm\sqrt{(900\ \text{m}/\text{s})^2 - (533\ \text{m}/\text{s})^2} = \pm725\ \text{m}/\text{s}.$

(b) $\vec{B}(0,0.2\ \text{m},0) = \frac{\mu_o}{4\pi}\frac{q\vec{v}\times\hat{r}}{r^2} = \frac{\mu_o}{4\pi}\frac{q}{r^2}\begin{vmatrix}\hat{i} & \hat{j} & \hat{k}\\ v_x & v_y & 0\\ 0 & 1 & 0\end{vmatrix} = +\frac{\mu_o}{4\pi}\frac{q}{r^2}v_x\hat{k}$

$\Rightarrow B(0,0.2\ \text{m},0) = -\frac{\mu_o}{4\pi}\frac{(-6.0\times10^{-3}\ \text{C})}{(0.20\ \text{m})^2}(\pm725\ \text{m}/\text{s}) = \pm1.09\times10^{-5}\ \text{T}.$

29-41: Choose a cube of edge length L, with one face on the y-z plane. Then:
$0 = \oint\vec{B}\cdot d\vec{A} = \iint_{x=L}\vec{B}\cdot d\vec{A} = \iint_{x=L}\frac{B_o x}{a}\hat{i}\cdot d\vec{A} = \frac{B_o L}{a}\iint_{x=L}dA = \frac{B_o L^3}{a} \Rightarrow B_o = 0$, so the only possible field is a zero field.

29-42: The magnetic field of charge q at the location of charge q' is into the page.
$\vec{F} = q'\vec{v}'\times\vec{B} = (q'v')\hat{j}\times\frac{\mu_o}{4\pi}\frac{q\vec{v}\times\hat{r}}{r^2} = (q'v')\hat{j}\times\left(\frac{\mu_o}{4\pi}\frac{qv}{r^2}\right)(-\hat{k}) = \left(\frac{\mu_o}{4\pi}\frac{qq'vv'\sin\theta}{r^2}\right)(-\hat{i})$

$\Rightarrow \vec{F} = -\left(\frac{\mu_o}{4\pi}\frac{(5.0\times10^{-6}\ \text{C})(-3.0\times10^{-6}\ \text{C})(7.5\times10^4\ \text{m/s})(3.2\times10^4\ \text{m/s})}{(0.500\ \text{m})^2}\left(\frac{0.3}{0.5}\right)\right)\hat{i}$

$\Rightarrow \vec{F} = (8.64\times10^{-9}\ \text{N})\hat{i}.$

29-43: $F = qvB = qv\left(\frac{\mu_o I}{2\pi r}\right) = \frac{\mu_o}{2\pi}\frac{(1.60\times10^{-19}\ \text{C})(4.00\times10^{4}\ \text{m/s})(1.50\ \text{A})}{(0.0800\ \text{m})} = 2.40\times10^{-20}\ \text{N}.$

Let the current run left to right, the electron move in the same direction, below the wire, then the magnetic field at the electron is into the page, and the electron feels a force downwards, away from the wire, by the right hand rule (remember the electron is negative).

29-44: (a)

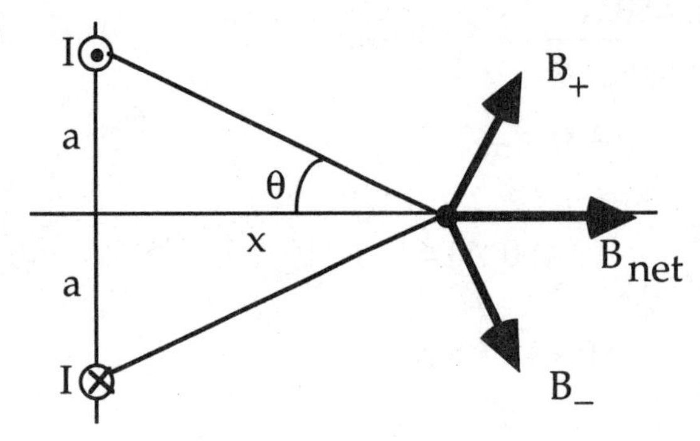

(b) At a position on the x-axis:

$$B_{net} = 2\frac{\mu_o I}{2\pi r}\sin\theta = \frac{\mu_o I}{\pi\sqrt{x^2+a^2}}\frac{a}{\sqrt{x^2+a^2}}$$

$$\Rightarrow B_{net} = \frac{\mu_o I a}{\pi(x^2+a^2)},$$ in the positive x-direction, as shown at left.

(d) The magnetic field is a maximum at the origin, x = 0.

(e) When $x >> a,\ B \approx \frac{\mu_o I a}{\pi x^2}.$

(c)

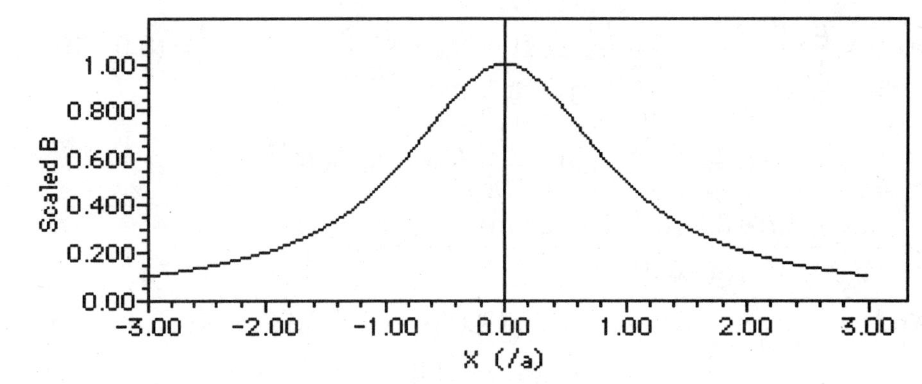

29-45: (a)

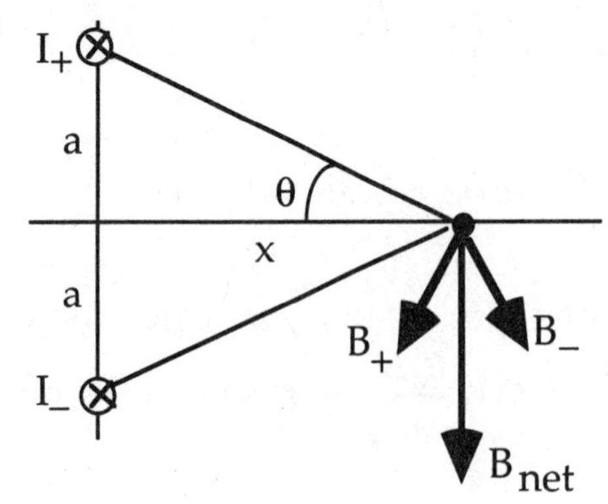

(b) At a position on the x-axis:

$$B_{net} = 2\frac{\mu_o I}{2\pi r}\cos\theta = \frac{\mu_o I}{\pi\sqrt{x^2+a^2}}\frac{x}{\sqrt{x^2+a^2}}$$

$$\Rightarrow B_{net} = \frac{\mu_o I x}{\pi(x^2+a^2)},$$ in the negative y-direction, as shown at left.

(c)

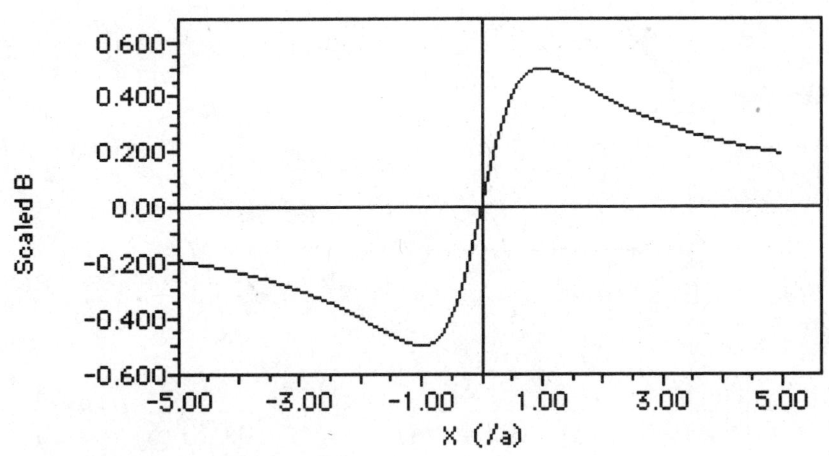

(d) The magnetic field is a maximum when:

$$\frac{dB}{dx} = 0 = \frac{C}{x^2 + a^2} - \frac{2Cx^2}{(x^2 + a^2)^2} \Rightarrow (x^2 + a^2) = 2x^2 \Rightarrow x = \pm a$$

(e) When $x >> a$, $B \approx \frac{\mu_o I}{\pi x}$, which is just like a wire carrying current 2I.

29-46: (a) Wire carrying current into the page, so it feels a force downwards from the other wires, as shown at right.

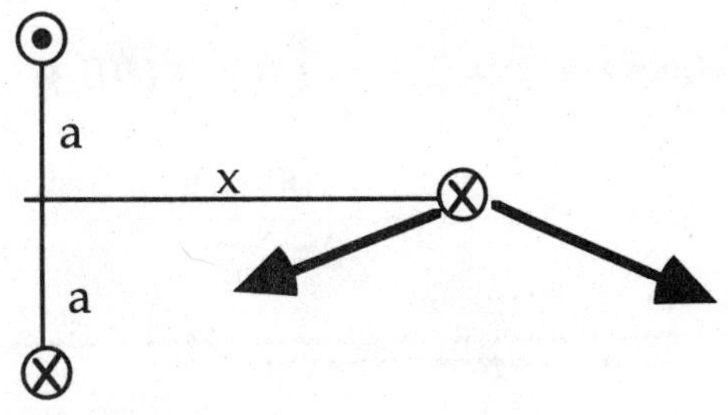

$$\frac{F}{L} = IB = I\left(\frac{\mu_o Ia}{\pi(x^2 + a^2)}\right)$$

$$\Rightarrow \frac{F}{L} = \frac{\mu_o (9.00\text{ A})^2 (0.300\text{ m})}{\pi((0.400\text{ m})^2 + (0.300\text{ m})^2)} = 3.89 \times 10^{-5}\text{ N/m}.$$

(b) If the wire carries current out of the page then the forces felt will be the opposite of part (a). Thus the force will be 3.89×10^{-5} N/m, upward.

29-47: (a) If the magnetic field at point P is zero, then from Figure 29-40 the current I_2 must out of the page, in order to cancel the field from I_1. Also:

$$B_1 = B_2 \Rightarrow \frac{\mu_o I_1}{2\pi r_1} = \frac{\mu_o I_2}{2\pi r_2} \Rightarrow I_2 = I_1 \frac{r_2}{r_1} = (6.00\text{ A})\frac{(0.500\text{ m})}{(1.50\text{ m})} = 2.00\text{ A}.$$

(b) Given the currents, the field at Q points to the right and has magnitude

$$B_Q = \frac{\mu_o}{2\pi}\left(\frac{I_1}{r_1} + \frac{I_2}{r_2}\right) = \frac{\mu_o}{2\pi}\left(\frac{6.00\text{ A}}{0.500\text{ m}} + \frac{2.00\text{ A}}{1.50\text{ m}}\right) = 2.13 \times 10^{-6}\text{ T}.$$

(c) The magnitude of the field at S is given by the sum of the squares of the two fields because they are at right angles. So:

$$B_S = \sqrt{B_1^2 + B_2^2} = \frac{\mu_o}{2\pi}\sqrt{\left(\frac{I_1}{r_1}\right)^2 + \left(\frac{I_2}{r_2}\right)^2} = \frac{\mu_o}{2\pi}\sqrt{\left(\frac{6.00\text{ A}}{0.60\text{ m}}\right)^2 + \left(\frac{2.00\text{ A}}{0.80\text{ m}}\right)^2} = 2.1 \times 10^{-6}\text{ T}.$$

29-48: (a)

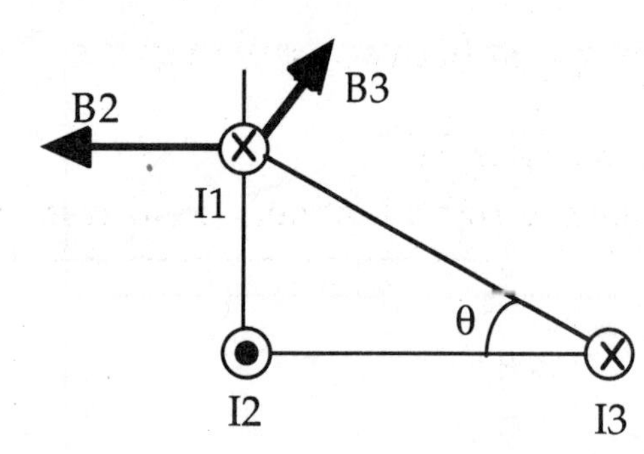

(b) $\vec{B}_2 = -\left(\frac{\mu_0 I_2}{2\pi r_2}\right)\hat{i}$

$\vec{B}_3 = \left(\frac{\mu_0 I_3}{2\pi r_3}\right)(\sin\theta\,\hat{i} + \cos\theta\,\hat{j})$

And so

$\vec{B} = \left(\frac{\mu_0}{2\pi}\right)\left(\left(-\frac{I_2}{r_2} + \frac{I_3}{r_3}\sin\theta\right)\hat{i} + \frac{I_3}{r_3}\cos\theta\,\hat{j}\right)$

$$\Rightarrow \vec{B} = \left(\frac{\mu_0}{2\pi}\right)\left(\left(-\frac{I_2}{(0.030\text{ m})} + \frac{I_3}{(0.040\text{ m})}(0.6)\right)\hat{i} + \frac{I_3}{(0.040\text{ m})}(0.8)\hat{j}\right)$$

$$\Rightarrow \vec{B} = \left(\frac{\mu_0}{2\pi}\right)\left((15I_3 - 33I_2)\,\hat{i} + (20I_3)\hat{j}\right).$$

(c) The actual direction of the force on the first wire cannot be completely determined without the other current magnitudes, but it must have magnitude $F = I_1 lB$, and be in the xy plane.

29-49: The forces on the top and bottom segments cancel, leaving the left and right sides: $\vec{F} = \vec{F}_l + \vec{F}_r = -(IlB_l)\hat{i} + (IlB_r)\hat{i} = Il\left(-\frac{\mu_0 I_{wire}}{2\pi r_l} + \frac{\mu_0 I_{wire}}{2\pi r_r}\right)\hat{i} = \frac{\mu_0 IlI_{wire}}{2\pi}\left(\frac{1}{r_r} - \frac{1}{r_l}\right)\hat{i}$

$$\Rightarrow \vec{F} = \frac{\mu_0(5.00\text{ A})(0.200\text{ m})(14.0\text{ A})}{2\pi}\left(\frac{1}{0.100\text{ m}} - \frac{1}{0.026\text{ m}}\right)\hat{i} = -(7.97\times10^{-5}\text{ N})\hat{i}.$$

29-50:

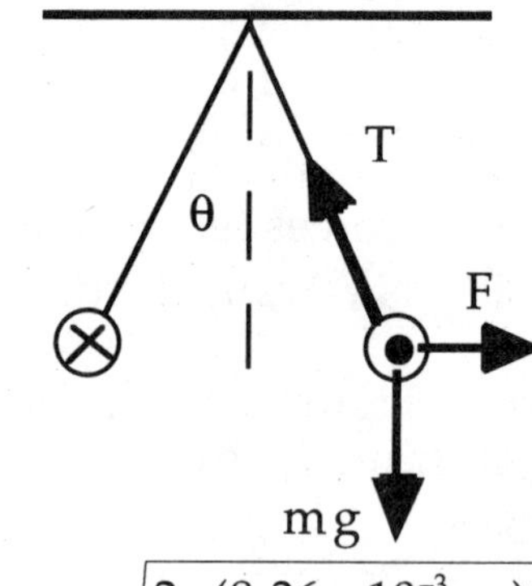

The wires are in equilibrium, so:
x: $F = T\sin\theta$ and *y*: $T\cos\theta = mg$

$\Rightarrow F = IlB = T\sin\theta = mg\tan\theta \Rightarrow I = \frac{mg\tan\theta}{lB}$.

But $B = \frac{\mu_0 I}{2\pi r} \Rightarrow I = \frac{2\pi rmg\tan\theta}{l\mu_0 I} \Rightarrow I = \sqrt{\frac{2\pi rmg\tan\theta}{l\mu_0}}$

And $r = [2(0.0400\text{ m})\sin(6.00°)] = 8.36\times10^{-3}$ m.

$$\Rightarrow I = \sqrt{\frac{2\pi(8.36\times10^{-3}\text{ m})(0.0375\text{ kg/m})(9.80\text{ m/s}^2)\tan(6.00°)}{\mu_0}} = 40.2\text{ A}.$$

29-51: Recall for a single loop: $B = \frac{\mu_0 Ia^2}{2(x^2 + a^2)^{3/2}}$. Here we have two loops, each of N turns, and measuring the field along the x-axis from between them means that the "x" in the formula is different for each case:

Left coil: $x \to x + \frac{a}{2} \Rightarrow B_l = \frac{\mu_0 NIa^2}{2((x + a/2)^2 + a^2)^{3/2}}$.

Right coil: $x \to x - \frac{a}{2} \Rightarrow B_r = \frac{\mu_0 NIa^2}{2((x - a/2)^2 + a^2)^{3/2}}$.

So the total field at a point x from the point between them is:

$$B=\frac{\mu_o NIa^2}{2}\left(\frac{1}{((x+a/2)^2+a^2)^{3/2}}+\frac{1}{((x-a/2)^2+a^2)^{3/2}}\right).$$

(b) Below left: Total magnetic field. Below right: Magnetic field from right coil.

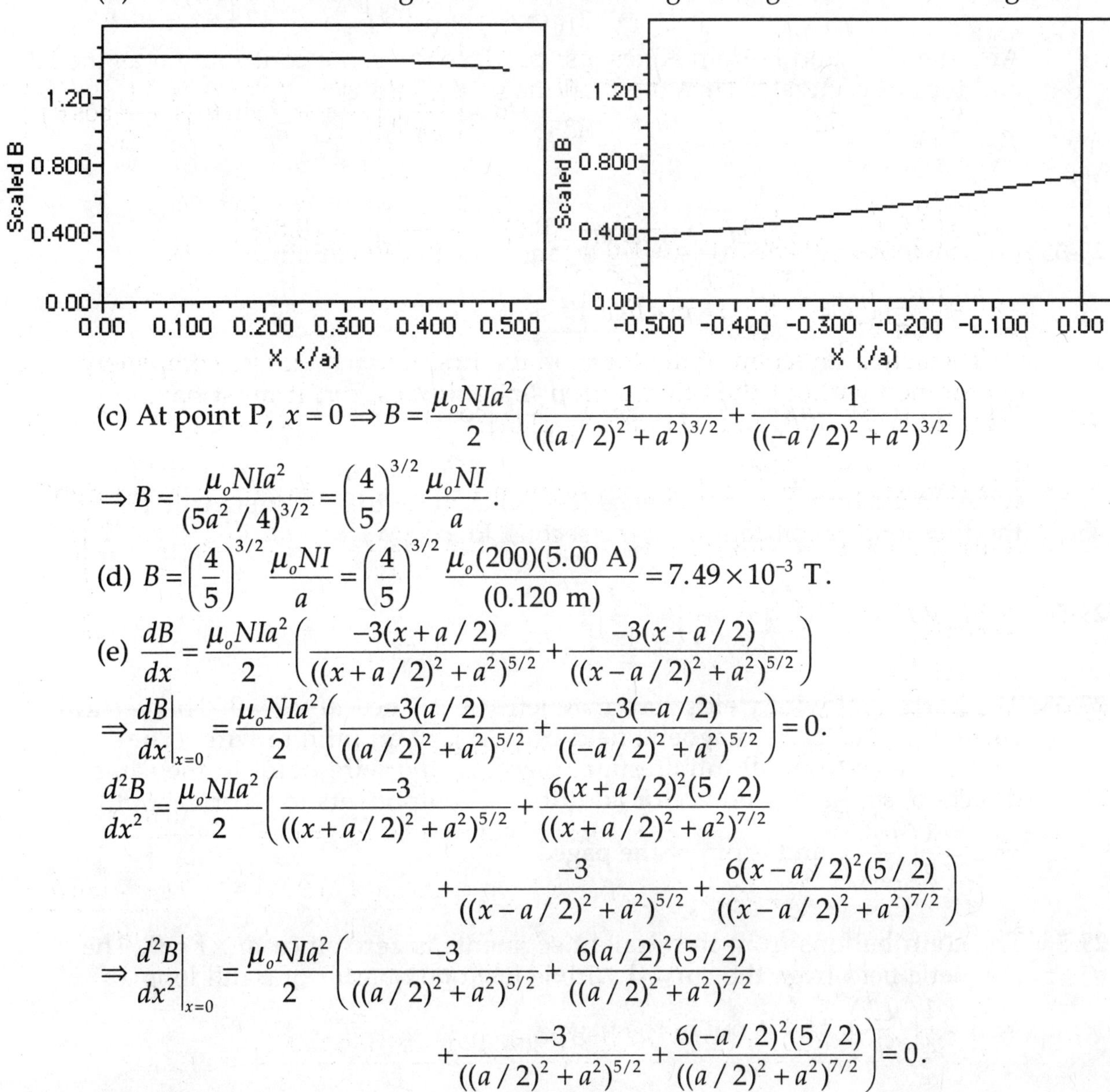

(c) At point P, $x=0 \Rightarrow B=\frac{\mu_o NIa^2}{2}\left(\frac{1}{((a/2)^2+a^2)^{3/2}}+\frac{1}{((-a/2)^2+a^2)^{3/2}}\right)$

$$\Rightarrow B=\frac{\mu_o NIa^2}{(5a^2/4)^{3/2}}=\left(\frac{4}{5}\right)^{3/2}\frac{\mu_o NI}{a}.$$

(d) $B=\left(\frac{4}{5}\right)^{3/2}\frac{\mu_o NI}{a}=\left(\frac{4}{5}\right)^{3/2}\frac{\mu_o(200)(5.00\text{ A})}{(0.120\text{ m})}=7.49\times10^{-3}\text{ T}.$

(e) $\frac{dB}{dx}=\frac{\mu_o NIa^2}{2}\left(\frac{-3(x+a/2)}{((x+a/2)^2+a^2)^{5/2}}+\frac{-3(x-a/2)}{((x-a/2)^2+a^2)^{5/2}}\right)$

$$\Rightarrow \left.\frac{dB}{dx}\right|_{x=0}=\frac{\mu_o NIa^2}{2}\left(\frac{-3(a/2)}{((a/2)^2+a^2)^{5/2}}+\frac{-3(-a/2)}{((-a/2)^2+a^2)^{5/2}}\right)=0.$$

$$\frac{d^2B}{dx^2}=\frac{\mu_o NIa^2}{2}\left(\frac{-3}{((x+a/2)^2+a^2)^{5/2}}+\frac{6(x+a/2)^2(5/2)}{((x+a/2)^2+a^2)^{7/2}}+\frac{-3}{((x-a/2)^2+a^2)^{5/2}}+\frac{6(x-a/2)^2(5/2)}{((x-a/2)^2+a^2)^{7/2}}\right)$$

$$\Rightarrow \left.\frac{d^2B}{dx^2}\right|_{x=0}=\frac{\mu_o NIa^2}{2}\left(\frac{-3}{((a/2)^2+a^2)^{5/2}}+\frac{6(a/2)^2(5/2)}{((a/2)^2+a^2)^{7/2}}+\frac{-3}{((a/2)^2+a^2)^{5/2}}+\frac{6(-a/2)^2(5/2)}{((a/2)^2+a^2)^{7/2}}\right)=0.$$

Since both first and second derivatives are zero, the field can only be changing very slowly.

29-52: A wire of length l produces a field $B=\frac{\mu_o I}{4\pi}\frac{l}{x\sqrt{x^2+(l/2)^2}}$. Here all edges produce a field into the page so we can just add them up:

$$x = a/2 \text{ and } l = b \Rightarrow B_{left} = \frac{\mu_0 I}{4\pi} \frac{b}{(a/2)\sqrt{(a/2)^2 + (b/2)^2}} = \frac{\mu_0 I}{\pi}\left(\frac{b}{a}\right)\frac{1}{\sqrt{a^2+b^2}}.$$

$$x = b/2 \text{ and } l = a \Rightarrow B_{top} = \frac{\mu_0 I}{4\pi} \frac{a}{(b/2)\sqrt{(b/2)^2 + (a/2)^2}} = \frac{\mu_0 I}{\pi}\left(\frac{a}{b}\right)\frac{1}{\sqrt{a^2+b^2}}.$$

And the right and bottom edges just produce the same contribution as the left and top, respectively. Thus the total magnetic field is:

$$B = \frac{2\mu_0 I}{\pi}\left(\frac{b}{a} + \frac{a}{b}\right)\frac{1}{\sqrt{a^2+b^2}} = \frac{2\mu_0 I}{\pi ab}\sqrt{a^2+b^2}.$$

29-53: (a) $x >> a \Rightarrow B = \frac{\mu_0 I a^2}{2(x^2+a^2)^{3/2}} \approx \frac{\mu_0 I a^2}{2x^{3/2}}$ and $|\vec{\tau}| = |\vec{\mu} \times \vec{B}| = \mu B \sin\phi$

$$\Rightarrow \tau = (I'A')\left(\frac{\mu_0 I a^2}{2x^3}\right)\sin\phi = \frac{\mu_0 \pi I I' a^2 a'^2 \sin\phi}{2x^3}.$$

(b) $U = -\vec{\mu}\cdot\vec{B} = -\mu B\cos\phi = -(I'\pi a'^2)\left(\frac{\mu_0 I a^2}{2x^3}\right)\cos\phi = -\frac{\mu_0 \pi I I' a^2 a'^2 \cos\phi}{2x^3}.$

(c) Having $x >> a$ allows us to simplify the form of the magnetic field, whereas assuming $x >> a'$ means we can assume that the magnetic field from the first loop is constant over the second loop.

29-54: $B = B_a - B_b = \frac{1}{2}\left(\frac{\mu_0 I}{2}\right)\left(\frac{1}{a} - \frac{1}{b}\right) = \frac{\mu_0 I}{4a}\left(1 - \frac{a}{b}\right).$

29-55: The horizontal wire yields zero magnetic field since $d\vec{l} \times \vec{r} = 0$. The vertical current provides the magnetic field of HALF of an infinite wire. (The contributions from all infinitesimal pieces of the wire point in the same direction, so there is no vector addition or components to worry about.)

$\Rightarrow B = \frac{1}{2}\left(\frac{\mu_0 I}{2\pi a}\right)$, and is out of the page.

29-56: The contributions from the straight segments is zero since $d\vec{l} \times \vec{r} = 0$. The magnetic field from the curved wire is just one quarter of a full loop:

$\Rightarrow B = \frac{1}{4}\left(\frac{\mu_0 I}{2R}\right)$, and is out of the page.

29-57: If there is a magnetic field component in the z-direction, it must be constant because of the symmetry of the wire. Therefore the contribution to a surface integral over a closed cylinder, encompassing a long straight wire will be zero: no flux through the barrel of the cylinder, and equal but opposite flux through the ends. The radial field will have no contribution through the ends, but through the barrel:

$$0 = \oint_S \vec{B}\cdot d\vec{A} = \oint_S \vec{B}_r \cdot d\vec{A} = \int_{barrel} \vec{B}_r \cdot d\vec{A} = \int_{barrel} B_r dA = B_r A_{barrel} = 0 \Rightarrow B_r = 0.$$

29-58: (a) $r < a \Rightarrow I_{encl} = 0 \Rightarrow B = 0$.

(b) $a < r < b \Rightarrow I_{encl} = I\left(\dfrac{A_{a\to r}}{A_{a\to b}}\right) = I\left(\dfrac{\pi(r^2 - a^2)}{\pi(b^2 - a^2)}\right) = I\dfrac{(r^2 - a^2)}{(b^2 - a^2)}$

$\Rightarrow \oint \vec{B}\cdot d\vec{l} = B2\pi r = \mu_o I\dfrac{(r^2 - a^2)}{(b^2 - a^2)} \Rightarrow B = \dfrac{\mu_o I}{2\pi r}\dfrac{(r^2 - a^2)}{(b^2 - a^2)}$.

(c) $r > b \Rightarrow I_{encl} = I \Rightarrow \oint \vec{B}\cdot d\vec{l} = B2\pi r = \mu_o I \Rightarrow B = \dfrac{\mu_o I}{2\pi r}$.

29-59: (a) $I = \int_S J\,dA = \int_S \alpha r\, r dr d\theta = \alpha 2\pi \int_0^R r^2 dr = \dfrac{2\pi\alpha R^3}{3} \Rightarrow \alpha = \dfrac{3I}{2\pi R^3}$.

(b) (i) $r \le R \Rightarrow I_{encl} = \dfrac{3I}{2\pi R^3}\int_S r^2 dr d\theta = \dfrac{3I}{2\pi R^3}2\pi\int_0^r r^2 dr = I\dfrac{r^3}{R^3}$

$\Rightarrow \oint \vec{B}\cdot d\vec{l} = B2\pi r = \mu_o I_{encl} = \mu_o\left(I\dfrac{r^3}{R^3}\right) \Rightarrow B = \dfrac{\mu_o I r^2}{2\pi R^3}$.

(ii) $r \ge R \Rightarrow I_{encl} = I \Rightarrow \oint \vec{B}\cdot d\vec{l} = B2\pi r = \mu_o I_{encl} = \mu_o I \Rightarrow B = \dfrac{\mu_o I}{2\pi r}$.

29-60: (a) $r < a \Rightarrow I_{encl} = I\left(\dfrac{A_r}{A_a}\right) = I\left(\dfrac{r^2}{a^2}\right) \Rightarrow \oint \vec{B}\cdot d\vec{l} = B2\pi r = \mu_o I_{encl} = \mu_o I\left(\dfrac{r^2}{a^2}\right) \Rightarrow B = \dfrac{\mu_o I r}{2\pi a^2}$.

When $r = a$, $B = \dfrac{\mu_o I}{2\pi a}$ which is just what was found from **Ex. (29-28), part (a)**.

(b) $b < r < c \Rightarrow I_{encl} = I - I\left(\dfrac{A_{b\to r}}{A_{b\to c}}\right) = I\left(1 - \dfrac{r^2 - b^2}{c^2 - b^2}\right)$

$\Rightarrow \oint \vec{B}\cdot d\vec{l} = B2\pi r = \mu_o I\left(1 - \dfrac{r^2 - b^2}{c^2 - b^2}\right) = \mu_o I\left(\dfrac{c^2 - r^2}{c^2 - b^2}\right) \Rightarrow B = \dfrac{\mu_o I}{2\pi r}\left(\dfrac{c^2 - r^2}{c^2 - b^2}\right)$.

When $r = b$, $B = \dfrac{\mu_o I}{2\pi b}$, just as in **Ex. (29-28), part (a)**, and at $r = c$, $B = 0$, just as in **Ex. (29-28), part (b)**.

29-61: (a) Below the sheet, all the magnetic field contributions from different wires add up to produce a magnetic field that points in the positive x-direction. (Components in the z-direction cancel.) Using Ampere's Law, where we use the fact that the field is anti-symmetrical above and below the current sheet, and that the legs of the path perpendicular provide nothing to the integral:

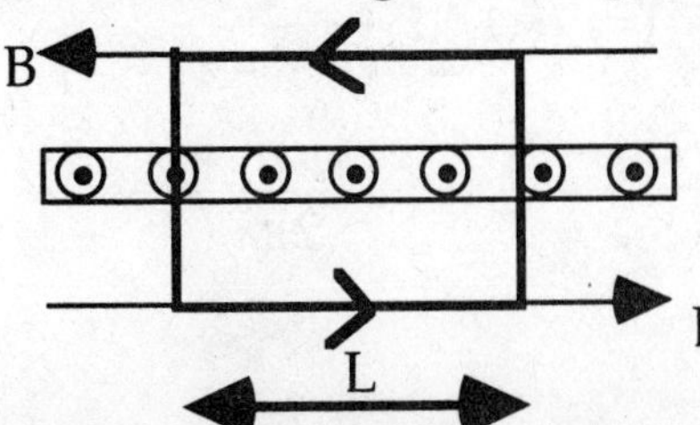

So, at a distance a beneath the sheet the magnetic field is:

$I_{encl} = nLI \Rightarrow \oint \vec{B}\cdot d\vec{l} = B2L = \mu_o nLI \Rightarrow B = \dfrac{\mu_o nI}{2}$, in the positive x-direction.

(Note there is no dependence on a.)

(b) The field has the same magnitude above the sheet, but points in the negative x-direction.

29-62: Two infinite sheets, as in **Pr. (29-61)**, are placed one above the other, with their currents opposite.

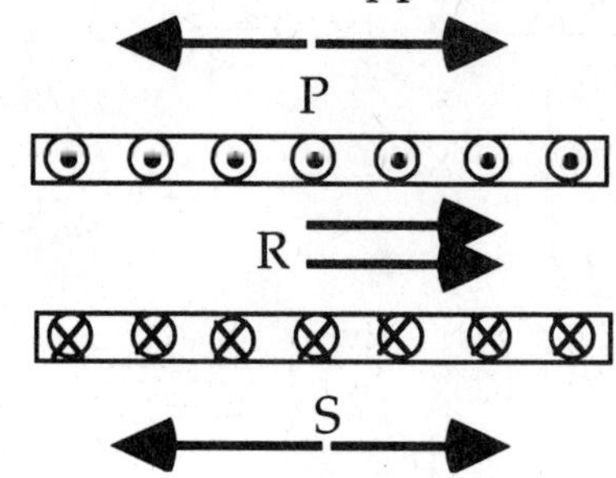

(a) Above the two sheets, the fields cancel (since there is no dependence upon the distance from the sheets).
(b) In between the sheets the two fields add up to yield $B = \mu_o nI$, to the right.
(c) Below the two sheets, their fields again cancel (since there is no dependence upon the distance from the sheets).

29-63: (a) $I = \int_S \vec{J} \cdot d\vec{A} = \frac{2I_o}{\pi a^2} \int_S \left(1 - \frac{r^2}{a^2}\right) r dr d\theta = \frac{2I_o}{\pi a^2} 2\pi \int_0^a \left(r - \frac{r^3}{a^2}\right) dr = \frac{4I_o}{a^2}\left(\frac{r^2}{2} - \frac{r^4}{4a^2}\right)\Bigg|_0^a$

$\Rightarrow I = \frac{4I_o}{a^2}\left(\frac{a^2}{2} - \frac{a^4}{4a^2}\right) = I_o.$

(b) For $r \ge a \Rightarrow \oint \vec{B} \cdot d\vec{l} = B2\pi r = \mu_o I_{encl} = \mu_o I_o \Rightarrow B = \frac{\mu_o I_o}{2\pi r}.$

(c) For $r \le a \Rightarrow I_{encl} = \int_S \vec{J} \cdot d\vec{A} = \frac{2I_o}{\pi a^2} \int_S \left(1 - \frac{r'^2}{a^2}\right) r' dr' d\theta = \frac{2I_o}{\pi a^2} 2\pi \int_0^r \left(r' - \frac{r'^3}{a^2}\right) dr'$

$\Rightarrow I_{encl} = \frac{4I_o}{a^2}\left(\frac{r'^2}{2} - \frac{r'^4}{4a^2}\right)\Bigg|_0^r = 2I_o \frac{r^2}{a^2}\left(1 - \frac{r^2}{2a^2}\right).$

(d) For $r \le a \Rightarrow \oint \vec{B} \cdot d\vec{l} = B2\pi r = \mu_o I_{encl} = 2\mu_o I_o \frac{r^2}{a^2}\left(1 - \frac{r^2}{2a^2}\right) \Rightarrow B = \frac{\mu_o I_o r}{\pi a^2}\left(1 - \frac{r^2}{2a^2}\right).$

At $r = a$, $B = \frac{\mu_o I_o}{2\pi a}$ for both parts (b) and (d).

29-64: (a) $I_o = \int_S \vec{J} \cdot d\vec{A} = \int_S \left(\frac{b}{r} e^{(r-a)/\delta}\right) r dr d\theta = 2\pi b \int_0^a e^{(r-a)/\delta} dr = 2\pi b \delta e^{(r-a)/\delta}\Big|_0^a = 2\pi b \delta\left(1 - e^{-a/\delta}\right)$

$\Rightarrow I_o = 2\pi(400\ \text{A/m})(0.050\ \text{m})\left(1 - e^{-(0.10/0.050)}\right) = 109\ \text{A}.$

(b) For $r \ge a \Rightarrow \oint \vec{B} \cdot d\vec{l} = B2\pi r = \mu_o I_{encl} = \mu_o I_o \Rightarrow B = \frac{\mu_o I_o}{2\pi r}.$

(c) $r \le a \Rightarrow I(r) = \int_S \vec{J} \cdot d\vec{A} = \int_S \left(\frac{b}{r'} e^{(r'-a)/\delta}\right) r' dr' d\theta = 2\pi b \int_0^r e^{(r'-a)/\delta} dr = 2\pi b \delta e^{(r'-a)/\delta}\Big|_0^r$

$\Rightarrow I(r) = 2\pi b \delta\left(e^{(r-a)/\delta} - e^{-a/\delta}\right) = 2\pi b \delta e^{-a/\delta}\left(e^{r/\delta} - 1\right) \Rightarrow I(r) = I_o \frac{(e^{r/\delta} - 1)}{(e^{a/\delta} - 1)}.$

(d) For $r \le a \Rightarrow \oint \vec{B} \cdot d\vec{l} = B(r)2\pi r = \mu_o I_{encl} = \mu_o I_o \frac{(e^{r/\delta} - 1)}{(e^{a/\delta} - 1)} \Rightarrow B = \frac{\mu_o I_o}{2\pi r} \frac{(e^{r/\delta} - 1)}{(e^{a/\delta} - 1)}.$

(e) At $r = \delta = 0.05 \text{ m} \Rightarrow B = \frac{\mu_o I_o}{2\pi\delta}\frac{(e-1)}{(e^{a/\delta}-1)} = \frac{\mu_o (109 \text{ A})}{2\pi(0.05 \text{ m})}\frac{(e-1)}{(e^{10/5}-1)} = 1.17\times10^{-4} \text{ T}.$

At $r = a = 0.10 \text{ m} \Rightarrow B = \frac{\mu_o I_o}{2\pi\delta}\frac{(e^{a/\delta}-1)}{(e^{a/\delta}-1)} = \frac{\mu_o (109 \text{ A})}{2\pi(0.10 \text{ m})} = 2.18\times10^{-4} \text{ T}.$

At $r = 2a = 0.20 \text{ m} \Rightarrow B = \frac{\mu_o I_o}{2\pi r} = \frac{\mu_o (109 \text{ A})}{2\pi(0.20 \text{ m})} = 1.09\times10^{-4} \text{ T}.$

29-65: $\int_{-\infty}^{\infty} B_x dx = \int_{-\infty}^{\infty} \frac{\mu_0 I a^2}{2(x^2+a^2)^{3/2}} dx = \frac{\mu_0 I}{2}\int_{-\infty}^{\infty}\frac{1}{((x/a)^2+1)^{3/2}} d(x/a) = \frac{\mu_0 I}{2}\int_{-\infty}^{\infty}\frac{dz}{(z^2+1)^{3/2}}$

$\Rightarrow \int_{-\infty}^{\infty} B_x dx = \frac{\mu_0 I}{2}\int_{-\pi/2}^{\pi/2}\cos\theta d\theta = \frac{\mu_0 I}{2}(\sin\theta)\Big|_{-\pi/2}^{\pi/2} = \mu_0 I$, where we used the substitution $z = \tan\theta$ to go from the first to second line.

This is just what Ampere's Law tells us to expect if we imagine the loop runs along the x-axis, closing on itself at infinity: $\oint \vec{B}\cdot d\vec{l} = \mu_0 I$.

29-66: The microscopic magnetic moments of an initially unmagnetised ferromagnetic material experiences torques from a magnet that align the magnetic domains with the external field, so they are attracted to the magnet. For a paramagnetic material, the same attraction occurs because the magnetic moments align themselves parallel to the external field.

For a diamagnetic material, the magnetic moments align anti-parallel to the external field so it is like two magnets repelling each other.

(b) The magnet can just pick up the iron cube so the force it exerts is:

$F_{Fe} = m_{Fe}g = \rho_{Fe}a^3 g = (7.8\times10^3 \text{ kg/m}^3)(0.010 \text{ m})^3(9.8 \text{ m/s}^2) = 0.0764 \text{ N}.$

But $F_{Fe} = IaB = \frac{\mu_{Fe}B}{a} = 0.0764 \text{ N} \Rightarrow \frac{B}{a} = \frac{0.0764 \text{ N}}{\mu_{Fe}}.$

So if the magnet tries to lift the aluminum cube of the same dimensions as the iron block, then the upward force felt by the cube is:

$F_{Al} = \frac{\mu_{Al}B}{a} = \frac{\mu_{Al}}{\mu_{Fe}} 0.0764 \text{ N} = \frac{K_{Al}}{K_{Fe}} 0.0764 \text{ N} = \frac{1.000022}{1400} 0.0764 \text{ N} = 5.46\times10^{-5} \text{ N}.$

But the weight of the aluminum cube is:

$W = m_{Al}g = \rho_{Al}a^3 g = (2.7\times10^3 \text{ kg/m}^3)(0.010 \text{ m})^3(9.8 \text{ m/s}^2) = 0.0265 \text{ N}.$

So the ratio of the magnetic force on the aluminum cube to the weight of the cube is $\frac{5.46\times10^{-5} \text{ N}}{0.0265 \text{ N}} = 2.1\times10^{-3}$, and the magnet cannot lift it.

(c) If the magnet tries to lift a silver cube of the same dimensions as the iron block, then the DOWNWARD force felt by the cube is:

$F_{Al} = \frac{\mu_{Ag}B}{a} = \frac{\mu_{Ag}}{\mu_{Fe}} 0.0764 \text{ N} = \frac{K_{Ag}}{K_{Fe}} 0.0764 \text{ N} = \frac{(1.00 - 2.6\times10^{-5})}{1400} 0.0764 \text{ N}$

$= -5.46\times10^{-5} \text{ N}.$

But the weight of the silver cube is:

$W = m_{Ag}g = \rho_{Ag}a^3 g = (10.5\times10^3 \text{ kg/m}^3)(0.010 \text{ m})^3(9.8 \text{ m/s}^2) = 0.103 \text{ N}.$

So the ratio of the magnetic force on the silver cube to the weight of the cube is $\dfrac{5.46\times10^{-5}\ \text{N}}{0.103\ \text{N}}=5.3\times10^{-4}$, and the magnet's effect would not be noticeable.

29-67: $M_{Fe}=(\mu_{atom\,of\,Fe})(\#\,Fe\ \text{atoms}/\text{m}^3)=(\mu_{atom\,of\,Fe})N_A(\#\,Fe\ \text{moles}/\text{m}^3)$

$$\Rightarrow M_{Fe}=(\mu_{atom\,of\,Fe})N_A\frac{\rho_{Fe}}{m_{mol}(\text{Fe})}\Rightarrow \mu_{atom\,of\,Fe}=\frac{M_{Fe}m_{mol}(\text{Fe})}{N_A\rho_{Fe}}$$

$$\Rightarrow \mu_{atom\,of\,Fe}=\frac{(8.0\times10^4\ \text{A/m})(0.0558\ \text{kg/mol})}{(6.02\times10^{23}\ \text{atoms/mol})(7.8\times10^3\ \text{kg/m}^3)}=9.51\times10^{-25}\ \text{Am}^2.$$

$$\Rightarrow \mu_{atom\,of\,Fe}=\frac{9.51\times10^{-25}\ \text{Am}^2}{9.27\times10^{-24}\ \text{Am}^2}\mu_B=0.103\mu_B.$$

29-68: (a) $j_c(\text{max})=\dfrac{E_o}{\rho}=\dfrac{0.3000\ \text{V/m}}{2300\ \Omega\text{m}}=1.30\times10^{-4}\ \text{A/m}^2.$

(b) $j_D(\text{max})=\varepsilon_o\dfrac{dE}{dt}=\varepsilon_o\omega E_o=2\pi\varepsilon_o fE_o=2\pi\varepsilon_o(60\ \text{Hz})(0.3000\ \text{V/m})$

$\Rightarrow j_D(\text{max})=1.00\times10^{-9}\ \text{A/m}^2.$

(c) If $j_c=j_D\Rightarrow\dfrac{E_o}{\rho}=\omega\varepsilon_oE_o\Rightarrow\omega=\dfrac{1}{\rho\varepsilon_o}=4.91\times10^7\ \text{rad/s}$

$\Rightarrow f=\dfrac{\omega}{2\pi}=\dfrac{4.91\times10^7\ \text{rad/s}}{2\pi}=7.82\times10^6\ \text{Hz}.$

(d) The two current densities are out of phase by 90° because one has a sine function and the other has a cosine, so the displacement current leads the conduction current by 90°.

29-69: (a) $j_c=\dfrac{I}{A}=\dfrac{V}{AR}=\dfrac{VA}{Ad\rho}=\dfrac{V}{d\rho}=\dfrac{q}{Cd\rho}=\dfrac{qd}{K\varepsilon_oAd\rho}$ and $RC=\dfrac{\rho d}{A}\dfrac{K\varepsilon_oA}{d}=K\varepsilon_o\rho.$

$\Rightarrow j_c(t)=\dfrac{q}{K\varepsilon_oA\rho}=\dfrac{Q_o}{K\varepsilon_oA\rho}e^{-t/RC}=\dfrac{Q_o}{K\varepsilon_oA\rho}e^{-t/K\varepsilon_o\rho}.$

(b) $j_D(t)=K\varepsilon_o\dfrac{dE}{dt}=K\varepsilon_o\dfrac{d(\rho j_c)}{dt}=K\varepsilon_o\rho\dfrac{Q_o}{K\varepsilon_oA\rho}\dfrac{d(e^{-t/K\varepsilon_o\rho})}{dt}=-\dfrac{Q_o}{K\varepsilon_oA\rho}e^{-t/K\varepsilon_o\rho}=-j_c(t).$

29-70: The amount of charge on a length Δx of the belt is:

$\Delta Q=L\Delta x\sigma\Rightarrow I=\dfrac{\Delta Q}{\Delta t}=L\dfrac{\Delta x}{\Delta t}\sigma=Lv\sigma.$

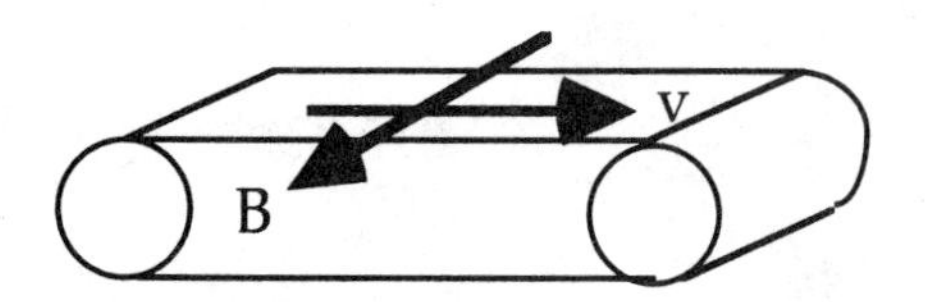

Approximating the belt as an infinite sheet: $B=\dfrac{\mu_oI}{2L}=\dfrac{\mu_ov\sigma}{2}$, out of the page, as shown at left.

29-71: The charge on a ring of radius r is $q=\sigma A=\sigma 2\pi rdr=\dfrac{2Qrdr}{a^2}$. If the disk rotates at n turns per second, then the current from that ring is:

$$I = \frac{\Delta q}{\Delta t} = nq = \frac{2Qnrdr}{a^2} \Rightarrow dB = \frac{\mu_o I}{2r} = \frac{\mu_o}{2r}\frac{2Qnrdr}{a^2} = \frac{\mu_o nQdr}{a^2}.$$

So we integrate out from the center to the edge of the disk to find:

$$B = \int_0^a dB = \int_0^a \frac{\mu_o nQdr}{a^2} = \frac{\mu_o nQ}{a}.$$

29-72: There are two parts to the magnetic field: that from the half loop and that from the straight wire segment running from -a to a.

$$B_x(ring) = \frac{1}{2}B_{loop} = -\frac{\mu_o Ia^2}{4(x^2+a^2)^{3/2}}$$

$$dB_y(ring) = dB\sin\theta\sin\phi = \frac{\mu_o I}{4\pi}\frac{dl}{(x^2+a^2)}\frac{x}{(x^2+a^2)^{1/2}}\sin\phi = \frac{\mu_o Iax\sin\phi d\phi}{4\pi(x^2+a^2)^{3/2}}$$

$$\Rightarrow B_y(ring) = \int_0^\pi dB_y(ring) = \int_0^\pi \frac{\mu_o Iax\sin\phi\, d\phi}{4\pi(x^2+a^2)^{3/2}} = \frac{\mu_o Iax}{4\pi(x^2+a^2)^{3/2}}\cos\phi\Big|_0^\pi = -\frac{\mu_o Iax}{2\pi(x^2+a^2)^{3/2}}.$$

$B_y(rod) = \dfrac{\mu_o Ia}{2\pi x(x^2+a^2)^{1/2}}$, using Eq. (29-8). So the total field components are:

$$B_x = -\frac{\mu_o Ia^2}{4(x^2+a^2)^{3/2}} \text{ and } B_y = \frac{\mu_o Ia}{2\pi x(x^2+a^2)^{1/2}}\left(1+\frac{x^2}{x^2+a^2}\right) = \frac{\mu_o Ia^3}{2\pi x(x^2+a^2)^{3/2}}.$$

29-73: (a) The magnetic force per unit length between two parallel, long wires is:

$$\frac{F}{L} = IB = \frac{\mu_o}{2\pi d}I^2 = \frac{\mu_o}{2\pi d}\left(\frac{I_o}{\sqrt{2}}\right)^2 = \frac{\mu_o}{4\pi d}\left(\frac{V}{R}\right)^2 = \frac{\mu_o}{4\pi d}\left(\frac{Q_o}{RC}\right)^2$$

$$\frac{F}{L} = \frac{m}{L}a = \lambda a = \frac{\mu_o}{4\pi d}\left(\frac{Q_o}{RC}\right)^2 \Rightarrow a = \frac{\mu_o Q_o^{\,2}}{4\pi\lambda\, dR^2C^2} \Rightarrow v_o = at = aRC = \frac{\mu_o Q_o^{\,2}}{4\pi\lambda\, dRC}.$$

(b) $$v_o = \frac{\mu_o(CV)^2}{4\pi\lambda\, dRC} = \frac{\mu_o CV^2}{4\pi\lambda\, dR} = \frac{\mu_o(1.0\times10^{-6}\text{ F})(2500\text{ V})^2}{4\pi(1.75\times10^{-3}\text{ kg/m})(0.02\text{ m})(0.032\,\Omega)} = 0.558\text{ m/s}.$$

(c) Height that the wire reaches above the original height:

$$\frac{1}{2}mv_o^{\,2} = mgh \Rightarrow h = \frac{v_o^{\,2}}{2g} = \frac{(0.558\text{ m/s})^2}{2(9.80\text{ m/s}^2)} = 0.0159\text{ m}.$$

Chapter 30: Electromagnetic Induction

30-1: $I = \frac{\mathcal{E}}{R} = -\frac{1}{R}\frac{d\Phi_B}{dt} = -\frac{A}{R}\frac{dB}{dt} = -\frac{0.0900\ \text{m}^2}{0.300\ \Omega}(-0.190\ \text{T/s}) = 0.0570\ \text{A}.$

30-2: $I = \frac{NA}{R}\frac{dB}{dt} \Rightarrow \frac{dB}{dt} = \frac{RI}{NA} = \frac{(40.0\ \Omega)(0.240\ \text{A})}{(500)\pi(0.0300\ \text{m})^2} = 6.79\ \text{T/s}.$

30-3: $\Phi_{B_f} = NBA,$ and $\Phi_{B_i} = NBA\cos 45° \Rightarrow \Delta\Phi_B = NBA(1-\cos 45°)$

$\Rightarrow \mathcal{E} = -\frac{\Delta\Phi_B}{\Delta t} = -\frac{NBA(1-\cos 45°)}{\Delta t} = -\frac{(50)(0.975\ \text{T})(0.120\ \text{m})(0.25\ \text{m})(1-\cos 45°)}{0.0800\ \text{s}}$

$\Rightarrow |\mathcal{E}| = 5.35\ \text{V}.$

30-4: $|\mathcal{E}| = \frac{\Delta\Phi_B}{\Delta t} = \frac{NBA}{\Delta t} = \frac{(400)(6.0\times10^{-5}\ \text{T})(1.6\times10^{-3}\ \text{m}^2)}{0.020\ \text{s}} = 1.9\times10^{-3}\ \text{V}.$

30-5: (a) $\mathcal{E} = \frac{1}{2}BR^2\omega \Rightarrow \omega = \frac{2\mathcal{E}}{BR^2} = \frac{2(5.00\ \text{V})}{(1.20\ \text{T})(0.320\ \text{m})^2} = 81.4\ \text{rad/s} = 13.0\ \text{rev/s}.$

(b) $P = IV = \tau\omega \Rightarrow \tau = \frac{IV}{\omega} = \frac{(200\ \text{A})(5.00\ \text{V})}{81.4\ \text{rad/s}} = 12.3\ \text{N}\cdot\text{m}.$

30-6: $\mathcal{E} = -\frac{d\Phi_B}{dt} = -\frac{d}{dt}(NBA\cos\omega t) = NBA\omega\sin\omega t \Rightarrow \mathcal{E}_{max} = NBA\omega$

$\Rightarrow \omega = \frac{\mathcal{E}_{max}}{NBA} = \frac{3.00\times10^{-2}\ \text{V}}{(80)(0.0250\ \text{T})(0.120\ \text{m})^2} = 1.04\ \text{rad/s}.$

30-7: $\mathcal{E} = \frac{\Delta\Phi_B}{\Delta t} = \frac{NBA}{\Delta t} = IR = \left(\frac{Q}{\Delta t}\right)R \Rightarrow QR = NBA \Rightarrow Q = \frac{NBA}{R}.$

30-8: From **Ex. (30-7)**, $Q = \frac{NBA}{R} = \frac{(60)(1.80\ \text{T})(1.50\times10^{-4}\ \text{m}^2)}{9.00\ \Omega + 16.0\ \Omega} = 6.48\times10^{-4}\ \text{C}.$

30-9: From **Ex. (30-7)**,

$Q = \frac{NBA}{R} \Rightarrow B = \frac{QR}{NA} = \frac{(7.28\times10^{-5}\ \text{C})(50.0\ \Omega + 30.0\ \Omega)}{(160)(4.00\times10^{-4}\ \text{m}^2)} = 0.0910\ \text{T}.$

30-10: (a) $\mathcal{E} = \frac{Nd\Phi_B}{dt} = NA\frac{d}{dt}(B) = NA\frac{d}{dt}\left((0.0100\ \text{T/s})t + (2.00\times10^{-4}\ \text{T/s}^3)t^3\right)$

$\Rightarrow \mathcal{E} = NA\left((0.010\ \text{T/s}) + (6.0\times10^{-4}\ \text{T/s})t^2\right) = 5.03\times10^{-3}\ \text{V} + (3.02\times10^{-4}\ \text{V/s}^2)t^2.$

(b) At $t = 10\ \text{s} \Rightarrow \mathcal{E} = 7.04\times10^{-3}\ \text{V} + (3.02\times10^{-4}\ \text{V/s}^2)(10\ \text{s})^2 = +0.0352\ \text{V}$

$\Rightarrow I = \frac{\mathcal{E}}{R} = \frac{0.0352\ \text{V}}{500\ \Omega} = 7.04\times10^{-5}\ \text{A}.$

30-11: From Example 30-5,

$$\mathcal{E}_{av} = \frac{2N\omega BA}{\pi} = \frac{2(500)(14\ \text{rev/s})(2\pi\ \text{rad/rev})(0.20\ \text{T})(0.10\ \text{m})^2}{\pi} = 56\ \text{V}$$

30-12: (a) $\mathcal{E} = \frac{d\Phi_B}{dt} = \frac{d}{dt}(NBA\cos\omega t) = NBA\omega\sin\omega t$ and 1200 rev/min = 20 rev/s, so:

$\Rightarrow \mathcal{E}_{max} = NBA\omega = (200)(0.080\ \text{T})\pi(0.030\ \text{m})^2(20\ \text{rev/s})(2\pi\ \text{rad/rev}) = 5.68\ \text{V}.$

(b) Average $\mathcal{E} = \frac{2}{\pi}\mathcal{E}_{max} = \frac{2}{\pi}5.68\ \text{V} = 3.62\ \text{V}.$

30-13: (a) If the magnetic field is increasing into the page, the induced magnetic field must oppose that change and point opposite the external field's direction, thus requiring a counter-clockwise current in the loop.
(b) If the magnetic field is decreasing into the page, the induced magnetic field must oppose that change and point in the external field's direction, thus requiring a clockwise current in the loop.
(c) If the magnetic field is constant, there is no changing flux, and therefore no induced current in the loop.

30-14: (a) When the switch is opened, the magnetic field to the right decreases. Therefore the second coil's induced current produces its own field to the right. That means that the current must pass through the resistor from point a to point b.
(b) If coil B is moved closer to coil A, more flux passes through it toward the right. Therefore the induced current must produce its own magnetic field to the left to oppose the increased flux. That means that the current must pass through the resistor from point b to point a.
(c) If the variable resistor R is decreased, then more current flows through coil A, and so a stronger magnetic field is produced, leading to more flux to the right through coil B. Therefore the induced current must produce its own magnetic field to the left to oppose the increased flux. That means that the current must pass through the resistor from point b to point a.

30-15: (a) With current passing from $a \to b$ and increasing, the magnetic field becomes stronger to the left, so the induced field points right, and the induced current must flow from right to left through the resistor.
(b) If the current passes from $b \to a$, and is decreasing, then there is less magnetic field pointing right, so the induced field points right, and the induced current must flow from right to left through the resistor.
(c) If the current passes from $b \to a$, and is increasing, then there is more magnetic field pointing right, so the induced field points left, and the induced current must flow from left to right through the resistor.

30-16: (a) Using Equation (30-6): $\mathcal{E} = vBL \Rightarrow B = \frac{\mathcal{E}}{vL} = \frac{2.00\ \text{V}}{(6.00\ \text{m/s})(0.190\ \text{m})} = 1.75\ \text{T}.$
(b) Point a is at a higher potential than point b, because there are more positive charges there.

30-17: $[vBL] = \left[\frac{\text{m}}{\text{s}}\text{Tm}\right] = \left[\frac{\text{m}}{\text{s}}\frac{\text{N}\cdot\text{s}}{\text{C}\cdot\text{m}}\text{m}\right] = \left[\frac{\text{N}\cdot\text{m}}{\text{C}}\right] = \left[\frac{J}{\text{C}}\right] = [V].$

30-18: (a) $\mathcal{E} = vBL = (6.00\ \text{m/s})(0.500\ \text{T})(0.055\ \text{m}) = 0.165\ \text{V}.$
(b) The potential difference between the ends of the rod is just the motional emf $V = 0.165\ \text{V}$.
(c) The positive charges are moved to end b, so b is at the higher potential.

30-19: (a) $\mathcal{E} = vBL \Rightarrow v = \frac{\mathcal{E}}{BL} = \frac{0.320\ \text{V}}{(1.20\ \text{T})(0.065\ \text{m})} = 4.10\ \text{m/s}.$

(b) $I = \frac{\mathcal{E}}{R} = \frac{0.320\ \text{V}}{0.800\ \Omega} = 0.400\ \text{A}.$

(c) $F = ILB = (0.400\ \text{A})(0.065\ \text{m})(1.20\ \text{T}) = 0.0312\ \text{N}$, to the left, since you must pull it to get the current to flow.

30-20: (a) $\mathcal{E} = vBL = (6.00\ \text{m/s})(0.300\ \text{T})(0.500\ \text{m}) = 0.900\ \text{V}.$
(b) The current flows counter-clockwise since its magnetic field must oppose the increasing flux through the loop.
(c) $F = ILB = \frac{\mathcal{E}LB}{R} = \frac{(0.900\ \text{V/m})(0.500\ \text{m})(0.300\ \text{T})}{2.00\ \Omega} = 0.0675\ \text{N}$, to the right
(d) $P_{mech} = Fv = (0.0675\ \text{N})(6.00\ \text{m/s}) = 0.405\ \text{W}.$
$P_{elec} = \frac{\mathcal{E}^2}{R} = \frac{(0.900\ \text{V})^2}{2.00\ \Omega} = 0.405\ \text{W}$. So both rates are equal.

30-21: For the loop pulled through the region of magnetic field,
(a)

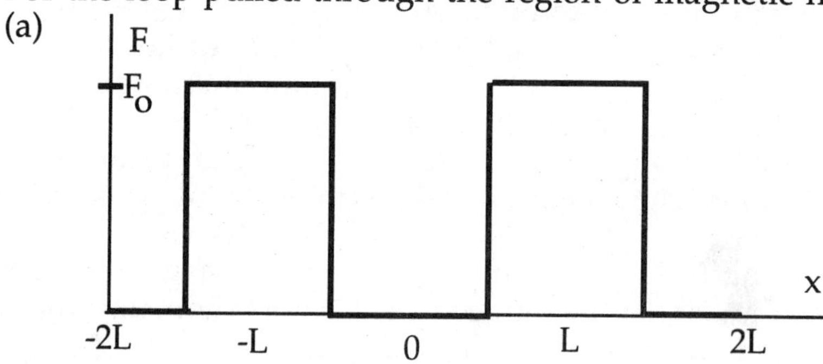

(b)

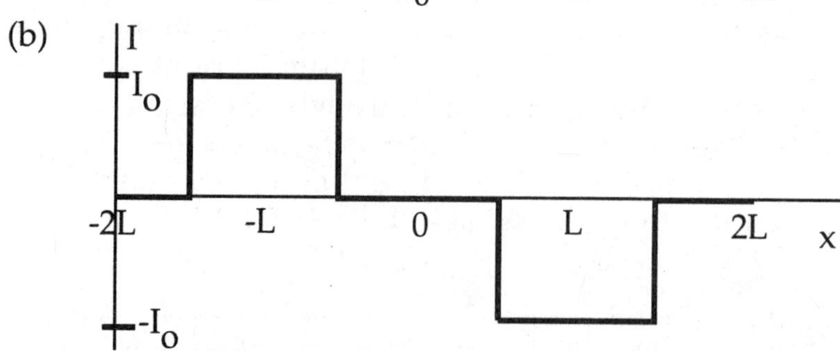

Where $\mathcal{E} = vBL = IR \Rightarrow I_o = \frac{vBL}{R}$ and $F_o = ILB = \frac{vB^2L^2}{R}$.

30-22: $\mathcal{E} = \frac{d\Phi_B}{dt} = \frac{d}{dt}(BA) = \frac{d}{dt}(\mu_o nIA) = \mu_o nA\frac{dI}{dt} \Rightarrow \frac{dI}{dt} = \frac{E \cdot 2\pi r}{\mu_o nA}$

$\Rightarrow \frac{dI}{dt} = \frac{(6.00\times10^{-6}\text{ V/m})2\pi(0.0300)}{\mu_o(500\text{ m}^{-1})\pi(0.0125\text{ m})^2} = 3.68\text{ A/s}.$

30-23: $\varepsilon = \frac{d\Phi_B}{dt} = \frac{d}{dt}(BA) = \frac{d}{dt}(\mu_o nIA) = \mu_o nA\frac{dI}{dt}$ and $\oint \vec{E}\cdot d\vec{l} = \varepsilon$

$\Rightarrow E = \frac{\mathcal{E}}{2\pi r} = \frac{\mu_o nA}{2\pi r}\frac{dI}{dt} = \frac{\mu_o nr}{2}\frac{dI}{dt}.$

(a) $r = 0.40\text{ cm} \Rightarrow E = \frac{\mu_o(800\text{ m}^{-1})(0.0040\text{ m})}{2}(50\text{ A/s}) = 1.01\times10^{-4}\text{ V/m}.$

(b) $r = 0.80\text{ cm} \Rightarrow E = \frac{\mu_o(800\text{ m}^{-1})(0.0080\text{ m})}{2}(50\text{ A/s}) = 2.01\times10^{-4}\text{ V/m}.$

30-24: $\mathcal{E} = -\frac{N\Delta\Phi_B}{\Delta t} = -\frac{NA(B_f - B_i)}{\Delta t} = \frac{NA\mu_o nI}{\Delta t} = \frac{\mu_o(8)(6.00\times10^{-4}\text{ m})(8000\text{ m}^{-1})(0.250\text{ A})}{0.0500\text{ s}}$

$\Rightarrow \mathcal{E} = 2.41\times10^{-4}\text{ V}.$

30-25: (a) The induced electric field lines are concentric circles since they cause the current to flow in circles.

(b) $E = \frac{1}{2\pi r}\mathcal{E} = \frac{1}{2\pi r}\frac{d\Phi_B}{dt} = \frac{1}{2\pi r}A\frac{dB}{dt} = \frac{r}{2}\frac{dB}{dt} = \frac{0.100\text{ m}}{2}(0.0450\text{ T/s})$

$\Rightarrow E = 2.25\times10^{-3}\text{ V/m}$, in the clockwise direction, since the induced magnetic field must reinforce the decreasing external magnetic field.

(c) $I = \frac{\mathcal{E}}{R} = \frac{\pi r^2}{R}\frac{dB}{dt} = \frac{\pi(0.100\text{ m})^2}{2.00\,\Omega}(0.0450\text{ T/s}) = 7.07\times10^{-4}\text{ A}.$

(d) $\varepsilon = IR = IR_{TOT}/2 = \frac{(7.07\times10^{-4}\text{A})(2.0\Omega)}{2} = 7.0\times10^{-4}\text{V}.$

(e) If the ring was cut and the ends separated slightly, then there would be a potential difference between the ends equal to the induced emf:

$\varepsilon = \pi r^2\frac{dB}{dt} = \pi(0.100\text{ m})^2(0.0450\text{ T/s}) = 1.41\times10^{-3}\text{ V}.$

30-26: (a) $\frac{d\Phi_B}{dt} = A\frac{dB}{dt} = \pi r_1^2\frac{dB}{dt}.$

(b) $E = \frac{1}{2\pi r_1}\frac{d\Phi_B}{dt} = \frac{\pi r_1^2}{2\pi r_1}\frac{dB}{dt} = \frac{r_1}{2}\frac{dB}{dt}.$

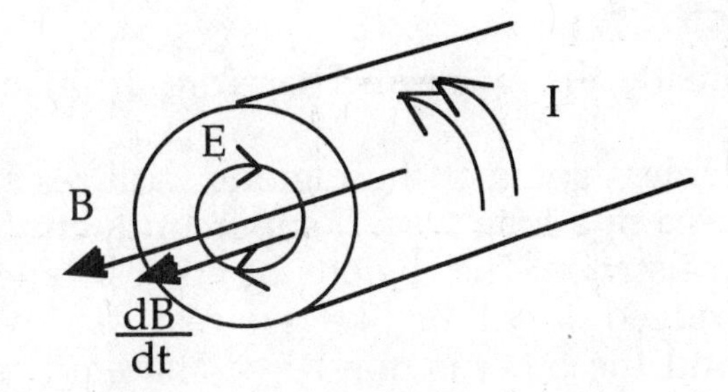

(c) All the flux is within r < R, so outside the solenoid

$$E = \frac{1}{2\pi r}\frac{d\Phi_B}{dt} = \frac{\pi R^2}{2\pi r}\frac{dB}{dt} = \frac{R^2}{2r}\frac{dB}{dt}.$$

(e) At $r = R/2$:

$$\Rightarrow \mathcal{E} = \frac{d\Phi_B}{dt} = \pi(R/2)^2\frac{dB}{dt} = \frac{\pi R^2}{4}\frac{dB}{dt}.$$

(f) At $r = R \Rightarrow \mathcal{E} = \frac{d\Phi_B}{dt} = \pi R^2\frac{dB}{dt}$.

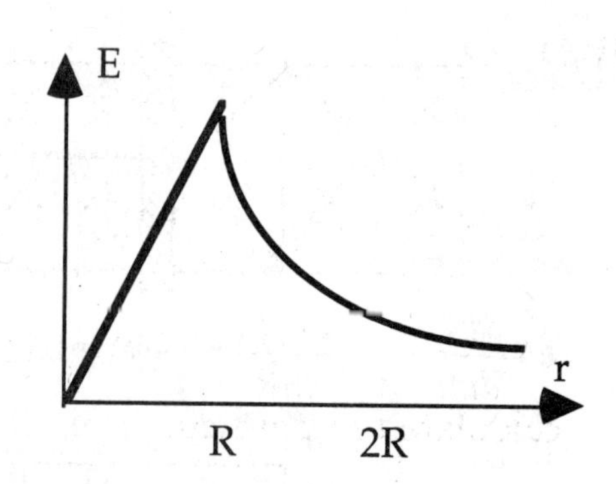

(g) At $r = 2R \Rightarrow \mathcal{E} = \frac{d\Phi_B}{dt} = \pi R^2\frac{dB}{dt}$. Note that the emf is independent of the distance from the center of the cylinder as long as one is outside it.

30-27: (a) For magnetic fields less than the critical field, there is no internal magnetic field, so:

Inside the superconductor: $\vec{B} = 0,\ \vec{M} = -\frac{\vec{B}_o}{\mu_o} = -\frac{(0.118\ \text{T})\hat{i}}{\mu_o} = -(9.39\times10^4\ \text{A/m})\hat{i}$.

Outside the superconductor: $\vec{B} = \vec{B}_o = -(0.118\ \text{T})\hat{i},\ \vec{M} = 0$.

(b) For magnetic fields greater than the critical field, $\chi = 0 \Rightarrow \vec{M} = 0$ both inside and outside the superconductor, and $\vec{B} = \vec{B}_o = (0.224\ \text{T})\hat{i}$, both inside and outside the superconductor.

30-28: (a) Just under $\vec{B}_{c1}$ (threshold of superconducting phase), the magnetic field in the material must be zero, and $\vec{M} = -\frac{\vec{B}_{c1}}{\mu_o} = -\frac{55\times10^{-3}\ \text{T}\hat{i}}{\mu_o} = -(4.38\times10^4\ \text{A/m})\hat{i}$.

(b) Just over $\vec{B}_{c2}$ (threshold of normal phase), there is zero magnetization, and $\vec{B} = \vec{B}_{c2} = (15.0\ \text{T})\hat{i}$.

30-29: In a superconductor there is no internal magnetic field, and so there is no changing flux and no induced emf, and no induced electric field.

$$0 = \oint_{\substack{\text{Inside}\\ \text{material}}} \vec{B}\cdot d\vec{l} = \mu_o I_{encl} = \mu_o(I_c + I_D) = \mu_o I_c \Rightarrow I_c = 0,$$ and so there is no current inside the material. Therefore, it must all be at the surface of the cylinder.

30-30: Unless some of the regions with resistance completely fill a cross-sectional area of a long type-II superconducting wire, there will still be no total resistance. The regions of no resistance provide the path for the current. Indeed, it will be like two resistors in parallel, where one has zero resistance and the other is non-zero. The equivalent resistance is still zero.

30-31: (a)

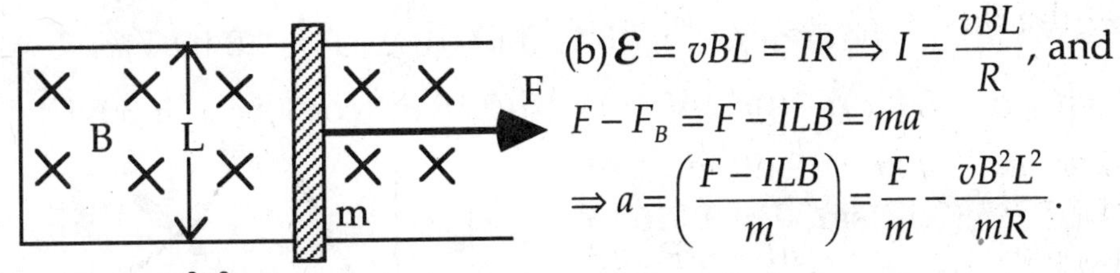

(b) $\mathcal{E} = vBL = IR \Rightarrow I = \frac{vBL}{R}$, and

$F - F_B = F - ILB = ma$

$\Rightarrow a = \left(\frac{F - ILB}{m}\right) = \frac{F}{m} - \frac{vB^2L^2}{mR}.$

$\Rightarrow \frac{dv}{dt} = \frac{F}{m} - \frac{vB^2L^2}{mR} \Rightarrow v(t) = v_t\left(1 - e^{-t(B^2L^2/mR)}\right)$, where v_t is the terminal velocity calculated in part (b).

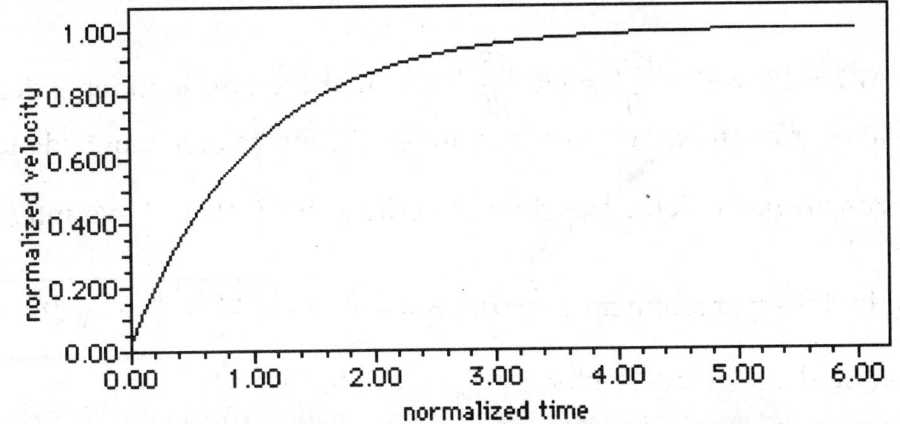

The terminal speed v_t occurs when the pulling force is equaled by the magnetic force: $F_B = ILB = \left(\frac{v_tLB}{R}\right)LB = \frac{v_tL^2B^2}{R} = F \Rightarrow v_t = \frac{FR}{L^2B^2}.$

30-32: Wire A: $\vec{v} \times \vec{B} = 0 \Rightarrow \mathcal{E} = 0.$
Wire C: $\mathcal{E} = vBL\sin\phi = (0.750 \text{ m/s})(0.300 \text{ T})(1.00 \text{ m})\sin 45° = 0.159 \text{ V}.$
Wire D: $\mathcal{E} = vBL\sin\phi = (0.750 \text{ m/s})(0.300 \text{ T})\sqrt{2}(1.00 \text{ m})\sin 45° = 0.225 \text{ V}.$

30-33: (a) $d\mathcal{E} = (\vec{v} \times \vec{B}) \cdot d\vec{r} = \omega rBdr \Rightarrow \mathcal{E} = \int_0^L \omega rBdr = \frac{1}{2}\omega L^2B = 0.0722 \text{ V}.$

(b) The potential difference between its ends is the same as the induced emf.

30-34: $P = \frac{V^2}{R} = \frac{\mathcal{E}^2}{R} = \frac{(vBL)^2}{R} \Rightarrow R = \frac{(vBL)^2}{P} = \frac{((4.0 \text{ m/s})(0.50 \text{ T})(3.0 \text{ m}))^2}{200 \text{ W}} = 0.18\ \Omega.$

30-35: Recall $v_{orbit} = \sqrt{\frac{Gm_E}{r}} = \sqrt{\frac{(6.67 \times 10^{-11} \text{ Nm}^2/\text{kg}^2)(5.97 \times 10^{24} \text{ kg})}{(6.38 \times 10^6 \text{ m} + 0.17 \times 10^6 \text{ m})}} = 7800 \text{ m/s}.$
Therefore: $\mathcal{E} = vBL = (7800 \text{ m/s})(5.0 \times 10^{-5} \text{ T})(500 \text{ m}) = 195 \text{ V}.$

30-36: (a) $\mathcal{E} = (\vec{v} \times \vec{B}) \cdot \vec{L} = ((2.5 \text{ m/s})\hat{i} \times ((0.12 \text{ T})\hat{i} - (0.22 \text{ T})\hat{j} - (0.09 \text{ T})\hat{k}) \cdot \vec{L}$

$\Rightarrow \mathcal{E} = \left((0.255 \text{ V/m})\hat{j} - (0.55 \text{ V/m})\hat{k}\right) \cdot \left((0.15 \text{ m})(\cos 30° \hat{i} + \sin 30° \hat{j})\right)$

$\Rightarrow \mathcal{E} = (0.225)(0.15)\sin 30° = 0.0169 \text{ V}.$

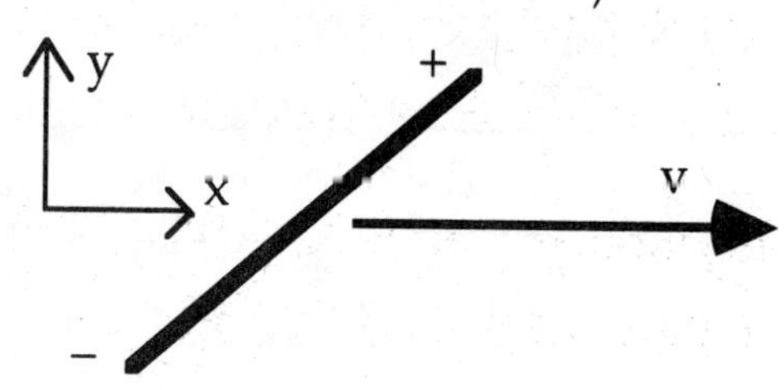

30-37: (a)

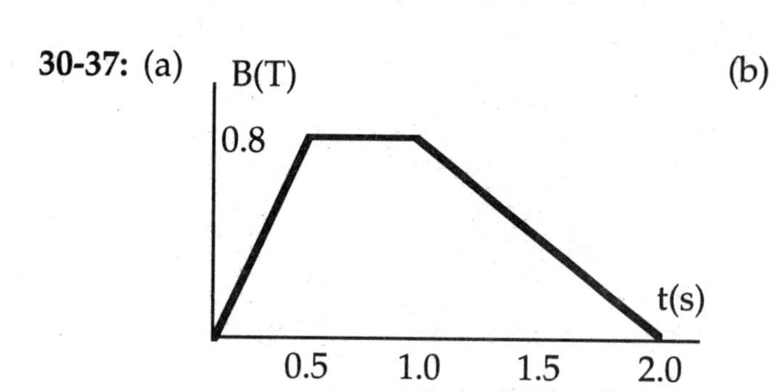

(b)

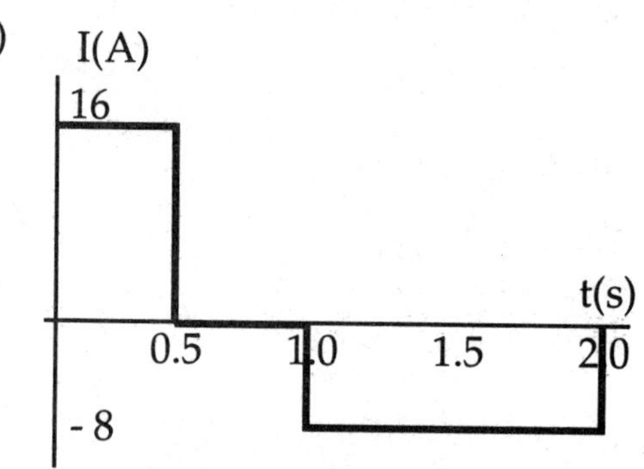

(c) $E_{\text{max}} = \dfrac{\mathcal{E}}{N2\pi r} = \dfrac{1}{2\pi r} NA \dfrac{dB}{dt} = \dfrac{1}{2\pi r} \pi r^2 \dfrac{dB}{dt} = \dfrac{r}{2} \dfrac{dB}{dt} = \dfrac{(0.50 \text{ m})}{2} \dfrac{0.80 \text{ T}}{0.50 \text{ s}} = 0.4 \text{ V}.$

30-38: (a) $I = \dfrac{\varepsilon}{R} = \dfrac{1}{R}\dfrac{d\Phi_B}{dt} = \dfrac{1}{R}\dfrac{d(BA\cos\omega t)}{dt} = \dfrac{BA\omega \sin\omega t}{R}.$

(b) $P = I^2 R = \dfrac{B^2 A^2 \omega^2 \sin^2 \omega t}{R}.$

(c) $\mu = IA = \dfrac{BA^2 \omega \sin \omega t}{R}.$

(d) $\tau = \mu B \sin\phi = \mu B \sin \omega t = \dfrac{B^2 A^2 \omega \sin^2 \omega t}{R}.$

(e) $P = \tau\omega = \dfrac{B^2 A^2 \omega^2 \sin^2 \omega t}{R}$, which is the same as part (b).

30-39: (a) $\Phi_B = BA = \dfrac{\mu_0 i}{2a}\pi a^2 = \dfrac{\mu_0 i \pi a}{2}.$

(b) $\mathcal{E} = -\dfrac{d\Phi_B}{dt} = iR \Rightarrow -\dfrac{d}{dt}\left(\dfrac{\mu_0 i \pi a}{2}\right) = -\dfrac{\mu_0 \pi a}{2}\dfrac{di}{dt} = iR \Rightarrow \dfrac{di}{dt} = -i\dfrac{2R}{\mu_0 \pi a}$

$\Rightarrow \dfrac{di}{dt} = -i\dfrac{2R}{\mu_0 \pi a}.$

(c) Solving $\dfrac{di}{i} = -dt\dfrac{2R}{\mu_0 \pi a}$ for $i(t)$ yields $i(t) = i_0 e^{-t(2R/\mu_0 \pi a)}.$

(d) We want $i(t) = i_0(0.010) = i_0 e^{-t(2R/\mu_0 \pi a)} \Rightarrow \ln(0.010) = -t(2R/\mu_0 \pi a)$

$\Rightarrow t = -\dfrac{\mu_0 \pi a}{2R}\ln(0.010) = -\dfrac{\mu_0 \pi (0.50 \text{ m})}{2(0.10\ \Omega)}\ln(0.010) = 4.55 \times 10^{-5} \text{ s}.$

(e) We can ignore the self-induced currents because it takes only a very short time for them to die out.

30-40: (a) Rotating about the y-axis:

$$\varepsilon_{\max} = \frac{d\Phi_B}{dt} = \omega BA = (25.0 \text{ rad/s})(0.300 \text{ T})(4.00\times 10^{-2} \text{ m}) = 0.300 \text{ V}.$$

(b) Rotating about the x-axis: $\frac{d\Phi_B}{dt} = 0 \Rightarrow \mathcal{E} = 0.$

(c) Rotating about the z-axis:

$$\mathcal{E}_{\max} = \frac{d\Phi_B}{dt} = \omega BA = (25.0 \text{ rad / s})(0.300 \text{ T})(4.00\times 10^{-2} \text{ m}) = 0.300 \text{ V}.$$

30-41: (a) $\mathcal{E} = -\frac{\Delta\Phi_B}{\Delta t} = -B\frac{\Delta A}{\Delta t} = -B\frac{-\pi r^2}{\Delta t} = (1.50 \text{ T})\frac{\pi(0.084/2 \text{ m})^2}{0.200 \text{ s}} = 0.0416 \text{ V}.$

(b) Since the flux through the loop is decreasing, the induced current must produce a field that goes into the page. Therefore the current flows from point a through the resistor to point b.

30-42: (a) When $I = i \Rightarrow B = \frac{\mu_o i}{2\pi r}.$

(b) $d\Phi_B = BdA = \frac{\mu_o i}{2\pi r} L dr.$

(c) $\Phi_B = \int_a^b d\Phi_B = \frac{\mu_o iL}{2\pi}\int_a^b \frac{dr}{r} = \frac{\mu_o iL}{2\pi}\ln(b/a).$

(d) $\mathcal{E} = \frac{d\Phi_B}{dt} = \frac{\mu_o L}{2\pi}\ln(b/a)\frac{di}{dt}.$

(e) $\mathcal{E} = \frac{\mu_o(0.200 \text{ m})}{2\pi}\ln(0.300/0.100)(8.50 \text{ A/s}) = 3.74\times 10^{-7} \text{ V}.$

30-43: At point a: $\mathcal{E} = \frac{d\Phi_B}{dt} = A\frac{dB}{dt} = \pi r^2\frac{dB}{dt}$ and $F = qE = q\frac{\mathcal{E}}{2\pi r} = \frac{qr}{2}\frac{dB}{dt}$, to the left.
At point b, the field is the same magnitude as at a since they are the same distance from the center. So $F = \frac{qr}{2}\frac{dB}{dt}$, but upward.
At point c, there is no force by symmetry arguments: one cannot have one direction picked out over any other, so the force must be zero.

30-44: (a) $|\mathcal{E}| = \frac{d\Phi_B}{dt} = A\frac{d}{dt}\left(B_o e^{-t/\tau}\right) = -\frac{\pi a^2 B_o}{\tau}e^{-t/\tau} \Rightarrow i = \frac{|\mathcal{E}|}{R} = \frac{\pi a^2 B_o}{R\tau}e^{-t/\tau}.$

(b) $\Phi_B = AB_o e^{-t/\tau} = \pi a^2 B_o e^{-t/\tau} = \pi(0.20 \text{ m})^2(0.15 \text{ T})(e^{-0/\tau}) = 0.0188 \text{ T}\cdot\text{m}^2.$

And $i = \frac{\pi a^2 B_o}{R\tau}e^{-t/\tau} = \frac{\pi(0.20 \text{ m})^2(0.15 \text{ T})}{(0.050\,\Omega)(0.03 \text{ s})} = 12.6 \text{ A}.$

(c) the same calculation, but with t = 0.060 s yields:
$\Phi_B = 0.0188 \text{ T}\cdot\text{m}^2$, and $i = 6.28 \text{ A}.$

30-45: (a) $\mathcal{E} = -\dfrac{d\Phi_B}{dt} = -\dfrac{d}{dt}(AB\cos\omega t) = \omega AB_o e^{-t/\tau}\sin\omega t - AB_o\dfrac{de^{-t/\tau}}{dt}\cos\omega t$

$\Rightarrow \mathcal{E} = AB_o e^{-t/\tau}\left(\omega\sin\omega t + \dfrac{\cos\omega t}{\tau}\right) = \dfrac{AB_o e^{-t/\tau}}{\tau}\cos\omega t(\omega\tau\tan\omega t + 1).$

(b) At $t = 0 \Rightarrow \mathcal{E} = AB_o\left(\dfrac{1}{\tau}\right) \Rightarrow i = \dfrac{\mathcal{E}}{R} = \dfrac{AB_o}{\tau R} = \dfrac{\pi(0.20\text{ m})^2(0.15\text{ T})}{(0.030\text{ s})(0.050\ \Omega)} = 12.6\text{ A}.$

(c) The current is zero when the induced emf is zero:

$\Rightarrow \tan\omega t = -\dfrac{1}{\omega\tau} = -\dfrac{1}{(400\text{ rad/s})(0.030\text{ s})} = -\dfrac{1}{12} \Rightarrow \omega t = 3.06\text{ rad}$

$\Rightarrow t = \dfrac{3.06\text{ rad}}{400\text{ rad/s}} = 7.65\times10^{-3}\text{ s}.$

30-46: (a) $\Phi_B = BA = B_o\pi r_o^2(1 - 3(t/t_o)^2 + 2(t/t_o)^3).$

(b) $\varepsilon = -\dfrac{d\Phi_B}{dt} = B_o\pi r_o^2\dfrac{d}{dt}(1 - 3(t/t_o)^2 + 2(t/t_o)^3) = B_o\pi r_o^2(-6(t/t_o) + 6(t/t_o)^2)$

$\Rightarrow \mathcal{E} = -\dfrac{6B_o\pi r_o^2}{t_o}\left(\left(\dfrac{t}{t_o}\right)^2 - \left(\dfrac{t}{t_o}\right)\right)$, so at $t = 5.0\times10^{-3}$ s,

$\mathcal{E} = -\dfrac{6B_o\pi(0.0420\text{m})^2}{0.010\text{ s}}\left(\left(\dfrac{5.0\times10^{-3}\text{ s}}{0.010\text{ s}}\right)^2 - \left(\dfrac{5.0\times10^{-3}\text{ s}}{0.010\text{ s}}\right)\right) = 0.0665\text{ V},$

counterclockwise.

(c) $i = \dfrac{\mathcal{E}}{R_{total}} \Rightarrow R_{total} = r + R = \dfrac{\mathcal{E}}{i} \Rightarrow r = \dfrac{0.0655\text{ V}}{3.0\times10^{-3}\text{ A}} - 12\ \Omega = 10.2\ \Omega.$

(d) Evaluating the emf at $t = 1.21\times10^{-2}$ s, using the equations of part (b): $\mathcal{E} = -0.0676$ V, and the current flows clockwise, from b to a through the resistor.

(e) $\mathcal{E} = 0 \Rightarrow 0 = \left(\left(\dfrac{t}{t_o}\right)^2 - \left(\dfrac{t}{t_o}\right)\right) \Rightarrow 1 = \dfrac{t}{t_o} \Rightarrow t = t_o = 0.010\text{ s}.$

30-47: (a) $d\mathcal{E} = (\vec{v}\times\vec{B})\cdot d\vec{r} = vB\,dr = \dfrac{\mu_o Iv}{2\pi r}dr \Rightarrow \mathcal{E} = \dfrac{\mu_o Iv}{2\pi}\displaystyle\int_d^{d+L}\dfrac{dr}{r} = \dfrac{\mu_o Iv}{2\pi}\ln\left(\dfrac{d+L}{d}\right).$

(b) The magnetic force is strongest at the top end, closest to the current carrying wire. Therefore, the top end, point a, is the higher potential since the force on positive charges is greatest there, leading to more positives gathering at that end.

(c) If the single bar was replaced by a rectangular loop, the edges parallel to the wire would have no emf induced, but the edges perpendicular to the wire will have an emf induced, just as in part (b). However, no current will flow because each edge will have its highest potential closest to the current carrying wire. It would be like having two batteries of opposite polarity connected in a loop.

30-48: The change in current in the primary coil leads to a change in the flux passing through the secondary, and a resulting flow of charge.

$$\Phi_B = BA = \mu_o niA \Rightarrow \frac{d\Phi_B}{dt} = \mu_o nA\frac{di}{dt} \Rightarrow \mathcal{E} = N_2\frac{d\Phi_B}{dt} = N_2\mu_o nA\frac{di}{dt}$$

$$\Rightarrow i_2 = \frac{\mathcal{E}}{R} = \frac{\mu_o N_2 nA}{R}\frac{di}{dt}$$

$$\Rightarrow Q = i_2 dt = \frac{\mu_o N_2 nA}{R}\Delta i = \frac{\mu_o (20)(900/0.50\ \text{m})\pi(0.040\ \text{m})^2}{30.0\ \Omega}(2.0\ \text{A}) = 1.52\times10^{-5}\ \text{C}.$$

(b)

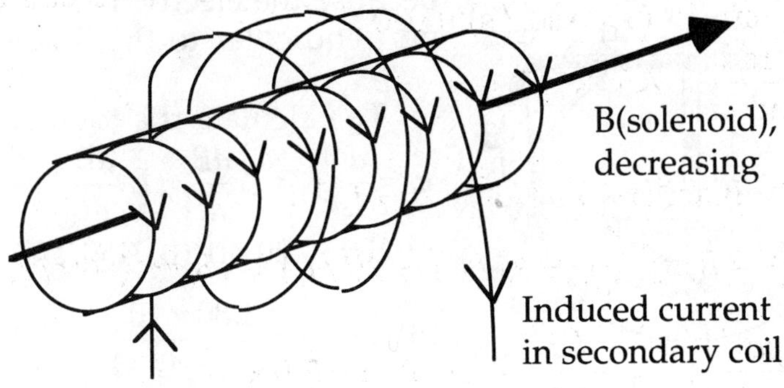

30-49: (a) $I = \frac{\mathcal{E}}{R} = \frac{vBa}{R} \Rightarrow F = IaB = \frac{vB^2a^2}{R}$.

(b) $F = ma = m\frac{dv}{dt} = \frac{vB^2a^2}{R} \Rightarrow \int_{v_o}^{v}\frac{dv'}{v'} = \frac{B^2a^2}{mR}\int_0^t dt' \Rightarrow v = v_o e^{-t(B^2a^2/mR)} = \frac{dx}{dt}$

$$\Rightarrow \int_0^x dx' = v_o\int_0^\infty e^{-t'(B^2a^2/mR)}dt' \Rightarrow x = -\frac{mRv_o}{B^2a^2}e^{-t'(B^2a^2/mR)}\Big|_0^\infty = \frac{mRv_o}{B^2a^2}.$$

30-50: (a)

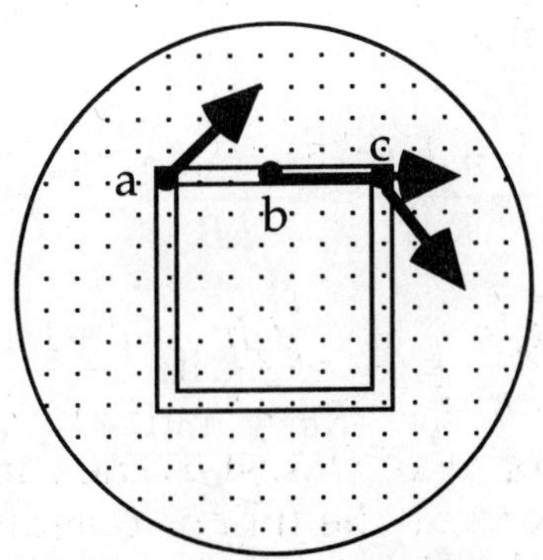

(b) To work out the amount of the electric field that is in the direction of the loop at a general position, we will use the geometry shown in the diagram at right.

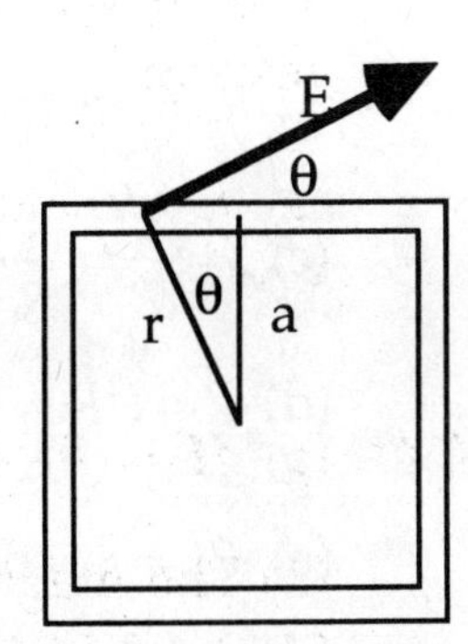

$$E_{loop} = E\cos\theta \text{ but } E = \frac{\mathcal{E}}{2\pi r} = \frac{\mathcal{E}}{2\pi(a/\cos\theta)} = \frac{\mathcal{E}\cos\theta}{2\pi a}$$

$$\Rightarrow E_{loop} = \frac{\mathcal{E}\cos^2\theta}{2\pi a} \text{ but } \mathcal{E} = \frac{d\Phi_B}{dt} = A\frac{dB}{dt} = \pi r^2\frac{dB}{dt} = \frac{\pi a^2}{\cos^2\theta}\frac{dB}{dt}$$

$\Rightarrow E_{loop} = \frac{\pi a^2}{2\pi a}\frac{dB}{dt} = \frac{a}{2}\frac{dB}{dt}$, which is exactly the value for a ring, obtained in **Ex. (30-25)**, and has no dependence on the part of the loop we pick.

(c) $I = \dfrac{\mathcal{E}}{R} = \dfrac{A}{R}\dfrac{dB}{dt} = \dfrac{L^2}{R}\dfrac{dB}{dt} = \dfrac{(0.20\text{ m})^2(0.045\text{ T/s})}{1.45\ \Omega} = 1.24\times10^{-3}\text{ A}.$

(d) $\mathcal{E}_{ab} = \dfrac{1}{8}\mathcal{E} = \dfrac{1}{8}L^2\dfrac{dB}{dt} = \dfrac{(0.20\text{ m})^2(0.045\text{ T/s})}{8} = 2.25\times10^{-4}\text{ V}.$

But there is potential drop $V = IR = -2.25\times10^{-4}V$, so the potential difference is zero.

30-51: (a)

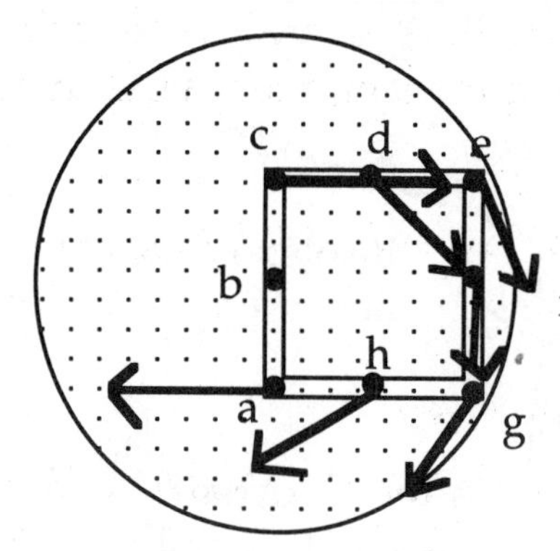

(b) The induced emf on the side ac is zero, because the electric field is always perpendicular to the line ac.

(c) To calculate the total emf in the loop,
$\mathcal{E} = \dfrac{d\Phi_B}{dt} = A\dfrac{dB}{dt} = L^2\dfrac{dB}{dt}$
$\Rightarrow \mathcal{E} = (0.20\text{ m})^2(0.0450\text{ T/s}) = 1.80\times10^{-3}\ V.$

(d) $I = \dfrac{\mathcal{E}}{R} = \dfrac{1.80\times10^{-3}\text{ V}}{1.45\ \Omega} = 1.24\times10^{-3}\text{ A}.$

(e) Since the loop is uniform, the resistance in length ac is one quarter of the total resistance. Therefore the potential difference between a and c is:
$V_{ac} = IR_{ac} = (1.24\times10^{-3}\text{ A})(1.45\ \Omega/4) = 4.50\times10^{-4}\text{ V}$, and the point a is at a higher potential since the current is flowing from a to c.

30-52: (a) As the bar starts to slide, the flux is decreasing, so the current flows to increase the flux, which means it flows from a to b.
(b) The magnetic force on the bar must eventually equal that of gravity.
$F_B = iLB = \dfrac{LB}{R}\mathcal{E} = \dfrac{LB}{R}\dfrac{d\Phi_B}{dt} = \dfrac{LB}{R}B\dfrac{dA}{dt} = \dfrac{LB^2}{R}(vL\cos\phi) = \dfrac{vL^2B^2}{R}\cos\phi$
$\Rightarrow F_g = mg\tan\phi = \dfrac{v_tL^2B^2}{R}\cos\phi \Rightarrow v_t = \dfrac{Rmg\tan\phi}{L^2B^2\cos\phi}.$

(c) $i = \dfrac{\mathcal{E}}{R} = \dfrac{1}{R}\dfrac{d\Phi_B}{dt} = \dfrac{1}{R}B\dfrac{dA}{dt} = \dfrac{B}{R}(vL\cos\phi) = \dfrac{vLB\cos\phi}{R} = \dfrac{mg\tan\phi}{LB}.$

(d) $P = i^2R = \dfrac{Rm^2g^2\tan^2\phi}{L^2B^2}.$

(e) $P_g = Fv\cos(90°-\phi) = mg\left(\dfrac{Rmg\tan\phi}{L^2B^2\cos\phi}\right)\sin\phi \Rightarrow P_g = \dfrac{Rm^2g^2\tan^2\phi}{L^2B^2}$, which is the same as found in part (d).

Chapter 31: Inductance

31-1: 1 H = 1 Wb/A = 1 Tm²/A = 1 Nm/A² = 1 J/A² = 1 (J/AC) s = 1 (V/A) s = 1 Ωs.

31-2: (a) $M = \varepsilon_2/(di/dt) = 7.3\times10^{-3}\,\text{V}/(-0.0850\text{ A/s}) = 0.0859$ H.
(b) $N_2 = 5,\ i_1 = 1.60\text{ A},\ \Rightarrow \Phi_{B_2} = i_1 M/N_2 = (1.60\text{ A})(0.0859\text{ H})/5 = 0.0275$ Wb.
(c) $di_2/dt = 0.05$ A/s and $\varepsilon_1 = M di_2/dt = (0.0859\text{ H})(0.05\text{ A/s}) = 4.29$ mV.

31-3: (a) $\varepsilon_2 = M(di_1/dt) = (0.275\text{ H})(0.050\text{ A/s}) = 0.0138$ V, and is constant.
(b) If the second coil has the same changing current, then the induced voltage is the same and $\varepsilon_1 = 0.0138$ V.

31-4: For a toroidal solenoid, $M = N_1\Phi_{B_2}/i_1$, and $\Phi_{B_2} = \mu_o N_2 i_2 A/2\pi r$. So, $M = \mu_o A N_1 N_2/2\pi r$.

31-5: (a) $M = N_2\Phi_{B_2}/i_1 = (300)(0.0280\text{ Wb})/(3.55\text{ A}) = 2.37$ H.
(b) When $i_2 = 1.60$ A, $\Phi_{B_1} = i_2 M/N_1 = (1.60\text{ A})(2.37\text{ H})/(800) = 4.73\times10^{-3}$ Wb.

31-6: The units of $[N\Phi/i] = \Omega\text{s} = \text{Vs/A} = \text{V/(A/s)} = [\mathcal{E}/(di/dt)]$.

31-7: (a) $L = \mathcal{E}/(di/dt) = (0.0180\text{ V})/(0.0600\text{ A/s}) = 0.300$ H.
(b) $\Phi_B = iL/N = (0.800\text{ A})(0.300\text{ H})/(300) = 8.00\times10^{-4}$ Wb.

31-8: For a toroidal solenoid, $L = N\Phi_B/i = \mathcal{E}/(di/dt)$. So solving for N we have:

$$N = \mathcal{E}i/\Phi_B(di/dt) = \frac{(8.40\times10^{-3}\text{ V})(1.25\text{ A})/}{(0.00375\text{ Wb})(0.0350\text{ A/s})} = 80\text{ turns.}$$

31-9: (a) $|\mathcal{E}| = L(di_1/dt) = (0.540\text{ H})(0.0300\text{ A/s}) = .0162$ V.
(b) Terminal a is at a higher potential since the coil pushes current through from b to a and if replaced by a battery it would have the + terminal at a.

31-10: (a) $L_\kappa = \kappa\mu_o N^2 A/2\pi r = \dfrac{(600\mu_o)(2000^2)(4.00\times10^{-5}\text{ m}^2)}{2\pi(0.0900\text{ m})} = 0.213$ H.
(b) Without the material, $L = \dfrac{1}{\kappa}L_\kappa = \dfrac{1}{600}(0.213\text{ H}) = 3.56\times10^{-4}$H.

31-11: For a long, straight solenoid: $L = N\Phi_B/i$ and $\Phi_B = \mu_o NiA/l \Rightarrow L = \mu_o N^2 A/l$.

31-12: (a) $U = Pt = (150\text{ W})(24\text{ h/day}\times3600\text{ s/h}) = 1.30\times10^7$ J.
(b) $U = \dfrac{1}{2}LI^2 \Rightarrow L = \dfrac{2U}{I^2} = \dfrac{2(1.30\times10^7\text{ J})}{(40.0\text{ A})^2} = 16200$ H.

31-13: (a) $u = \frac{U}{Vol} = \frac{B^2}{2\mu_o} \Rightarrow \text{Volume} = \frac{2\mu_o U}{B^2} = \frac{2\mu_o (3.60 \times 10^6 \text{ J})}{(0.500 \text{ T})^2} = 36.2 \text{ m}^3.$

(b) $B^2 = \frac{2\mu_o U}{\text{Vol}} = \frac{2\mu_o (3.60 \times 10^6 \text{ J})}{0.125 \text{ m}^3} = 72.4 \text{ T}^2 \Rightarrow B = 8.51 \text{ T}.$

31-14: (a) $U = \frac{1}{2} LI^2 = (16.0 \text{ H})(0.350 \text{ A})^2 / 2 = 0.980 \text{ J}.$

(b) $P = I^2 R = (0.350 \text{ A})^2 (200\ \Omega) = 24.5 \text{ W}.$

31-15: $U = \frac{1}{2} LI^2 = \frac{\mu_o N^2 A I^2}{4\pi r} \Rightarrow N = \sqrt{\frac{4\pi r U}{\mu_o A I^2}} = \sqrt{\frac{4\pi (0.120 \text{ m})(0.350 \text{ J})}{\mu_o (4.00 \times 10^{-4} \text{ m}^2)(12.5 \text{ A})^2}} = 2590 \text{ turns}.$

31-16: (a) $B = \frac{\mu_o N I}{2\pi r} = \frac{\mu_o (500)(3.50 \text{ A})}{2\pi (0.0780 \text{ m})} = 4.49 \text{ mT}.$

(b) From Eq. (31-10), $u = \frac{B^2}{2\mu_o} = \frac{(4.49 \times 10^{-3} \text{ T})^2}{2\mu_o} = 8.01 \text{ J/m}^3.$

(c) Volume $V = 2\pi r A = 2\pi (0.0780 \text{ m})(4.00 \times 10^{-6} \text{ m}^2) = 1.96 \times 10^{-6} \text{ m}^3.$

(d) $U = uV = (8.01 \text{ J/m}^3)(1.96 \times 10^{-6} \text{ m}^3) = 1.57 \times 10^{-5} \text{ J}.$

(e) $L = \frac{\mu_o N^2 A}{2\pi r} = \frac{\mu_o (500)^2 (4.00 \times 10^{-6} \text{ m}^2)}{2\pi (0.0780 \text{ m})} = 2.56 \times 10^{-6} \text{ H}.$

31-17: Starting with Eq. (31-9), follow exactly the same steps as in the text except that the magnetic permeability μ is used in place of μ_o.

31-18: (a) free space: $U = uV = \frac{B^2}{2\mu_o} V = \frac{(0.420 \text{ T})^2}{2\mu_o} (0.0250 \text{ m}^3) = 1760 \text{ J}.$

(b) material with $\kappa = 600 \Rightarrow U = uV = \frac{B^2}{2\kappa\mu_o} V = \frac{(0.420 \text{ T})^2}{2(600)\mu_o} (0.0250 \text{ m}^3) = 2.92 \text{ J}.$

31-19: Units of $L/R = \text{H}/\Omega = (\Omega \text{s})/\Omega = \text{s} =$ units of time.

31-20: When switch 1 is closed and switch 2 is open:

$$L\frac{di}{dt} + iR = 0 \Rightarrow \frac{di}{dt} = -i\frac{R}{L} \Rightarrow \int_{I_o}^{i} \frac{di'}{i'} = -\frac{R}{L}\int_0^t dt' \Rightarrow \ln(i/I_o) = -\frac{R}{L}t \Rightarrow i = I_o e^{-t(R/L)}.$$

31-21: (a) $\frac{di}{dt} = \frac{\mathcal{E} - iR}{L}.$ When $i = 0 \Rightarrow \frac{di}{dt} = \frac{12.0 \text{ V}}{3.00 \text{ H}} = 4.00 \text{ A/s}.$

(b) When $i = 1.00 \text{ A} \Rightarrow \frac{di}{dt} = \frac{12.0 \text{ V} - (1.00 \text{ A})(7.00\ \Omega)}{3.00 \text{ H}} = 1.67 \text{ A/s}.$

(c) At $t = 0.200 \text{ s} \Rightarrow i = \frac{\mathcal{E}}{R}(1 - e^{-(R/L)t}) = \frac{12.0 \text{ V}}{7.00\ \Omega}(1 - e^{-(7.00\ \Omega/3.00 \text{ H})(0.20\text{s})}) = 0.64 \text{ A}.$

(d) As $t \to \infty \Rightarrow i \to \frac{\mathcal{E}}{R} = \frac{12.0 \text{ V}}{7.00\ \Omega} = 1.71 \text{ A}.$

31-22: (a) $U = \frac{1}{2}LI^2 \Rightarrow I = \sqrt{\frac{2U}{L}} = \sqrt{\frac{2(0.210\text{ J})}{0.095\text{ H}}} = 2.10\text{ A}$

$\Rightarrow \mathcal{E} = IR = (2.10\text{ A})(320\ \Omega) = 673\text{ V}.$

(b) $i = Ie^{-(R/L)t}$ and $U = \frac{1}{2}Li^2 = \frac{1}{2}LI^2e^{-2(R/L)t} = \frac{1}{2}U_o = \frac{1}{2}\left(\frac{1}{2}LI^2\right) \Rightarrow e^{-2(R/L)t} = \frac{1}{2}$

$\Rightarrow t = -\frac{L}{2R}\ln\left(\frac{1}{2}\right) = -\frac{0.095\text{ H}}{2(320\ \Omega)}\ln\left(\frac{1}{2}\right) = 1.03 \times 10^{-4}\text{ s}.$

31-23: (a) $I_o = \frac{\mathcal{E}}{R} = \frac{120\text{ V}}{400\ \Omega} = 0.300\text{ A}.$

(b) $i = I_o e^{-(R/L)t} = (0.300\text{ A})e^{-(400\ \Omega/0.200\text{ H})(2.0\times10^{-4}\text{ s})} = 0.201\text{ A}.$

(c) $V_{cb} = V_{ab} = iR = (0.201\text{ A})(400\ \Omega) = 80.4\text{ V}$, and c is at the higher potential.

(d) $\frac{i}{I_o} = \frac{1}{2} = e^{-(R/L)t_{1/2}} \Rightarrow t_{1/2} = -\frac{L}{R}\ln\left(\frac{1}{2}\right) = 3.47 \times 10^{-4}\text{ s}.$

31-24: (a) At $t = 0 \Rightarrow v_{ab} = 0$ and $v_{bc} = 80\text{ V}$.

(b) As $t \to \infty \Rightarrow v_{ab} \to 80\text{ V}$ and $v_{bc} \to 0$.

(c) When $i = 0.050\text{ A} \Rightarrow v_{ab} = iR = 20\text{ V}$ and $v_{bc} = 80\text{ V} - 20\text{ V} = 60\text{ V}$.

31-25: (a) $P = \mathcal{E}i = \mathcal{E}I_o(1 - e^{-(R/L)t}) = \frac{\mathcal{E}^2}{R}(1 - e^{-(R/L)t}) = \frac{(12.0\text{ V})^2}{7.00\ \Omega}(1 - e^{-(7.00\ \Omega/3.00\text{ H})t})$

$\Rightarrow P = (20.6\text{ W})(1 - e^{-(2.33)t}).$

(b) $P_R = i^2R = \frac{\mathcal{E}^2}{R}(1 - e^{-(R/L)t})^2 = \frac{(12.0\text{ V})^2}{7.00\ \Omega}(1 - e^{-(7.00\ \Omega/3.00\text{ H})t})^2$

$\Rightarrow P_R = (20.6\text{ W})(1 - e^{-(2.33)t})^2.$

(c) $P_L = iL\frac{di}{dt} = \frac{\mathcal{E}}{R}(1 - e^{-(R/L)t})L\left(\frac{\mathcal{E}}{L}e^{-(R/L)t}\right) = \frac{\mathcal{E}^2}{R}(e^{-(R/L)t} - e^{-2(R/L)t})$

$\Rightarrow P_L = (20.6\text{ W})(e^{-(2.33)t} - e^{-(4.67)t}).$

(d) Note that if we expand the exponential in part (b), then parts (b) and (c) add to give part (a), and the total power delivered is dissipated in the resistor and inductor.

31-26: Equation (31-20) is $\frac{d^2q}{dt^2} + \frac{1}{LC}q = 0$. We will solve the equation using:

$q = Q\cos(\omega t + \phi) \Rightarrow \frac{dq}{dt} = -\omega Q\sin(\omega t + \phi) \Rightarrow \frac{d^2q}{dt^2} = -\omega^2 Q\cos(\omega t + \phi).$

$\Rightarrow \frac{d^2q}{dt^2} + \frac{1}{LC}q = -\omega^2 Q\cos(\omega t + \phi) + \frac{Q}{LC}\cos(\omega t + \phi) = 0 \Rightarrow \omega^2 = \frac{1}{LC} \Rightarrow \omega = \frac{1}{\sqrt{LC}}.$

31-27: $[LC] = \text{H}\cdot\text{F} = \text{H}\cdot\frac{\text{C}}{\text{V}} = \Omega\cdot\text{s}\cdot\frac{\text{C}}{\text{V}} = \frac{\Omega}{\text{V}}\cdot\frac{\text{C}}{\text{s}}\cdot\text{s}^2 = \frac{1}{\text{A}}\cdot\text{A}\cdot\text{s}^2 = \text{s}^2 \Rightarrow \left[\sqrt{LC}\right] = \text{s}.$

31-28: (a) $\omega = \frac{1}{\sqrt{LC}} = 2\pi f \Rightarrow L = \frac{1}{4\pi^2 f^2 C} = \frac{1}{4\pi^2 (5.4\times10^5)^2 (2.67\times10^{-11}\ \text{F})} = 3.25\times10^{-3}\ \text{H}.$

(b) $C_{\text{min}} = \frac{1}{4\pi^2 f_{\text{max}}^2 L} = \frac{1}{4\pi^2 (1.6\times10^6)^2 (3.25\times10^{-3}\ \text{H})} = 3.04\times10^{-12}\ \text{F}.$

31-29: (a) $T = \frac{2\pi}{\omega} = 2\pi\sqrt{LC} = 2\pi\sqrt{(3.00\ \text{H})(6.00\times10^{-4}\ \text{F})} = 0.267\ \text{s}.$

(b) $Q = CV = (6.00\times10^{-4}\ \text{F})(24.0\ \text{V}) = 0.0144\ \text{C}.$

(c) $U_o = \frac{1}{2}CV^2 = \frac{1}{2}(6.00\times10^{-4}\ \text{F})(24.0\ \text{V})^2 = 0.173\ \text{J}.$

(d) At $t = 0,\ q = Q = Q\cos(\omega t + \phi) \Rightarrow \phi = 0.$

$t = 0.0444\ \text{s},\ q = Q\cos(\omega t) = (0.0144\ \text{C})\cos\left(\frac{0.0444\ \text{s}}{\sqrt{(3.00\ \text{H})(6.00\times10^{-4}\ \text{F}}}\right) = 7.22\times10^{-3}\ \text{C}.$

(e) $t = 0.0444\ \text{s},\ i = \frac{dq}{dt} = -\omega Q\sin(\omega t)$

$\Rightarrow i = -\frac{0.0144\ \text{C}}{\sqrt{(3.00\ \text{H})(6.00\times10^{-4}\ \text{H})}}\sin\left(\frac{0.0444\ \text{s}}{\sqrt{(3.00\ \text{H})(6.00\times10^{-4}\ \text{H})}}\right) = -0.294\ \text{A}.$

(f) Capacitor: $U_C = \frac{q^2}{2C} = \frac{(7.21\times10^{-3}\ \text{C})^2}{2(6.00\times10^{-4}\ \text{F})} = 0.0433\ \text{J}.$

Inductor: $U_L = \frac{1}{2}Li^2 = \frac{1}{2}(3.00\ \text{H})(0.294\ \text{A})^2 = 0.129\ \text{J}.$

31-30: $\omega = \frac{1}{\sqrt{(0.225\ \text{H})(6.3\times10^{-6}\ \text{F})}} = 840\ \text{rad/s}$

(a) $i_{\text{max}} = \omega Q_{\text{max}} \Rightarrow Q_{\text{max}} = \frac{i_{\text{max}}}{\omega} = \frac{8.50\times10^{-3}\ \text{A}}{840\ \text{rad/s}} = 1.01\times10^{-5}\text{C}$

(b) $i = -\omega Q\sin(\omega t') \Rightarrow \omega t' = \arcsin\left(-\frac{i}{\omega Q}\right) = \arcsin\left(\frac{(5.0\times10^{-3}\ \text{A})}{(840\ \text{rad/s})(1.01\times10^{-5}\ \text{C})}\right)$

$\Rightarrow \omega t' = 0.629\ \text{rad} \Rightarrow q = Q\cos(\omega t') = ((1.01\times10^{-5}\ \text{C})\cos(0.629\ \text{rad}) = 8.18\times10^{-6}\ \text{C}.$

31-31: (a) $\frac{d^2q}{dt^2} + \frac{1}{LC}q = 0 \Rightarrow q = LC\frac{di}{dt} = (0.750\ \text{H})(1.80\times10^{-5}\ \text{F})(3.40\ \text{A/s}) = 4.59\times10^{-5}\ \text{C}.$

(b) $\mathcal{E} = \frac{q}{C} = \frac{4.20\times10^{-4}\ \text{C}}{1.80\times10^{-5}\ \text{F}} = 23.3\ \text{V}.$

31-32: $i_{\text{max}} = \omega Q_{\text{max}} \Rightarrow Q_{\text{max}} = \frac{i_{\text{max}}}{\omega} = i_{\text{max}}\sqrt{LC}$

$\Rightarrow Q_{\text{max}} = (2.00\ \text{A})\sqrt{(0.800\ \text{H})(5.00\times10^{-10}\ \text{F})} = 4.00\times10^{-5}\ \text{C}.$

$\Rightarrow U_{\text{max}} = \frac{Q_{\text{max}}^2}{2C} = \frac{(4.00\times10^{-5}\ \text{C})^2}{2(5.00\times10^{-10}\ \text{F})} = 1.60\ \text{J}.$

31-33: (a) When $R = 0,\ \omega_o = \dfrac{1}{\sqrt{LC}} = \dfrac{1}{\sqrt{(0.500\text{ H})(6.00\times10^{-4}\text{ F})}} = 57.7\text{ rad/s}.$

(b) We want $\dfrac{\omega}{\omega_o} = 0.9 \Rightarrow \dfrac{\left(1/LC - R^2/4L^2\right)}{1/LC} = 1 - \dfrac{R^2C}{4L} = (0.9)^2$

$$\Rightarrow R = \sqrt{\frac{4L}{C}(1-(0.9)^2)} = \sqrt{\frac{4(0.500\text{ H})(0.19)}{(6.00\times10^{-4}\text{ F})}} = 25.2\ \Omega.$$

31-34: $\left[\dfrac{L}{C}\right] = \dfrac{\text{H}}{\text{F}} = \dfrac{\Omega\cdot\text{s}}{\text{C/V}} = \dfrac{\Omega\cdot\text{V}}{\text{A}} = \Omega^2 \Rightarrow \left[\sqrt{\dfrac{L}{C}}\right] = \Omega.$

31-35: $\omega'^2 = \dfrac{1}{LC} - \dfrac{R^2}{4L^2} = \dfrac{1}{9LC} \Rightarrow R^2 = 4L^2\left(\dfrac{1}{LC} - \dfrac{1}{9LC}\right) \Rightarrow R = 2L\sqrt{\dfrac{1}{LC} - \dfrac{1}{9LC}}$

$$\Rightarrow R = 2(0.335\text{ H})\sqrt{\frac{1}{(0.335\text{ H})(8.5\times10^{-4}\text{ F})} - \frac{1}{9(0.335\text{ H})(8.5\times10^{-4}\text{ F})}} = 37.4\ \Omega.$$

31-36: (a) $q = Ae^{-(R/2L)t}\cos(\omega' t + \phi)$

$$\Rightarrow \frac{dq}{dt} = -A\frac{R}{2L}e^{-(R/2L)t}\cos(\omega' t + \phi) - \omega' Ae^{-(R/2L)t}\sin(\omega' t + \phi).$$

$$\Rightarrow \frac{d^2q}{dt^2} = A\left(\frac{R}{2L}\right)^2 e^{-(R/2L)t}\cos(\omega' t + \phi) + 2\omega' A\frac{R}{2L}e^{-(R/2L)t}\sin(\omega' t + \phi) - \omega'^2 Ae^{-(R/2L)t}\cos(\omega' t + \phi).$$

$$\Rightarrow \frac{d^2q}{dt^2} + \frac{R}{L}\frac{dq}{dt} + \frac{q}{LC} = q\left(\left(\frac{R}{2L}\right)^2 - \omega'^2 - \frac{R^2}{2L^2} + \frac{1}{LC}\right) = 0 \Rightarrow \omega'^2 = \frac{1}{LC} - \frac{R^2}{4L^2}.$$

(b) At $t = 0,\ q = Q,\ i = \dfrac{dq}{dt} = 0$:

$$\Rightarrow q = A\cos\phi = Q \text{ and } \frac{dq}{dt} = -\frac{R}{2L}A\cos\phi - \omega' A\sin\phi = 0$$

$$\Rightarrow A = \frac{Q}{\cos\phi} \text{ and } -\frac{QR}{2L} - \omega' Q\tan\phi = 0 \Rightarrow \tan\phi = -\frac{R}{2L\omega'} = -\frac{R}{2L\sqrt{1/LC - R^2/4L^2}}.$$

31-37: Subbing $x \to q,\ m \to L,\ b \to R,\ k \to \dfrac{1}{C}$, we find:

(a) Eq. (13-41): $\dfrac{d^2x}{dt^2} + \dfrac{b}{m}\dfrac{dx}{dt} + \dfrac{kx}{m} = 0 \ \to$ Eq. (31-27): $\dfrac{d^2q}{dt^2} + \dfrac{R}{L}\dfrac{dq}{dt} + \dfrac{q}{LC} = 0.$

(b) Eq. (13-43): $\omega' = \sqrt{\dfrac{k}{m} - \dfrac{b^2}{4m^2}} \ \to$ Eq. (31-29): $\omega' = \sqrt{\dfrac{1}{LC} - \dfrac{R^2}{4L^2}}.$

(c) Eq. (13-42): $x = Ae^{-(b/2m)t}\cos(\omega' t + \phi) \to$ Eq. (31-28): $q = Ae^{-(R/2L)t}\cos(\omega' t + \phi).$

31-38: (a) $\mathcal{E} = -L\dfrac{di}{dt} = -L\dfrac{d}{dt}((0.18\text{ A})\sin((120\pi/\text{s})t))$

$$\Rightarrow \mathcal{E} = -(0.80\text{ H})(0.18\text{ A})(120\pi)\cos((120\pi/\text{s})t) = -(54.3\text{ V})\cos((120\pi/\text{s})t).$$

(b) $\mathcal{E}_{max} = 54.3\text{ V}$; $i = 0$, since the emf and current are 90° out of phase.
(c) $i_{max} = 0.18\text{ A}$; $\mathcal{E} = 0$, since the emf and current are 90° out of phase.

31-39: (a) $\mathcal{E} = -L\frac{di}{dt} \Rightarrow L = \mathcal{E}/(di/dt) = (60.0\text{ V})/(5.00\text{ A/s}) = 12.0\text{ H}.$

(b) $\mathcal{E} = \frac{d\Phi}{dt} \Rightarrow \Phi_f - \Phi_i = \mathcal{E}\Delta t \Rightarrow \Phi_f = (60.0\text{ V})(10.0\text{ s}) = 600\text{ Wb}.$

(c) $P_L = Li\frac{di}{dt} = (12.0\text{ H})(50.0\text{ A})(5.00\text{ A/s}) = 3000\text{ W}.$

$P_R = i^2R = (50.0\text{ A})^2(25.0\ \Omega) = 62500\text{ W} \Rightarrow \frac{P_L}{P_R} = 0.0480.$

31-40: (a) $M = \frac{N_2}{I}\Phi_{B_2} = \frac{N_2}{I}\frac{A_2}{A_1}\Phi_{B_1} = \frac{N_2A_2}{IA_1}\frac{\mu N_1IA_1}{l_1} = \frac{\mu N_1N_2A_2}{l_1} = \frac{\mu N_1N_2\pi r_2^{\,2}}{l_1}.$

(b) $|\mathcal{E}_2| = N_2\frac{d\Phi_{B_2}}{dt} = N_2\frac{\mu N_1A_2}{l_1}\frac{di_1}{dt} = \frac{\mu N_1N_2\pi r_2^{\,2}}{l_1}\frac{di_1}{dt}.$

(c) $|\mathcal{E}_1| = M_{12}\frac{di_2}{dt} = M\frac{di_2}{dt} = \frac{\mu N_1N_2\pi r_2^{\,2}}{l_1}\frac{di_2}{dt}.$

31-41: (a)

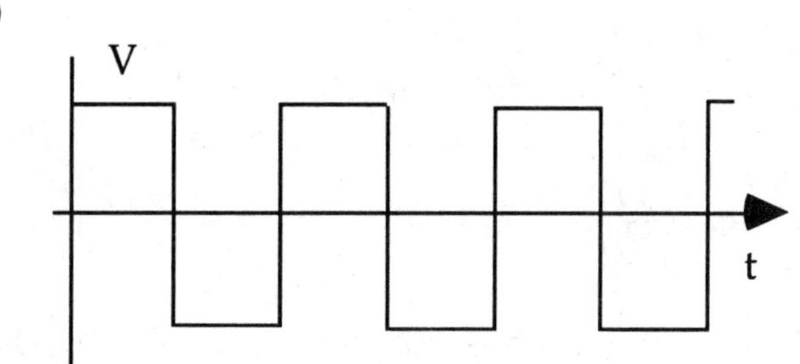

(b) Since the voltage is determined by the derivative of the current, the V versus t graph is indeed proportional to the derivative of the current graph.

31-42: (a) $\mathcal{E} = L\frac{di}{dt} = (7.50\times10^{-3}\text{ H})\frac{d}{dt}((0.380\text{ A})\sin(\pi t/0.0400\text{ s}))$

$\Rightarrow \mathcal{E}_{max} = (7.50\times10^{-3}\text{ H})(0.380\text{ A})\frac{\pi}{0.0400\text{ s}} = 0.224\text{ V}.$

(b) $\Phi_{B\max} = \frac{Li_{max}}{N} = \frac{(7.50\times10^{-3}\text{ H})(0.380\text{ A})}{300} = 9.50\times10^{-6}\text{ Wb}.$

(c) $\mathcal{E}(t) = -L\frac{di}{dt} = -(7.50\times10^{-3}\text{ H})(0.380\text{ A})(\pi/0.0400\text{ s})\cos(\pi t/0.0400\text{ s})$

$\Rightarrow \mathcal{E}(t) = -(0.224\text{ V})\cos((78.5\text{ s}^{-1})t) \Rightarrow \mathcal{E}(0.03\text{ s}) = -(0.224\text{ V})\cos((78.5\text{ s}^{-1})(0.03\text{ s}))$

$\Rightarrow \mathcal{E}(t) = 0.158\text{ V}.$

31-43: (a) $\Phi_B = \int_a^b B(hdr) = \int_a^b \left(\frac{\mu_o Ni}{2\pi r}\right)(hdr) = \frac{\mu_o Nih}{2\pi}\int_a^b \frac{dr}{r} = \frac{\mu_o Nih}{2\pi}\ln(b/a).$

(b) $L = \frac{N\Phi_B}{i} = \frac{\mu_o N^2 h}{2\pi}\ln(b/a).$

(c) $\ln(b/a) = \ln(1-(b-a)/a) \approx \frac{b-a}{a} + \frac{(b-a)^2}{2a^2} + \ldots \Rightarrow L \approx \frac{\mu_o N^2 h}{2\pi}\left(\frac{b-a}{a}\right).$

31-44: (a) $\oint \vec{B}\cdot d\vec{l} = \mu_o I_{encl} \Rightarrow B2\pi r = \mu_o i \Rightarrow B = \frac{\mu_o i}{2\pi r}.$

(b) $d\Phi_B = BdA = \frac{\mu_o i}{2\pi r} l dr.$

(c) $\Phi_B = \int_a^b d\Phi_B = \frac{\mu_o il}{2\pi}\int_a^b \frac{dr}{r} = \frac{\mu_o il}{2\pi}\ln(b/a).$

(d) $L = \frac{N\Phi_B}{i} = l\frac{\mu_o}{2\pi}\ln(b/a).$

(e) $U = \frac{1}{2}Li^2 = \frac{1}{2}l\frac{\mu_o}{2\pi}\ln(b/a)i^2 = \frac{\mu_o li^2}{4\pi}\ln(b/a).$

31-45: (a) $\oint \vec{B}\cdot d\vec{l} = \mu_o I_{encl} \Rightarrow B2\pi r = \mu_o i \Rightarrow B = \frac{\mu_o i}{2\pi r}.$

(b) $u = \frac{B^2}{2\mu_o} \Rightarrow dU = udV = u(l2\pi rdr) = \frac{1}{2\mu_o}\left(\frac{\mu_o i}{2\pi r}\right)^2 (l2\pi rdr) = \frac{\mu_o i^2 l}{4\pi r}dr.$

(c) $U = \int_a^b dU = \frac{\mu_o i^2 l}{4\pi}\int_a^b \frac{dr}{r} = \frac{\mu_o i^2 l}{4\pi}\ln(b/a).$

(d) $U = \frac{1}{2}Li^2 \Rightarrow L = \frac{2U}{i^2} = l\frac{\mu_o}{2\pi}\ln(b/a)$, which is the same as in **Pr. (31-44)**.

31-46: (a) $L_1 = \frac{N_1\Phi_{B_1}}{i_1} = \frac{N_1 A}{i_1}\left(\frac{\mu_o N_1 i_1}{2\pi r}\right) = \frac{\mu_o N_1{}^2 A}{2\pi r}, \; L_2 = \frac{N_2\Phi_{B_2}}{i_2} = \frac{N_2 A}{i_2}\left(\frac{\mu_o N_2 i_2}{2\pi r}\right) = \frac{\mu_o N_2{}^2 A}{2\pi r}.$

(b) $M^2 = \left(\frac{\mu_o N_1 N_2 A}{2\pi r}\right)^2 = \frac{\mu_o N_1{}^2 A}{2\pi r}\frac{\mu_o N_2{}^2 A}{2\pi r} = L_1 L_2.$

31-47: $u_B = u_E \Rightarrow \frac{\varepsilon_o E^2}{2} = \frac{B^2}{2\mu_o} \Rightarrow B = \sqrt{\varepsilon_o\mu_o E^2} = \sqrt{\varepsilon_o\mu_o}E = \sqrt{\varepsilon_o\mu_o}(900\text{ V/m}) = 3.00\times10^{-6}\text{ T}.$

31-48: (a) $R = \frac{V}{i_f} = \frac{36.0\text{ V}}{0.017\text{ A}} = 2120\,\Omega.$

(b) $i = i_f(1-e^{-(R/L)t}) \Rightarrow \frac{Rt}{L} = -\ln(1-i/i_f) \Rightarrow L = \frac{-Rt}{\ln(1-i/i_f)}$

$\Rightarrow L = \frac{-(2120\,\Omega)(3.65\times10^{-4}\text{ s})}{\ln(1-(8.5/17))} = 1.12\text{ H}.$

31-49: (a) After one time constant has passed:

$$i=\frac{\mathcal{E}}{R}(1-e^{-1})=\frac{12.0\text{ V}}{7.00\ \Omega}(1-e^{-1})=1.08\text{ A}\Rightarrow U=\frac{1}{2}Li^2=\frac{1}{2}(3.00\text{ H})(1.08\text{ A})^2=1.76\text{ J}.$$

Or, using **Pr. (31-25(c))**:

$$U=\int P_L dt=(20.6\text{ W})\int_0^{3/7}(e^{-(2.33)t}-e^{-(4.67)t})dt=(20.6\text{ W})\left(\frac{(1-e^{-1})}{2.33}-\frac{(1-e^{-2})}{4.67}\right)-1.76\text{ J}.$$

(b) $$U_{tot}=(20.6\text{ W})\int_0^{3/7}(1-e^{-(R/L)t})dt=(20.6\text{ W})\left(\frac{L}{R}+\frac{L}{R}(e^{-1}-1)\right)$$

$$\Rightarrow U_{tot}=(20.6\text{ W})\frac{3.00\text{ H}}{7.00\ \Omega}e^{-1}=3.24\text{ J}.$$

(c) $$U_R=(20.6\text{ W})\int_0^{3/7}(1-2e^{-(R/L)t}-2e^{-2(R/L)t})dt=(20.6\text{ W})\left(\frac{L}{R}+\frac{L}{R}(e^{-1}-1)-\frac{L}{2R}(e^{-2}-1)\right)$$

$$\Rightarrow U_R=(20.6\text{ W})\frac{3.00\text{ H}}{7.00\ \Omega}(0.168)=1.48\text{ J}.$$

The energy dissipated over the inductor (part (a)), plus the energy lost over the resistor (part (b)), sums to the total energy output (part (b)).

31-50: (a) $$U=\frac{1}{2}Li_o^{\,2}=\frac{1}{2}L\left(\frac{\mathcal{E}}{R}\right)^2=\frac{1}{2}(0.200\text{ H})\left(\frac{120\text{ V}}{400\ \Omega}\right)^2=9.00\times10^{-3}\text{ J}.$$

(b) $$i=\frac{\mathcal{E}}{R}e^{-(R/L)t}\Rightarrow\frac{di}{dt}=-\frac{R}{L}i\Rightarrow\frac{dU_L}{dt}=iL\frac{di}{dt}=-Ri^2=-\frac{\mathcal{E}^2}{R}e^{-2(R/L)t}$$

$$\Rightarrow\frac{dU_L}{dt}=-\frac{(120\text{ V})^2}{400\ \Omega}e^{-2(400/0.200)(2.00\times10^{-4})}=-16.2\text{ W}.$$

(c) In the resistor: $$\frac{dU_R}{dt}=i^2R=\frac{\mathcal{E}^2}{R}e^{-2(R/L)t}=\frac{(120\text{ V})^2}{400\ \Omega}e^{-2(400/0.200)(2.00\times10^{-4})}=16.2\text{ W}.$$

(d) $$P_R(t)=i^2R=\frac{\mathcal{E}^2}{R}e^{-2(R/L)t}$$

$$\Rightarrow U_R=\frac{\mathcal{E}^2}{R}\int_0^{\infty}e^{-2(R/L)t}dt=\frac{\mathcal{E}^2}{R}\frac{L}{2R}=\frac{(120\text{ V})^2(0.200\text{ H})}{2(400\ \Omega)^2}=9.00\times10^{-3}\text{ J},$$ which is the same as part (a).

31-51: Multiplying Eq. (31-27) by i, yields:

$$i^2R+Li\frac{di}{dt}-\frac{q}{C}i=i^2R+Li\frac{di}{dt}+\frac{q}{C}\frac{dq}{dt}=i^2R+\frac{d}{dt}\left(\frac{1}{2}Li^2\right)+\frac{d}{dt}\left(\frac{1}{2}\frac{q^2}{C}\right)=P_R+P_L+P_C=0.$$

That is, the rate of energy dissipation throughout the circuit must balance over all of the circuit elements.

31-52: (a) If $$t=\frac{3T}{8}\Rightarrow q=Q\cos(\omega t)=Q\cos\left(\frac{2\pi}{T}\frac{3T}{8}\right)=Q\cos\left(\frac{3\pi}{4}\right)=\frac{Q}{\sqrt{2}}$$

$$\Rightarrow i=\frac{1}{\sqrt{LC}}\left(\sqrt{Q^2-q^2}\right)=\frac{1}{\sqrt{LC}}\left(\sqrt{Q^2-Q^2/2}\right)=\sqrt{\frac{Q^2}{2LC}}$$

$$\Rightarrow U_E = \frac{1}{2}Li^2 = \frac{1}{2}L\frac{Q^2}{2LC} = \frac{1}{2}\frac{Q^2}{2C} = \frac{q^2}{2C} = U_B.$$

(b) The two energies are next equal when $q = \frac{Q}{\sqrt{2}} \Rightarrow \omega t = \frac{5\pi}{8} \Rightarrow t = \frac{5T}{8}$.

31-53: $U = \frac{1}{2}CV^2 \Rightarrow C = \frac{2U}{V^2} = \frac{2(2.00\times10^{-4}\ \text{J})}{(50.0\ \text{V})^2} = 1.60\times10^{-7}\ \text{F}.$

Also $\omega = \frac{1}{\sqrt{LC}} \Rightarrow L = \frac{1}{C\omega^2} = \frac{1}{(1.60\times10^{-7}\ \text{F})(8.00\times10^4\ \text{rad/s})^2} = 9.77\times10^{-4}\ \text{H}.$

31-54: (a) $V_{\max} = \frac{Q}{C} = \frac{5.00\times10^{-6}\ \text{C}}{4.00\times10^{-4}\ \text{F}} = 0.0125\ \text{V}.$

(b) $\frac{1}{2}Li_{\max}{}^2 = \frac{Q^2}{2C} \Rightarrow i_{\max} = \frac{Q}{\sqrt{LC}} = \frac{5.00\times10^{-6}\ \text{C}}{\sqrt{(0.0900\ \text{H})(4.00\times10^{-4}\ \text{F})}} = 8.33\times10^{-4}\ \text{A}.$

(c) $U_{\max} = \frac{1}{2}Li_{\max}{}^2 = \frac{1}{2}(0.0900\ \text{H})(8.33\times10^{-4}\ \text{A})^2 = 3.13\times10^{-8}\ \text{J}.$

(d) If $i = \frac{1}{2}i_{\max} \Rightarrow U = \frac{1}{4}U_{\max} = 7.81\times10^{-9}\ \text{J} \Rightarrow U_E = \frac{3}{4}U_{\max} = \frac{\left(\sqrt{\frac{3}{4}}Q\right)^2}{2C} = \frac{q^2}{2C}$

$\Rightarrow q = \sqrt{\frac{3}{4}}Q = 4.33\times10^{-6}\ \text{C}.$

31-55: (a) At $t = 0$, all the current passes through the resistor R_1, so the voltage v_{ab} is the total voltage of 120 V.
(b) Point a is at a higher potential than point b.
(c) v_{cd} = 120 V since there is no current through R_2.
(d) Point c is at a higher potential than point b.
(e) After a long time, the switch is opened, and the inductor initially

maintains the current of $i_{R_2} = \frac{\mathcal{E}}{R_2} = \frac{120\ \text{V}}{50.0\ \Omega} = 2.40\ \text{A}$. Therefore the potential between a and b is $v_{ab} = iR_1 = (2.40\ \text{A})(30.0\ \Omega) = 72.0\ \text{V}$.
(f) Point b is at a higher potential than point a.
(g) $v_{cd} = i(R_1 + R_2) = (2.40\ \text{A})(30\ \Omega + 50\ \Omega) = 192\ \text{V}$.
(h) Point d is at a higher potential than point c.

31-56: (a) Switch is closed, then at some later time:

$\frac{di}{dt} = 4.00\ \text{A/s} \Rightarrow v_{cd} = L\frac{di}{dt} = (8.00\ \text{H})(4.00\ \text{A/s}) = 32.0\ \text{V}.$

The top circuit loop: $90\ \text{V} = i_1R_1 \Rightarrow i_1 = \frac{90\ \text{V}}{40\ \Omega} = 2.25\ \text{A}.$

The bottom loop: $90\ \text{V} - i_2R_2 - 32\ \text{V} = 0 \Rightarrow i_2 = \frac{58\ \text{V}}{30\ \Omega} = 1.93\ \text{A}.$

(b) After a long time: $i_2 = \frac{90\ \text{V}}{30\ \Omega} = 3.00\ \text{A}$, and immediately when the switch is opened, the inductor maintains this current, so $i_1 = i_2 = 3.00\ \text{A}$.

31-57: (a) Immediately after S_1 is closed, $i_o = 0,\ v_{ac} = 0,$ and $v_{cb} = 36.0\ \text{V}$, since the inductor stops the current flow.

(b) After a long time, $i_o = \dfrac{\mathcal{E}}{R_o + R} = \dfrac{36.0\ \text{V}}{50\ \Omega + 150\ \Omega} = 0.180\ \text{A},$

$v_{ac} = i_o R_o = (0.18\ \text{A})(50\ \Omega) = 9.00\ \text{V},$ and $v_{cb} = 36.0\ \text{V} - 9.00\ \text{V} = 27.0\ \text{V}.$

(c) $i(t) = \dfrac{\varepsilon}{R_{total}}\left(1 - e^{-(R/L)t}\right) \Rightarrow i(t) = (0.180\ \text{A})\left(1 - e^{-50t}\right),$

$v_{ac}(t) = i(t)R_o = (9.00\ \text{V})\left(1 - e^{-50t}\right)$ and

$v_{ac}(t) = \mathcal{E} - i(t)R_o = 36.0\ \text{V} - (9.00\ \text{V})\left(1 - e^{-50t}\right) = (9.00\ \text{V})\left(3 + e^{-50t}\right).$

Below are the graphs of current and voltage found above.

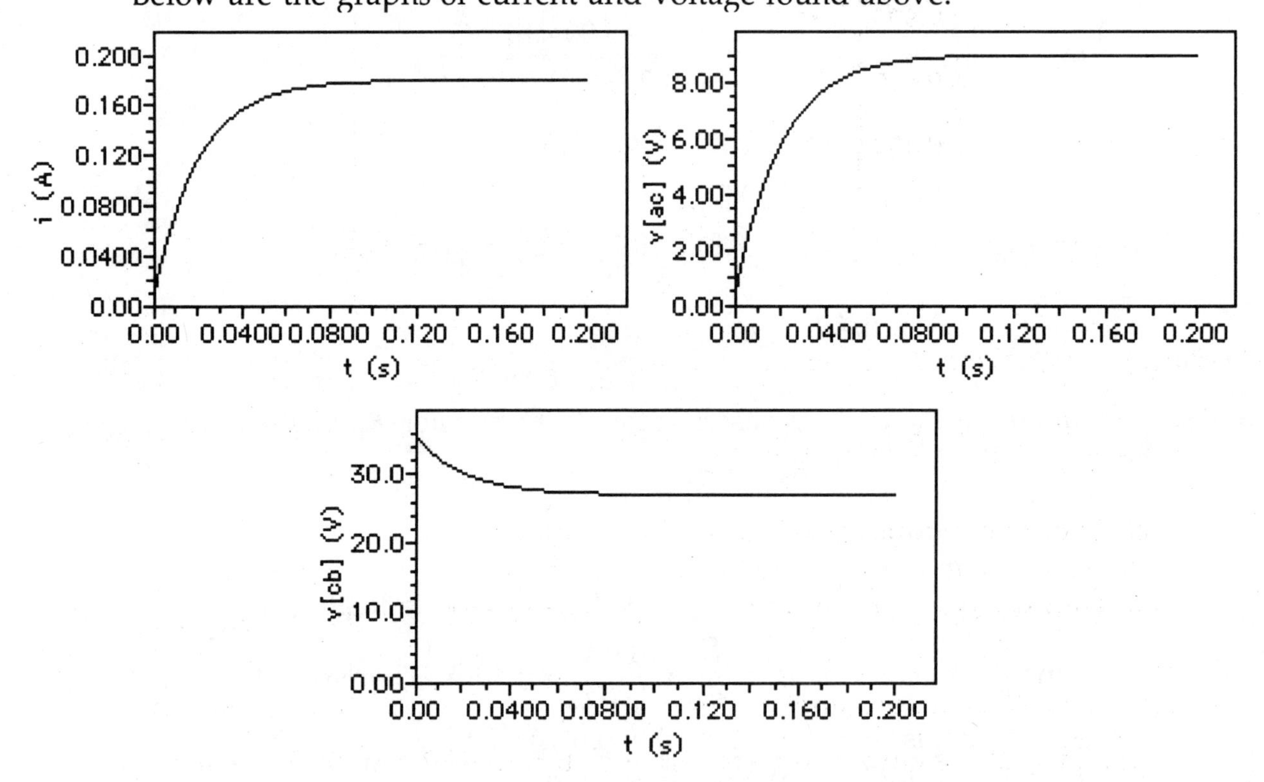

31-58: (a) Immediately after S_2 is closed, the inductor maintains the current $i = 0.180\text{A}$ through R. The Kirchoff's Rules around the outside of the circuit yield: $\mathcal{E} + \mathcal{E}_L - iR - i_o R_o = 36.0\ \text{V} + (0.18)(150) - (0.18)(150) - i_o(50) = 0$

$\Rightarrow i_o = \dfrac{36\ \text{V}}{50\ \Omega} = 0.720\ \text{A},\ v_{ac} = (0.72\ \text{A})(50\ \text{V}) = 36.0\ \text{V}$ and $v_{cb} = 0.$

(b) After a long time, $v_{ac} = 36.0\ \text{V},$ and $v_{cb} = 0.$ Thus $i_o = \dfrac{\mathcal{E}}{R_o} = \dfrac{36.0\ \text{V}}{50\ \Omega} = 0.720\ \text{A},$

$i_R = 0,$ and $i_{S_2} = 0.720\ \text{A}.$

(c) $i_o = 0.720\ \text{A},\ i_R(t) = \dfrac{\varepsilon}{R_{total}} e^{-(R/L)t} \Rightarrow i_R(t) = (0.180\ \text{A})e^{-50t}$, and

$i_{S_2}(t) = (0.720\ \text{A}) - (0.180\ \text{A})e^{-50t} = (0.180\ \text{A})\left(4 - e^{-50t}\right)$

Below are the graphs of the currents found above.

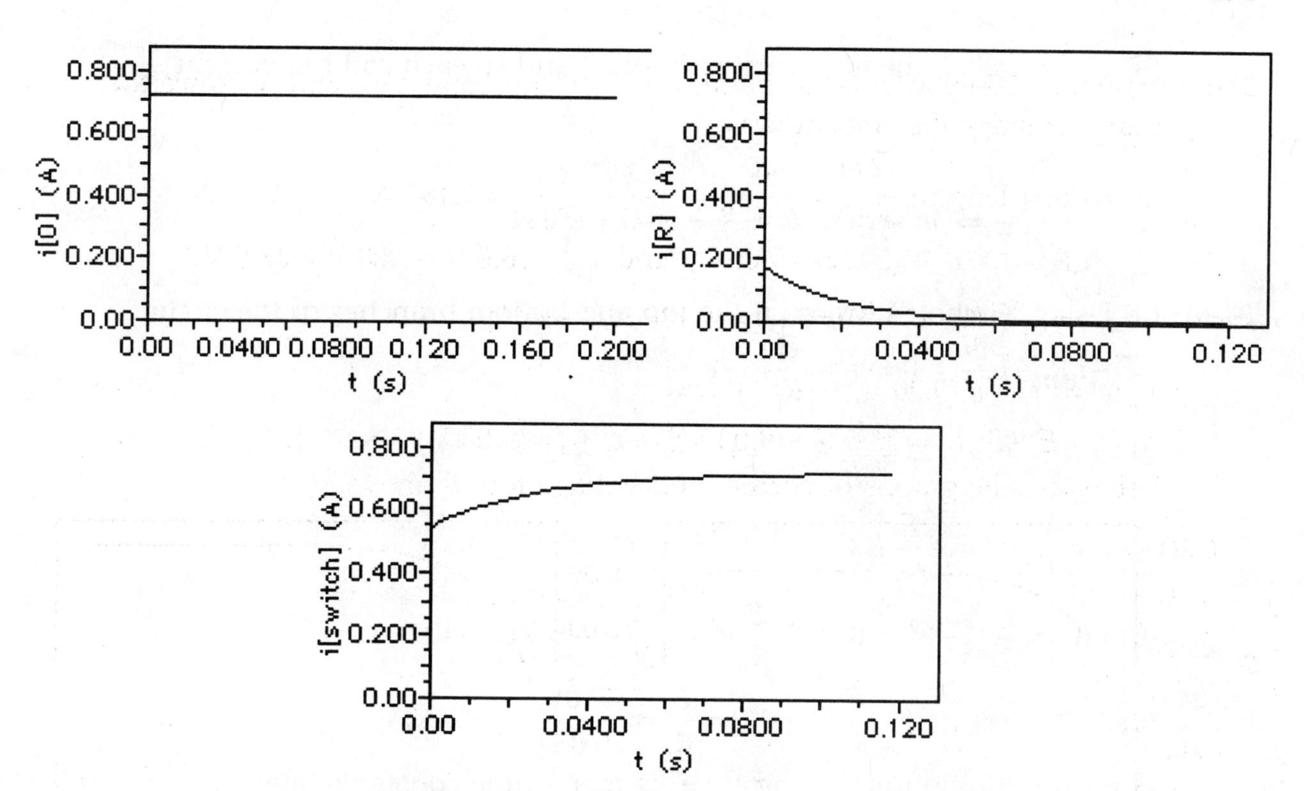

31-59: (a) Using Kirchoff's Rules: $\mathcal{E} - i_1R_1 = 0 \Rightarrow i_1 = \dfrac{\mathcal{E}}{R_1}$, and

$$\mathcal{E} - L\frac{di_2}{dt} - i_2R_2 = 0 \Rightarrow i_2 = \frac{\mathcal{E}}{R_2}(1 - e^{-(R_2/L)t}).$$

(b) After a long time, $i_1 = \dfrac{\mathcal{E}}{R_1}$ still, and $i_2 = \dfrac{\mathcal{E}}{R_2}$.

(c) After the switch is opened, $i_1 = i_2 = \dfrac{\varepsilon}{R_2}e^{-(R_1+R_2/L)t}$, and the current drops off.

(d) A 60 W light bulb implies $R = \dfrac{V^2}{P} = \dfrac{(120\ \text{V})^2}{60\ \text{W}} = 240\ \Omega$. If the switch is opened, and the current is to fall from 0.90 A to 0.30 A in 0.20 s, then:

$$i_2 = (0.900\ \text{A})e^{-(R_1+R_2/L)t} \Rightarrow 0.300\ \text{A} = (0.900\ \text{A})e^{-((240\ \Omega+R_2)/52.0\ \text{H})(0.20\ \text{s})}$$

$$\Rightarrow \frac{52.0\ \text{H}}{0.200\ \text{s}}\ln(3.00) = 240\ \Omega + R_2 \Rightarrow R_2 = 45.6\ \Omega.$$

$$\Rightarrow \mathcal{E} = i_2R_2 = (0.900\ \text{A})(240\ \Omega) = 41.1\ \text{V}.$$

(e) Before the switch is opened, $i_o = \dfrac{\mathcal{E}}{R_1} = \dfrac{41.1\ \text{V}}{240\ \Omega} = 0.171\ \text{A}$

31-60: (a) If in series, each inductor provides $\mathcal{E} = (L+M)\dfrac{di}{dt}$, but for equal coils $M = L$, each provides $\mathcal{E} = 2L\dfrac{di}{dt} \Rightarrow \mathcal{E}_{total} = 4L\dfrac{di}{dt} \Rightarrow L_{eff} = 4L$.

(b) If in parallel, the total current is still i and so each coil has current $\frac{i}{2}$.

Thus: $\mathcal{E}_{total} = (L+L)\frac{1}{2}\frac{di}{dt} = L\frac{di}{dt} \Rightarrow L_{eff} = L.$

(c) $\omega = \frac{1}{\sqrt{LC}} \Rightarrow$ in series: $\omega = \frac{1}{\sqrt{4LC}} = \frac{1}{2\sqrt{LC}}$

31-61: (a) Using Kirchoff's Rules on the top and bottom branches of the circuit:

$$\mathcal{E} - i_1 R_1 - L\frac{di_1}{dt} = 0 \Rightarrow i_1 = \frac{\mathcal{E}}{R_1}(1 - e^{-(R_1/L)t}).$$

$$\mathcal{E} - i_2 R_2 - \frac{q_2}{C} = 0 \Rightarrow -\frac{di_2}{dt}R_2 - \frac{i_2}{C} = 0 \Rightarrow i_2 = \frac{\mathcal{E}}{R_2}e^{-(1/R_2C)t}$$

$$\Rightarrow q_2 = \int_0^t i_2\,dt' = -\frac{\mathcal{E}}{R_2}R_2Ce^{-(1/R_2C)t'}\Big|_0^t = \mathcal{E}C(1 - e^{-(1/R_2C)t}).$$

(b) $i_1(0) = \frac{\mathcal{E}}{R_1}(1 - e^0) = 0,\ i_2 = \frac{\mathcal{E}}{R_2}e^0 = \frac{48.0\text{ V}}{5000\ \Omega} = 9.60 \times 10^{-3}\text{ A}.$

(c) As $t \to \infty$: $i_1(\infty) = \frac{\mathcal{E}}{R_1}(1 - e^{-\infty}) = \frac{\mathcal{E}}{R_1} = \frac{48.0\text{V}}{25.0\ \Omega} = 1.92\text{ A},\ i_2 = \frac{\mathcal{E}}{R_2}e^{-\infty} = 0.$

A good definition of a "long time" is many time constants later.

(d) $i_1 = i_2 \Rightarrow \frac{\mathcal{E}}{R_1}(1 - e^{-(R_1/L)t}) = \frac{\mathcal{E}}{R_2}e^{-(1/R_2C)t} \Rightarrow (1 - e^{-(R_1/L)t}) = \frac{R_1}{R_2}e^{-(1/R_2C)t}.$

Expanding the exponentials like $e^x = 1 + x + \frac{x^2}{2} + \frac{x^3}{3!} + \ldots$, we find:

$$\frac{R_1}{L}t - \frac{1}{2}\left(\frac{R_1}{L}\right)^2 t^2 + \ldots = \frac{R_1}{R_2}\left(1 - \frac{t}{RC} + \frac{t^2}{2R^2C^2} - \ldots\right) \Rightarrow t\left(\frac{R_1}{L} + \frac{R_1}{R_2{}^2C}\right) + O(t^2) + \ldots = \frac{R_1}{R_2},\text{ if}$$

we have assumed that $t \ll 1$. Therefore:

$$\Rightarrow t \approx \frac{1}{R_2}\left(\frac{1}{(1/L) + (1/R_2{}^2C)}\right) = \left(\frac{LR_2C}{L + R_2{}^2C}\right)$$

$$\Rightarrow t = \left(\frac{(8.0\text{ H})(5000\ \Omega)(2.0 \times 10^{-5}\text{ F})}{8.0\text{ H} + (5000\ \Omega)^2(2.0 \times 10^{-5}\text{ F})}\right) = 1.6 \times 10^{-3}\text{ s}.$$

(e) At $t = 1.57 \times 10^{-3}$ s: $i_1 = \frac{\mathcal{E}}{R_1}(1 - e^{-(R_1/L)t}) = \frac{48\text{ V}}{25\ \Omega}(1 - e^{-(25/8)t}) = 9.4 \times 10^{-3}\text{ A}.$

(f) We want to know when the current is half its final value. We note that the current i_2 is very small to begin with, and just gets smaller, so we ignore it and find: $i_{1/2} = 0.960\text{ A} = i_1 = \frac{\mathcal{E}}{R_1}(1 - e^{-(R_1/L)t}) = (1.92\text{ A})(1 - e^{-(R_1/L)t}).$

$$\Rightarrow e^{-(R_1/L)t} = 0.500 \Rightarrow t = -\frac{L}{R_1}\ln(0.5) = \frac{8.0\text{ H}}{25\ \Omega}\ln(0.5) = 0.22\text{ s}.$$

31-62: (a) Using Kirchoff's Rules on the left and right branches:

Left: $\mathcal{E} - (i_1 + i_2)R - L\frac{di_1}{dt} = 0 \Rightarrow R(i_1 + i_2) + L\frac{di_1}{dt} = \mathcal{E}.$

Right: $\mathcal{E} - (i_1 + i_2)R - \frac{q_2}{C} = 0 \Rightarrow R(i_1 + i_2) + \frac{q_2}{C} = \mathcal{E}$.

(b) Initially, with the switch just closed, $i_1 = 0$, $i_2 = \frac{\mathcal{E}}{R}$ and $q_2 = 0$.

(c) The substitution of the solutions into the circuit equations to show that they satisfy the equations is a somewhat tedious exercise in book-keeping that is left to the reader.

We will show that the initial conditions are satisfied:

At $t = 0$, $q_2 = \frac{\mathcal{E}}{\omega R} e^{-\beta t} \sin(\omega t) = \frac{\mathcal{E}}{\omega R} \sin(0) = 0$,

$$i_1(t) = \frac{\mathcal{E}}{R}\left(1 - e^{-\beta t}\left[(2\omega RC)^{-1} \sin(\omega t) + \cos(\omega t)\right]\right) \Rightarrow i_1(0) = \frac{\mathcal{E}}{R}\left(1 - \left[\cos(0)\right]\right) = 0.$$

(d) When does i_2 first equal zero?

$$i_2(t) = 0 = \frac{\mathcal{E}}{R} e^{-bt}\left[-(2\omega RC)^{-1} \sin(\omega t) + \cos(\omega t)\right] \Rightarrow -(2\omega RC)^{-1} \tan(\omega t) + 1 = 0$$

$$\Rightarrow \tan(\omega t) = +2\omega RC = +2(40 \text{ rad/s})(1000\ \Omega)(6.25 \times 10^{-6} \text{ F}) = +0.5.$$

$$\Rightarrow \omega t = \arctan(+0.5) = +0.464 \Rightarrow t = \frac{0.464}{40.0 \text{ rad/s}} = 0.0116 \text{ s}.$$

31-63: (a) $\Phi_B = BA = B_L A_L + B_{Air} A_{Air} = \frac{\mu_o N i}{W}((D - d)W) + \frac{K\mu_o N i}{W}(dW) = \mu_o N i[(D - d) + Kd]$

$$\Rightarrow L = \frac{N\Phi_B}{i} = \mu_o N^2[(D - d) + Kd] = L_o - L_o \frac{d}{D} + L_f \frac{d}{D} = L_o + \left(\frac{L_f - L_o}{D}\right)d$$

$$\Rightarrow d = \left(\frac{L - L_o}{L_f - L_o}\right)D, \text{ where } L_o = \mu_o N^2 D, \text{ and } L_f = K\mu_o N^2 D.$$

(b) Using $K = \chi_m + 1$ we can find the inductance for any height $L = L_o\left(1 + \chi_m \frac{d}{D}\right)$.

Height of Fluid	Inductance of Liquid Oxygen	Inductance of Mercury
$d = \frac{D}{4}$	0.63024 H	0.63000 H
$d = \frac{D}{2}$	0.63048 H	0.62999 H
$d = \frac{3D}{4}$	0.63072 H	0.62999 H
$d = D$	0.63096 H	0.62998 H

Where we used the values $\chi_m(O_2) = 1.52 \times 10^{-3}$ and $\chi_m(Hg) = -2.9 \times 10^{-5}$.

(d) The volume gauge is much better for the liquid oxygen than the mercury because there is an easily detectable spread of values for the liquid oxygen, but not for the mercury.

Chapter 32: Alternating Current

32-1: (a) Since the voltage is sinusoidal, the average is zero.

(b) $V_{rms} = \dfrac{V}{\sqrt{2}} = \dfrac{90\text{ V}}{\sqrt{2}} = 63.6\text{ V}.$

32-2: (a) $I = \sqrt{2}I_{rms} = \sqrt{2}(1.40\text{ A}) = 1.98\text{ A}.$ (b) $I_{rav} = \dfrac{2}{\pi}I = \dfrac{2}{\pi}(1.98\text{ A}) = 1.26\text{ A}.$

(c) The root mean square voltage is always greater than the rectified average, because squaring the current before averaging, then square-rooting to get the root mean square value will always give a larger value than just averaging.

32-3: (a) $X_L = \omega L = 2\pi fL = 2\pi(50\text{ Hz})(2.00\text{ H}) = 628\ \Omega.$

(b) $X_L = \omega L = 2\pi fL \Rightarrow L = \dfrac{X_L}{2\pi f} = \dfrac{2.00\ \Omega}{2\pi(50\text{ Hz})} = 6.37 \times 10^{-3}\text{ H}.$

(c) $X_C = \dfrac{1}{\omega C} = \dfrac{1}{2\pi fC} = \dfrac{1}{2\pi(50\text{ Hz})(2.0 \times 10^{-6}\text{ F})} = 1590\ \Omega.$

(d) $X_C = \dfrac{1}{2\pi fC} \Rightarrow C = \dfrac{1}{2\pi f X_C} = \dfrac{1}{2\pi(50\text{ Hz})(2.00\ \Omega)} = 1.59 \times 10^{-3}\text{ F}.$

32-4: (a) $X_L = \omega L = 2\pi fL = 2\pi(60\text{ Hz})(0.80\text{ H}) = 302\ \Omega.$ If $f = 600\text{ Hz},\ X_L = 3020\ \Omega.$

(b) $X_C = \dfrac{1}{\omega C} = \dfrac{1}{2\pi fC} = \dfrac{1}{2\pi(60\text{ Hz})(7.00 \times 10^{-6}\text{ F})} = 379\ \Omega.$ If $f = 600\text{ Hz},\ X_C = 37.9\ \Omega.$

(c) $X_C = X_L \Rightarrow \dfrac{1}{\omega C} = \omega L \Rightarrow \omega = \dfrac{1}{\sqrt{LC}} = \dfrac{1}{\sqrt{(0.800\text{ H})(7.00 \times 10^{-6}\text{ Hz})}} = 422\text{ Hz}.$

32-5: (a) $V = IX_L = I\omega L \Rightarrow I = \dfrac{V}{\omega L} = \dfrac{80.0\text{ V}}{(100\text{ rad/s})(4.00\text{ H})} = 0.200\text{ A}.$

(b) $I = \dfrac{V}{\omega L} = \dfrac{80.0\text{ V}}{(1000\text{ rad/s})(4.00\text{ H})} = 0.0200\text{ A}.$

(c) $I = \dfrac{V}{\omega L} = \dfrac{80.0\text{ V}}{(10{,}000\text{ rad/s})(4.00\text{ H})} = 0.00200\text{ A}.$

(d)

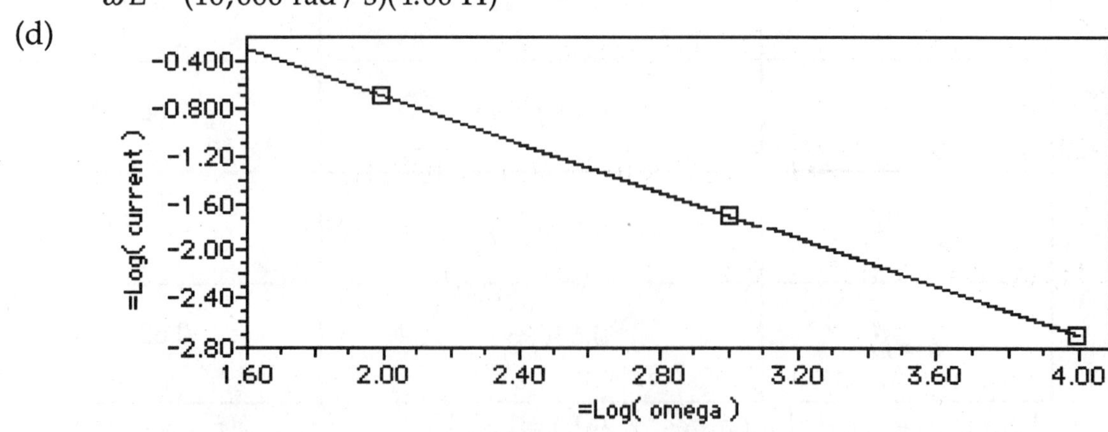

32-6: (a) $V = IX_C = \dfrac{I}{\omega C} \Rightarrow I = V\omega C = (80.0\text{ V})(100\text{ rad/s})(4.00 \times 10^{-6}\text{ F}) = 0.0320\text{ A}.$

(b) $I = V\omega C = (80.0\text{ V})(1000\text{ rad / s})(4.00 \times 10^{-6}\text{ F}) = 0.320\text{ A}.$

(c) $I = V\omega C = (80.0\text{ V})(10{,}000\text{ rad / s})(4.00 \times 10^{-6}\text{ F}) = 3.20\text{ A}.$

(d)

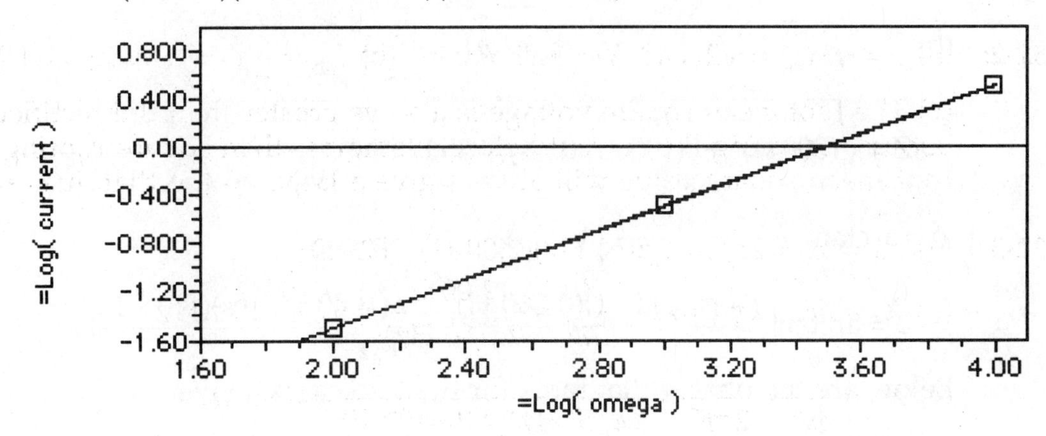

32-7: (a) $i = \dfrac{v}{R} = \dfrac{(2.50\text{ V})\cos((950\text{ rad/s})t)}{300\ \Omega} = (8.33 \times 10^{-3}\text{ A})\cos((950\text{ rad/s})t).$

(b) $X_L = \omega L = (950\text{ rad / s})(0.800\text{ H}) = 760\ \Omega.$

(c) $v_L = L\dfrac{di}{dt} = -(\omega L)(8.33 \times 10^{-3}\text{ A})\sin((950\text{ rad / s})t) = -(6.33\text{ V})\sin((950\text{ rad / s})t).$

32-8: (a) $X_c = \dfrac{1}{\omega C} = \dfrac{1}{(300\text{ rad/s})(6.00 \times 10^{-6}\text{ F})} = 556\ \Omega.$

(b) To find the voltage across the resistor we need to know the current, which can be found from the capacitor (remembering that it is out of phase by 90° from the capacitor's voltage).

$i = \dfrac{v_C}{X_C} = \dfrac{v\cos(\omega t)}{X_C} = \dfrac{(8.50\text{ V})\cos((300\text{ rad/s})t)}{556\ \Omega} = (0.0153\text{ A})\cos((300\text{ rad/s})t)$

$\Rightarrow v_R = iR = (0.0153\text{ A})(400\ \Omega)\cos((300\text{ rad/s})t) = (6.12\text{ V})\cos((300\text{ rad/s})t).$

32-9: $V_C = \dfrac{I}{\omega C} \Rightarrow C = \dfrac{I}{\omega V_C} = \dfrac{(1.40\text{ A})}{2\pi(60\text{ Hz})(170\text{ V})} = 2.18 \times 10^{-5}\text{ F}.$

32-10: $V_L = I\omega L \Rightarrow \omega = \dfrac{V_L}{IL} = \dfrac{(24.0\text{ V})}{(4.00 \times 10^{-3}\text{ A})(6.00 \times 10^{-4}\text{ H})} = 1.00 \times 10^{7}\text{ Hz}.$

32-11: (a) $Z = \sqrt{R^2 + (X_L - X_C)^2} = \sqrt{R^2 + (\omega L - \dfrac{1}{\omega C})^2}$

$$\Rightarrow Z = \sqrt{(300\ \Omega)^2 + \left((500\text{ rad / s})(0.250\text{ H}) - \dfrac{1}{(500\text{ rad / s})(8.00 \times 10^{-6}\text{ F})}\right)^2} = 325\ \Omega.$$

$$\phi = \arctan\left(\frac{\omega L - 1/(\omega C)}{R}\right)$$

$$\Rightarrow \phi = \arctan\left(\frac{(500\ \text{rad/s})(0.250\ \text{H}) - 1/(500\ \text{rad/s})(8.0\times10^{-6}\ \text{F})}{300\ \Omega}\right) = -22.6°.$$

(b) $Z = \sqrt{R^2 + (X_L - X_C)^2} = \sqrt{R^2 + (\omega L - \frac{1}{\omega C})^2}$

$$\Rightarrow Z = \sqrt{(300\ \Omega)^2 + \left((1000\ \text{rad/s})(0.250\ \text{H}) - \frac{1}{(1000\ \text{rad/s})(8.00\times10^{-6}\ \text{F})}\right)^2} = 325\ \Omega.$$

$$\phi = \arctan\left(\frac{\omega L - 1/(\omega C)}{R}\right)$$

$$\Rightarrow \phi = \arctan\left(\frac{(1000\ \text{rad/s})(0.250\ \text{H}) - 1/(1000\ \text{rad/s})(8.0\times10^{-6}\ \text{F})}{300\ \Omega}\right) = +22.6°.$$

Below are the phasor diagrams for the two cases above.

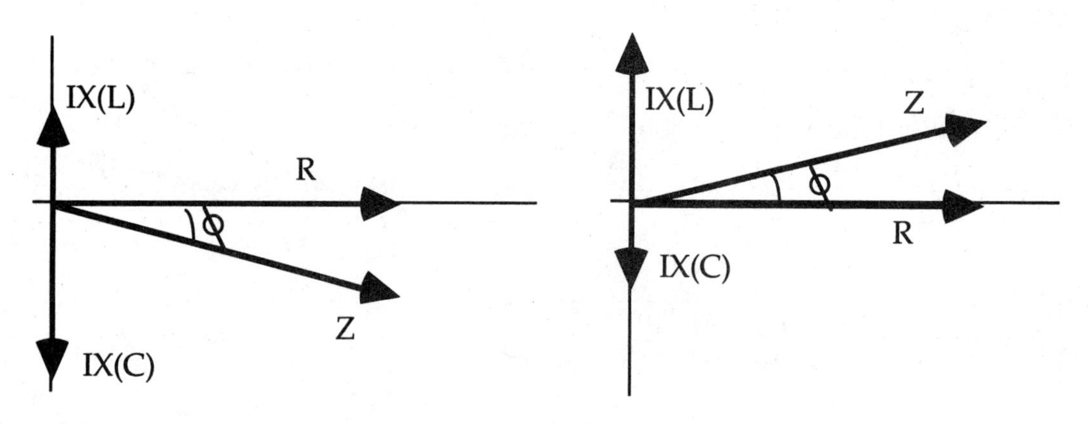

32-12: Using $Z = \sqrt{R^2 + (\omega L - \frac{1}{\omega C})^2}$ and $\phi = \arctan\left(\frac{\omega L - 1/(\omega C)}{R}\right)$, along with the values $R = 400\ \Omega$, $L = 0.900\ \text{H}$, and $C = 2.50\times10^{-6}\ \text{F}$:

(a) $\omega = 1000\ \text{rad/s}$: $Z = 640\ \Omega$, $\phi = +51.3°$;
$\omega = 750\ \text{rad/s}$: $Z = 408\ \Omega$, $\phi = +19.5°$;
$\omega = 500\ \text{rad/s}$: $Z = 532\ \Omega$, $\phi = -41.2°$.

(b) The current increases at first, then decreases as the frequency decreases from 1000 rad/s to 500 rad/s.

(c) The phase angle was calculated in part (a) for all frequencies.

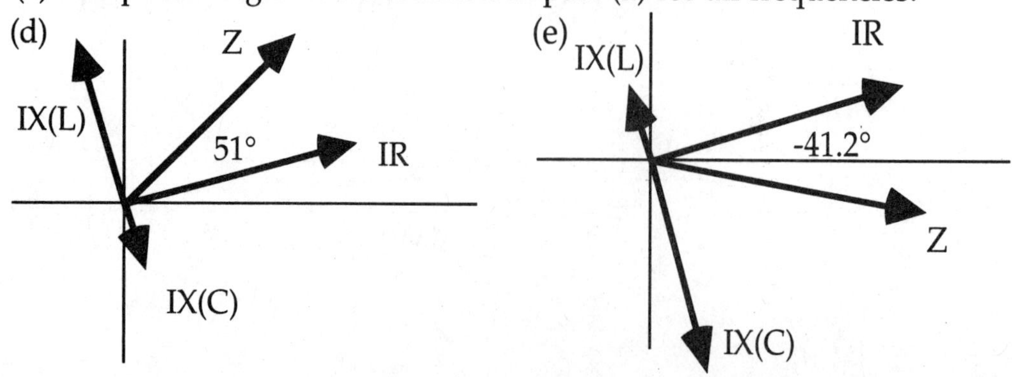

32-13: (a) If $\omega = \omega_o = \dfrac{1}{\sqrt{LC}}$

$\Rightarrow X = \omega L - \dfrac{1}{\omega C}$

$\Rightarrow X = \dfrac{L}{\sqrt{LC}} - \dfrac{1}{C/\sqrt{LC}} = 0.$

(b) When $\omega > \omega_o \Rightarrow X > 0.$

(c) When $\omega < \omega_o \Rightarrow X < 0.$

(d) At right is the graph of X against ω.

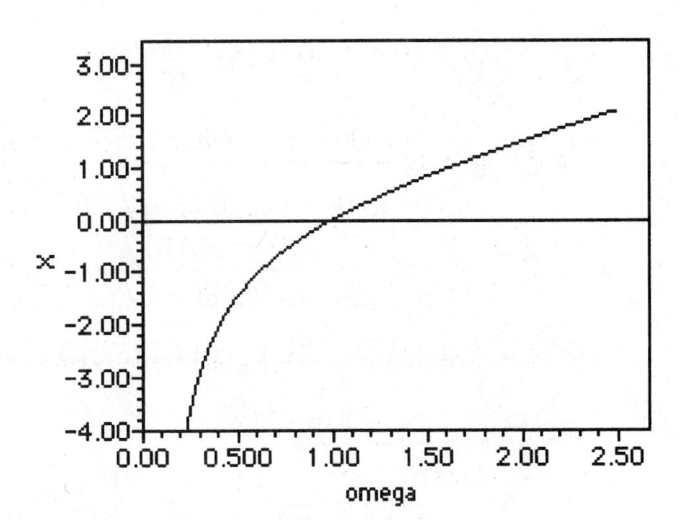

32-14: (a) $Z = \sqrt{R^2 + (\omega L)^2} = \sqrt{(300\ \Omega)^2 + ((200\ \text{rad/s})(0.600\ \text{H}))^2} = 323\ \Omega.$

(b) $I = \dfrac{V}{Z} = \dfrac{70.0\ \text{V}}{323\ \Omega} = 0.217\ \text{A}.$

(c) $V_R = IR = (0.217\ \text{A})(300\ \Omega) = 65.0\ \text{V};$

$V_L = I\omega L$

$= (0.217\ \text{A})(200\ \text{rad/s})(0.600\ \text{H})$

$\Rightarrow V_L = 26.0\ \text{V}.$

(e)

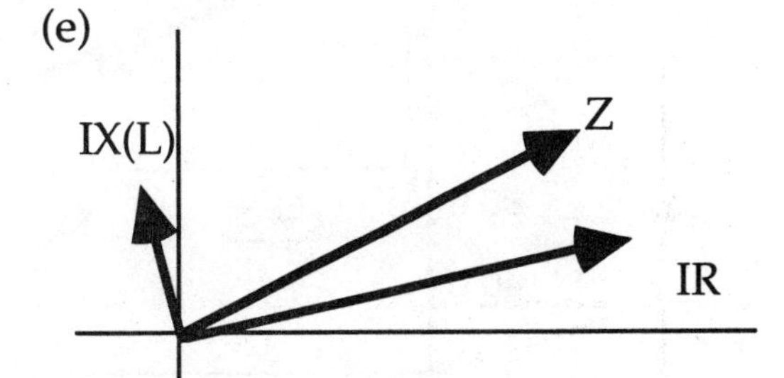

(d) $\phi = \arctan\left(\dfrac{v_L}{v_R}\right) = \arctan\left(\dfrac{26.0\ \text{V}}{65.0\ \text{V}}\right) = 21.8°$, and the voltage leads the current.

32-15: (a) $Z = \sqrt{R^2 + (1/\omega C)^2} = \sqrt{(300\ \Omega)^2 + 1/\left((200\ \text{rad/s})(4.0\times10^{-6}\ \text{F})\right)^2} = 1290\ \Omega.$

(b) $I = \dfrac{V}{Z} = \dfrac{70.0\ \text{V}}{1590\ \Omega} = 0.0545\ \text{A}.$

(c) $V_R = IR = (0.0545\ \text{A})(300\ \Omega) = 16.3\ \text{V};$

$V_C = \dfrac{I}{\omega C} = \dfrac{(0.0545\ \text{A})}{(200\ \text{rad/s})(4.0\times10^{-6}\ \text{F})} = 68.1\ \text{V}.$

(d) $\phi = \arctan\left(\dfrac{V_C}{V_R}\right) = \arctan\left(\dfrac{68.1\ \text{V}}{16.3\ \text{V}}\right) = -76.5°$, and the voltage lags the current.

(e)

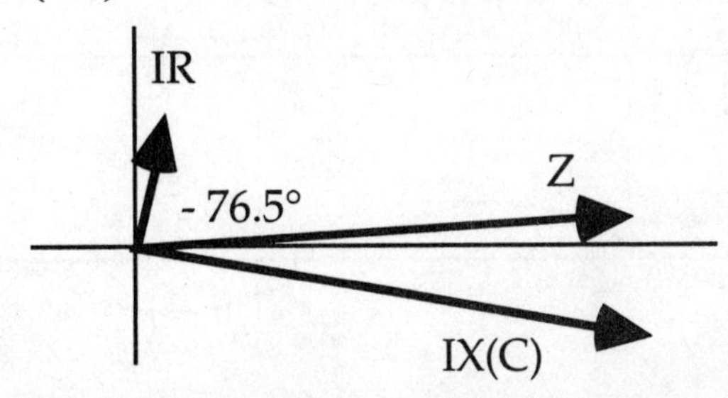

32-16: (a) $Z = (\omega L - 1/\omega C) = (200 \text{ rad/s})(0.600 \text{ H}) - \dfrac{1}{(200 \text{ rad/s})(4.0 \times 10^{-6} \text{ F})} = 1130\ \Omega.$

(b) $I = \dfrac{V}{Z} = \dfrac{70.0 \text{ V}}{1130\ \Omega} = 0.0619 \text{ A}.$

(c) $V_L = I\omega L = (0.0619 \text{ A})(200 \text{ rad/s})(0.600 \text{ H}) = 7.43 \text{ V}$

$V_C = \dfrac{I}{\omega C} = \dfrac{(0.0619 \text{ A})}{(200 \text{ rad/s})(4.0 \times 10^{-6} \text{ F})} = 77.4 \text{ V}.$

(d) $\phi = \arctan\left(\dfrac{V_L - V_C}{V_R}\right) = \arctan(-\infty) = -90.0°$, and the voltage lags the current.

(e)

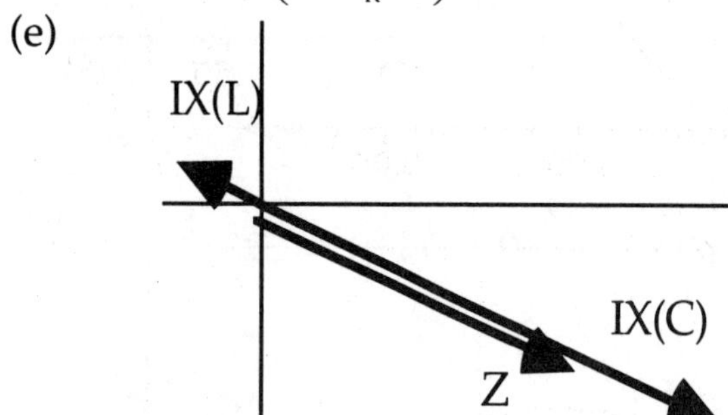

32-17: (a)

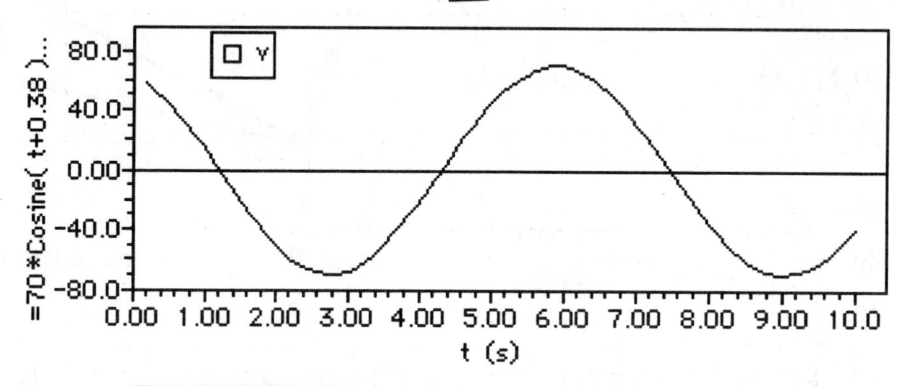

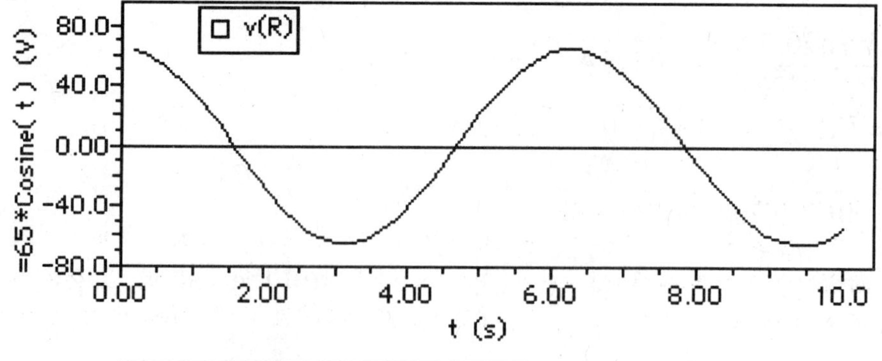

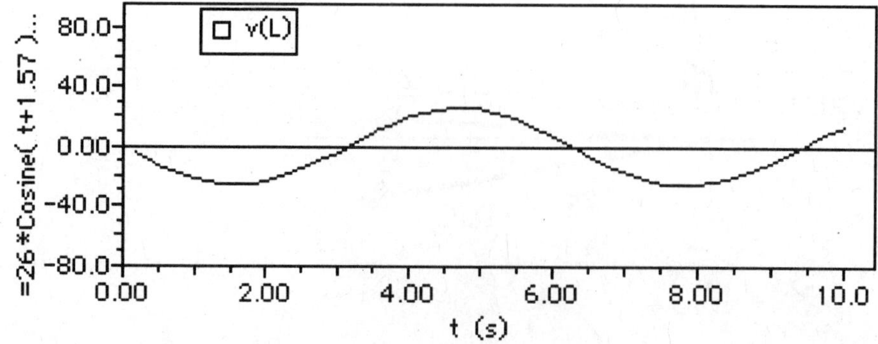

(b) The different voltages are:
$v = (70.0\text{ V})\cos(200t + 21.8°)$, $v_R = (65.0\text{ V})\cos(200t)$, $v_L = (26.0\text{ V})\cos(200t + 90°)$
At t = 20 ms: $v = -22.8$ V, $v_R = -42.5$ V, $v_L = 19.7$ V. Note $v_R + v_L = v$.
(c) At t = 40 ms: $v = -35.2$ V, $v_R = -9.5$ V, $v_L = -25.7$ V. Note $v_R + v_L = v$.

32-18: (a)

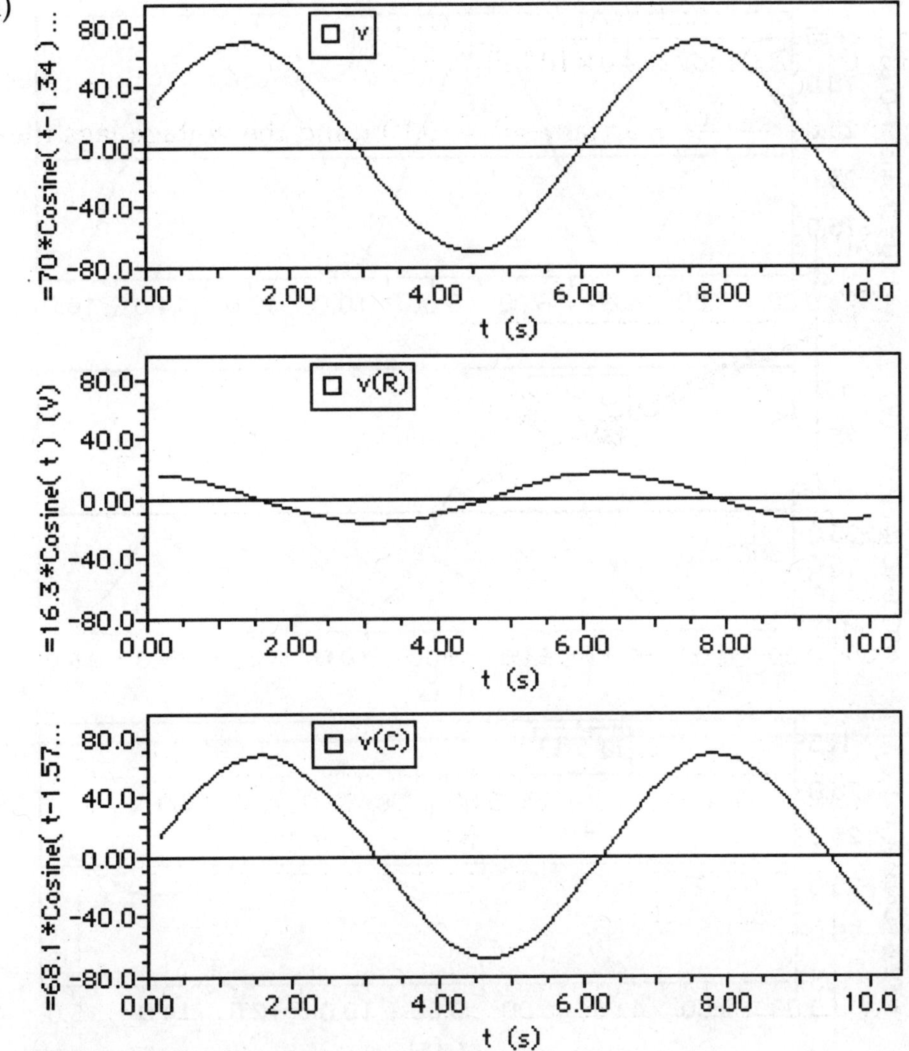

(b) The different voltages are:
$v = (70.0\text{ V})\cos(200t - 76.5°)$, $v_R = (16.3\text{ V})\cos(200t)$, $v_L = (68.1\text{ V})\cos(200t - 90°)$
At t = 20 ms: $v = -62.2$ V, $v_R = -10.7$ V, $v_C = -51.5$ V. Note $v_R + v_C = v$.
(c) At t = 40 ms: $v = 65.0$ V, $v_R = -2.4$ V, $v_C = 67.4$ V. Note $v_R + v_C = v$.

32-19: (a) $Z = \sqrt{R^2 + (\omega L - 1/\omega C)^2}$

$$\Rightarrow Z = \sqrt{(300\ \Omega)^2 + \left((400\text{ rad/s})(0.25\text{ H}) - 1/\left((400\text{ rad/s})(8.0\times10^{-6}\text{ F})\right)\right)^2} = 368\ \Omega.$$

$$\Rightarrow I = \frac{V}{Z} = \frac{120\text{ V}}{368\ \Omega} = 0.326\text{ A}.$$

(b) $\phi = \arctan\left(\dfrac{\omega L - 1/\omega C}{R}\right) = \arctan\left(\dfrac{100\ \Omega - 312\ \Omega}{300\ \Omega}\right) = -35.3°$, and the voltage lags the current.

(c) $V_R = IR = (0.326\,\text{A})(300\,\Omega) = 97.8$ V;

$V_L = I\omega L = (0.326\,\text{A})(400\text{ rad/s})(0.250\text{ H}) = 32.6$ V;

$$V_C = \frac{I}{\omega C} = \frac{(0.326\,\text{A})}{(400\text{ rad/s})(8.0\times10^{-6}\text{ F})} = 102\text{ V}.$$

32-20: (a)

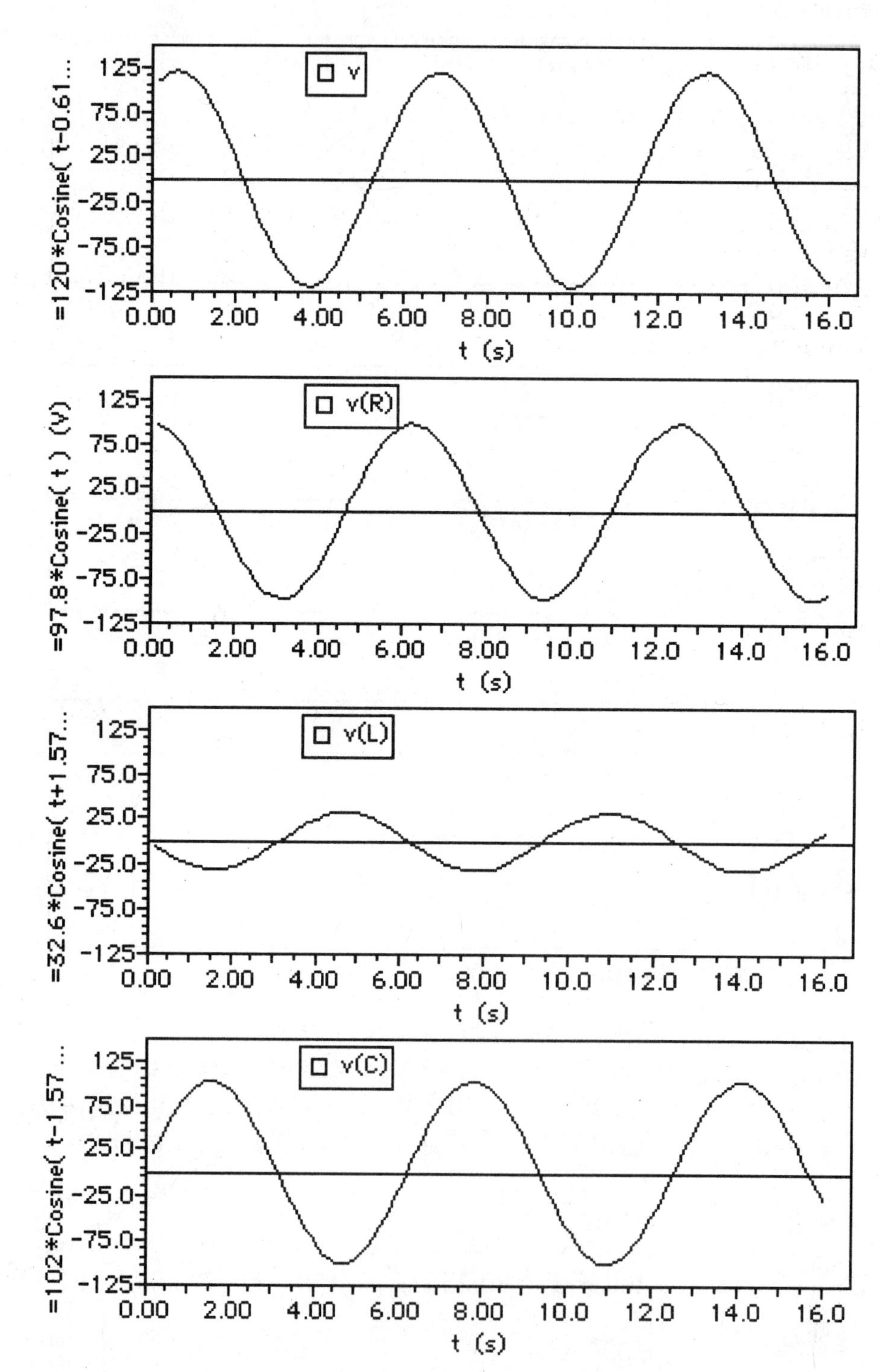

The different voltages plotted above are: $v = (120\text{ V})\cos(400t - 0.616)$, $v_R = (97.8\text{ V})\cos(200t)$, $v_L = (32.6\text{ V})\cos(200t + 90°)$ $v_C = (102\text{ V})\cos(200t - 90°)$.

(b) At t = 10 ms: $v = -116$ V, $v_R = -63.9$ V, $v_L = 24.7$ V, $v_C = -77.2$ V.

(c) At t = 25 ms: $v = -120\text{ V},\ v_R = -82.1\text{ V},\ v_L = 17.7\text{ V},\ v_C = -55.5\text{ V}.$
In both parts (b) and (c), note that the total voltage equals the sum of the other voltages at the given instant.

32-21: (a) From **Ex. (32-11)**:
$$P = \frac{1}{2}I^2 Z\cos\phi = I_{rms}{}^2 Z\cos\phi = (0.325\text{ A})^2(325\ \Omega)\cos(22.6°) = 31.7\text{ W}.$$
(b) $P_{av(R)} = I_{rms}{}^2 R = (0.325\text{ A})^2(300\ \Omega) = 31.7\text{ W}.$
(c) The voltage and current for the inductor are out of phase by 90°, so the power is zero.
(d) The voltage and current for the capacitor are out of phase by 90°, so the power is zero.
(e) The total power equals the sum of the powers over the circuit elements.

32-22: (a) The power factor equals:
$$\cos\phi = \frac{R}{Z} = \frac{R}{\sqrt{R^2 + (\omega L)^2}} = \frac{(300\ \Omega)}{\sqrt{(300\ \Omega)^2 + ((50\text{ rad/s})(4.30\text{ H}))^2}} = 0.813.$$
(b) $$P_{av} = \frac{1}{2}\frac{V^2}{Z}\cos\phi = \frac{1}{2}\frac{(180\text{ V})^2}{\sqrt{(300\ \Omega)^2 + ((50\text{ rad/s})(4.30\text{ H}))^2}}(0.813) = 35.7\text{ W}.$$

32-23: (a) Using the phasor diagram at right we can see:
$$\cos\phi = \frac{IR}{I\sqrt{R^2 + X_L{}^2 - X_C{}^2}} = \frac{R}{Z}.$$
(b) $$P_{av} = \frac{1}{2}\frac{V^2}{Z}\cos\phi = \frac{V_{rms}{}^2}{Z}\cos\phi$$
$$\Rightarrow P_{av} = \frac{V_{rms}{}^2}{Z}\frac{R}{Z} = I_{rms}{}^2 R.$$

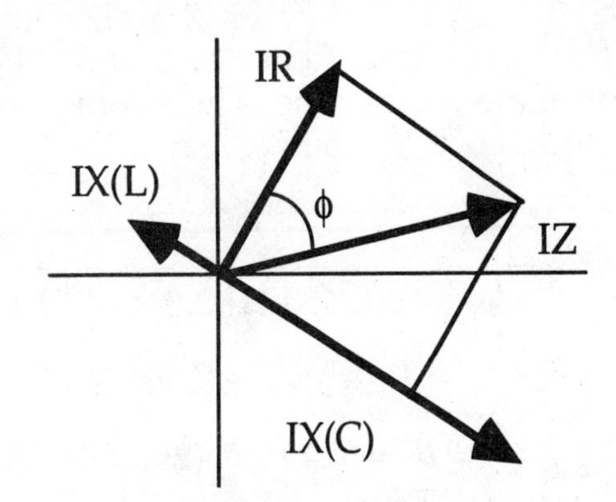

32-24: $$P_{av} = \frac{V_{rms}{}^2}{Z^2}R = \frac{(160\text{ V})^2}{(150\ \Omega)^2}(110\ \Omega) = 125\text{ W}.$$

32-25: (a) At resonance:
$$\omega_o = \frac{1}{\sqrt{LC}} = \frac{1}{\sqrt{(0.900\text{ H})(2.5\times 10^{-6}\text{ F})}}.$$
$$\Rightarrow \omega_o = 667\text{ rad/s}.$$

(b)

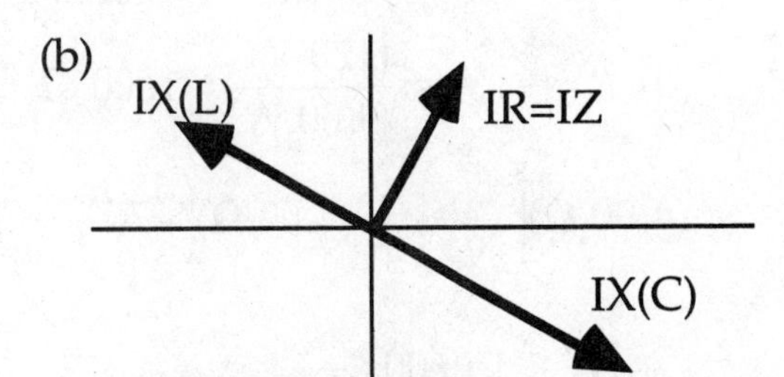

(c) $$V_1 = V_{rms(source)} = \frac{V}{\sqrt{2}} = \frac{90.0\text{ V}}{\sqrt{2}} = 63.6\text{V},\ I_{rms} = \frac{V_{rms}}{Z} = \frac{V_{rms}}{R} = \frac{63.6\text{ V}}{400\ \Omega} = 0.159\text{ A}.$$
$$V_2 = I_{rms}\omega_o L = (0.159\text{ A})(667\text{ Hz})(0.900\text{ H}) = 95.5\text{ V}.$$
$$V_3 = \frac{I_{rms}}{\omega_o C} = \frac{(0.159\text{ A})}{(667\text{ Hz})(2.5\times 10^{-6}\text{ F})} = 95.5\text{ V} = V_2,$$

$V_4 = 0$, since the capacitor and inductor's volteges cancel each other.

$$V_5 = V_{rms(source)} = \frac{V}{\sqrt{2}} = \frac{90.0\ \text{V}}{\sqrt{2}} = 63.6\text{V}.$$

(d) If the resistance is changed, that has no affect upon the resonance frequency: $\omega_o = 667$ rad / s.

(e) $$I_{rms} = \frac{V_{rms}}{Z} = \frac{V_{rms}}{R} = \frac{63.6\ \text{V}}{100\ \Omega} = 0.636\ \text{A}$$

32-26: (a) $$\omega_o = \frac{1}{\sqrt{LC}} = \frac{1}{\sqrt{(0.400\ \text{H})(8.0\times 10^{-6}\ \text{F})}} = 559\ \text{rad/s}.$$

(b) $I = 0.325$ A at resonance, so: $R = Z = \dfrac{V}{I} = \dfrac{240\ \text{V}}{0.325\ \text{A}} = 738\ \Omega.$

(c) At resonance:

$$V_{peak}(R) = 240\ \text{V},\ V_{peak}(L) = V_{peak}(C) = I\omega L = (0.325\text{A})(559\ \text{rad/s})(0.40\ \text{H}) = 72.7\ \text{V}.$$

32-27: (a) At resonance, the power factor is equal to one, because the impedance of the circuit is exactly equal to the resistance, so $\dfrac{R}{Z} = 1.$

(b) Average power: $$P_{av} = \frac{V_{rms}^{\ 2}}{R} = \frac{1}{2}\frac{(300\ \text{V})^2}{300\ \Omega} = 150\ \text{W}.$$

(c) If the capacitor is changed, and then resonance is again attained, the power factor again equals one. The average power still has no dependence on the capacitor, so $P_{av} = 150$ W again.

32-28: (a) $$\omega_o = \frac{1}{\sqrt{LC}} = \frac{1}{\sqrt{(0.400\ \text{H})(2.0\times 10^{-8}\ \text{F})}} = 11{,}200\ \text{rad/s}.$$

(b) $$V_C = \frac{I}{\omega C} \Rightarrow I = V_C\omega C = (650\ \text{V})(11{,}200\ \text{rad/s})(2.00\times 10^{-8}\ \text{F}) = 0.146\ \text{A}$$

$$\Rightarrow V_{\max(source)} = IR = (0.146\ \text{A})(250\ \Omega) = 36.4\ \text{V}.$$

32-29: (a) $\dfrac{N_1}{N_2} = \dfrac{120}{6} = 20.$ (b) $I_{rms} = \dfrac{V_{rms}}{R} = \dfrac{6.00\ \text{V}}{4.00\ \Omega} = 1.50\ \text{A}.$

(c) $P_{av} = I_{rms}V_{rms} = (1.50\ \text{A})(6.00\ \text{V}) = 9.00\ \text{W}.$

(d) $R = \dfrac{V_{rms}^{\ 2}}{P} = \dfrac{(120\ \text{V})^2}{9.00\ \text{W}} = 1600\ \Omega$, and note that this is the same as

$$(4.00\ \Omega)\left(\frac{N_1}{N_2}\right)^2 = (4.00\ \Omega)\left(\frac{120}{6}\right)^2 = 1600\ \Omega.$$

32-30: (a) $\dfrac{N_2}{N_1} = \dfrac{15{,}600}{120} = 130.$

(b) $P = I_2V_2 = (0.0100\ \text{A})(15{,}600\ \text{V}) = 156\ \text{W}.$

(c) $I_1 = I_2\dfrac{N_2}{N_1} = (0.0100\ \text{A})(130) = 1.30\ \text{A}.$

32-31: (a) $R_1 = R_2\left(\frac{N_1}{N_2}\right)^2 \Rightarrow \frac{N_1}{N_2} = \sqrt{\frac{R_1}{R_2}} = \sqrt{\frac{6400\ \Omega}{16.0\ \Omega}} = 20.$

(b) $V_2 = V_1\left(\frac{N_2}{N_1}\right) = (100\text{V})\frac{1}{20} = 5.00\ \text{V}.$

32-32: (a) $X_{L_2} = \omega_2 L = 2\omega_1 L = 2\left(\frac{1}{\omega_1 C}\right) = 2\left(\frac{2}{\omega_2 C}\right) = 4X_{C_2} \Rightarrow \frac{X_{L_2}}{X_{C_2}} = 4$, and so the inductor's reactance is greater than that of the capacitor.

(b) $X_{L_3} = \omega_3 L = \frac{\omega_1 L}{3} = \left(\frac{1}{3\omega_1 C}\right) = \left(\frac{1}{9\omega_3 C}\right) = \frac{1}{9}X_{C_3} \Rightarrow \frac{X_{L_2}}{X_{C_2}} = \frac{1}{9}$, and so the capacitor's reactance is greater than that of the inductor.

32-33: (a) $I_{rav} = 0$ when $\omega t = (n + 1/2)\pi \Rightarrow t_1 = \frac{\pi}{2\omega},\ t_2 = \frac{3\pi}{2\omega} \Rightarrow t_2 - t_1 = \frac{\pi}{\omega}.$

(b) $\int_{t_1}^{t_2} i\,dt = \int_{t_1}^{t_2} I\cos(\omega t)dt = \frac{I}{\omega}\sin(\omega t)\Big|_{t_1}^{t_2} = \frac{I}{\omega}[\sin(3\pi/2) - \sin(\pi/2)] = -\frac{2I}{\omega}$

(c) So, $I_{rav}(t_2 - t_1) = \frac{2I}{\omega} \Rightarrow I_{rav} = \frac{\omega}{\pi}\frac{2I}{\omega} = \frac{2I}{\pi}.$

32-34: (a) $Z_{tweeter} = \sqrt{R^2 + (1/\omega C)^2}$ (b) $Z_{woofer} = \sqrt{R^2 + (\omega L)^2}$

(c) If $Z_{tweeter} = Z_{woofer}$, then the current splits evenly through each branch.

(d) At the cross-over point, where currents are equal:

$R^2 + (1/\omega C)^2 = R^2 + (\omega L)^2 \Rightarrow \omega = \frac{1}{\sqrt{LC}}.$

32-35: $\phi = \arctan\left(\frac{\omega L}{R}\right) \Rightarrow L = \frac{R}{\omega}\tan\phi = \frac{R}{2\pi f}\tan\phi = \left(\frac{40.0\ \Omega}{2\pi(100\ \text{Hz})}\right)\tan(38.4°) = 0.0505\ \text{H}.$

32-36: (a) If $\omega = 500$ rad/s: $Z = \sqrt{R^2 + (\omega L - 1/\omega C)^2}$

$\Rightarrow Z = \sqrt{(300\ \Omega)^2 + \left((500\ \text{rad/s})(0.60\ \text{H}) - 1/\left((500\ \text{rad/s})(4.0\times10^{-6}\ \text{F})\right)\right)^2} = 361\ \Omega.$

$\Rightarrow I = \frac{V}{Z} = \frac{70\ \text{V}}{361\ \Omega} = 0.194\ \text{A} \Rightarrow I_{rms} = \frac{I}{\sqrt{2}} = 0.137\ \text{A}.$

So, $V_1 = I_{rms}R = (0.137\ \text{A})(300\ \Omega) = 41.2\text{V},$

$V_2 = I_{rms}X_L = I_{rms}\omega L = (0.137\ \text{A})(500\ \text{rad/s})(0.600\ \text{H}) = 41.2\text{V},$

$V_3 = I_{rms}X_C = \frac{I_{rms}}{\omega C} = \frac{(0.137\ \text{A})}{(500\ \text{rad/s})(4.0\times10^{-6}\ \text{F})} = 68.7\text{V},$

$V_4 = V_3 - V_2 = 68.7\text{V} - 41.2\text{V} = 27.5\ \text{V}$, and

$V_5 = \mathcal{E}_{rms} = 49.5\text{V}.$

(b) If $\omega = 1000$ rad/s, using the same steps as above in part (a):

$Z = 461\ \Omega,\ V_1 = 32.2\text{V},\ V_2 = 64.4\text{V},\ V_3 = 26.8\text{V},\ V_4 = 37.5\text{V},\ V_5 = 49.5\text{V}.$

32-37: (a) $V_C = IX_C \Rightarrow I = \frac{V_C}{X_C} = \frac{720 \text{ V}}{600 \ \Omega} = 1.20 \text{ A}.$

(b) $Z = \frac{V}{I} = \frac{240 \text{ V}}{1.20 \text{ A}} = 200 \ \Omega.$

(c) $Z = \sqrt{R^2 + (X_L - X_C)^2} \Rightarrow X_L = X_C \pm \sqrt{Z^2 - R^2} = 600 \ \Omega \pm \sqrt{(200 \ \Omega)^2 - (120 \ \Omega)^2}$

$\Rightarrow X_L = 440 \ \Omega \text{ or } 760 \ \Omega.$

32-38: $Z = \sqrt{R^2 + X_L^2} = \sqrt{(500 \ \Omega)^2 + (300 \ \Omega)^2} = 583 \ \Omega.$

$P_{av} = \frac{V_{rms}^2}{Z}\frac{R}{Z} \Rightarrow V_{rms} = Z\sqrt{\frac{P_{av}}{R}} = (583 \ \Omega)\sqrt{\frac{900 \text{ W}}{500 \ \Omega}} = 782 \text{ V}.$

32-39: $P_{av} = I_{rms}^2 Z \frac{R}{Z} \Rightarrow R = \frac{P_{av}}{I_{rms}^2} = \frac{16.0 \text{ W}}{(0.500 \text{ A})^2} = 64.0 \ \Omega$

$\Rightarrow Z = \sqrt{R^2 + X_L^2} = \sqrt{(64.0 \ \Omega)^2 + (25.0 \ \Omega)^2} = 68.7 \ \Omega.$

32-40: (a) $P_{av} = \frac{V_{rms}^2}{Z}\cos\phi \Rightarrow Z = \frac{V_{rms}^2 \cos\phi}{P_{av}} = \frac{(110 \text{ V})^2(0.480)}{(280 \text{ W})} = 20.7 \ \Omega$

$\Rightarrow R = Z\cos\phi = (20.7 \Omega)(0.480) = 9.96 \ \Omega.$

(b) $Z = \sqrt{R^2 + X_L^2} \Rightarrow X_L = \sqrt{Z^2 - R^2} = \sqrt{(20.7 \ \Omega)^2 - (9.96 \ \Omega)^2} = 18.1 \ \Omega$. But at resonance, the inductive and capacitive reactances equal each other. So:

$X_C = \frac{1}{\omega C} \Rightarrow C = \frac{1}{\omega X_C} = \frac{1}{2\pi f X_C} = \frac{1}{2\pi(60.0 \text{ Hz})(18.1 \ \Omega)} = 1.46 \times 10^{-4} \text{ F}.$

(c) At resonance, $P = \frac{V^2}{R} = \frac{(110 \text{ V})^2}{9.96 \ \Omega} = 1210 \text{ W}.$

32-41: (a) If the original voltage was lagging the circuit current, the addition of an inductor will help it "catch up", since a pure LR circuit would have the voltage leading. This will increase the power factor, because it is largest when the current and voltage are in phase.

(b) Since the voltage is lagging, we have an RC circuit, and if the power factor is to be made equal to one, we need an inductor such that $X_L = X_C$. So:

$R = 0.630Z = 0.630(50.0 \ \Omega) = 31.5 \ \Omega$

$\Rightarrow Z = \sqrt{R^2 + X_C^2} \Rightarrow X_C = \sqrt{Z^2 - R^2} = \sqrt{(50 \ \Omega)^2 + (31.5 \ \Omega)^2} = 38.8 \ \Omega.$

$X_L = X_C = 38.8 \ \Omega = \omega L \Rightarrow L = \frac{X_C}{\omega} = \frac{38.8 \ \Omega}{2\pi(60 \text{ Hz})} = 0.103 \text{ H}.$

32-42: (a) $\cos\phi = \frac{R}{Z} = \frac{R}{\sqrt{R^2 + (X_L - X_C)^2}} = \frac{300 \ \Omega}{\sqrt{(300 \ \Omega)^2 + (500 \ \Omega - 700 \ \Omega)^2}} = \frac{300 \ \Omega}{361 \ \Omega} = 0.832.$

(b) $P_{av} = \frac{V_{rms}^2}{Z}\cos\phi \Rightarrow V_{rms} = \sqrt{\frac{P_{av}Z}{\cos\phi}} = \sqrt{\frac{(60 \text{ W})(361 \ \Omega)}{0.832}} = 161 \text{ V}.$

32-43: (a) $\tan\phi = \dfrac{X_L - X_C}{R} \Rightarrow X_L = X_C + R\tan\phi = 400\ \Omega + (200\ \Omega)\tan(-32°) = 275\ \Omega.$

(b) $P_{av} = I_{rms}{}^2 R \Rightarrow I_{rms} = \sqrt{\dfrac{P_{av}}{R}} = \sqrt{\dfrac{(210\ \text{W})}{(200\ \Omega)}} = 1.02\ \text{A}.$

(c) $V_{rms} = I_{rms}Z = I_{rms}\sqrt{R^2 + (X_L - X_C)^2}$

$\Rightarrow V_{rms} = (1.02\ \text{A})\sqrt{(200\ \Omega)^2 + (275\ \Omega - 400\ \Omega)} = 242\ \text{V}.$

32-44: (a) Since the voltage drop between any two points must always be equal, the parallel LRC sircuit must have equal potential drops over the capacitor, inductor and resistor, so $v_R = v_L = v_C = v$. Also, the sum of currents entering any junction must equal the current leaving the junction. Therefore, the sum of the currents in the branches must equal the current through the source: $i = i_R + i_L + i_C$.

(b) $i_R = \dfrac{v}{R}$ is always in phase with the voltage. $i_L = \dfrac{v}{\omega L}$ lags the voltage by 90°, and $i_C = v\omega C$ leads the voltage by 90°.

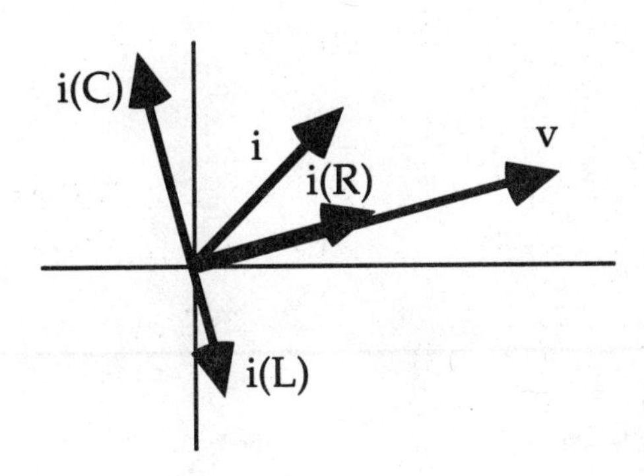

(c) From the diagram,

$$I^2 = I_R{}^2 + (I_C - I_L)^2 = \left(\frac{V}{R}\right)^2 + \left(V\omega C - \frac{V}{\omega L}\right)^2$$

(d) From (c): $I = V\sqrt{\dfrac{1}{R^2} + \left(\omega C - \dfrac{1}{\omega L}\right)^2}$. But

$$I = \frac{V}{Z} \Rightarrow \frac{1}{Z} = \sqrt{\frac{1}{R^2} + \left(\omega C - \frac{1}{\omega L}\right)^2}.$$

32-45: At resonance, the total current is a minimum, and $I_C = I_L$.

(a) $I_C = I_L \Rightarrow V\omega_o C = \dfrac{V}{\omega_o L} \Rightarrow \omega_o C = \dfrac{1}{\omega_o L} \Rightarrow \omega_o = \dfrac{1}{\sqrt{LC}}.$

(b) At $\omega = \omega_o$, $Z = R$, since the inductor and capacitor terms cancel.

(c) At $\omega = \omega_o$, I and V are in phase, so the phase angle is zero.

32-46: (a) $I_R = \dfrac{V}{R} = \dfrac{368\ \text{V}}{700\ \Omega} = 0.526\ \text{A}.$

(b) $I_C = V\omega C = (368\ \text{V})(450\ \text{rad/s})(2.0 \times 10^{-6}\ \text{F}) = 0.331\ \text{A}.$

(c) $\phi = \arctan\left(\dfrac{I_C}{I_R}\right) = \arctan\left(\dfrac{0.331\ \text{A}}{0.526\ \text{A}}\right) = 32.2°$, leading the voltage.

(d) $I = \sqrt{I_R{}^2 + I_C{}^2} = \sqrt{(0.331\ \text{A})^2 + (0.526\ \text{A})^2} = 0.621\ \text{A}.$

32-47: (a) $\omega_o = \dfrac{1}{\sqrt{LC}} = \dfrac{1}{\sqrt{(0.500\text{ H})(6.00\times10^{-6}\text{ F})}} = 1830\text{ Hz}.$

(b)

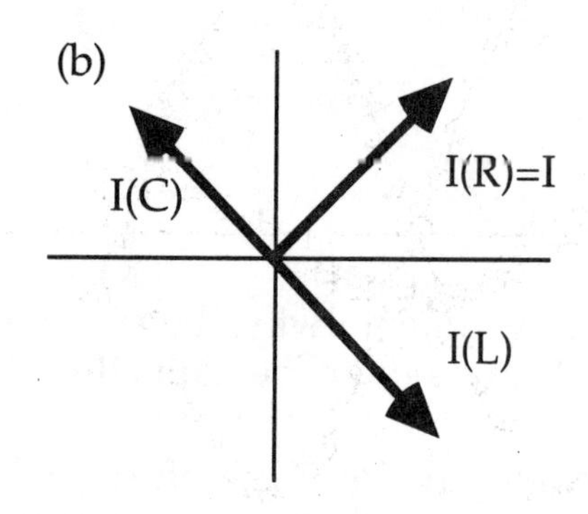

(c) At the resonant frequency $Z = R = 400\ \Omega.$
$\Rightarrow I = \dfrac{V}{R} = \dfrac{120\text{ V}}{400\ \Omega} = 0.300\text{ A}.$

(d) $I_R = I = 0.300\text{ A},$
$I_c = I_L = V\omega C = (120\text{ V})(1830\text{ Hz})(6.00\times10^{-6}\text{F})$
$\Rightarrow I_c = I_L = 0.131\text{ A}.$

32-48: (a) For $\omega = 800\text{ rad/s},\ Z = \sqrt{R^2 + (\omega L - 1/\omega C)^2}$

$$\Rightarrow Z = \sqrt{(500\ \Omega)^2 + \left((800\text{ rad/s})(2.0\text{ H}) - 1/\left((800\text{ rad/s})(5.0\times10^{-7}\text{ F})\right)\right)^2} = 1030\ \Omega.$$

$$\Rightarrow I = \frac{V}{Z} = \frac{100\text{ V}}{1030\ \Omega} = 0.0971\text{A} \Rightarrow V_R = IR = (0.0971\text{A})(500\ \Omega) = 48.6\text{ V}.$$

$$V_C = \frac{I}{\omega C} = \frac{0.0971\text{ A}}{(800\text{ rad/s})(5.0\times10^{-7}\text{F})} = 243\text{ V}.$$

$$V_L = I\omega L = (0.0971\text{ A})(800\text{ rad/s})(2.00\text{ H}) = 155\text{ V}.$$

Also note $\phi = \arctan\left(\dfrac{\omega L - 1/(\omega C)}{R}\right) = -60.9°.$

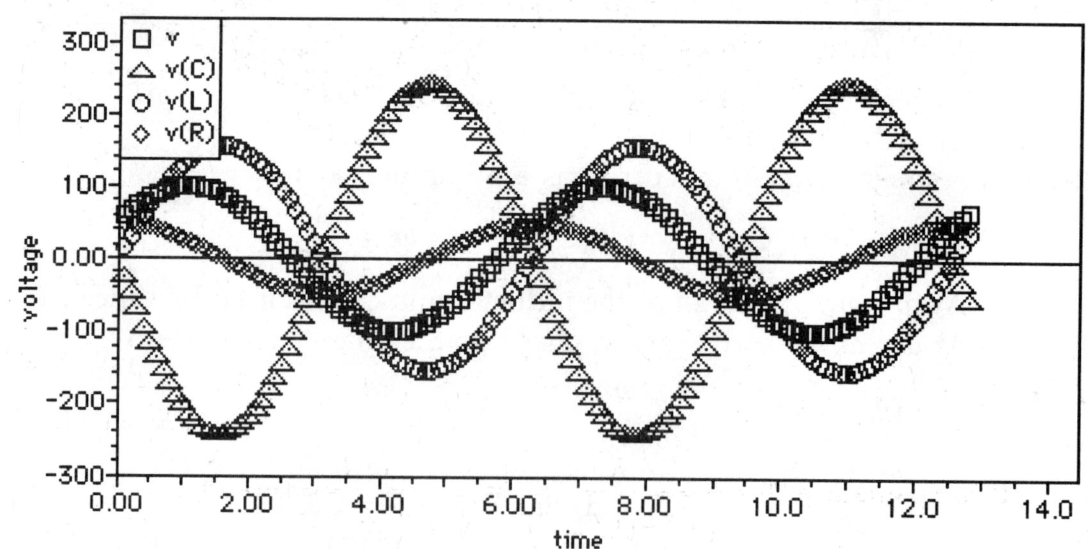

(b) Repeating exactly the same calculations as above for $\omega = 1000\text{ rad/s}$:
$Z = R = 500\ \Omega;\ \phi = 0;\ I = 0.200\text{A};\ V_R = V = 100\text{ V};\ V_C = V_L = 400\text{ V}.$

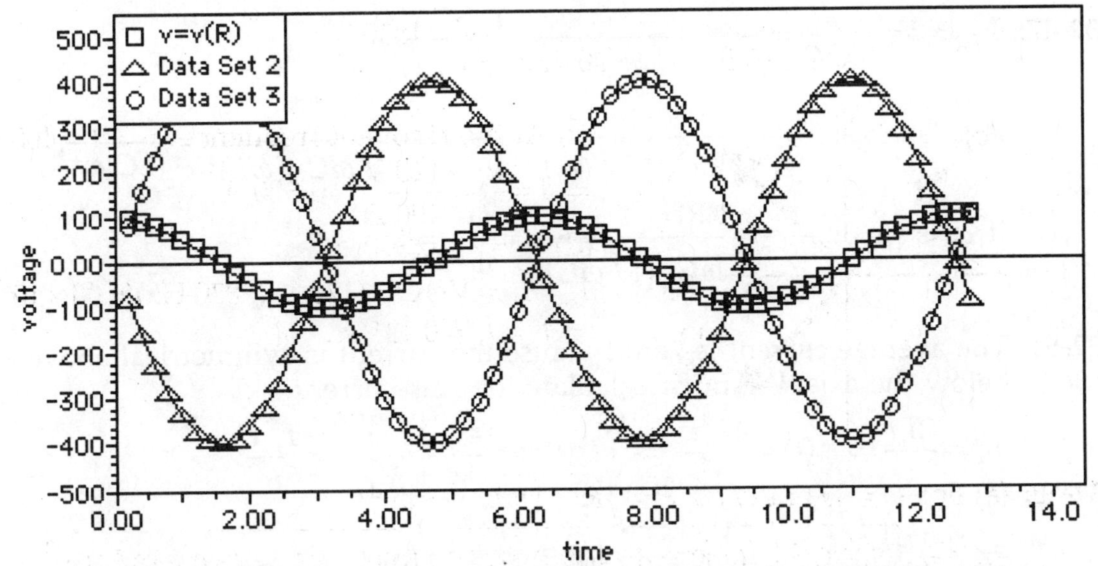

(c) Repeating exactly the same calculations as part (a) for $\omega = 1250 \text{ rad/s}$:
$Z = R = 1030\ \Omega;\ \phi = +60.9°;\ I = 0.0971\text{ A};\ V_R = 48.6\text{ V};\ V_C = 155\text{ V};\ V_L = 243\text{ V}$.

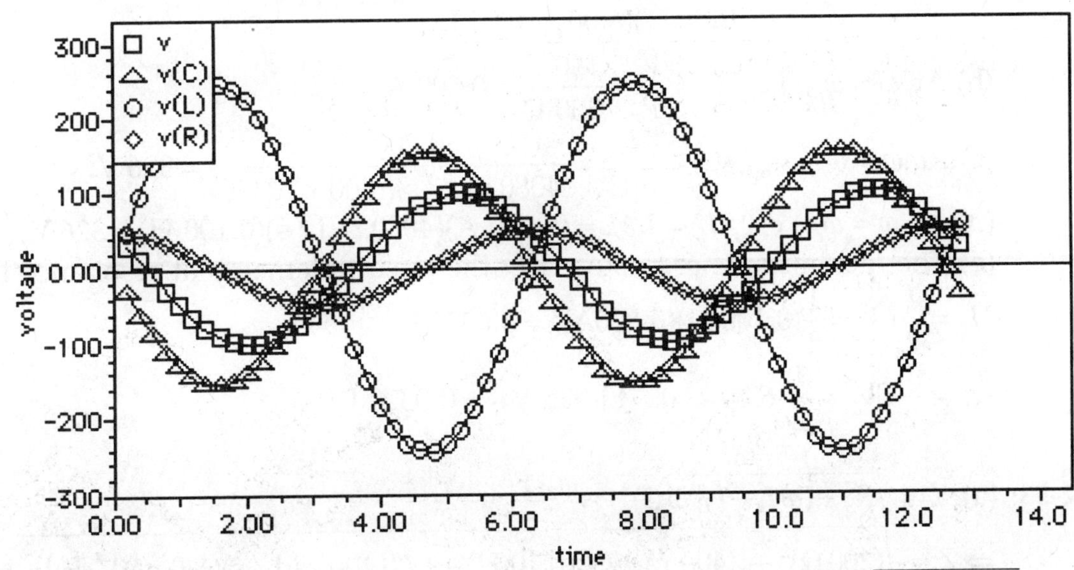

32-49: $V_{out} = \sqrt{V_R{}^2 + V_L{}^2} = I\sqrt{R^2 + (\omega L)^2} = \dfrac{V_s}{Z}\sqrt{R^2 + (\omega L)^2} \Rightarrow \dfrac{V_{out}}{V_s} = \dfrac{\sqrt{R^2 + (\omega L)^2}}{\sqrt{R^2 + (\omega L - 1/\omega C)^2}}$.

If ω is small: $\dfrac{V_{out}}{V_s} \approx \dfrac{R}{\sqrt{R^2 + (1/\omega C)^2}} = \dfrac{\omega R}{\sqrt{\omega^2 R^2 + (1/C)^2}} \approx \omega RC$.

If ω is large: $\dfrac{V_{out}}{V_s} \approx \dfrac{\sqrt{(\omega L)^2}}{\sqrt{(\omega L)^2}} = 1$.

32-50: $V_{out} = V_C = \dfrac{I}{\omega C} \Rightarrow \dfrac{V_{out}}{V_s} = \dfrac{1}{\omega C\sqrt{R^2 + (\omega L - 1/\omega C)^2}}.$

If ω is large: $\dfrac{V_{out}}{V_s} = \dfrac{1}{\omega C\sqrt{R^2 + (\omega L - 1/\omega C)^2}} \approx \dfrac{1}{\omega C\sqrt{(\omega L)^2}} = \dfrac{1}{(LC)\omega^2}.$

If ω is small: $\dfrac{V_{out}}{V_s} \approx \dfrac{1}{\omega C\sqrt{(1/\omega C)^2}} = \dfrac{\omega C}{\omega C} = 1.$

32-51: The average current is zero because the current is symmetrical above and below the axis. We must calculate the rms-current:

$$I(t) = \frac{2I_o t}{\tau} \Rightarrow I^2(t) = \frac{4I_o^2 t^2}{\tau^2} \Rightarrow \int_0^{\tau/2} I^2(t)dt = \frac{4I_o^2}{\tau^2}\left[\frac{t^3}{3}\right]_0^{\tau/2} = \frac{I_o^2 \tau}{6}.$$

$$\Rightarrow \langle I^2 \rangle = \left(\frac{I_o^2 \tau}{6}\right) \bigg/ \left(\frac{\tau}{2}\right) = \frac{I_o^2}{3} \Rightarrow I_{rms} = \sqrt{\frac{I_o^2}{3}} = \frac{I_o}{\sqrt{3}}.$$

32-52: (a) $\omega_o = \dfrac{1}{\sqrt{LC}} = \dfrac{1}{\sqrt{(0.100\text{ H})(6.00 \times 10^{-7}\text{ F})}} = 4080\text{ rad/s}.$

(b) At $\omega = \omega_o$, $I_{\max}(R) = \dfrac{V}{R} = \dfrac{160\text{ V}}{200\ \Omega} = 0.800\text{ A}.$

(c) At $\omega = \omega_o$, $V_{\max}(C) = \dfrac{I}{\omega C} = \dfrac{0.800\text{ A}}{(4080\text{ rad/s})(6.00 \times 10^{-7}\text{ F})} = 326\text{ V}.$

(d) At $\omega = \omega_o$, $V_{\max}(L) = I\omega L = (0.800\text{ A})(4080\text{ rad/s})(0.100\text{ H}) = 326\text{ V}.$

(e) The maximum energy stored in the circuit elements at resonance is:

$U_L = \dfrac{1}{2}LI^2 = \dfrac{1}{2}(0.100\text{ H})(0.800\text{A})^2 = 0.0320\text{ J}.$

$U_C = \dfrac{1}{2}CV^2 = \dfrac{1}{2}(6.00 \times 10^{-7}\text{ F})(326\text{ V})^2 = 0.0320\text{ J}.$

32-53: (a) $Z = \sqrt{R^2 + (\omega L - 1/\omega C)^2}$

$\Rightarrow Z = \sqrt{(200\ \Omega)^2 + \left((400\text{ rad/s})(0.10\text{ H}) - 1/\left((400\text{ rad/s})(6.0 \times 10^{-7}\text{ F})\right)\right)^2} = 4130\ \Omega$

$\Rightarrow I_{\max} = \dfrac{V}{Z} = \dfrac{160\text{ V}}{4130\ \Omega} = 0.0387\text{ A}.$

(b) $V_{\max}(C) = \dfrac{I_{\max}}{\omega C} = \dfrac{0.0387\text{ A}}{(400\text{ rad/s})(6.00 \times 10^{-7}\text{F})} = 161\text{ V}.$

(c) $V_{\max}(L) = I_{\max}\omega L = (0.0387\text{ A})(400\text{ rad/s})(0.100\text{ H}) = 1.55\text{ V}.$

(d) $U_{\max}(C) = \dfrac{1}{2}CV^2 = \dfrac{1}{2}(6.00 \times 10^{-7}\text{ F})(161\text{ V})^2 = 7.81 \times 10^{-3}\text{ J}.$

$U_{\max}(L) = \dfrac{1}{2}LI^2 = \dfrac{1}{2}(0.100\text{ H})(0.0387\text{ A})^2 = 7.50 \times 10^{-5}\text{ J}.$

32-54: (a) $\omega_o = \frac{1}{\sqrt{LC}} = \frac{1}{\sqrt{(2.25\text{ H})(8.00\times10^{-7}\text{ F})}} = 745\text{ rad/s}$

(b) $Z = \sqrt{R^2 + (\omega L - 1/\omega C)^2}$

$\Rightarrow Z = \sqrt{(400\ \Omega)^2 + \left((745\text{ rad/s})(2.25\text{ H}) - 1/\left((745\text{ rad/s})(8.0\times10^{-7}\text{ F})\right)\right)^2} = 400\ \Omega.$

$\Rightarrow I_{rms} = \frac{V_{rms}}{Z} = \frac{120\text{ V}}{400\ \Omega} = 0.300\text{ A}.$

(c) We want

$I = \frac{1}{2}I_{rms} = \frac{V_{rms}}{Z} = \frac{V_{rms}}{\sqrt{R^2 + (\omega L - 1/\omega C)^2}} \Rightarrow R^2 + (\omega L - 1/\omega C)^2 = \frac{4V_{rms}^{\ 2}}{I_{rms}^{\ 2}}$

$\Rightarrow \omega^2 L^2 + \frac{1}{\omega^2 C^2} - \frac{2L}{C} + R^2 - \frac{4V_{rms}^{\ 2}}{I_{rms}^{\ 2}} = 0$

$\Rightarrow (\omega^2)^2 L^2 + \omega^2\left(R^2 - \frac{2L}{C} - \frac{4V_{rms}^{\ 2}}{I_{rms}^{\ 2}}\right) + \frac{1}{C^2} = 0.$

Substituting in the values for this problem, the equation becomes:

$(\omega^2)^2(5.06) + \omega^2(-6.11\times10^6) + 1.56\times10^{12} = 0.$

Solving this quadratic equation in ω^2 we find

$\omega^2 = 3.69\times10^5\text{ rad}^2/\text{s}^2$ or $8.37\times10^5\text{ rad}^2/\text{s}^2 \Rightarrow \omega = 607\text{ rad/s}$ or 915 rad/s.

(d) Making the same substitutions but for $R = 4.00\ \Omega$ and $40.0\ \Omega$, they both yield the same frequencies: $\omega_1 = 588\text{ rad/s}$ and $\omega_2 = 945\text{ rad/s}$.

Therefore, they both have a resonance width $|\omega_1 - \omega_2| = 357\text{ rad/s}$.

For $R = 400\ \Omega$, the resonance width is $|\omega_1 - \omega_2| = 308\text{ rad/s}$.

32-55: (a) $I = \frac{V}{Z} = \frac{V}{\sqrt{R^2 + (\omega L - 1/\omega C)^2}}.$

(b) $P_{av} = \frac{1}{2}I^2 R = \frac{1}{2}\left(\frac{V}{Z}\right)^2 R = \frac{V^2 R/2}{R^2 + (\omega L - 1/\omega C)^2}.$

(c) The average power and the current amplitude are both greatest when the denominator is smallest, which occurs for $\omega_o L = \frac{1}{\omega_o C} \Rightarrow \omega_o = \frac{1}{\sqrt{LC}}.$

(d) $P_{av} = \frac{(100\text{ V})^2(200\ \Omega)/2}{(200\ \Omega)^2 + (\omega(2.00\text{ H}) - 1/\omega(5.00\times10^{-6}\text{ F}))^2}$

$\Rightarrow P_{av} = \frac{25\omega^2}{40{,}000\omega^2 + (2\omega^2 - 2{,}000{,}000)^2}.$

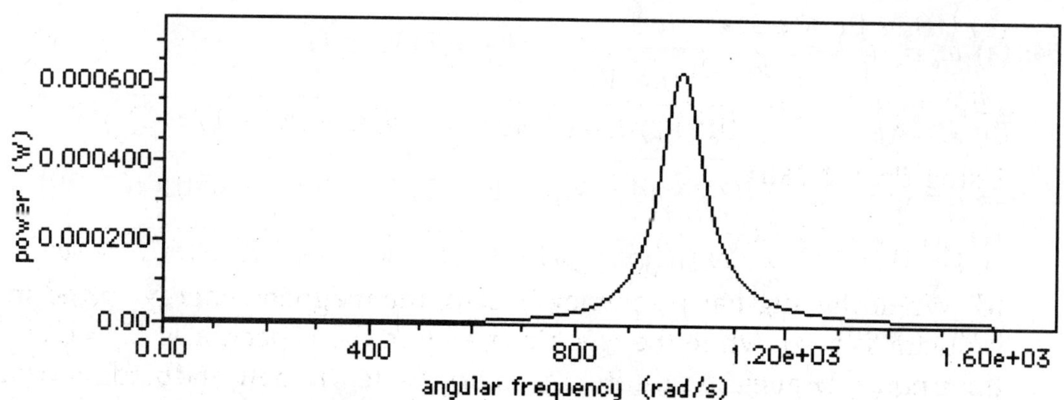

Note that as the angular frequency goes to zero, the power and current are zero, just as they are when the angular frequency goes to infinity. This graph exhibits the same strongly peaked nature as the light red curve in Fig. (32-14).

32-56: (a) $V_L = I\omega L = \dfrac{V\omega L}{Z} = \dfrac{V\omega L}{\sqrt{R^2 + (\omega L - 1/\omega C)^2}}$.

(b) $V_C = \dfrac{I}{\omega C} = \dfrac{V}{\omega C Z} = \dfrac{V}{\omega C\sqrt{R^2 + (\omega L - 1/\omega C)^2}}$.

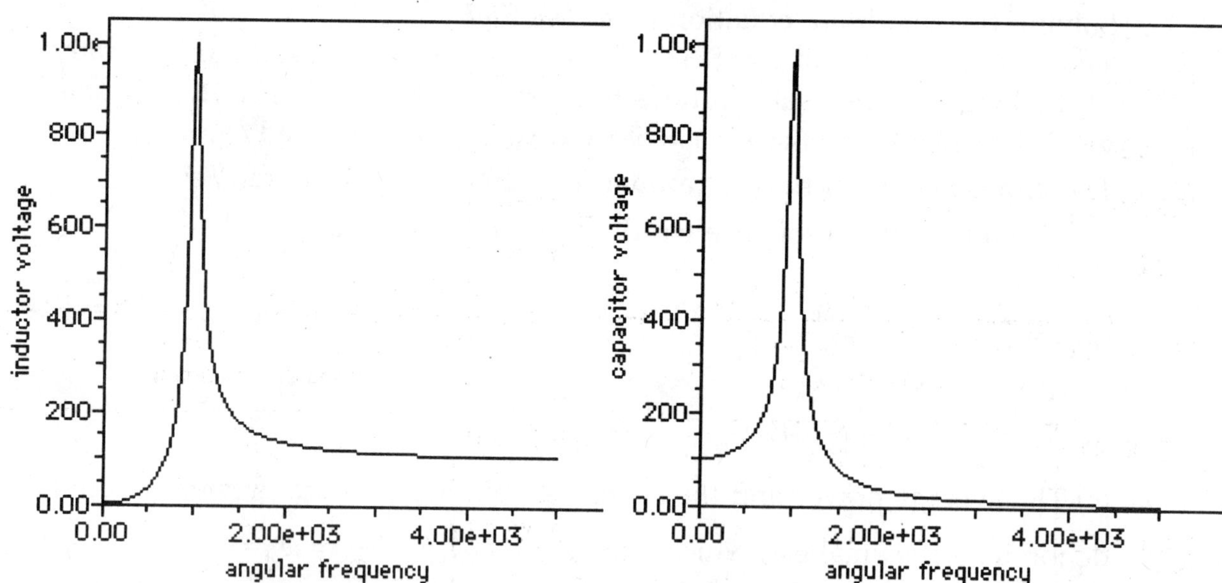

(d) When the angular frequency is zero, the inductor has zero voltage while the capacitor has a voltage of 100 V (equal to the total source voltage). At very high frequencies, the capacitor voltage goes to zero, while the inductor's voltage goes to 100 V. At resonance, $\omega_o = \dfrac{1}{\sqrt{LC}} = 1000 \text{ rad/s}$, the two voltages are equal, and are a maximum, 1000 V.

32-57: (a) $U_B = \dfrac{1}{2}Li^2 \Rightarrow \langle U_B \rangle = \dfrac{1}{2}L\langle i^2 \rangle = \dfrac{1}{2}LI_{rms}{}^2 = \dfrac{1}{2}L\left(\dfrac{I}{\sqrt{2}}\right)^2 = \dfrac{1}{4}LI^2.$

$U_E = \dfrac{1}{2}Cv^2 \Rightarrow \langle U_E \rangle = \dfrac{1}{2}C\langle v^2 \rangle = \dfrac{1}{2}CV_{rms}{}^2 = \dfrac{1}{2}C\left(\dfrac{V}{\sqrt{2}}\right)^2 = \dfrac{1}{4}CV^2.$

(b) Using **Pr. (32-55a)**:

$$\langle U_B \rangle = \frac{1}{4}LI^2 = \frac{1}{4}L\left(\frac{V}{\sqrt{R^2+(\omega L-1/\omega C)^2}}\right)^2 = \frac{LV^2}{4\left(R^2+(\omega L-1/\omega C)^2\right)}.$$

Using **Pr. (32-56b)**:

$$\langle U_E \rangle = \frac{1}{4}CV_C{}^2 = \frac{1}{4}C\frac{V^2}{\omega^2 C^2\left(R^2+(\omega L-1/\omega C)^2\right)} = \frac{V^2}{4\omega^2 C\left(R^2+(\omega L-1/\omega C)^2\right)}.$$

(d) When the angular frequency is zero, the magnetic energy stored in the inductor is zero, while the electric energy in the capacitor is $U_E = CV^2/4$. As the frequency goes to infinity, the energy soted in both inductor and capacitor go to zero. The energies equal each other at the resonant frequency where $\omega_o = \frac{1}{\sqrt{LC}}$ and $U_B = U_E = \frac{LV^2}{4R^2}$.

(c) Below are the graphs of the magnetic and electric energies, the top two showing the general features, while the bottom two show the details close to angular frequency equal to zero.

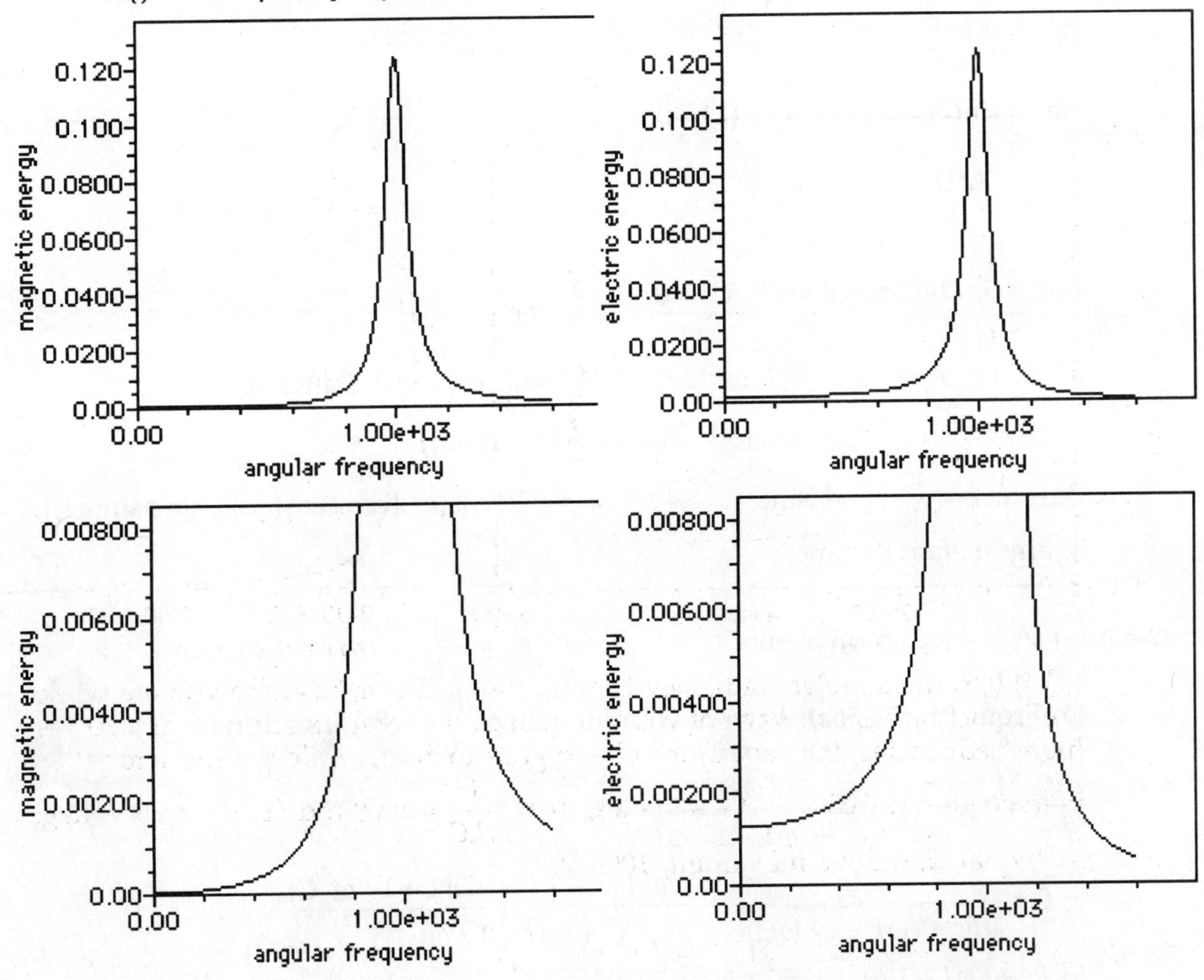

32-58: We want $P_{av}(\omega_1) =$ maximum, $P_{av}(\omega_2) = 0.01P_{av}(\omega_1)$. Maximum power implies $\omega_o = \frac{1}{\sqrt{LC}} \Rightarrow C = \frac{1}{L\omega^2} = \frac{1}{(1.0\times10^{-6}\text{ H})(6.0\times10^{8}\text{ rad/s})^2} = 2.78\times10^{-12}$ F.

$$P_{av}(\omega_2) = 0.01P_{av}(\omega_1) \Rightarrow \frac{V^2R/2}{R^2 + (\omega L - 1/\omega C)^2} = \frac{1}{100}\left(\frac{V^2}{2R}\right)$$

$$\Rightarrow 100R^2 = R^2 + (\omega L - 1/\omega C)^2 \Rightarrow R = \sqrt{\frac{(\omega L - 1/\omega C)^2}{99}} = \frac{(\omega L - 1/\omega C)}{\sqrt{99}}$$

$$\Rightarrow R = \frac{1}{\sqrt{99}}\left((5.99\times10^{8}\text{ rad/s})(1.00\times10^{-6}\text{ H}) - \frac{1}{(5.99\times10^{8}\text{ rad/s})(2.78\times10^{-12}\text{ F})}\right)$$

$\Rightarrow R = 0.153\ \Omega$.

32-59: (a) $p_R = i^2R = I^2\cos^2(\omega t)R = V_RI\cos^2(\omega t)) = \frac{1}{2}V_RI(1+\cos(2\omega t))$

$$\Rightarrow P_{av}(R) = \frac{1}{T}\int_0^T p_R dt = \frac{V_RI}{2T}\int_0^T (1+\cos(2\omega t))dt = \frac{V_RI}{2T}[t]_0^T = \frac{1}{2}V_RI\,.$$

(b) $p_L = Li\frac{di}{dt} = -\omega LI^2\cos(\omega t)\sin(\omega t) = -\frac{1}{2}V_LI\sin(2\omega t)$.

But $\int_0^T \sin(2\omega t)dt = 0 \Rightarrow P_{av}(L) = 0$.

(c) $p_C = \frac{dU}{dt} = \frac{d}{dt}\left(\frac{q^2}{2C}\right) = \frac{q}{C}i = v_Ci = V_CI\sin(\omega t)\cos(\omega t) = \frac{1}{2}V_CI\sin(2\omega t)$.

But $\int_0^T \sin(2\omega t)dt = 0 \Rightarrow P_{av}(C) = 0$.

(d) $p = p_R + p_L + p_C = V_RI\cos^2(\omega t) - \frac{1}{2}V_LI\sin(2\omega t) + \frac{1}{2}V_CI\sin(2\omega t)$

$\Rightarrow p = I\cos(\omega t)(V_R\cos(\omega t) - V_L\sin(\omega t) + V_C\sin(\omega t))$.

But $\cos\phi = \frac{V_R}{V}$ and $\sin\phi = \frac{V_L - V_C}{V} \Rightarrow p = VI\cos(\omega t)(\cos\phi\cos(\omega t) - \sin\phi\sin(\omega t))$, at any instant of time.

32-60: (a) $V_R =$ maximum when $V_C = V_L \Rightarrow \omega = \omega_o = \frac{1}{\sqrt{LC}}$.

(b) From **Pr. (32-56a)**, $V_L =$ maximum when $\frac{dV_L}{d\omega} = 0$. Therefore:

$$\frac{dV_L}{d\omega} = 0 = \frac{d}{d\omega}\left(\frac{V\omega L}{\sqrt{R^2 + (\omega L - 1/\omega C)^2}}\right)$$

$$\Rightarrow 0 = \frac{VL}{\sqrt{R^2 + (\omega L - 1/\omega C)^2}} - \frac{V\omega^2L(L - 1/\omega^2C)(L + 1/\omega^2C)}{(R^2 + (\omega L - 1/\omega C)^2)^{3/2}}$$

$$\Rightarrow R^2 + (\omega L - 1/\omega C)^2 = \omega^2(L^2 - 1/\omega^4C^2)$$

$$\Rightarrow R^2 + \frac{1}{\omega^2C^2} - \frac{2L}{C} = -\frac{1}{\omega^2C^2} \Rightarrow \frac{1}{\omega^2} = LC - \frac{R^2C^2}{2} \Rightarrow \omega = \frac{1}{\sqrt{LC - R^2C^2/2}}.$$

(c) From **Pr. (32-56b)**, V_C = maximum when $\frac{dV_C}{d\omega}=0$, Therefore:

$$\frac{dV_C}{d\omega}=0=\frac{d}{d\omega}\left(\frac{V}{\omega C\sqrt{R^2+(\omega L-1/\omega C)^2}}\right)$$

$$\Rightarrow 0=-\frac{V}{\omega^2 C\sqrt{R^2+(\omega L-1/\omega C)^2}}-\frac{V(L-1/\omega^2 C)(L+1/\omega^2 C)}{C\left(R^2+(\omega L-1/\omega C)^2\right)^{3/2}}$$

$$\Rightarrow R^2+(\omega L-1/\omega C)^2=-\omega^2(L^2-1/\omega^4 C^2)$$

$$\Rightarrow R^2+\omega^2 L^2-\frac{2L}{C}=-\omega^2 L^2\Rightarrow \omega=\sqrt{\frac{1}{LC}-\frac{R^2}{2L^2}}.$$

32-61: (a) From the current phasors we know that $Z=\sqrt{R^2+(\omega L-1/\omega C)^2}$

$$\Rightarrow Z=\sqrt{(400\ \Omega)^2+\left((1000\text{ rad/s})(0.50\text{ H})-\frac{1}{(1000\text{ rad/s})(1.25\times10^{-6}\text{ F})}\right)^2}=500\ \Omega.$$

$$\Rightarrow I=\frac{V}{Z}=\frac{200\text{ V}}{500\ \Omega}=0.400\text{ A}.$$

(b) $\phi=\arctan\left(\frac{\omega L-1/(\omega C)}{R}\right)$

$$\Rightarrow \phi=\arctan\left(\frac{(1000\text{ rad/s})(0.500\text{ H})-1/(1000\text{ rad/s})(1.25\times10^{-6}\text{ F})}{400\ \Omega}\right)=+36.9^\circ.$$

(c) $Z_{cpx}=R+i\left(\omega L-\frac{1}{\omega C}\right)$

$$\Rightarrow Z_{cpx}=400\ \Omega-i\left((1000\text{ rad/s})(0.50\text{ H})-\frac{1}{(1000\text{ rad/s})(1.25\times10^{-6}\text{ F})}\right)$$

$$=400\ \Omega-300\ \Omega i$$

$$\Rightarrow Z=\sqrt{(400\ \Omega)^2+(-300\ \Omega)^2}=500\ \Omega.$$

(d) $I_{cpx}=\frac{V}{Z_{cpx}}=\frac{200\text{ V}}{(400\text{ - }300\text{i})\ \Omega}=\left(\frac{8+6i}{25}\right)\text{A}\Rightarrow I=\sqrt{\left(\frac{8+6i}{25}\right)\left(\frac{8-6i}{25}\right)}=0.400\text{ A}.$

(e) $\tan\phi=\frac{\text{Im}(I_{cpx})}{\text{Re}(I_{cpx})}=\frac{6/25}{8/25}=0.75\Rightarrow\phi=+36.9^\circ.$

(f) $V_{R_{cpx}}=I_{cpx}R=\left(\frac{8+6i}{25}\right)(400\ \Omega)=(128+96i)\text{V}.$

$V_{L_{cpx}}=iI_{cpx}\omega L=i\left(\frac{8+6i}{25}\right)(1000\text{ rad/s})(0.500\text{ H})=(-120+160i)\text{V}.$

$V_{L_{cpx}}=i\frac{I_{cpx}}{\omega C}=i\left(\frac{8+6i}{25}\right)\frac{1}{(1000\text{ rad/s})(1.25\times10^{-6}\text{ F})}=(+192-256i)\text{V}.$

(g) $V_{cpx}=V_{R_{cpx}}+V_{L_{cpx}}+V_{L_{cpx}}=(128+96i)\text{V}+(-120+160i)\text{V}+(192-256i)\text{V}=200\text{ V}.$

Chapter 33: Electromagnetic Waves

33-1: $d = c\Delta t = (3.0\times10^{8}\ \text{m/s})(6.0\times10^{-7}\ \text{s}) = 180\ \text{m}.$

33-2: (a) $t = \dfrac{d}{c} = \dfrac{1.49\times10^{11}\ \text{m}}{3.00\times10^{8}\ \text{m/s}} = 497\ \text{s} = 8\ \text{minutes}\ 17\ \text{seconds}.$

(b) Light travel time is:

$$26\ \text{years} = (26\ \text{years})\frac{(365\ \text{days})}{(1\ \text{year})}\frac{(24\ \text{hours})}{(1\ \text{day})}\frac{(3600\ \text{s})}{(1\ \text{hour})} = 8.21\times10^{8}\ \text{s}$$

$d = ct = (3.0\times10^{8}\ \text{m/s})(8.21\times10^{8}\ \text{s}) = 2.5\times10^{17}\ \text{m} = 2.5\times10^{14}\ \text{km}.$

33-3: (a) $\lambda = \dfrac{c}{f} = \dfrac{3.00\times10^{8}\ \text{m/s}}{9.09\times10^{7}\ \text{Hz}} = 3.30\ \text{m}.$

(b) $E_{max} = cB_{max} = (3.00\times10^{8}\ \text{m/s})(5.56\times10^{-11}\ \text{T}) = 0.0167\ \text{V/m}.$

33-4: $B_{max} = \dfrac{E_{max}}{c} = \dfrac{4.75\times10^{-3}\ \text{V/m}}{3.00\times10^{8}\ \text{m/s}} = 1.58\times10^{-11}\ \text{T}.$

So $\dfrac{B_{max}}{B_{earth}} = \dfrac{1.58\times10^{-11}\ \text{T}}{5\times10^{-5}\ \text{T}} = 3\times10^{-7}$, and thus B_{max} is much weaker than B_{earth}.

33-5: $\vec{B}(y,t) = B_{max}\sin(\omega t - ky)\hat{i} = B_{max}\sin\left(2\pi f\left(t - \frac{y}{c}\right)\right)\hat{i}$

$\Rightarrow \vec{B}(y,t) = (7.30\times10^{-4}\ \text{T})\sin\left(2\pi(4.00\times10^{13}\ \text{Hz})\left(t - \dfrac{y}{(3.00\times10^{8}\ \text{m/s})}\right)\right)\hat{i}$

$\Rightarrow \vec{B}(y,t) = (7.30\times10^{-4}\ \text{T})\sin\left((2.51\times10^{14}\ \text{rad/s})t - (8.38\times10^{5}\ \text{m}^{-1})y\right)\hat{i}.$

$\vec{E}(y,t) = (B(y,t)\hat{i})\times(c\hat{j})$

$\Rightarrow \vec{E}(y,t) = (2.19\times10^{5}\ \text{V/m})\sin\left((2.51\times10^{14}\ \text{rad/s})t - (8.38\times10^{5}\ \text{m}^{-1})y\right)\hat{k}.$

33-6: (a) $f = \dfrac{c}{\lambda} = \dfrac{3.00\times10^{8}\ \text{m/s}}{2.75\times10^{-7}\ \text{m}} = 1.09\ \times10^{15}\ \text{Hz}.$

(b) $B_{max} = \dfrac{E_{max}}{c} = \dfrac{5.80\times10^{-3}\ \text{V/m}}{3.00\times10^{8}\ \text{m/s}} = 1.93\times10^{-11}\ \text{T}.$

(c) The electric field is in the y-direction, and the wave is propagating in the -z-direction. So the magnetic field is in the x-direction, since $\hat{S}\propto\vec{E}\times\vec{B}$. Thus:

$\vec{E}(z,t) = E_{max}\sin(\omega t - kz)\hat{j} = E_{max}\sin\left(2\pi f\left(t - \frac{z}{c}\right)\right)\hat{j}$

$\Rightarrow \vec{E}(z,t) = (5.80\times10^{-3}\ \text{V/m})\sin\left(2\pi(1.09\times10^{15}\ \text{Hz})\left(t - \dfrac{z}{3.00\times10^{8}\ \text{m/s}}\right)\right)\hat{j}$

$\Rightarrow \vec{E}(z,t) = (5.80\times10^{-3}\ \text{V/m})\sin\left((6.85\times10^{15}\ \text{rad/s})t - (2.28\times10^{7}\ \text{m}^{-1})z\right)\hat{j}.$

And $\vec{B}(z,t) = \dfrac{E(z,t)}{c}\hat{i} = (1.93\times10^{-11}\ \text{T})\sin\left((6.85\times10^{15}\ \text{rad/s})t - (2.28\times10^{7}\ \text{m}^{-1})z\right)\hat{i}.$

33-7: (a) $\omega = 2\pi f = \frac{2\pi c}{\lambda} \Rightarrow \lambda = \frac{2\pi c}{\omega} = \frac{2\pi(3.00\times10^{8}\ \text{m/s})}{(1.45\times10^{14}\ \text{rad/s})} = 1.30\times10^{-5}\ \text{m}.$

(b) Since the electric field is in the -z-direction, and the wave is propagating in the +x-direction, then the magnetic field is in the +y-direction ($\hat{S} \propto \vec{E}\times\vec{B}$). So:

$$\vec{B}(x,t) = \frac{E(x,t)}{c}\hat{j} = \frac{E_o}{c}\sin(\omega t - kx)\hat{j} = \frac{E_o}{c}\sin(\omega t - \frac{\omega}{c}x)\hat{j}$$

$$\Rightarrow \vec{B}(x,t) = \left(\frac{2.30\times10^{5}\ \text{V/m}}{3.00\times10^{8}\ \text{m/s}}\right)\sin\left((1.45\times10^{16}\ \text{rad/s})t - \frac{(1.45\times10^{16}\ \text{rad/s})}{(3.00\times10^{8}\ \text{m/s})}x\right)\hat{j}$$

$$\Rightarrow \vec{B}(x,t) = (7.67\times10^{-4}\ \text{T})\sin\left((1.45\times10^{16}\ \text{rad/s})t - (4.83\times10^{5}\ \text{m}^{-1})x\right)\hat{j}.$$

33-8: (a) $k = \frac{2\pi}{\lambda} = \frac{2\pi f}{c} \Rightarrow f = \frac{kc}{2\pi} = \frac{(7.45\times10^{4}\ \text{rad/m})(3.0\times10^{8}\ \text{m/s})}{2\pi} = 3.56\times10^{12}\ \text{Hz}.$

(b) Since the magnetic field is in the +x-direction, and the wave is propagating in the -y-direction, then the electric field is in the -z-direction ($\hat{S} \propto \vec{E}\times\vec{B}$). So:

$$\vec{E}(y,t) = -cB(y,t)\hat{k} = -cB_o\sin(2\pi f t + ky)\hat{k}$$

$$\Rightarrow \vec{E}(y,t) = -\left(c(4.38\times10^{-8}\ \text{T})\right)\sin\left((2.24\times10^{13}\ \text{rad/s})t + (7.45\times10^{4}\ \text{rad/m})y\right)\hat{k}$$

$$\Rightarrow \vec{E}(y,t) = -(13.1\ \text{V/m})\sin\left((2.24\times10^{13}\ \text{rad/s})t + (7.45\times10^{4}\ \text{rad/m})y\right)\hat{k}.$$

33-9: $S = \frac{\varepsilon_o}{\sqrt{\varepsilon_o\mu_o}}E^2 = \sqrt{\frac{\varepsilon_o}{\mu_o}}E^2 = \sqrt{\frac{\varepsilon_o}{\mu_o}}Ec\frac{E}{c} = c\sqrt{\frac{\varepsilon_o}{\mu_o}}EB = \frac{1}{\sqrt{\varepsilon_o\mu_o}}\sqrt{\frac{\varepsilon_o}{\mu_o}}EB = \frac{EB}{\mu_o}.$

33-10: Recall that $\hat{S} \propto \vec{E}\times\vec{B}$, so:

(a) $\hat{S} = \hat{i}\times(-\hat{j}) = -\hat{k}.$

(b) $\hat{S} = \hat{j}\times\hat{i} = -\hat{k}.$

(c) $\hat{S} = (-\hat{k})\times(-\hat{i}) = \hat{j}.$

(d) $\hat{S} = \hat{i}\times(-\hat{k}) = \hat{j}.$

33-11: (a) $B_{\text{max}} = \frac{E_{\text{max}}}{c} = \frac{0.0800\ \text{V/m}}{3.00\times10^{8}\ \text{m/s}} = 2.67\times10^{-10}\ \text{T}.$

(b) Note that the antenna radiates only above the ground, so the area in calculating the power is that of half a sphere.

$$P = S\cdot A = \left(\frac{E_{\text{max}}B_{\text{max}}}{2\mu_o}\right)(2\pi r^2) = \left(\frac{(0.080\ \text{V/m})(2.67\times10^{-10}\ \text{T})}{2\mu_o}\right)(2\pi(5.0\times10^{4}\ \text{m})^2)$$

$$\Rightarrow P = 1.33\times10^{5}\ \text{W}.$$

(c) $P \propto E^2 \propto \frac{1}{r^2} \Rightarrow$ if $E \to \frac{1}{2}E$, we need $r \to 2r = 2(50\ \text{km}) = 100\ \text{km}$.

33-12: (a) The electric field is in the -y-direction, and the magnetic field is in the +z-direction, so $\hat{S} = \hat{E}\times\hat{B} = (-\hat{j})\times\hat{k} = -\hat{i}$. That is, the Poynting vector is in the -x-direction.

(b) $S(x,t) = \frac{E(x,t)B(x,t)}{\mu_o} = -\frac{E_{\text{max}}B_{\text{max}}}{\mu_o}\sin^2(\omega t + kx) = -\frac{E_{\text{max}}B_{\text{max}}}{2\mu_o}\left(1 - \cos(2(\omega t + kx))\right).$

But over one period, the cosine function averages to zero, so we have:

$$|S_{av}| = \frac{E_{\max} B_{\max}}{2\mu_o}.$$

33-13: $P - S_{av} A - \frac{E_{\max}{}^2}{2c\mu_o} \cdot (4\pi r^2) \rightarrow E_{\max} - \sqrt{\frac{Pc\mu_o}{2\pi r^2}}$

$$\Rightarrow E_{\max} = \sqrt{\frac{(75.0 \text{ W})(3.00 \times 10^8 \text{ m/s})\mu_o}{2\pi(3.00 \text{ m})^2}} = 22.4 \text{ V/m}.$$

$$\Rightarrow B_{\max} = \frac{E_{\max}}{c} = \frac{22.4 \text{ V/m}}{3.00 \times 10^8 \text{ m/s}} = 7.45 \times 10^{-8} \text{ T}.$$

33-14: (a) $f = \frac{c}{\lambda} = \frac{3.00 \times 10^8 \text{m/s}}{0.0520 \text{ m}} = 5.77 \times 10^9 \text{ Hz}.$

(b) $B_{\max} = \frac{E_{\max}}{c} = \frac{0.0450 \text{ V/m}}{3.00 \times 10^8 \text{ m/s}} = 1.50 \times 10^{-10} \text{ T}.$

(c) $I = S_{av} = \frac{EB}{2\mu_o} = \frac{(0.0450 \text{ V/m})(1.50 \times 10^{-10} \text{ T})}{2\mu_o} = 2.69 \times 10^{-6} \text{ W/m}^2.$

33-15: (a) The momentum density $\frac{dp}{dV} = \frac{S_{av}}{c^2} = \frac{780 \text{ W/m}^2}{(3.0 \times 10^8 \text{ m/s})^2} = 8.7 \times 10^{-15} \text{ kg/m}^2 \cdot s.$

(b) The momentum flow rate $\frac{1}{A}\frac{dp}{dt} = \frac{S_{av}}{c} = \frac{780 \text{ W/m}^2}{3.0 \times 10^8 \text{ m/s}} = 2.6 \times 10^{-6} \text{ Pa}.$

33-16: (a) The momentum density $\frac{dp}{dV} = \frac{S_{av}}{c^2} = \frac{750 \text{ W/m}^2}{(3.0 \times 10^8 \text{ m/s})^2} = 8.3 \times 10^{-15} \text{ kg/m}^2 \cdot s.$

(b) The momentum flow rate $\frac{1}{A}\frac{dp}{dt} = \frac{S_{av}}{c} = \frac{750 \text{ W/m}^2}{3.0 \times 10^8 \text{ m/s}} = 2.5 \times 10^{-6} \text{ Pa}.$

33-17: (a) Reflecting light: $p_{rad} = \frac{1}{A}\frac{dp}{dt} = \frac{2S_{av}}{c} = \frac{2(935 \text{ W/m}^2)}{3.0 \times 10^8 \text{ m/s}} = 6.23 \times 10^{-6} \text{ Pa}$

$$\Rightarrow p_{rad} = \frac{6.23 \times 10^{-6} \text{ Pa}}{1.013 \times 10^5 \text{ Pa/atm}} = 6.15 \times 10^{-11} \text{ atm}.$$

(b) Absorbed light: $p_{rad} = \frac{1}{A}\frac{dp}{dt} = \frac{S_{av}}{c} = \frac{935 \text{ W/m}^2}{3.0 \times 10^8 \text{ m/s}} = 3.12 \times 10^{-6} \text{ Pa}$

$$\Rightarrow p_{rad} = \frac{3.12 \times 10^{-6} \text{ Pa}}{1.013 \times 10^5 \text{ Pa/atm}} = 3.08 \times 10^{-11} \text{ atm}.$$

33-18: $u_E = \frac{1}{2}\varepsilon E^2 = \frac{1}{2}\varepsilon(vB)^2 = \frac{1}{2}\varepsilon\left(\frac{1}{\sqrt{\varepsilon\mu}}\right)^2 B^2 = \frac{1}{2}\varepsilon\frac{1}{\varepsilon\mu}B^2 = \frac{B^2}{2\mu} = u_B.$

33-19: $E = vB = \dfrac{B}{\sqrt{\varepsilon\mu}} = \dfrac{B}{\sqrt{K_E \varepsilon_o K_B \mu_o}} = \dfrac{cB}{\sqrt{K_E K_B}}$

$\Rightarrow E = \dfrac{(3.00\times10^8 \text{ m/s})(4.80\times10^{-9} \text{ T})}{\sqrt{(1.83)(1.36)}} = 0.913 \text{ V/m}.$

33-20: (a) $v = \dfrac{c}{\sqrt{K_E K_B}} = \dfrac{(3.00\times10^8 \text{ m/s})}{\sqrt{(3.75)(5.30)}} = 6.73\times10^7 \text{ m/s}.$

(b) $\lambda = \dfrac{v}{f} = \dfrac{6.73\times10^7 \text{ m/s}}{75.0 \text{ Hz}} = 8.97\times10^5 \text{ m}.$

(c) $B = \dfrac{E}{v} = \dfrac{8.40\times10^{-3} \text{ V/m}}{6.73\times10^7 \text{ m/s}} = 1.25\times10^{-10} \text{ T}.$

(d) $I = \dfrac{EB}{2K_B\mu_o} = \dfrac{(8.40\times10^{-3} \text{ V/m})(1.25\times10^{-10} \text{ T})}{2(5.30)\mu_o} = 7.87\times10^{-8} \text{ W/m}^2.$

33-21: (a) $\lambda = \dfrac{v}{f} = \dfrac{2.48\times10^8 \text{ m/s}}{2.00\times10^7 \text{ Hz}} = 12.4 \text{ m}.$

(b) $\lambda_o = \dfrac{c}{f} = \dfrac{3.00\times10^8 \text{ m/s}}{2.00\times10^7 \text{ Hz}} = 15.0 \text{ m}.$

(c) $n = \dfrac{c}{v} = \dfrac{3.00\times10^8 \text{ m/s}}{2.48\times10^8 \text{ m/s}} = 1.21.$

(d) $v = \dfrac{c}{\sqrt{K_E}} \Rightarrow K_E = \dfrac{c^2}{v^2} = n^2 = (1.21)^2 = 1.46.$

33-22: (a) $v = f\lambda = (4.20\times10^7 \text{ Hz})(5.25 \text{ m}) = 2.21\times10^8 \text{ m/s}.$

(b) $K_E = \dfrac{c^2}{v^2} = \dfrac{(3.00\times10^8 \text{ m/s})^2}{(2.21\times10^8 \text{ m/s})^2} = 1.85.$

(c) $I = \dfrac{E_{\max}^{\ 2}}{2\mu_o v} \Rightarrow E_{\max} = \sqrt{2\mu_o v I}$

$\Rightarrow E_{\max} = \sqrt{2\mu_o(2.21\times10^8 \text{ m/s})(3.60\times10^{-6} \text{ W/m}^2)} = 0.0447 \text{ V/m}.$

And $B_{\max} = \dfrac{E_{\max}}{v} = \dfrac{0.0447 \text{ V/m}}{2.21\times10^8 \text{ m/s}} = 2.02\times10^{-10} \text{ T}.$

33-23: (a) $\Delta x = \dfrac{\lambda}{2} \Rightarrow \lambda = 2\Delta x = 2(4.00 \text{ mm}) = 8.00 \text{ mm}.$

(b) $\Delta x_E = \Delta x_B = 4.00 \text{ mm}.$

(c) $v = f\lambda = (2.00\times10^{10} \text{ Hz})(8.00\times10^{-3} \text{ m}) = 1.60\times10^8 \text{ m/s}.$

33-24: (a) $\Delta x = \dfrac{\lambda}{2} = \dfrac{c}{2f} = \dfrac{3.00\times10^8 \text{ m/s}}{2(7.40\times10^{14} \text{ Hz})} = 2.03\times10^{-7} \text{ m}.$

(b) The distance between the electric and magnetic nodal planes is one quarter of a wavelength $= \dfrac{\lambda}{4} = \dfrac{\Delta x}{2} = \dfrac{2.03\times10^{-7} \text{ m}}{2} = 1.01\times10^{-7} \text{ m}.$

33-25: (a) The node-antinode distance $= \frac{\lambda}{4} = \frac{v}{4f} = \frac{2.40\times10^8\ \text{m/s}}{4(1.70\times10^{11}\ \text{Hz})} = 3.53\times10^{-4}\ \text{m}$.

(b) The distance between the electric and magnetic antinodes is one quarter of a wavelength $= \frac{\lambda}{4} = \frac{v}{4f} = \frac{2.40\times10^8\ \text{m/s}}{4(1.70\times10^{11}\ \text{Hz})} = 3.53\times10^{-4}\ \text{m}$.

(c) The distance between the electric and magnetic nodes is also one quarter of a wavelength $= \frac{\lambda}{4} = \frac{v}{4f} = \frac{2.40\times10^8\ \text{m/s}}{4(1.70\times10^{11}\ \text{Hz})} = 3.53\times10^{-4}\ \text{m}$.

33-26: $\Delta x_{nodes} = \frac{\lambda}{2} = \frac{c}{2f} = \frac{3.00\times10^8\ \text{m/s}}{2(6.00\times10^8\ \text{Hz})} = 0.250\ \text{m} = 25.0\ \text{cm}$. There must be nodes at the planes, which are 75.0 cm apart, and there are two nodes between the planes, each 25.0 cm from a plane. It is at these nodes that a point charge will remain at rest, since the electric fields there are zero.

33-27: (a) $\frac{\partial^2 E(x,t)}{\partial x^2} = \frac{\partial^2}{\partial x^2}(-2E_{max}\sin kx\cos\omega t) = \frac{\partial}{\partial x}(-2kE_{max}\cos kx\cos\omega t)$

$\Rightarrow \frac{\partial^2 E(x,t)}{\partial x^2} = 2k^2E_{max}\sin kx\cos\omega t = \frac{\omega^2}{c^2}2E_{max}\sin kx\cos\omega t = \varepsilon_o\mu_o\frac{\partial^2 E(x,t)}{\partial t^2}$.

Similarly: $\frac{\partial^2 B(x,t)}{\partial x^2} = \frac{\partial^2}{\partial x^2}(2B_{max}\cos kx\sin\omega t) = \frac{\partial}{\partial x}(-2kB_{max}\sin kx\sin\omega t)$

$\Rightarrow \frac{\partial^2 B(x,t)}{\partial x^2} = 2k^2B_{max}\cos kx\sin\omega t = \frac{\omega^2}{c^2}2B_{max}\cos kx\sin\omega t = \varepsilon_o\mu_o\frac{\partial^2 B(x,t)}{\partial t^2}$.

(b) $\frac{\partial E(x,t)}{\partial x} = \frac{\partial}{\partial x}(-2E_{max}\sin kx\cos\omega t) = -2kE_{max}\cos kx\cos\omega t$

$\Rightarrow \frac{\partial E(x,t)}{\partial x} = -\frac{\omega}{c}2E_{max}\cos kx\cos\omega t = -\omega 2\frac{E_{max}}{c}\cos kx\cos\omega t = -\omega 2B_{max}\cos kx\cos\omega t$

$\Rightarrow \frac{\partial E(x,t)}{\partial x} = -\frac{\partial}{\partial t}(2B_{max}\cos kx\sin\omega t) = -\frac{\partial B(x,t)}{\partial t}$.

Similarly: $-\frac{\partial B(x,t)}{\partial x} = \frac{\partial}{\partial x}(-2B_{max}\cos kx\sin\omega t) = 2kB_{max}\sin kx\sin\omega t$

$\Rightarrow -\frac{\partial B(x,t)}{\partial x} = -\frac{\omega}{c}2B_{max}\sin kx\sin\omega t = -\frac{\omega}{c^2}2cB_{max}\sin kx\sin\omega t$

$\Rightarrow -\frac{\partial B(x,t)}{\partial x} = \varepsilon_o\mu_o\omega 2E_{max}\sin kx\sin\omega t = \varepsilon_o\mu_o\frac{\partial}{\partial t}(-2E_{max}\sin kx\cos\omega t) = \varepsilon_o\mu_o\frac{\partial E(x,t)}{\partial t}$.

33-28: (a) $f = \frac{c}{\lambda} = \frac{3.0\times10^8\ \text{m/s}}{1000\ \text{m}} = 3.0\times10^5\ \text{Hz}$.

(b) $f = \frac{c}{\lambda} = \frac{3.0\times10^8\ \text{m/s}}{1.0\ \text{m}} = 3.0\times10^8\ \text{Hz}$.

(c) $f = \frac{c}{\lambda} = \frac{3.0\times10^8\ \text{m/s}}{1.0\times10^{-6}\ \text{m}} = 3.0\times10^{14}\ \text{Hz}$.

(d) $f = \frac{c}{\lambda} = \frac{3.0\times10^8\ \text{m/s}}{1.0\times10^{-9}\ \text{m}} = 3.0\times10^{17}\ \text{Hz}$.

33-29: (a) x-rays: $\lambda = \dfrac{c}{f} = \dfrac{3.00 \times 10^{8}\ \text{m/s}}{4.00 \times 10^{18}\ \text{Hz}} = 7.50 \times 10^{-11}\ \text{m} = 0.0750\ \text{nm}.$

(b) blue light: $\lambda = \dfrac{c}{f} = \dfrac{3.00 \times 10^{8}\ \text{m/s}}{6.50 \times 10^{14}\ \text{Hz}} = 4.62 \times 10^{-7}\ \text{m} = 462\ \text{nm}.$

33-30: Using a Gaussian surface such that the front surface is ahead of the wave front (no electric or magnetic fields) and the back face is behind the wave front (as shown at right), we have:

$$\oint \vec{E} \cdot d\vec{A} = E_x A = \frac{Q_{encl}}{\varepsilon_o} = 0 \Rightarrow E_x = 0.$$

$$\oint \vec{B} \cdot d\vec{A} = B_x A = 0 \Rightarrow B_x = 0.$$

So the wave must be transverse, since there are no components of the electric or magnetic field in the direction of propagation.

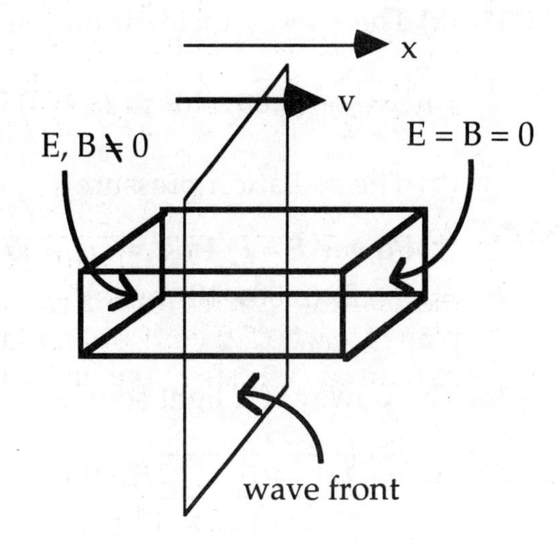

33-31: From Eq. (33-12): $\dfrac{\partial}{\partial t}\left(\dfrac{\partial E_y(x,t)}{\partial x}\right) = \dfrac{\partial}{\partial t}\left(-\dfrac{\partial B_z(x,t)}{\partial t}\right) = -\dfrac{\partial^2 B_z(x,t)}{\partial t^2}$

But also from Eq. (33-14): $-\dfrac{\partial}{\partial x}\left(\dfrac{\partial B_z(x,t)}{\partial x}\right) = \dfrac{\partial}{\partial x}\left(\varepsilon_o \mu_o \dfrac{\partial E_y(x,t)}{\partial t}\right) = -\varepsilon_o \mu_o \dfrac{\partial^2 B_z(x,t)}{\partial t^2}$

$$\Rightarrow \frac{\partial^2 B_z(x,t)}{\partial x^2} = \varepsilon_o \mu_o \frac{\partial^2 B_z(x,t)}{\partial t^2}.$$

33-32: Assume $\vec{E} = E_{\max}\hat{j}\sin(\omega t - kx)$ and $\vec{B} = B_{\max}\hat{k}\sin(\omega t - kx + \phi)$, with $-\pi < \phi < \pi$.
Then Eq. (33-12) implies:

$$\frac{\partial E}{\partial x} = -\frac{\partial B}{\partial t} \Rightarrow -kE_{\max}\cos(\omega t - kx) = -\omega B_{\max}\cos(\omega t - kx + \phi) \Rightarrow \phi = 0.$$

$$\Rightarrow kE_{\max} = \omega B_{\max} \Rightarrow E_{\max} = \frac{\omega}{k}B_{\max} = \frac{2\pi f}{2\pi/\lambda}B_{\max} = f\lambda B_{\max} = cB_{\max}.$$

Similarly for Eq. (33-14):

$$-\frac{\partial B}{\partial x} = \varepsilon_o \mu_o \frac{\partial E}{\partial t} \Rightarrow -kB_{\max}\cos(\omega t - kx + \phi) = \varepsilon_o \mu_o \omega E_{\max}\cos(\omega t - kx) \Rightarrow \phi = 0.$$

$$\Rightarrow kB_{\max} = \varepsilon_o \mu_o \omega E_{\max} \Rightarrow B_{\max} = \frac{\varepsilon_o \mu_o \omega}{k}E_{\max} = \frac{2\pi f}{c^2 2\pi/\lambda}E_{\max} = \frac{f\lambda}{c^2}E_{\max} = \frac{1}{c}E_{\max}.$$

33-33: $E(x,t) = E_{\max}\sin(\omega t - kx) \Rightarrow u_E = \dfrac{1}{2}\varepsilon_o E^2 = \dfrac{1}{2}\varepsilon_o E_{\max}{}^2 \sin^2(\omega t - kx)$

$$\Rightarrow u_E = \frac{\varepsilon_o c^2}{2}\left(\frac{E_{\max}}{c}\right)^2 \sin^2(\omega t - kx) = \frac{1}{2\mu_o}B_{\max}{}^2 \sin^2(\omega t - kx) = \frac{B^2}{2\mu_o} = u_B.$$

33-34: (a) $I = \frac{1}{2}\varepsilon_o c E_{\max}{}^2 \Rightarrow E_{\max} = \sqrt{\frac{2I}{\varepsilon_o c}} = \sqrt{\frac{2(1400\ \text{W}/\text{m}^2)}{\varepsilon_o(3.0\times10^8\ \text{m}/\text{s})}} = 1000\ \text{V}/\text{m}.$

(b) $P = IA = (1400\ \text{W}/\text{m}^2)(4\pi(1.5\times10^{11}\ \text{m})^2) = 4.0\times10^{26}\ \text{W}.$

33-35: (a) The energy incident on the mirror is $E = Pt = IAt = \frac{1}{2}\varepsilon_o cE^2 At$

$\Rightarrow E = \frac{1}{2}\varepsilon_o(3.00\times10^8\ \text{m}/\text{s})(0.0350\ \text{V}/\text{m})^2(6.00\times10^{-4}\ \text{m}^2)(1.00\ \text{s}) = 9.76\times10^{-10}\ \text{J}.$

(b) The radiation pressure $p_{rad} = \frac{2I}{c} = \varepsilon_o E^2 = \varepsilon_o(0.0350\ \text{V}/\text{m})^2 = 1.08\times10^{-14}\ \text{Pa}.$

(c) Power $P = I\cdot 4\pi R^2 = cp_{rad}2\pi R^2$

$\Rightarrow P = 2\pi(3.00\times10^8\ \text{m}/\text{s})(1.08\times10^{-14}\ \text{Pa})(4.00\ \text{m})^2 = 3.27\times10^{-4}\ \text{W}.$

33-36: The power of a light source is $P = IA = \left(\frac{1}{2}\varepsilon_o cE_{\max}{}^2\right)\cdot 4\pi r^2 \Rightarrow r = \sqrt{\frac{P}{2\pi\varepsilon_o cE_{\max}{}^2}}$

$\Rightarrow r = \sqrt{\frac{75.0\ \text{W}}{2\pi\varepsilon_o(3.00\times10^8\ \text{m}/\text{s})(0.430\ \text{V}/\text{m})^2}} = 156\ \text{m}.$

33-37: (a) The laser intensity $I = \frac{P}{A} = \frac{4P}{\pi D^2} = \frac{4(3.60\times10^{-3}\ \text{W})}{\pi(4.00\times10^{-3}\ \text{m})^2} = 286\ \text{W}/\text{m}^2.$

But $I = \frac{1}{2}\varepsilon_o cE^2 \Rightarrow E = \sqrt{\frac{2I}{\varepsilon_o c}} = \sqrt{\frac{2(286\ \text{W}/\text{m}^2)}{\varepsilon_o(3.00\times10^8\text{m}/\text{s})}} = 465\ \text{V}/\text{m}.$

And $B = \frac{E}{c} = \frac{465\ \text{V}/\text{m}}{3.00\times10^8\ \text{m}/\text{s}} = 1.55\times10^{-6}\ \text{T}.$

(b) $u_B = u_E = \frac{1}{4}\varepsilon_o E^2 = \frac{1}{4}\varepsilon_o(465\ \text{V}/\text{m})^2 = 4.78\times10^{-7}\ \text{J}/\text{m}^3.$

(c) In three meters of the laser beam, the total energy is:

$E_{tot} = u_{tot}\text{Vol} = 2u_E(AL) = 2u_E\pi D^2 L/4$

$\Rightarrow E_{tot} = 2(4.78\times10^{-7}\ \text{J/m}^3)\pi(4.00\times10^{-3}\ \text{m})^2(0.500\ \text{m})/4 = 6.00\times10^{-12}\ \text{J}$

33-38: (a) $f = \frac{c}{\lambda} = \frac{3.00\times10^8\ \text{m}/\text{s}}{0.0425\ \text{m}} = 7.06\times10^9\ \text{Hz}.$

(b) $B_{\max} = \frac{E_{\max}}{c} = \frac{0.840\ \text{V}/\text{m}}{3.00\times10^8\ \text{m}/\text{s}} = 2.80\times10^{-9}\ \text{T}.$

(c) $I = \frac{1}{2}\varepsilon_o cE_{\max}{}^2 = \frac{1}{2}\varepsilon_o(3.00\times10^8\text{m}/\text{s})(0.840\ \text{V}/\text{m})^2 = 9.37\times10^{-4}\ \text{W}/\text{m}^2.$

(d) $F = pA = \frac{IA}{c} = \frac{EBA}{2\mu_o c} = \frac{(0.840\ \text{V}/\text{m})(2.80\times10^{-9}\ \text{T})(0.500\ \text{m}^2)}{2\mu_o(3.00\times10^8\text{m}/\text{s})} = 1.56\times10^{-12}\ \text{N}.$

33-39: (a) At the sun's surface:

$$P = IA \Rightarrow I = \frac{P}{A} = \frac{P}{4\pi R^2} = \frac{3.9\times 10^{26}\ \text{W}}{4\pi(6.96\times 10^{8}\ \text{m})^2} = 6.4\times 10^{7}\ \text{W/m}^2$$

$$\Rightarrow p_{rad} = \frac{I}{c} = \frac{6.4\times 10^{7}\ \text{W/m}^2}{3.00\times 10^{8}\ \text{m/s}} = 0.21\ \text{Pa}.$$

Half-way out from the sun's center, the intensity is 4 times more intense, and so is the radiation pressure: $p_{rad}(R_{sun}/2) = 0.85\ \text{Pa}$.

At the top of the earth's atmosphere, the measured sunlight intensity is 1400 W/m^2 = 5×10^{-6} Pa, which is about 100,000 times less than the values above.

(b) The gas pressure in at the sun's surface is 50,000 times greater than the radiation pressure, and half-way out of the sun the gas pressure is believed to be about 6×10^{13} times greater than the radiation pressure. Therefore it is reasonable to ignore radiation pressure when modeling the sun's interior structure.

33-40: (a) $\vec{S}(x,t) = \frac{E_{max}B_{max}}{2\mu_o}(1-\cos 2(\omega t - kx))\hat{i} \Rightarrow S(x,t) < 0 \Rightarrow \cos 2(\omega t - kx) > 1$, which never happens. So the Poynting vector is always positive, which makes sense since the direction of wave propagation by definition is the direction of energy flow.

(b)

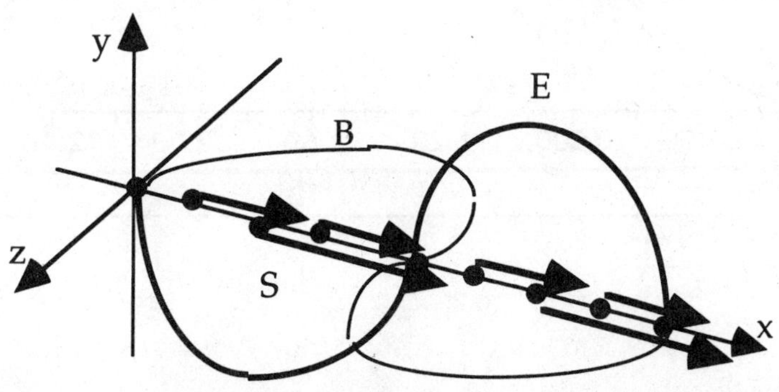

33-41: (a) $B = \mu_o ni \Rightarrow \frac{dB}{dt} = \mu_o n\frac{di}{dt} \Rightarrow \frac{d\Phi_B}{dt} = \frac{dB}{dt}A = \mu_o nA\frac{di}{dt}$.

So, $\oint \vec{E}\cdot d\vec{l} = -\frac{d\Phi_B}{dt} \Rightarrow E2\pi r = -\mu_o nA\frac{di}{dt} = -\mu_o n\pi r^2\frac{di}{dt}$

$$\Rightarrow E = -\frac{\mu_o nr}{2}\frac{di}{dt}.$$

(b) The direction of the Poynting vector is radially inward, since the magnetic field is along the solenoid's axis and the electric field is circumferential. It's magnitude $S = \frac{EB}{\mu_o} = \frac{\mu_o n^2 ri}{2}\frac{di}{dt}$.

(c) $u = \frac{B^2}{2\mu_o} = \frac{(\mu_o ni)^2}{2\mu_o} = \frac{\mu_o n^2 i^2}{2} \Rightarrow U = u(lA) = ul\pi a^2 = \frac{\mu_o \pi n^2 i^2 la^2}{2}$.

But also $U = \frac{Li^2}{2} \Rightarrow Li = \frac{2U}{i} = \frac{\mu_o \pi n^2 i^2 l a^2}{i} = \mu_o \pi n^2 i l a^2$, and so the rate of energy increase due to the increasing current is given by $P = Li\frac{di}{dt} = \mu_o \pi n^2 i l a^2 \frac{di}{dt}$.

(d) The in-flow of electromagnetic energy through a cylindrical surface located at the solenoid coils is $\iint \vec{S} \cdot d\vec{A} = S 2\pi a l = \frac{\mu_o n^2 a i}{2}\frac{di}{dt} \cdot 2\pi a l = \mu_o \pi n^2 i l a^2 \frac{di}{dt}$.

(e) The values from parts (c) and (d) are identical for the flow of energy, and hence we can consider the energy stored in a current carrying solenoid as having entered through its cylindrical walls while the current was attaining its steady-state value.

33-42: (a) The energy density, as a function of x, fro the equations for the electric and magnetic fields of Eqs. (33-38) and (33-39) is given by:
$u = \varepsilon_o E^2 = 4\varepsilon_o E_{\max}{}^2 \sin^2 kx \cos^2 \omega t$

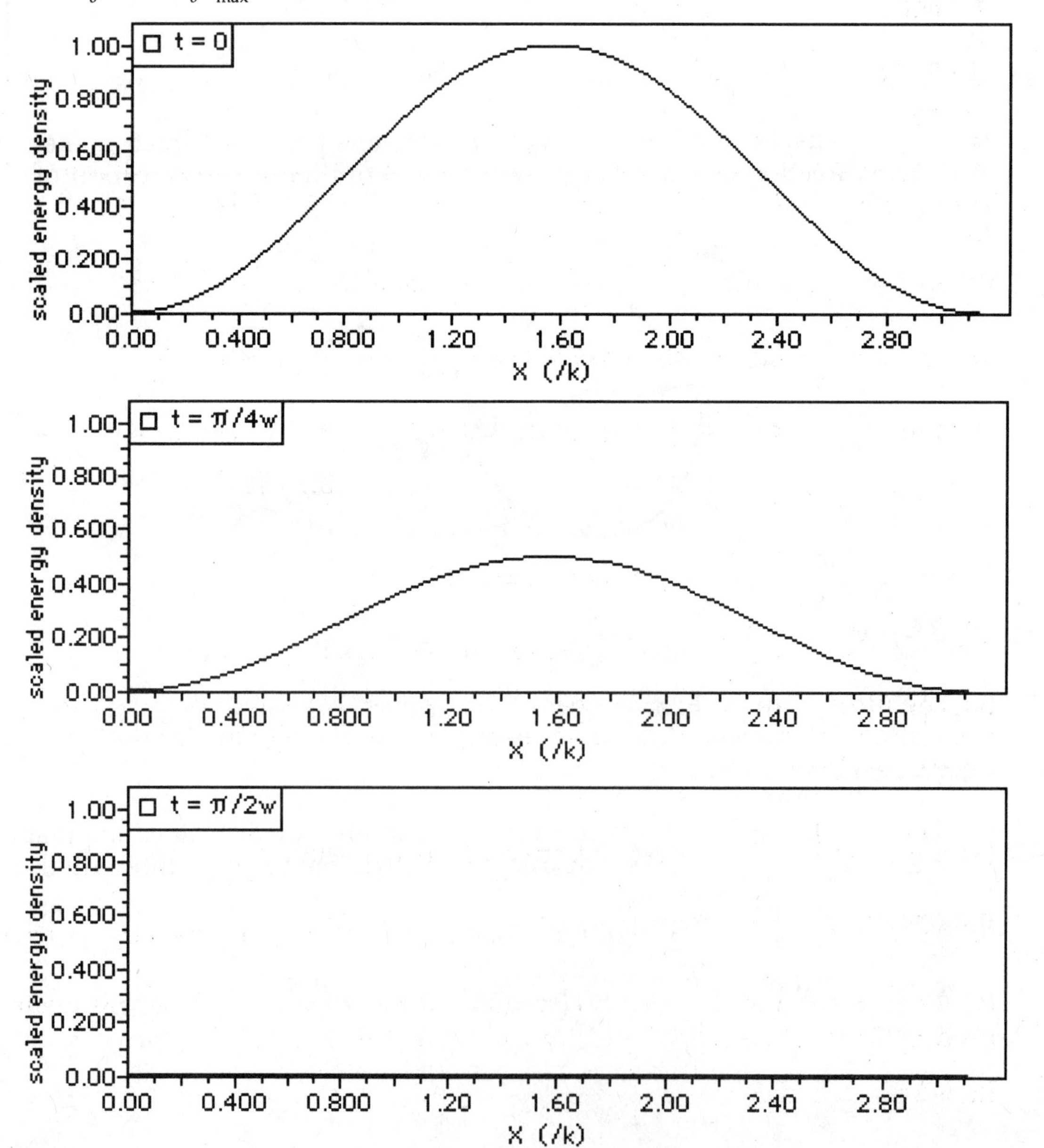

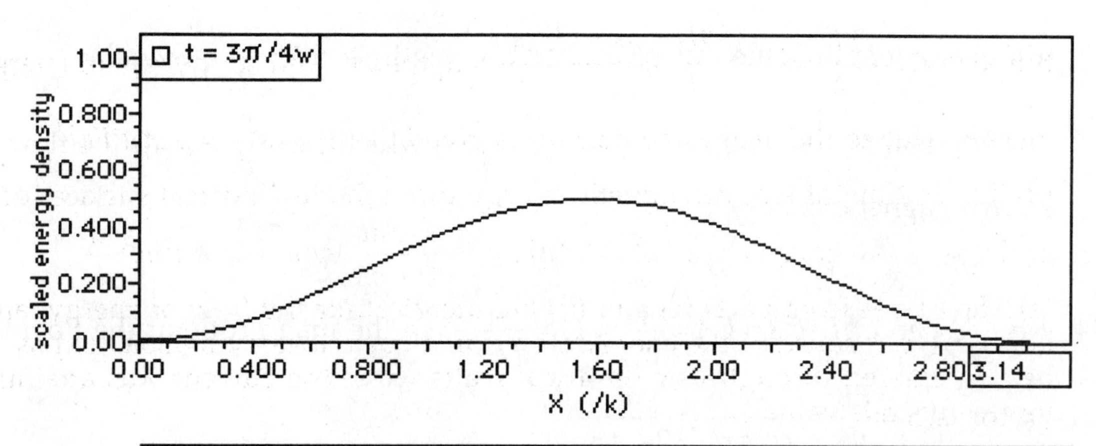

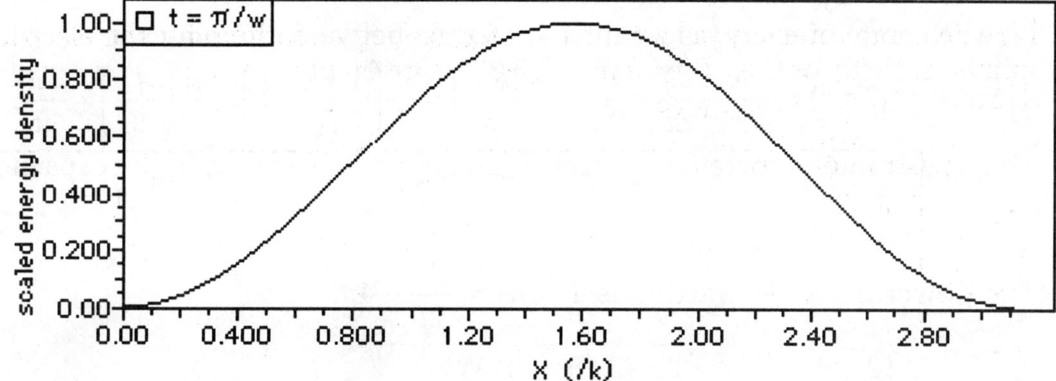

(b) At $t=\dfrac{\pi}{4\omega}$, $\cos\omega t=\cos\dfrac{\pi}{4}=\dfrac{1}{\sqrt{2}}$ and $\sin\omega t=\sin\dfrac{\pi}{4}=\dfrac{1}{\sqrt{2}}$.

For $0<x<\dfrac{\pi}{2k}$, $\sin kx>0,\ \cos kx>0\Rightarrow\hat{S}=\hat{E}\times\hat{B}=-\hat{j}\times\hat{k}=-\hat{\imath}$.

And for $\dfrac{\pi}{2k}<x<\dfrac{\pi}{k}$, $\sin kx>0,\ \cos kx<0\Rightarrow\hat{S}=\hat{E}\times\hat{B}=-\hat{j}\times-\hat{k}=\hat{\imath}$.

At $t=\dfrac{3\pi}{4\omega}$, $\cos\omega t=\cos\dfrac{3\pi}{4}=-\dfrac{1}{\sqrt{2}}$ and $\sin\omega t=\sin\dfrac{3\pi}{4}=\dfrac{1}{\sqrt{2}}$.

For $0<x<\dfrac{\pi}{2k}$, $\sin kx>0,\ \cos kx>0\Rightarrow\hat{S}=\hat{E}\times\hat{B}=\hat{j}\times\hat{k}=\hat{\imath}$.

And for $\dfrac{\pi}{2k}<x<\dfrac{\pi}{k}$, $\sin kx>0,\ \cos kx<0\Rightarrow\hat{S}=\hat{E}\times\hat{B}=\hat{j}\times-\hat{k}=-\hat{\imath}$.

(c) The plots from part (a) can be interpreted as two waves passing through each other in opposite directions, adding constructively at certain times, and destructively at others.

33-43: (a) $E=\rho J=\dfrac{\rho I}{A}=\dfrac{\rho I}{\pi a^2}$, in the direction of the current.

(b) $\oint\vec{B}.d\vec{l}=\mu_o I\Rightarrow B=\dfrac{\mu_o I}{2\pi a}$, counterclockwise when looking into the current.

(c) The direction of the Poynting vector $\hat{S}=\hat{E}\times\hat{B}=\hat{k}\times\hat{\phi}=-\hat{\rho}$, where we have used cylindrical coordinates, with the current in the z-direction.

Its magnitude is $S=\dfrac{EB}{\mu_o}=\dfrac{1}{\mu_o}\dfrac{\rho I}{\pi a^2}\dfrac{\mu_o I}{2\pi a}=\dfrac{\rho I^2}{2\pi^2 a^3}$.

(d) Over a length l, the rate of energy flowing in is $SA = \frac{\rho I^2}{2\pi^2 a^3} 2\pi a l = \frac{\rho l I^2}{\pi a^2}$.

The thermal power loss is $I^2 R = I^2 \frac{\rho l}{A} = \frac{\rho l I^2}{\pi a^2}$, which exactly equals the flow of electromagnetic energy.

33-44: $B = \frac{\mu_o i}{2\pi r}$, and $\oint_S \vec{E} \cdot d\vec{A} = EA = \frac{q}{\varepsilon_o} \Rightarrow E = \frac{q}{\pi \varepsilon_o r^2}$, so the magnitude of the Poynting vector is $S = \frac{EB}{\mu_o} = \frac{qi}{2\varepsilon_o \pi r^3} = \frac{q}{2\varepsilon_o \pi r^3} \frac{dq}{dt}$.

Now, the rate of energy flow into the region between the plates is:

$$\iint \vec{S} . d\vec{A} = S(2\pi r l) = \frac{lq}{\varepsilon_o \pi r^2} \frac{dq}{dt} = \frac{1}{2} \frac{l}{\varepsilon_o \pi r^2} \frac{d(q^2)}{dt} = \frac{d}{dt}\left(\frac{1}{2} \frac{l}{\varepsilon_o A} q^2\right) = \frac{d}{dt}\left(\frac{q^2}{2C}\right) = \frac{dU}{dt}.$$

This is just rate of increase in electrostatic energy U stored in the capacitor.

33-45: The power from the antenna is $P = IA = \frac{cB_{max}^{\ 2}}{2\mu_o} 4\pi r^2$. So

$$\Rightarrow B_{max} = \sqrt{\frac{2\mu_o P}{4\pi r^2 c}} = \sqrt{\frac{2\mu_o (2.75 \times 10^6 \text{ W})}{4\pi (500 \text{ m})^2 (3.00 \times 10^8 \text{ m/s})}} = 8.56 \times 10^{-8} \text{ T}$$

$$\Rightarrow \frac{dB}{dt} = \omega B_{max} = 2\pi f B_{max} = 2\pi (1.0 \times 10^7 \text{ Hz})(8.56 \times 10^{-8} \text{ T}) = 5.38 \text{ T/s}$$

$$\Rightarrow \mathcal{E} = -\frac{d\Phi}{dt} = -A\frac{dB}{dt} = \frac{\pi D^2}{4} \frac{dB}{dt} = \frac{\pi (0.400 \text{ m})^2 (5.38 \text{ T/s})}{4} = 0.676 \text{ V}.$$

33-46: $I = \frac{P}{A} = \frac{1}{2} \varepsilon_o c E^2 \Rightarrow E = \sqrt{\frac{2I}{\varepsilon_o c}} = \sqrt{\frac{2(62.5 \text{ W/m}^2)}{\varepsilon_o (3.00 \times 10^8 \text{ m/s})}} = 217 \text{ V/m}.$

33-47: The force and acceleration that the astronaut will experience by turning on the flashlight and pointing it away from the shuttle is calculated from the Poynting vector magnitude from the flashlight:

$F = \frac{dp}{dt} = \frac{SA}{c} = \frac{P}{c} = ma \Rightarrow a = \frac{P}{mc}$. Knowing the distance to travel, initial speed and the acceleration enables the time of travel to be calculated:

$\Delta x = v_o t + \frac{1}{2} a t^2$. But $v_o = 0$:

$$\Rightarrow t = \sqrt{\frac{2\Delta x}{a}} = \sqrt{\frac{2\Delta x \, mc}{P}} = \sqrt{\frac{2(12 \text{ m})(200 \text{ kg})(3.00 \times 10^8 \text{ m/s})}{100 \text{ W}}} = 1.2 \times 10^5 \text{ s} = 33 \text{ hrs}.$$

33-48: $P = IA \Rightarrow I = \frac{P}{A} = \frac{1}{2}\varepsilon_o cE^2 \Rightarrow E = \sqrt{\frac{2P}{A\varepsilon_o c}} = \sqrt{\frac{2Vi}{A\varepsilon_o c}}$

$$\Rightarrow E = \sqrt{\frac{2Vi}{A\varepsilon_o c}} = \sqrt{\frac{2(5.00\times10^5\ \text{V})(1000\ \text{A})}{(100\ \text{m}^2)\varepsilon_o(3.00\times10^8\ \text{m/s})}} = 6.14\times10^4\ \text{V/m}.$$

And $B = \frac{E}{c} = \frac{6.14\times10^4\ \text{V/m}}{3.00\times10^8\ \text{m/s}} = 2.05\times10^{-4}\ \text{T}.$

33-49: (a) $F_G = \frac{GM_S m}{r^2} = \frac{GM_S}{r^2}\cdot\frac{4\pi R^3\rho}{3} = \frac{4\pi GM_S R^3\rho}{3r^2}.$

(b) Assuming that the sun's radiation is intercepted by the particle's cross-section, we can write the force on the particle as:

$$F = \frac{IA}{c} = \frac{L}{4\pi r^2}\cdot\frac{\pi R^2}{c} = \frac{LR^2}{4cr^2}.$$

(c) So if the force of gravity and the force from the radiation pressure on a particle from the sun are equal, we can solve for the particle's radius:

$$F_G = F \Rightarrow \frac{4\pi GM_S R^3\rho}{3r^2} = \frac{LR^2}{4cr^2} \Rightarrow R = \frac{3L}{16\pi GM_S\rho c}$$

$$\Rightarrow R = \frac{3(3.9\times10^{26}\ \text{W})}{16\pi(6.7\times10^{-11}\ \text{N}\cdot\text{m}^2/\text{kg}^2)(2.0\times10^{30}\ \text{kg})(3000\ \text{kg/m}^3)(3.0\times10^8\ \text{m/s})}$$

$$\Rightarrow R = 1.9\times10^{-7}\ \text{m}.$$

(d) If the particle has a radius smaller than that found in part (c), then the radiation pressure overcomes the gravitational force and results in an acceleration away from the sun, thus removing all such particles from the solar system.

33-50: (a) The momentum transfer is always greatest when reflecting surfaces are used (consider a ball colliding with a wall - the wall exerts a greater force if the ball rebounds rather than sticks). So in solar sailing one would want to use a reflecting sail.

(b) The equation for repulsion comes from balancing the gravitational force and the force from the radiation pressure. As seen in **Pr. (33-49)**, the latter is:

$F_{rad} = \frac{2LA}{4\pi r^2 c}$. Thus: $F_G = F_{rad} \Rightarrow \frac{GM_S m}{r^2} = \frac{2LA}{4\pi r^2 c} \Rightarrow A = \frac{4\pi GM_S mc}{2L}$

$$\Rightarrow A = \frac{4\pi(6.7\times10^{-11}\ \text{N}\cdot\text{m}^2/\text{kg}^2)(2.0\times10^{30}\ \text{kg})(5000\ \text{kg})(3.0\times10^8\ \text{m/s})}{(2)\,3.9\times10^{26}\ \text{W}}$$

$$\Rightarrow A = 3.2\times10^6\ \text{m}^2 = 3.2\ \text{km}^2 = \frac{3.2\ \text{km}^2}{(1.6\ \text{km/mile})^2} = 1.2\ \text{square miles}.$$

(c) This answer is independent of the distance from the sun since both the gravitational force and the radiation pressure go down like one over the distance squared, and thus the distance cancels out of the problem.

33-51: (a) $\left[\frac{q^2a^2}{6\pi\varepsilon_o c^3}\right] = \frac{\text{C}^2(\text{m/s}^2)^2}{(\text{C}^2/\text{N}\cdot\text{m}^2)(\text{m/s})^3} = \frac{\text{Nm}}{\text{s}} = \frac{\text{J}}{\text{s}} = \text{W} = \left[\frac{dE}{dt}\right].$

(b) For a proton moving in a circle, the acceleration can be rewritten:

$$a = \frac{v^2}{R} = \frac{\frac{1}{2}mv^2}{\frac{1}{2}mR} = \frac{2(5.0\times10^6 \text{ eV})(1.6\times10^{-19} \text{ J/eV})}{(1.67\times10^{-27} \text{ kg})(0.85 \text{ m})} = 1.1\times10^{15} \text{ m/s}^2.$$

The rate at which it emits energy because of its acceleration is:

$$\frac{dE}{dt} = \frac{q^2a^2}{6\pi\varepsilon_0 c^3} = \frac{(1.6\times10^{-19}\text{C})^2(1.1\times10^{15} \text{ m/s}^2)^2}{6\pi\varepsilon_0(3.0\times10^8 \text{ m/s})^3} = 7.2\times10^{-24} \text{ J/s} = 4.5\times10^{-5} \text{ eV/s}.$$

So the fraction of its energy that it radiates every second is:

$$\frac{(dE/dt)(1 \text{ s})}{E} = \frac{4.5\times10^{-5} \text{ eV}}{5.0\times10^6 \text{ eV}} = 9.0\times10^{-12}.$$

(c) Carrying out the same calculations as in part (b), but now for an electron at the same speed and radius. That means the electron's acceleration is the same as the proton, and thus so is the rate at which it emits energy, since they also have the same charge. However, the electron's initial energy differs from the proton's by the ratio of their masses:

$$E_e = E_p\frac{m_e}{m_p} = (5.0\times10^6 \text{ eV})\frac{(9.11\times10^{-31} \text{ kg})}{(1.67\times10^{-271} \text{ kg})} = 2730 \text{ eV}.$$

So the fraction of its energy that it radiates every second is:

$$\frac{(dE/dt)(1 \text{ s})}{E} = \frac{4.5\times10^{-5} \text{ eV}}{2730 \text{ eV}} = 1.7\times10^{-8}.$$

33-52: For the electron in the classical hydrogen atom, its acceleration is:

$$a = \frac{v^2}{R} = \frac{\frac{1}{2}mv^2}{\frac{1}{2}mR} = \frac{2(13.6 \text{ eV})(1.60\times10^{-19} \text{ J/eV})}{(9.11\times10^{-31} \text{ kg})(5.29\times10^{-11} \text{ m})} = 9.03\times10^{22} \text{ m/s}^2.$$

Then using the formula for the rate of energy emission given in **Pr. (33-51)**:

$$\frac{dE}{dt} = \frac{q^2a^2}{6\pi\varepsilon_0 c^3} = \frac{(1.60\times10^{-19}\text{C})^2(9.03\times10^{22} \text{ m/s}^2)^2}{6\pi\varepsilon_0(3.00\times10^8 \text{ m/s})^3}$$

$\Rightarrow \frac{dE}{dt} = 4.64\times10^{-8} \text{ J/s} = 2.89\times10^{11} \text{ eV/s}$, which means that the electron would almost immediately lose all its energy!

Chapter 34: The Nature and Propagation of Light

34-1: $\theta_b = \arcsin\left(\dfrac{n_a}{n_b}\sin\theta_a\right) = \arcsin\left(\dfrac{1.80}{1.52}\sin 29.0°\right) = 35.0°.$

34-2: (a) $\theta_{water} = \arcsin\left(\dfrac{n_{air}}{n_{water}}\sin\theta_{air}\right) = \arcsin\left(\dfrac{1.00}{1.33}\sin 43.0°\right) = 30.8°.$

(b) This calculation has no dependence on the glass because we can omit that step in the chain: $n_{air}\sin\theta_{air} = n_{glass}\sin\theta_{glass} = n_{water}\sin\theta_{water}$.

34-3: (a) Incident and reflected angles are always equal $\Rightarrow \theta'_r = \theta'_a = 37.5°$.

(b) $\theta'_b = \dfrac{\pi}{2} - \theta_b = \dfrac{\pi}{2} - \arcsin\left(\dfrac{n_a}{n_b}\sin\theta_a\right) = \dfrac{\pi}{2} - \arcsin\left(\dfrac{1.00}{1.52}\sin 52.5°\right) = 58.5°.$

34-4: $\lambda_{plastic} = \dfrac{c}{fn} = \dfrac{3.00\times10^{8}\text{ m/s}}{(5.00\times10^{14}\text{ Hz})(1.65)} = 3.64\times10^{-7}\text{ m}.$

$\lambda_{vacuum} = \dfrac{c}{f} = \dfrac{3.00\times10^{8}\text{ m/s}}{5.00\times10^{14}\text{ Hz}} = 6.00\times10^{-7}\text{ m}.$

34-5: (a) $n = \dfrac{c}{v} = \dfrac{3.00\times10^{8}\text{ m/s}}{1.84\times10^{8}\text{ m/s}} = 1.63.$

(b) $\lambda_o = n\lambda = (1.63)(5.42\times10^{-7}\text{ m}) = 8.84\times10^{-7}\text{ m}.$

34-6: $\theta_b = \arcsin\left(\dfrac{n_a}{n_b}\sin\theta_a\right) = \arcsin\left(\dfrac{1.33}{1.52}\sin 40.0°\right) = 34.2°$. But this is the angle from the normal to the surface, so the angle from the vertical is an additional 20° because of the tilt of the surface. Therefore the angle is 54.2°.

34-7: (a) $v = \dfrac{c}{n} = \dfrac{3.00\times10^{8}\text{ m/s}}{1.62} = 1.85\times10^{8}\text{ m/s}.$

(b) $\lambda = \dfrac{\lambda_o}{n} = \dfrac{(5.00\times10^{-7}\text{ m})}{1.62} = 3.09\times10^{-7}\text{ m}.$

34-8: $\lambda_{water}n_{water} = \lambda_{CS_2}n_{CS_2} \Rightarrow \lambda_{CS_2} = \dfrac{\lambda_{water}n_{water}}{n_{CS_2}} = \dfrac{(4.44\times10^{-7}\text{ m})(1.333)}{1.628} = 3.64\times10^{-7}\text{ m}.$

34-9: As shown below, the angle between the beams and the prism is A/2 and the angle between the beams and the vertical is A, so the total angle between the two beams is 2A.

34-9 cont:

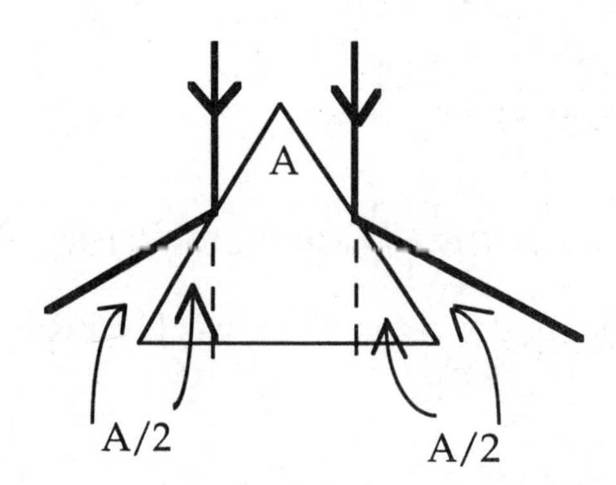

34-10: Rotating a mirror by an angle θ while keeping the incoming beam constant leads to an increase in the incident angle ϕ by θ. Therefore the angle between incoming and outgoing beams becomes $2\theta + 2\phi$, where an additional deflection of 2θ arose from the mirror rotation.

34-11: (a) The slower the speed of the wave, the larger the index of refraction - so air has a larger index of refraction than water.

(b) $\theta_{crit} = \arcsin\left(\frac{n_b}{n_a}\right) = \arcsin\left(\frac{v_{air}}{v_{water}}\right) = \arcsin\left(\frac{344\ \text{m/s}}{1320\ \text{m/s}}\right) = 15.1°.$

34-12: $\theta_{crit} = \arcsin\left(\frac{n_b}{n_a}\right) = \arcsin\left(\frac{1.00}{1.66}\right) = 37.0°.$

34-13: (a) Going from the liquid into air: $\frac{n_b}{n_a} = \sin\theta_{crit} \Rightarrow n_a = \frac{1.00}{\sin 37.0°} = 1.66.$

So: $\theta_b = \arcsin\left(\frac{n_a}{n_b}\sin\theta_a\right) = \arcsin\left(\frac{1.66}{1.00}\sin 24.0°\right) = 42.5°.$

(b) Going from air into the liquid:

$\theta_b = \arcsin\left(\frac{n_a}{n_b}\sin\theta_a\right) = \arcsin\left(\frac{1.00}{1.66}\sin 24.0°\right) = 14.2°.$

34-14: $\theta_{crit} = \arcsin\left(\frac{n_b}{n_a}\right) = \arcsin\left(\frac{1.00}{1.33}\right) = 48.6°.$

So the largest diameter circle from which light can emerge is:
$D = 2d\tan\theta_{crit} = 2(0.540\ \text{m})\tan 48.6° = 1.23\ \text{m}$, where d is the depth of the light source.

34-15: (a) Through the first filter: $I_1 = \frac{1}{2}I_o.$

The second filter: $I_2 = \frac{1}{2}I_o\cos^2(52.0°) = 0.190\,I_o.$

(b) The light is linearly polarized.

34-16: (a) $I = I_{max}\cos^2\phi \Rightarrow I = I_{max}\cos^2(22.5°) = 0.854\,I_{max}$.
(b) $I = I_{max}\cos^2\phi \Rightarrow I = I_{max}\cos^2(45.0°) = 0.500\,I_{max}$.
(c) $I = I_{max}\cos^2\phi \Rightarrow I = I_{max}\cos^2(67.5°) = 0.146\,I_{max}$.

34-17: (a) $I_1 = \frac{1}{2}I_o,\ I_2 = \frac{1}{2}I_o\cos^2(30.0°) = 0.375\,I_o,\ I_3 = I_2\cos^2(60.0°) = 0.0938\,I_o$.
(b) $I_1 = \frac{1}{2}I_o,\ I_2 = \frac{1}{2}I_o\cos^2(90.0°) = 0$.

34-18: From the picture at right,
$\theta_r = 37.0°$, and so:
$$n_b = n_a\frac{\sin\theta_a}{\sin\theta_b} = 1.33\frac{\sin 53°}{\sin 37°} = 1.77.$$

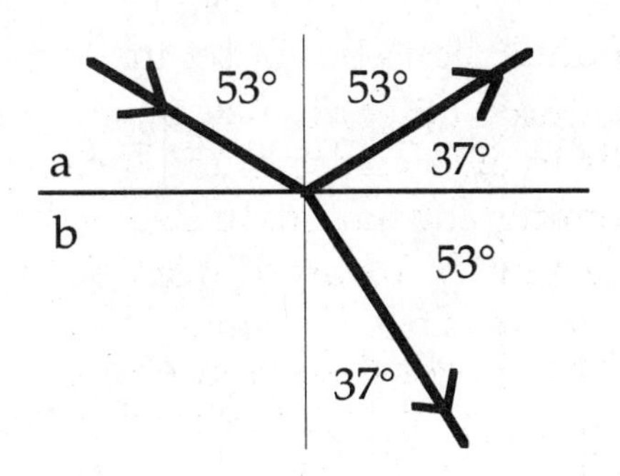

34-19: (a) $\tan\theta_p = \frac{n_b}{n_a} = \tan 56.8° = 1.53 \Rightarrow n_b = 1.53$.
(b) $\theta_b = \arcsin\left(\frac{n_a}{n_b}\sin\theta_a\right) = \arcsin\left(\frac{1.00}{1.53}\sin 56.8°\right) = 33.2°$.

34-20: (a) $\tan\theta_p = \frac{n_b}{n_a} \Rightarrow \theta_p = \arctan\left(\frac{n_b}{n_a}\right) = \arctan\left(\frac{1.33}{1.00}\right) = 53.1°$, so 36.9° from horizontal.
(b) The electric field is polarized perpendicular to the plane of incidence, so is parallel to the surface of the water.

34-21: (a) $\tan\theta_p = \frac{n_b}{n_a} \Rightarrow n_a = \frac{n_b}{\tan\theta_p} = \frac{1.00}{\tan 35.2°} = 1.42$.
(b) $\theta_b = \arcsin\left(\frac{n_a}{n_b}\sin\theta_a\right) = \arcsin\left(\frac{1.42}{1.00}\sin 35.2°\right) = 54.8°$.

34-22: (a) All the electric field is in the plane perpendicular to the propagation direction, and maximum intensity through the filters is at 90° to the filter orientation for the case of minimum intensity. Therefore rotating the second filter by 90° when the situation originally showed the maximum intensity means one ends with a dark cell.
(b) If filter P_1 is rotated by 90°, then the electric field oscillates in the direction pointing toward the P_2 filter, and hence no intensity passes through the second filter: see a dark cell.

(c) Even if P2 is rotated back to its original position, the new plane of oscillation of the electric field, determined by the first filter, allows zero intensity to pass through the second filter.

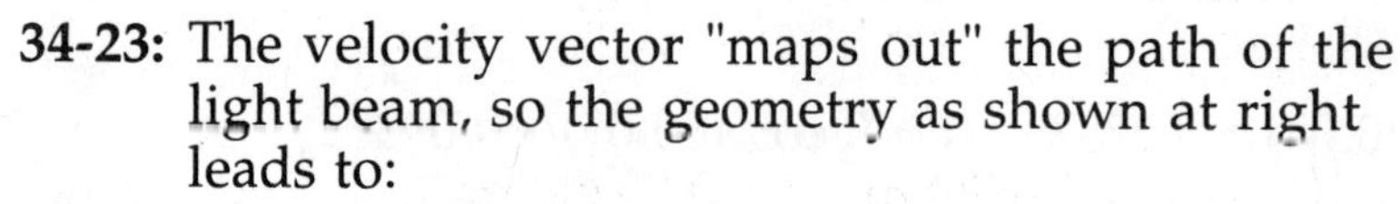

34-23: The velocity vector "maps out" the path of the light beam, so the geometry as shown at right leads to:

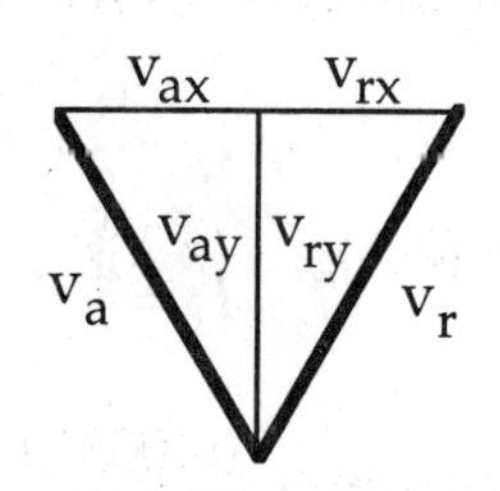

$v_a = v_r$ and $\theta_a = \theta_r$

$$\Rightarrow \arccos\left(\frac{v_{a_y}}{v_a}\right) = \arccos\left(\frac{v_{r_y}}{v_r}\right) \Rightarrow v_{a_y} = -v_{r_y}, \text{ with}$$

the minus sign chosen by inspection. Similarly,

$$\Rightarrow \arcsin\left(\frac{v_{a_x}}{v_a}\right) = \arcsin\left(\frac{v_{r_x}}{v_r}\right) \Rightarrow v_{a_x} = v_{r_x}.$$

34-24: $n_a \sin\theta_a = n_b \sin\theta_b = n_b \sin\left(\frac{\theta_a}{2}\right)$

$$\Rightarrow (1.00)\sin\theta_a = \sin 2\left(\frac{\theta_a}{2}\right) = 2\sin\left(\frac{\theta_a}{2}\right)\cos\left(\frac{\theta_a}{2}\right) = (1.62)\sin\left(\frac{\theta_a}{2}\right)$$

$$\Rightarrow 2\cos\left(\frac{\theta_a}{2}\right) = (1.62) \Rightarrow \theta_a = 2\arccos\left(\frac{1.62}{2}\right) = 71.8°.$$

34-25: Consider three mirrors, M_1 in the (x,y)-plane, M_2 in the (y,z)-plane and M_3 in the (x,z)-plane. A light ray bouncing from M_1 changes the sign of the z-component of the velocity, bouncing from M_2 changes the x-component, and from M_3 changes the y-component. Thus the velocity, and hence also the path, of the light beam flips by 180°.

34-26: (a) $\theta_b = \arcsin\left(\frac{n_a}{n_b}\sin\theta_a\right) = \arcsin\left(\frac{v_b}{v_a}\sin\theta_a\right) = \arcsin\left(\frac{1480}{344}\sin 9.73°\right) = 46.6°.$

(b) $\theta_{crit} = \arcsin\left(\frac{v_a}{v_b}\right) = \arcsin\left(\frac{344}{1480}\right) = 13.4°.$

34-27: $\theta_b = 90° - \arcsin\left(\frac{n_a}{n_b}\right) = 90° - \arcsin\left(\frac{1.00}{1.31}\right) = 40.2°.$

But $n_a \sin\theta_a = n_b \sin\theta_b \Rightarrow \theta_a = \arcsin\left(\frac{n_b \sin\theta_b}{n_a}\right) = \arcsin\left(\frac{1.31\sin(40.2°)}{1.00}\right) = 57.8°.$

34-28: $\theta_a = \arctan\left(\frac{1.5 \text{ m}}{1.2 \text{ m}}\right) = 51° \Rightarrow \theta_b = \arcsin\left(\frac{n_a}{n_b}\sin\theta_a\right) = \arcsin\left(\frac{1.00}{1.33}\sin 51°\right) = 36°.$

So the distance along the bottom of the pool from directly below where the light enters to where it hits the bottom is:

$x = (4.0 \text{ m})\tan\theta_b = (4.0 \text{ m})\tan 36° = 2.9 \text{ m}.$

$\Rightarrow x_{total} = 1.5 \text{ m} + x = 1.5 \text{ m} + 2.9 \text{ m} = 4.4 \text{ m}.$

34-29: $\theta_a = \arctan\left(\dfrac{8.0\text{ cm}}{16.0\text{ cm}}\right) = 27°$ and $\theta_b = \arctan\left(\dfrac{4.0\text{ cm}}{16.0\text{ cm}}\right) = 14°$.

So, $n_a \sin\theta_a = n_b \sin\theta_b \Rightarrow n_b = \left(\dfrac{n_a \sin\theta_a}{\sin\theta_b}\right) = \left(\dfrac{1.00 \sin 27°}{\sin 14°}\right) = 1.8.$

34-30: $\#\lambda = (\#\lambda)_{air} + (\#\lambda)_{glass} = \dfrac{d_{air}}{\lambda} + \dfrac{d_{glass}}{\lambda} n = \dfrac{0.0270\text{ m}}{4.50\times10^{-7}\text{ m}} + \dfrac{0.00300\text{ m}}{4.50\times10^{-7}\text{ m}}(1.50) = 70{,}000.$

34-31: $\theta_{crit} = \arctan\left(\dfrac{(0.00524\text{ m})/2}{0.00320\text{ m}}\right) = 39.3° = \arcsin\left(\dfrac{n_b}{n_a}\right) = \arcsin\left(\dfrac{1.0}{n}\right)$

$\Rightarrow n = \dfrac{1}{\sin 39.3°} = 1.58.$

34-32: $n_a \sin\theta_a = n_b \sin\theta_b \Rightarrow n_a = \left(\dfrac{n_b \sin\theta_b}{\sin\theta_a}\right) = \left(\dfrac{1.33 \sin 90°}{\sin 45°}\right) = 1.9.$

34-33: $n_a \sin\theta_a = n_b \sin\theta_b \Rightarrow \theta_b = \arcsin\left(\dfrac{n_a \sin\theta_a}{n_b}\right) = \arcsin\left(\dfrac{1.58 \sin(30.0°)}{1.00}\right) = 52.2°.$

So the angle below the horizontal is $\theta_b - 30° = 52.2° - 30.0° = 22.2°$, and thus the angle between the two emerging beams is 44.4°.

34-34: $n_a \sin\theta_a = n_b \sin\theta_b \Rightarrow n_b = \left(\dfrac{n_b \sin\theta_b}{\sin\theta_a}\right) = \left(\dfrac{1.62 \sin 60°}{\sin 90°}\right) = 1.40.$

34-35: $n_a \sin\theta_a = n_b \sin\theta_b \Rightarrow n_b = \left(\dfrac{n_b \sin\theta_b}{\sin\theta_a}\right) = \left(\dfrac{1.60 \sin 51.0°}{\sin 90°}\right) = 1.24.$

34-36: The beam of light will emerge at the same angle as it entered the fluid as seen by following what happens via Snell's Law at each of the interfaces. That is, the emergent beam is at 38.0° from the normal.

34-37: The critical angle at the water-air interface is given by:

$\theta_{crit} = \arcsin\left(\dfrac{n_a \sin 90°}{n_w}\right) = \arcsin\left(\dfrac{1.000}{1.333}\right) = 48.61°.$

This is the angle with which the beam must enter the water, so:

$n_g \sin\theta_g = n_w \sin\theta_w \Rightarrow \theta_g = \arcsin\left(\dfrac{n_w \sin\theta_w}{n_g}\right) = \arcsin\left(\dfrac{1.333 \sin(48.61°)}{1.473}\right) = 42.76°.$

(Also note that the same answer is obtained by ignoring the water and considering the critical angle to simply be between the glycerin and air:

$\theta_{crit} = \arcsin\left(\dfrac{n_a \sin 90°}{n_g}\right) = \arcsin\left(\dfrac{1.000}{1.473}\right) = 42.76°.$)

34-38: (a) In air: $\theta_p = \arctan\left(\frac{n_b}{n_a}\right) = \arctan\left(\frac{1.62}{1.00}\right) = 58.3°$.

(b) In water: $\theta_p = \arctan\left(\frac{n_b}{n_a}\right) = \arctan\left(\frac{1.62}{1.33}\right) = 50.6°$

34-39: (a) $I = \frac{1}{2}I_o \cos^2\theta \cos^2(90° - \theta) = \frac{1}{2}I_o(\cos\theta \sin\theta)^2 = \frac{1}{8}I_o \sin^2 2\theta$.
(b) For maximum transmission, we need $2\theta = 90°$, so $\theta = 45°$.

34-40: A quarter-wave plate shifts the phase of the light by $\theta = 90°$. Circularly polarized light is out of phase by 90°, so the use of a quarter-wave plate will bring it back into phase, resulting in linearly polarized light.

34-41: Both *l*-leucine and *d*-glutamic acid exhibit linear relationships between concentration and rotation angle. The dependence for *l*-leucine is:
Rotation angle(°) = (−0.11° 100 ml / g)C(g / 100 ml), and for *d*-glutamic acid is:
Rotation angle(°) = (0.124° 100 ml / g)C(g / 100 ml).

34-42: (a) A birefringent material has different speeds (or equivalently wavelengths) in two different directions, so:
$\lambda_1 = \frac{\lambda_o}{n_1}$ and $\lambda_2 = \frac{\lambda_o}{n_2} \Rightarrow \frac{D}{\lambda_1} = \frac{D}{\lambda_2} + \frac{1}{4} \Rightarrow \frac{n_1 D}{\lambda_0} = \frac{n_2 D}{\lambda_0} + \frac{1}{4} \Rightarrow D = \frac{\lambda_o}{4(n_1 - n_2)}$.
(b) $D = \frac{\lambda_o}{4(n_1 - n_2)} = \frac{5.90 \times 10^{-7}\ \text{m}}{4(1.658 - 1.486)} = 8.58 \times 10^{-7}\ \text{m}$.

34-43: (a) The maximum intensity from the table is at $\theta = 35°$, so the polarized component of the wave is in that direction (or else we would not have maximum intensity at that angle).
(b) At $\theta = 40°$: $I = 24.8\ \text{W/m}^2 = \frac{1}{2}I_o + I_p \cos^2(40° - 35°)$
$\Rightarrow 24.8\ \text{W/m}^2 = 0.500 I_o + 0.996 I_p$ (1).
At $\theta = 120°$: $I = 5.2\ \text{W/m}^2 = \frac{1}{2}I_o + I_p \cos^2(120° - 35°)$
$\Rightarrow 5.2\ \text{W/m}^2 = 0.500 I_o + 7.60 \times 10^{-3} I_p$ (2).
Solving equations (1) and (2) we find:
$\Rightarrow 19.6\ \text{W/m}^2 = 0.989 I_p \Rightarrow I_p = 19.8\ \text{W/m}^2$.
Then if one subs this back into equation (1), we find:
$5.049 = 0.500 I_o \Rightarrow I_o = 10.1\ \text{W/m}^2$.

34-44: (a) n decreases with increasing λ, so n is smaller for red than for blue. So beam a is the red one.
(b) The separation of the emerging beams is given by some elementary geometry. $x = x_r - x_v = d\tan\theta_r - d\tan\theta_v \Rightarrow d = \frac{x}{\tan\theta_r - \tan\theta_v}$, where x is the

vertical beam separation as they emerge from the glass $x = \dfrac{1.0 \text{ mm}}{\sin 20°} = 2.92 \text{ mm}$.

From the ray geometry, we also have;

$$\theta_r = \arcsin\left(\frac{1.00}{1.61\tan 70°}\right) = 35.7° \text{ and } \theta_v = \arcsin\left(\frac{1.00}{1.66\tan 70°}\right) = 34.5°, \text{ so:}$$

$$d = \frac{x}{\tan\theta_r - \tan\theta_v} = \frac{2.92 \text{ mm}}{\tan 35.7° - \tan 34.5°} = 9 \text{ cm}.$$

34-45: (a) For sunlight entering the earth's atmosphere from the sun BELOW the horizon, we can calculate the angle δ as follows:
$n_a \sin\theta_a = n_b \sin\theta_b \Rightarrow (1.00)\sin\theta_a = n\sin\theta_b$, where $n_b = n$ is the atmosphere's index of refraction. But the geometry of the situation tells us:

$$\sin\theta_b = \frac{R}{R+h} \Rightarrow \sin\theta_a = \frac{nR}{R+h} \Rightarrow \delta = \theta_a - \theta_b = \arcsin\left(\frac{nR}{R+h}\right) - \arcsin\left(\frac{R}{R+h}\right).$$

(b) $$\delta = \arcsin\left(\frac{(1.0003)(6.4\times 10^6 \text{ m})}{6.4\times 10^6 \text{ m} + 2.0\times 10^4 \text{ m}}\right) - \arcsin\left(\frac{6.4\times 10^6 \text{ m}}{6.4\times 10^6 \text{ m} + 2.0\times 10^4 \text{ m}}\right)$$

$\Rightarrow \delta = 0.22°$. This is about the same as the angular radius of the sun, 0.25°.

34-46: (a) The distance traveled by the light ray is the sum of the two diagonal segments: $d = \left(x^2 + y_1^2\right)^{1/2} + \left((l-x)^2 + y_2^2\right)^{1/2}$

Then the time taken to travel that distance is just:

$$t = \frac{d}{c} = \frac{\left(x^2 + y_1^2\right)^{1/2} + \left((l-x)^2 + y_2^2\right)^{1/2}}{c}$$

(b) Taking the derivative with respect to x of the time and setting it to zero yields:

$$\frac{dt}{dx} = \frac{1}{c}\frac{d}{dt}\left[\left(x^2 + y_1^2\right)^{1/2} + \left((l-x)^2 + y_2^2\right)^{1/2}\right]$$

$$\Rightarrow \frac{dt}{dx} = \frac{1}{c}\left[x\left(x^2 + y_1^2\right)^{-1/2} - (l-x)\left((l-x)^2 + y_2^2\right)^{-1/2}\right] = 0$$

$$\Rightarrow \frac{x}{\sqrt{x^2 + y_1^2}} = \frac{(l-x)}{\sqrt{(l-x)^2 + y_2^2}} \Rightarrow \sin\theta_1 = \sin\theta_2 \Rightarrow \theta_1 = \theta_2.$$

34-47: (a) The time taken to travel from point A to point B is just:

$$t = \frac{d_1}{v_1} + \frac{d_2}{v_2} = \frac{\sqrt{h_1^2 + x^2}}{v_1} + \frac{\sqrt{h_2^2 + (l-x)^2}}{v_2}$$

Taking the derivative with respect to x of the time and setting it to zero yields:

$$\frac{dt}{dx} = 0 = \frac{d}{dt}\left[\frac{\sqrt{h_1^2 + x^2}}{v_1} + \frac{\sqrt{h_2^2 + (l-x)^2}}{v_2}\right] = \frac{x}{v_1\sqrt{h_1^2 + x^2}} - \frac{(l-x)}{v_2\sqrt{h_2^2 + (l-x)^2}}$$

But $v_1 = \dfrac{c}{n_1}$ and $v_2 = \dfrac{c}{n_2} \Rightarrow \dfrac{n_1 x}{\sqrt{h_1^2 + x^2}} = \dfrac{n_2(l-x)}{\sqrt{h_2^2 + (l-x)^2}} \Rightarrow n_1\sin\theta_1 = n_2\sin\theta_2$.

34-48: (a) To let the most light possible through N polarizers, with a total rotation of 90°, we need as little shift from one polarizer to the next. That is, the angle between successive polarizers should be constant and equal to $\frac{\pi}{2N}$. Then:

$$I_1 = I_o \cos^2\left(\frac{\pi}{2N}\right),\ I_2 = I_o \cos^4\left(\frac{\pi}{2N}\right), \ldots \Rightarrow I = I_N = I_o \cos^{2N}\left(\frac{\pi}{2N}\right).$$

(b) If $n >> 1$, $\cos^n\theta = \left(1 - \frac{\theta^2}{2} + \ldots\right)^n = 1 - \frac{n}{2}\theta^2 + \ldots$

$$\Rightarrow \cos^{2N}\left(\frac{\pi}{2N}\right) \approx 1 - \frac{(2N)}{2}\left(\frac{\pi}{2N}\right)^2 = 1 - \frac{\pi^2}{4N} \approx 1, \text{ for large } N.$$

34-49: (a) $n_a \sin\theta_a = n_b \sin\theta_b \Rightarrow \sin\theta_a = n_b \sin\frac{A}{2}$.

But $\theta_a = \frac{A}{2} + \alpha \Rightarrow \sin\left(\frac{A}{2} + \alpha\right) = \sin\frac{A + 2\alpha}{2} = n \sin\frac{A}{2}$.

At each face of the prism the deviation is α, so $2\alpha = \delta \Rightarrow \sin\frac{A + \delta}{2} = n\sin\frac{A}{2}$.

(b) From part (a), $\delta = 2\arcsin\left(n\sin\frac{A}{2}\right) - A$

$$\Rightarrow \delta = 2\arcsin\left((1.62)\sin\frac{60.0°}{2}\right) - 60.0° = 48.2°.$$

(c) If two colors have different indices of refraction for the glass, then the deflection angles for them will differ:

$$\delta_{red} = 2\arcsin\left((1.60)\sin\frac{60.0°}{2}\right) - 60.0° = 46.3°$$

$$\delta_{violet} = 2\arcsin\left((1.64)\sin\frac{60.0°}{2}\right) - 60.0° = 50.2° \Rightarrow \Delta\delta = 50.2° - 46.3° = 3.9°.$$

34-50: (a) For light in air incident on a parallel-faced plate, Snell's Law yields:
$n\sin\theta_a = n'\sin\theta'_b = n'\sin\theta_b = n\sin\theta'_a \Rightarrow \sin\theta_a = \sin\theta'_a \Rightarrow \theta_a = \theta'_a$.

(b) Adding more plates just adds extra steps in the middle of the above equation which always cancel out. The requirement of parallel faces ensures that the angle $\theta'_n = \theta_n$, and the chain of equations can continue.

(c) The lateral displacement of the beam can be calculated using geometry:

$$d = L\sin(\theta_a - \theta'_b) \text{ and } L = \frac{t}{\cos\theta'_b} \Rightarrow d = \frac{t\sin(\theta_a - \theta'_b)}{\cos\theta'_b}.$$

(d) $\theta'_b = \arcsin\left(\frac{n\sin\theta_a}{n'}\right) = \arcsin\left(\frac{\sin 60.0°}{1.66}\right) = 31.4°$

$$\Rightarrow d = \frac{(1.80\text{ cm})\sin(60.0° - 31.4°)}{\cos 31.4°} = 1.01\text{ cm}.$$

34-51: (a) Multiplying Eq. (1) by $\sin\beta$ and Eq. (2) by $\sin\alpha$ yields:

$$(1):\ \frac{x}{a}\sin\beta = \sin\omega t\cos\alpha\sin\beta - \cos\omega t\sin\alpha\sin\beta$$

(2): $\frac{y}{a}\sin\alpha = \sin\omega t\cos\beta\sin\alpha - \cos\omega t\sin\beta\sin\alpha$

Subtracting yields: $\frac{x\sin\beta - y\sin\alpha}{a} = \sin\omega t(\cos\alpha\sin\beta - \cos\beta\sin\alpha)$.

(b) Multiplying Eq. (1) by $\cos\beta$ and Eq. (2) by $\cos\alpha$ yields:

(1): $\frac{x}{a}\cos\beta = \sin\omega t\cos\alpha\cos\beta - \cos\omega t\sin\alpha\cos\beta$

(2): $\frac{y}{a}\cos\alpha = \sin\omega t\cos\beta\cos\alpha - \cos\omega t\sin\beta\cos\alpha$

Subtracting yields: $\frac{x\cos\beta - y\cos\alpha}{a} = -\cos\omega t(\sin\alpha\cos\beta - \sin\beta\cos\alpha)$.

(c) Squaring and adding the results of parts (a) and (b) yields:
$(x\sin\beta - y\sin\alpha)^2 + (x\cos\beta - y\cos\alpha)^2 = a^2(\sin\alpha\cos\beta - \sin\beta\cos\alpha)^2$

(d) Expanding the left hand side, we have:
$x^2(\sin^2\beta + \cos^2\beta) + y^2(\sin^2\alpha + \cos^2\alpha) - 2xy(\sin\alpha\sin\beta + \cos\alpha\cos\beta)$
$= x^2 + y^2 - 2xy(\sin\alpha\sin\beta + \cos\alpha\cos\beta) = x^2 + y^2 - 2xy\cos(\alpha - \beta)$.
The right hand side can be rewritten: $a^2(\sin\alpha\cos\beta - \sin\beta\cos\alpha)^2 = a^2\sin^2(\alpha - \beta)$.
Therefore: $x^2 + y^2 - 2xy\cos(\alpha - \beta) = a^2\sin^2(\alpha - \beta)$.
Or: $x^2 + y^2 - 2xy\cos\delta = a^2\sin^2\delta$, where $\delta = \alpha - \beta$.

(e) $\delta = 0$: $x^2 + y^2 - 2xy = (x - y)^2 = 0 \Rightarrow x = y$, which is a straight diagonal line.

$\delta = \frac{\pi}{4}$: $x^2 + y^2 - \sqrt{2}xy = \frac{a^2}{2}$, which is an ellipse.

$\delta = \frac{\pi}{2}$: $x^2 + y^2 = a^2$, which is a circle.

This pattern repeats for the remaining phase differences.

34-52: (a) By the symmetry of the triangles, $\theta_b^A = \theta_a^B$, and $\theta_a^C = \theta_r^B = \theta_a^B = \theta_b^A$.
Therefore, $\sin\theta_b^C = n\sin\theta_a^C = n\sin\theta_b^A = \sin\theta_a^A \Rightarrow \theta_b^C = \theta_a^A$.

(b) The total angular deflection of the ray is:
$\Delta = \theta_a^A - \theta_b^A + \pi - 2\theta_a^B + \theta_b^C - \theta_a^C = 2\theta_a^A - 4\theta_b^A + \pi$.

(c) From Snell's Law, $\sin\theta_a^A = n\sin\theta_b^A \Rightarrow \theta_b^A = \arcsin\left(\frac{1}{n}\sin\theta_a^A\right)$

$\Rightarrow \Delta = 2\theta_a^A - 4\theta_b^A + \pi = 2\theta_a^A - 4\arcsin\left(\frac{1}{n}\sin\theta_a^A\right) + \pi$.

(d) $\frac{d\Delta}{d\theta_a^A} = 0 = 2 - 4\frac{d}{d\theta_a^A}\left(\arcsin\left(\frac{1}{n}\sin\theta_a^A\right)\right) \Rightarrow 0 = 2 - \frac{4}{\sqrt{1 - \sin^2\theta_1 / n^2}}\cdot\left(\frac{\cos\theta_1}{n}\right)$

$\Rightarrow 4\left(1 - \frac{\sin^2\theta_1}{n^2}\right) = \left(\frac{16\cos^2\theta_1}{n^2}\right) \Rightarrow 4\cos^2\theta_1 = n^2 - 1 + \cos^2\theta_1$

$\Rightarrow 3\cos^2\theta_1 = n^2 - 1 \Rightarrow \cos^2\theta_1 = \frac{1}{3}(n^2 - 1)$.

(e) For violet: $\theta_1 = \arccos\left(\sqrt{\frac{1}{3}(n^2 - 1)}\right) = \arccos\left(\sqrt{\frac{1}{3}(1.342^2 - 1)}\right) = 58.89°$

$\Rightarrow \Delta_{violet} = 139.2° \Rightarrow \theta_{violet} = 40.8°$.

For red: $\theta_1 = \arccos\left(\sqrt{\frac{1}{3}(n^2-1)}\right) = \arccos\left(\sqrt{\frac{1}{3}(1.330^2-1)}\right) = 59.58°$

$\Rightarrow \Delta_{red} = 137.5° \Rightarrow \theta_{red} = 42.5°$.

Therefore the color which appears higher is red.

34-53: (a) For the secondary rainbow, we will follow similar steps to **CP. (34-51)**. The total angular deflection of the ray is:

$\Delta = \theta_a^A - \theta_b^A + \pi - 2\theta_b^A + \pi - 2\theta_b^A + \theta_a^A - \theta_b^A = 2\theta_a^A - 6\theta_b^A + 2\pi$, where we have used the fact from the previous problem that all the internal angles are equal and the two external equals are equal. Also using the Snell's Law relationship, we have: $\theta_b^A = \arcsin\left(\frac{1}{n}\sin\theta_a^A\right)$.

$\Rightarrow \Delta = 2\theta_a^A - 6\theta_b^A + 2\pi = 2\theta_a^A - 6\arcsin\left(\frac{1}{n}\sin\theta_a^A\right) + 2\pi$.

(b) $\frac{d\Delta}{d\theta_a^A} = 0 = 2 - 6\frac{d}{d\theta_a^A}\left(\arcsin\left(\frac{1}{n}\sin\theta_a^A\right)\right) \Rightarrow 0 = 2 - \frac{6}{\sqrt{1-\sin^2\theta_2/n^2}}\cdot\left(\frac{\cos\theta_2}{n}\right)$

$\Rightarrow n^2\left(1-\sin^2\theta_2/n^2\right) = \left(n^2 - 1 + \cos^2\theta_2\right) = 9\cos^2\theta_2 \Rightarrow \cos^2\theta_2 = \frac{1}{8}(n^2-1)$.

(c) For violet: $\theta_2 = \arccos\left(\sqrt{\frac{1}{8}(n^2-1)}\right) = \arccos\left(\sqrt{\frac{1}{8}(1.342^2-1)}\right) = 71.55°$

$\Rightarrow \Delta_{violet} = 233.2° \Rightarrow \theta_{violet} = 53.2°$.

For red: $\theta_2 = \arccos\left(\sqrt{\frac{1}{8}(n^2-1)}\right) = \arccos\left(\sqrt{\frac{1}{8}(1.330^2-1)}\right) = 71.94°$

$\Rightarrow \Delta_{red} = 230.1° \Rightarrow \theta_{red} = 50.1°$.

Therefore the color which appears higher is violet.

Chapter 35: Geometric Optics

35-1: A mirror does not change the height of the object in the image, nor does the distance from the mirror change. So, the image is 43.6 cm to the right of the mirror, and its height is 3.45 cm.

35-2: Using similar triangles,

$$\frac{h_{tree}}{h_{mirror}} = \frac{d_{tree}}{d_{mirror}} \Rightarrow h_{tree} = h_{mirror}\frac{d_{tree}}{d_{mirror}} = 5.00\text{ cm}\frac{36.0\text{ m} + 0.300\text{ m}}{0.300\text{ m}} = 6.05\text{ m}.$$

35-3: If up is the +y-direction and right is the +x-direction, then the object is at $(-x_o, -y_o)$, P_2' is at $(x_o, -y_o)$, and mirror 1 flips the y-values, so the image is at (x_o, y_o) which is P_3'.

35-4: (a) $m = -\frac{s'}{s} = -\frac{-35.0}{10.0} = 3.50$, where s' comes from part (b).

(b) $\frac{1}{s} + \frac{1}{s'} = \frac{1}{f} \Rightarrow \frac{1}{s'} = \frac{2}{28.0\text{ cm}} - \frac{1}{10.0\text{ cm}} \Rightarrow s' = -35.0\text{ cm}$. Since s' is negative, the image is virtual.

(c)

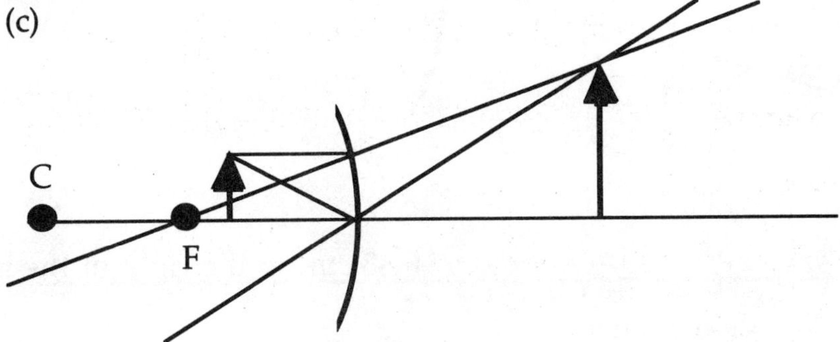

35-5: $\frac{1}{s} + \frac{1}{s'} = \frac{1}{f} \Rightarrow \frac{1}{s'} = \frac{1}{1.85\text{ m}} - \frac{1}{3.86 \times 10^{8}\text{ m}} \Rightarrow s' = -1.85\text{ m}$.

$\Rightarrow m = -\frac{1.85}{386{,}000} = -4.79 \times 10^{-9} \Rightarrow y' = my = (-4.79 \times 10^{-9})(3480\text{ km}) = 0.0167\text{ m}$.

35-6: (a) $f = \frac{R}{2} = \frac{21.0\text{ cm}}{2} = 10.5\text{ cm}$.

(b) If the spherical mirror is immersed in water, its focal length is unchanged - it just depends upon the physical geometry of the mirror.

35-7: (b) $\frac{1}{s} + \frac{1}{s'} = \frac{1}{f} \Rightarrow \frac{1}{s'} = \frac{2}{20\text{ cm}} - \frac{1}{18\text{ cm}} \Rightarrow s' = 22.5\text{ cm}$, to the left of the mirror.

$y' = -y\frac{s'}{s} = -(0.40\text{ cm})\frac{22.5\text{ cm}}{18.0\text{ cm}} = -0.500\text{ cm}$, and the image is inverted and real.

35-7: (a)

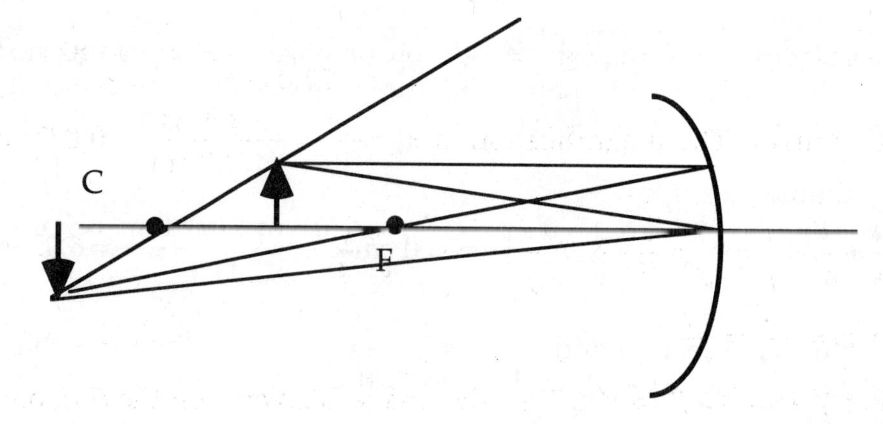

35-8: (a)

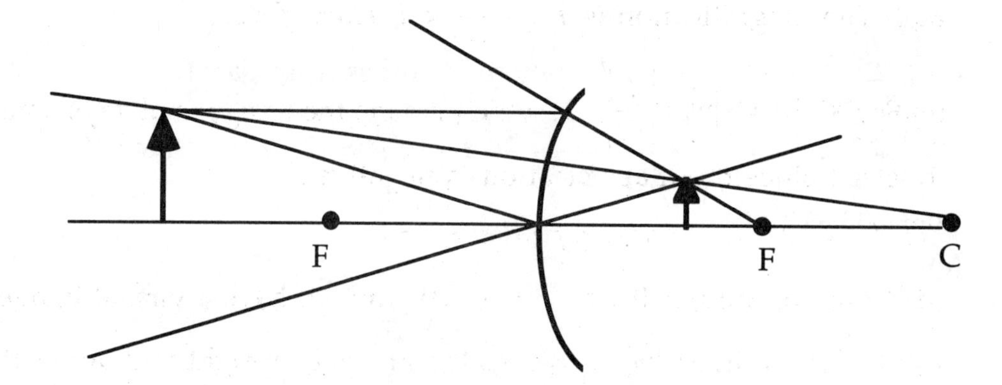

(b) $\frac{1}{s}+\frac{1}{s'}=\frac{1}{f}\Rightarrow\frac{1}{s'}=-\frac{2}{20\text{ cm}}-\frac{1}{18\text{ cm}}\Rightarrow s'=-6.43\text{ cm}$, to the right of the mirror.

$y'=-y\frac{s'}{s}=-(0.400\text{ cm})\frac{-6.43\text{ cm}}{18.0\text{ cm}}=0.143\text{ cm}$, and the image is upright and virtual.

35-9: (a)

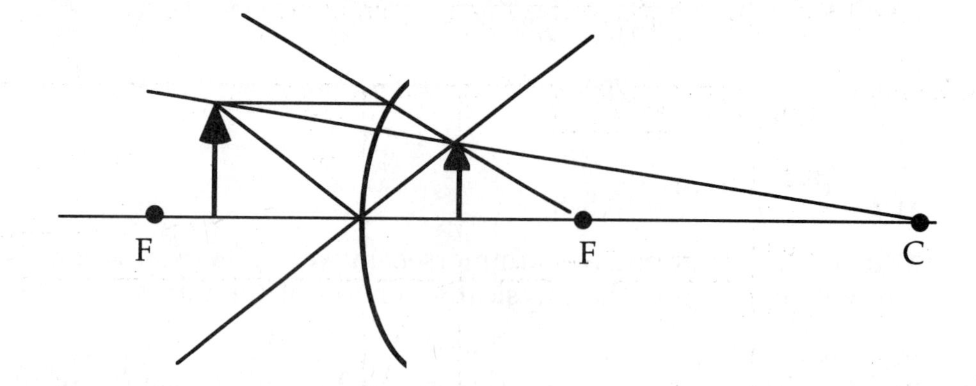

(b) $\frac{1}{s}+\frac{1}{s'}=\frac{1}{f}\Rightarrow\frac{1}{s'}=-\frac{2}{24\text{ cm}}-\frac{1}{9.0\text{ cm}}\Rightarrow s'=-5.14\text{ cm}$, to the right of the mirror.

$y'=-y\frac{s'}{s}=-(1.50\text{ cm})\frac{-5.14\text{ cm}}{9.0\text{ cm}}=0.857\text{ cm}$, and the image is upright and virtual.

35-10: $R = -4.00\text{ cm}$, $\frac{1}{s}+\frac{1}{s'}=\frac{1}{f} \Rightarrow \frac{1}{s'} = -\frac{2}{4.00\text{ cm}} - \frac{1}{21.0\text{ cm}} \Rightarrow s' = -1.83\text{ cm}$, to the left of the mirror. The magnification is $m = -\frac{s'}{s} = -\frac{-1.83\text{ cm}}{21.0\text{ cm}} = 0.0871\text{ cm}$.

35-11: (a) $\frac{1}{s}+\frac{1}{s'}=\frac{1}{f} \Rightarrow \frac{1}{s'}=\frac{1}{f}-\frac{1}{s}=\frac{s-f}{fs} \Rightarrow s'=\frac{sf}{s-f}$.

Also $m = -\frac{s'}{s} = \frac{f}{f-s}$.

(b) For $f>0,\ \ s>f \Rightarrow s'>0$, so the image is always on the outgoing side and is real. The magnification is $m=\frac{f}{f-s}<0$, since $f<s$.

(c) For $s \geq 2f \Rightarrow |m| < \left|\frac{f}{-f}\right| = 1$, which means the image is always smaller and inverted since the magnification is negative..

For $f<s<2f \Rightarrow 0<s-f<f \Rightarrow |m| > \frac{f}{f} = 1$.

(d) Concave mirror: $0<s<f \Rightarrow s'<0$, and we have a virtual image to the right of the mirror. $|m| > \frac{f}{f} = 1$, so the image is upright and larger than the object.

35-12: For a convex mirror, $f<0 \Rightarrow s' = \frac{sf}{s-f} = -\frac{s|f|}{s+|f|} < 0$. Therefore the image is always virtual. Also $m = \frac{f}{f-s} = \frac{-|f|}{-|f|-s} = \frac{|f|}{|f|+s} > 0$, so the image is erect, and $m<1$ since $|f|+s>|f|$, so the image is smaller.

35-13: (a)

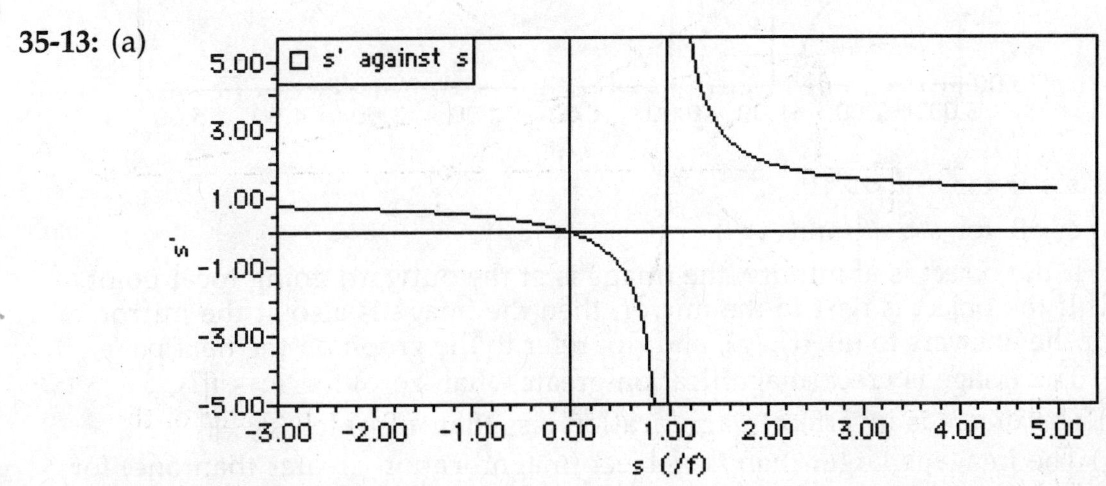

(b) $s'>0$ for $s>f,\ s<0$.

(c) $s'<0$ for $0<s<f$.

(d) If the object is just outside the focal point, then the image position approaches positive infinity.

(e) If the object is just inside the focal point, the image is at negative infinity, "behind" the mirror.
(f) If the object is at infinity, then the image is at the focal point.
(g) If the object is next to the mirror, then the image is also at the mirror.
(h)

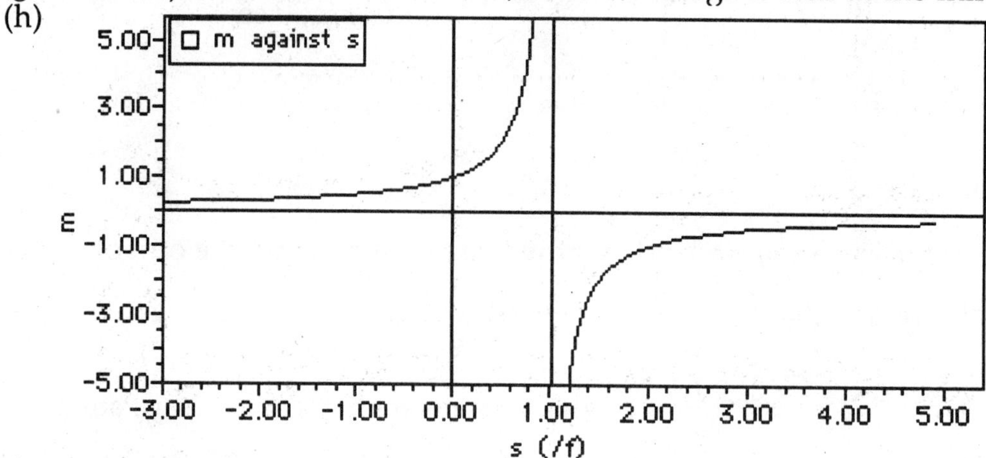

(i) The image is erect if $s < f$.
(j) The image is inverted if $s > f$
(k) The image is larger if $0 < s < 2f$.
(l) The image is smaller if $s > 2f$ or $s < 0$
(e) As the object is moved closer and closer to the focal point, the magnification INCREASES to infinite values

35-14: (a)

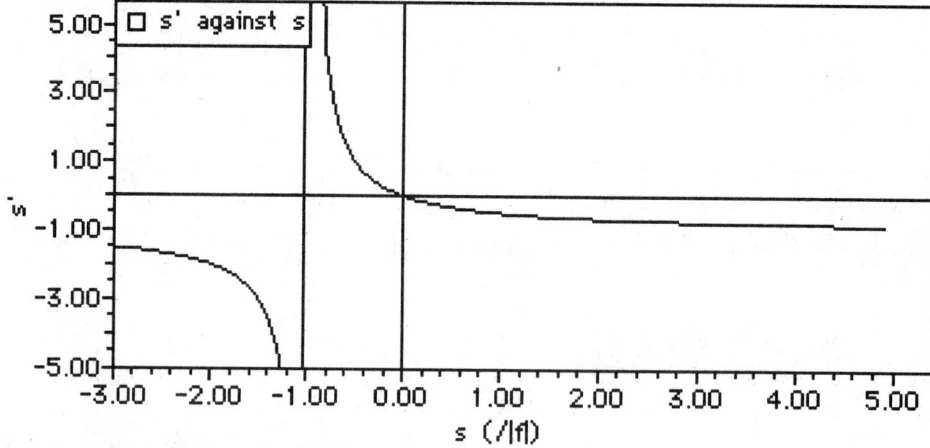

(a) $s' > 0$ for $-|f| < s < 0$.
(b) $s' < 0$ for $s < -|f|$ and $s > 0$.
(c) If the object is at infinity, the image is at the outward going focal point.
(d) If the object is next to the mirror, then the image is also at the mirror.
For the answers to (e), (f), (g), and (h), refer to the graph on the next page.
(e) The image is erect (magnification greater than zero) for $s > -|f|$.
(f) The image is inverted (magnification less than zero) for $s < -|f|$.
(g) The image is larger than the object (magnification greater than one) for $-2|f| < s < 0$.
(h) The image is smaller than the object (magnification less than one) for $s > 0$ and $s < -2|f|$.

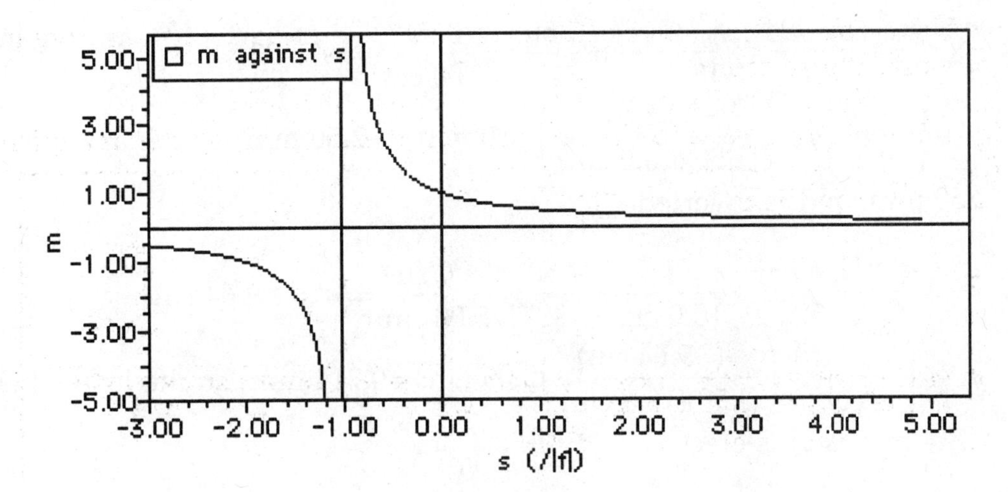

35-15: (a) For $R > 0$ and $n_a > n_b$, with $\theta_a = \alpha + \phi$ and $\theta_b = \phi + \beta$, we have:

$$n_b\theta_b = n_a\theta_a \Rightarrow \theta_b = \phi + \beta = \frac{n_a}{n_b}(\alpha + \phi) \Rightarrow n_a\alpha - n_b\beta = (n_b - n_a)\phi.$$

But $\alpha = \frac{h}{s}$, $\beta = \frac{h}{-s'}$, and $\phi = \frac{h}{R}$, so subbing them in one finds:

$$\frac{n_a}{s} + \frac{n_b}{s'} = \frac{(n_b - n_a)}{R}.$$

Also, the magnification calculation yields:

$$\tan\theta_a \frac{y}{s'} \text{ and } \tan\theta_b \frac{y'}{-s'} \Rightarrow \frac{n_a y}{s'} = \frac{n_b y'}{s'} \Rightarrow m = \frac{y'}{y} = -\frac{n_a s'}{n_b s}.$$

(b) For $R < 0$ and $n_a < n_b$, with $\theta_a = \alpha - \phi$ and $\theta_b = \beta - \phi$, we have:

$$n_b\beta - n_a\alpha = (n_b - n_a)\phi.$$

But $\alpha = \frac{h}{s}$, $\beta = \frac{h}{-s}$, and $\phi = \frac{h}{-R} \Rightarrow -\frac{n_a}{s} - \frac{n_b}{s'} = -\frac{(n_b - n_a)}{R}$, so subbing them in one finds: $\frac{n_a}{s} + \frac{n_b}{s'} = \frac{(n_b - n_a)}{R}$.

Also, the magnification calculation yields:

$$n_a \tan\theta_a \approx n_b \tan\theta_b \Rightarrow \frac{n_a y}{s} = -\frac{n_b y'}{s'} \Rightarrow m = \frac{y'}{y} = -\frac{n_a s'}{n_b s}.$$

35-16: (a) $\frac{n_a}{s} + \frac{n_b}{s'} = \frac{n_b - n_a}{R} \Rightarrow \frac{1}{\infty} + \frac{1.50}{s'} = \frac{0.50}{3.50 \text{ cm}} \Rightarrow s' = 10.5 \text{ cm}.$

(b) $\frac{n_a}{s} + \frac{n_b}{s'} = \frac{n_b - n_a}{R} \Rightarrow \frac{1}{16.0 \text{ cm}} + \frac{1.50}{s'} = \frac{0.50}{3.50 \text{ cm}} \Rightarrow s' = 18.7 \text{ cm}.$

(c) $\frac{n_a}{s} + \frac{n_b}{s'} = \frac{n_b - n_a}{R} \Rightarrow \frac{1}{4.00 \text{ cm}} + \frac{1.50}{s'} = \frac{0.50}{3.50 \text{ cm}} \Rightarrow s' = -14.0 \text{ cm}.$

35-17: $\frac{n_a}{s} + \frac{n_b}{s'} = \frac{n_b - n_a}{R} \Rightarrow \frac{n_a}{60.0 \text{ cm}} + \frac{1.50}{90.0 \text{ cm}} = \frac{1.50 - n_a}{3.50 \text{ cm}}$

$$\Rightarrow n_a\left(\frac{1}{60.0 \text{ cm}} + \frac{1}{3.50 \text{ cm}}\right) = \frac{-1.50}{90.0 \text{ cm}} + \frac{1.50}{3.50 \text{ cm}} = 0.412$$

$$\Rightarrow n_a = 1.36.$$

35-18: $\frac{n_a}{s}+\frac{n_b}{s'}=\frac{n_b-n_a}{R}\Rightarrow\frac{1}{18.0\text{ cm}}+\frac{1.50}{s'}=\frac{0.50}{5.00\text{ cm}}\Rightarrow s'=33.8\text{ cm}.$

$y'=\left(\frac{-n_a s'}{n_b s}\right)y=\left(\frac{-33.8\text{ cm}}{(1.50)(18.0\text{ cm})}\right)2.00\text{ mm}=-2.50\text{ mm}$, so the image height is 2.50 mm, and is inverted.

35-19: $\frac{n_a}{s}+\frac{n_b}{s'}=\frac{n_b-n_a}{R}\Rightarrow\frac{1}{18.0\text{ cm}}+\frac{1.50}{s'}=\frac{-0.50}{5.00\text{ cm}}\Rightarrow s'=-9.64\text{ cm}$

$y'=\left(\frac{-n_a s'}{n_b s}\right)y=\left(\frac{-(-9.64\text{ cm})}{(1.50)(18.0\text{ cm})}\right)2.00\text{ mm}=0.714\text{ mm}$, so the image height is 0.714 mm, and is erect.

35-20: (a) $\frac{n_a}{s}+\frac{n_b}{s'}=\frac{n_b-n_a}{R}\Rightarrow\frac{1.33}{17.0\text{ cm}}+\frac{1.00}{s'}=\frac{-0.33}{-17.0\text{ cm}}\Rightarrow s'=-17.0\text{ cm}$, so the fish appears to be at the center of the bowl.

$m=\left(\frac{-n_a s'}{n_b s}\right)=\left(\frac{-(1.33)(-17.0\text{ cm})}{(1.00)(17.0\text{ cm})}\right)=+1.33.$

(b) $\frac{n_a}{s}+\frac{n_b}{s'}=\frac{n_b-n_a}{R}\Rightarrow\frac{1.00}{\infty}+\frac{1.33}{s'}=\frac{0.33}{+17.0\text{ cm}}\Rightarrow s'=+68.5\text{ cm}$, which is outside the bowl.

35-21: $\frac{n_a}{s}+\frac{n_b}{s'}=0\Rightarrow\frac{1.309}{2.10\text{ cm}}+\frac{1.00}{s'}=0\Rightarrow s'=-1.60\text{ cm}.$

35-22: (a) $\frac{n_a}{s}+\frac{n_b}{s'}=0\Rightarrow\frac{1.33}{9.00\text{ cm}}+\frac{1.00}{s'}=0\Rightarrow s'=-6.77\text{ cm}$, so the fish appears 6.77 cm below the surface.

(b) $\frac{n_a}{s}+\frac{n_b}{s'}=0\Rightarrow\frac{1.33}{39.0\text{ cm}}+\frac{1.00}{s'}=0\Rightarrow s'=-29.3\text{ cm}$, so the image of the fish appears 29.3 cm below the surface.

35-23: (a) The lens equation is the same for both thin lenses and spherical mirrors, so the derivation of the equations in **Ex. (35-11)** is identical and one gets:

$\frac{1}{s}+\frac{1}{s'}=\frac{1}{f}\Rightarrow\frac{1}{s'}=\frac{1}{f}-\frac{1}{s}=\frac{s-f}{fs}\Rightarrow s'=\frac{sf}{s-f}$, and also $m=-\frac{s'}{s}=\frac{f}{f-s}$.

(b) Again, one gets exactly the same equations for a converging lens rather than a concave mirror because the equations are identical. The difference lies in the interpretation of the results. For a lens, the outgoing side is *not* that on which the object lies, unlike for a mirror. So for an object on the left side of the lens, a positive image distance means that the image is on the right of the lens, and a negative image distance means that the image is on the left side of the lens.

(c) Again, for **Ex. (35-12)** and **(35-14)**, the change from a convex mirror to a diverging lens changes nothing in the exercises, except for the interpretation of the location of the images, as explained in part (b) above.

35-24: $\dfrac{1}{s}+\dfrac{1}{s'}=\dfrac{1}{f}\Rightarrow\dfrac{1}{s}=\dfrac{1}{10.0\text{ cm}}-\dfrac{1}{14.0\text{ cm}}\Rightarrow s=35.0\text{ cm}.$

$m=-\dfrac{s'}{s}=-\dfrac{14.0}{35.0}=-0.400\Rightarrow y=\dfrac{y'}{m}=\dfrac{1.00\text{ cm}}{-0.400}=-2.50\text{ cm}$, so the object is 2.50 cm tall.

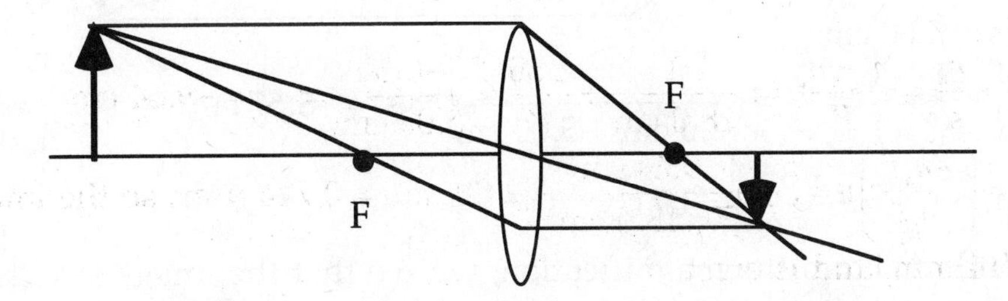

35-25: Note that a diverging lens cannot form a real image as stated in **Ex. (35-24).**

$\dfrac{1}{s}+\dfrac{1}{s'}=\dfrac{1}{f}\Rightarrow\dfrac{1}{s}=\dfrac{1}{-10.0\text{ cm}}-\dfrac{1}{14.0\text{ cm}}\Rightarrow s=-5.83\text{ cm}.$

$m=-\dfrac{s'}{s}=-\dfrac{14.0}{-5.83}=+2.40\Rightarrow y=\dfrac{y'}{m}=\dfrac{1.00\text{ cm}}{2.40}=0.417\text{ cm}.$

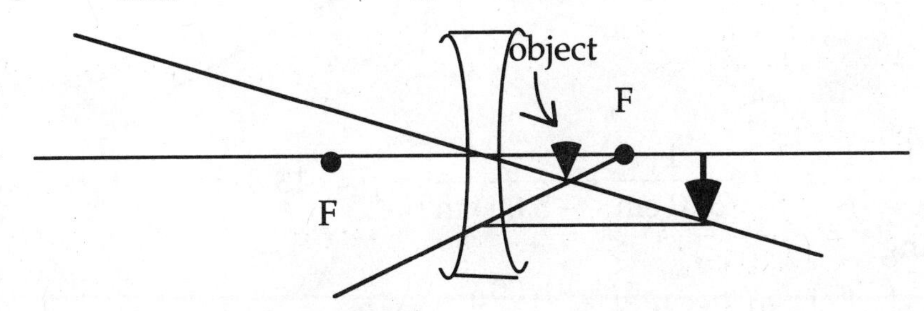

35-26: (a) $\dfrac{1}{s}+\dfrac{1}{s'}=\dfrac{1}{f}\Rightarrow\dfrac{1}{f}=\dfrac{1}{12.0\text{ cm}}+\dfrac{1}{25.0\text{ cm}}\Rightarrow f=8.11\text{ cm}$, converging.

(b) $y=y'\left(-\dfrac{s'}{s}\right)=(0.50\text{ cm})\left(-\dfrac{25.0}{12.0}\right)=-1.04\text{ cm}$, so the image is inverted.

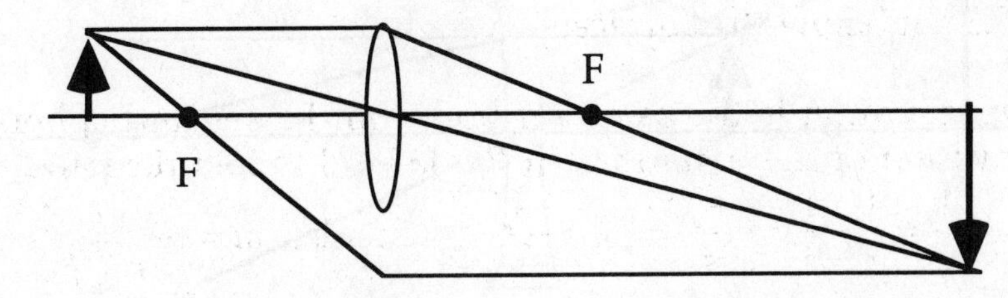

(c)

35-27: For s= 20 cm:

(a) $\dfrac{1}{s}+\dfrac{1}{s'}=\dfrac{1}{f}\Rightarrow\dfrac{1}{s'}=\dfrac{1}{12.0\text{ cm}}-\dfrac{1}{20.0\text{ cm}}\Rightarrow s'=30.0\text{ cm}.$

(b) $m=-\dfrac{s'}{s}=-\dfrac{30.0}{20.0}=-1.50.$

(c) and (d) From the magnification, we see that the image is real and inverted.

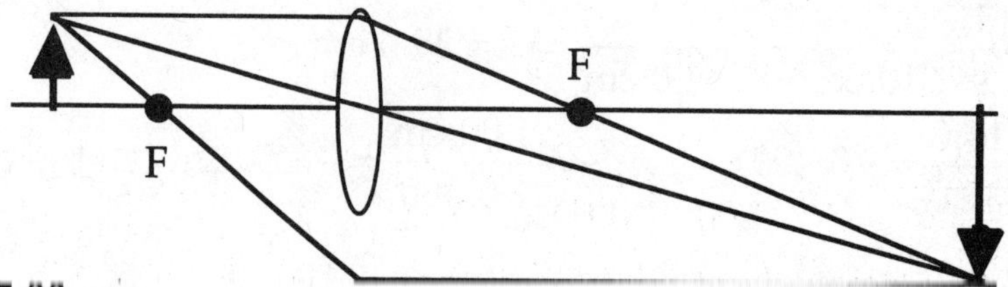

For s = 5.00 cm:

(a) $\frac{1}{s}+\frac{1}{s'}=\frac{1}{f}\Rightarrow\frac{1}{s'}=\frac{1}{12.0\text{ cm}}-\frac{1}{5.00\text{ cm}}\Rightarrow s'=-8.57\text{ cm}.$

(b) $m=-\frac{s'}{s}=-\frac{-8.57}{5.00}=1.71.$

(c) and (d) From the magnification, we see that the image is virtual and erect.

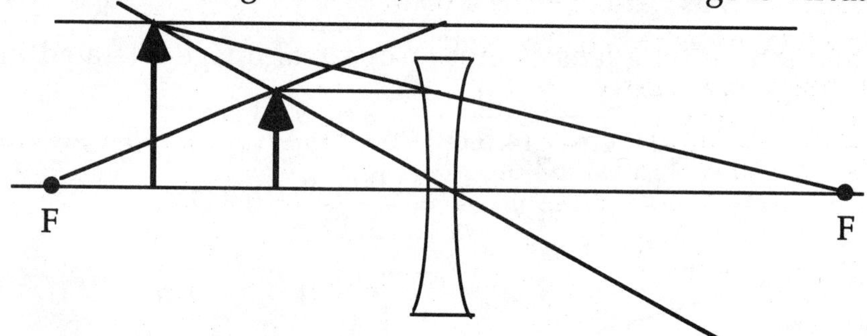

35-28: (a) $\frac{1}{s}+\frac{1}{s'}=\frac{1}{f}\Rightarrow\frac{1}{f}=\frac{1}{20.0\text{ cm}}+\frac{1}{-8.00\text{ cm}}\Rightarrow f=-13.3\text{ cm}$, and the lens is diverging.

(b) $y'=y\left(-\frac{s'}{s}\right)=(0.400\text{ cm})\left(-\frac{-8.00}{20.0}\right)=0.160\text{ cm}$, and is erect.

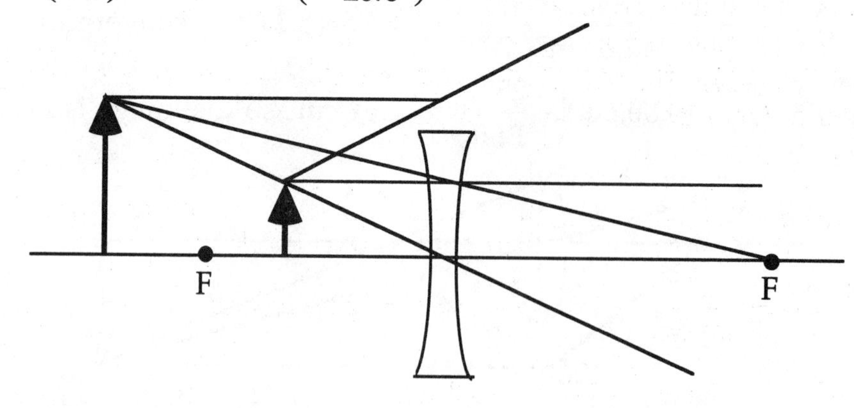

(c)

35-29: $m=\frac{y'}{y}=\frac{35.0}{5.00}=7.00=-\frac{s'}{s}$. Also:

$\frac{1}{s}+\frac{1}{s'}=\frac{1}{f}\Rightarrow\frac{1}{s}=\frac{1}{6.00\text{ cm}}-\frac{1}{s'}\Rightarrow\frac{s'}{s}=\frac{s'}{6.00\text{ cm}}-1=-7.00\Rightarrow s'=-36.0\text{ cm}$, (to the left).

$\Rightarrow\frac{1}{s}=\frac{1}{6.00\text{ cm}}-\frac{1}{-36.0\text{ cm}}\Rightarrow s=5.14\text{ cm}$, and the image is virtual (since s'<0).

35-30: $m = \frac{y'}{y} = -\frac{5.00}{40.0} = -\frac{s'}{s} \Rightarrow s = 8.00\,s'$. Also:

$$\frac{1}{s} + \frac{1}{s'} = \frac{1}{f} \Rightarrow \frac{1}{8.00\,s'} + \frac{1}{s'} = \frac{1}{7.00} \Rightarrow \frac{9.00}{8.00\,s'} = \frac{1}{7.00} \Rightarrow s' = 7.88 \text{ cm, (to the right).}$$

$\Rightarrow s = 8.00(7.88 \text{ cm}) = 63.0 \text{ cm}$, and the image is real (since s'>0).

35-31: $$\frac{1}{s} + \frac{1}{s'} = (n-1)\left(\frac{1}{R_1} - \frac{1}{R_2}\right) \Rightarrow \frac{1}{18.0 \text{ cm}} + \frac{1}{s'} = (0.48)\left(\frac{1}{4.00 \text{ cm}} - \frac{1}{2.50 \text{ cm}}\right)$$

$\Rightarrow s' = -7.84$ cm, (to the left of the lens).

35-32: (a) Given $s' = 3.75s$, and $s + s' = 3.00 \text{ m}$

$\Rightarrow 4.75s = 3.00 \text{ m} \Rightarrow s = 0.632 \text{ m}$ and $s' = 2.37 \text{ m}$.

(b) The image is inverted.

(c) $\frac{1}{f} = \frac{1}{s} + \frac{1}{s'} = \frac{1}{0.632 \text{ m}} + \frac{1}{2.37 \text{ m}} \Rightarrow f = 0.499 \text{ m}$, and the lens is converging.

35-33: (a) For the first lens: $\frac{1}{s} + \frac{1}{s'} = \frac{1}{f} \Rightarrow \frac{1}{18.0 \text{ cm}} + \frac{1}{s'} = \frac{1}{10.0 \text{ cm}} \Rightarrow s' = 22.5 \text{ cm}$.

So $m_1 = -\frac{22.5}{18.0} = -1.25$.

For the second lens: $s = 8.00 \text{ cm} - 22.5 \text{ cm} = -14.5 \text{ cm}$.

$$\frac{1}{s} + \frac{1}{s'} = \frac{1}{f} \Rightarrow \frac{1}{-14.5 \text{ cm}} + \frac{1}{s'} = -\frac{1}{10.0 \text{ cm}} \Rightarrow s' = -32.2 \text{ cm},\ m_2 = -\frac{-32.2}{-14.5} = -2.22.$$

So the image is 32.2 cm to the left of the second lens, and is therefore 24.2 cm to the left of the first lens.

(b) The final image is virtual.

(c) Since the magnification is $m = m_1 m_2 = (-1.25)(-2.22) = 2.78$, the final image is erect and has a height $y' = (2.78)(2.00 \text{ mm}) = 5.55 \text{ mm}$.

35-34: (a) $\frac{1}{f} = (n-1)\left(\frac{1}{R_1} - \frac{1}{R_2}\right) = (0.50)\left(\frac{1}{10.0 \text{ cm}} - \frac{1}{30.0 \text{ cm}}\right) \Rightarrow f = 30 \text{ cm}$.

$\Rightarrow \frac{1}{s} + \frac{1}{s'} = \frac{1}{f} \Rightarrow \frac{1}{40 \text{ cm}} + \frac{1}{s'} = \frac{1}{30 \text{ cm}} \Rightarrow s' = 120 \text{ cm}$, and

$y' = y\left(-\frac{s'}{s}\right) = (1.0 \text{ cm})\left(-\frac{120}{40}\right) = -3.0 \text{ cm}$.

(b) Adding a second identical lens 160 cm to the right of the first means that the first lens's image becomes an object for the second, a distance of 40 cm from that second lens. Note also that the magnification factor is also the same.

$\Rightarrow \frac{1}{s} + \frac{1}{s'} = \frac{1}{f} \Rightarrow \frac{1}{40 \text{ cm}} + \frac{1}{s'} = \frac{1}{30 \text{ cm}} \Rightarrow s' = 120 \text{ cm},\ y' = (-3.0 \text{ cm})(-3.0) = 9.0 \text{ cm}$,

and the image is erect.

(c) Putting an identical lens just 40 cm from the first means that the first lens's image becomes an object for the second, a distance of 80 cm to the right of the second lens.

$$\Rightarrow \frac{1}{s}+\frac{1}{s'}=\frac{1}{f} \Rightarrow \frac{1}{-80\text{ cm}}+\frac{1}{s'}=\frac{1}{30\text{ cm}} \Rightarrow s' = 21.8\text{ cm},$$

and $y' = (-3.0\text{ cm})\left(\frac{21.8}{80}\right) = -0.82\text{ cm}$, and the image is inverted.

35-35: $\frac{1}{f} = (n-1)\left(\frac{1}{R_1}-\frac{1}{R_2}\right) = (0.50)\left(\frac{1}{\pm 10.0\text{ cm}}-\frac{1}{\pm 20.0\text{ cm}}\right) \Rightarrow f = \pm 40\text{ cm},\ \pm 13.3\text{ cm}.$

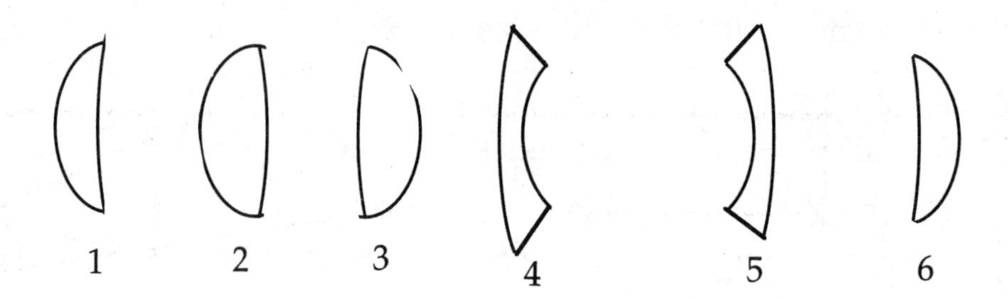

$f_1 = 40\text{ cm}; f_2 = 13.3\text{ cm}; f_3 = 13.3\text{ cm}; f_4 = -40\text{ cm}; f_5 = -40\text{ cm}; f_6 = 40\text{ cm}.$

35-36: We have a converging lens if the focal length is positive, which requires:
$\frac{1}{f} = (n-1)\left(\frac{1}{R_1}-\frac{1}{R_2}\right) > 0 \Rightarrow \left(\frac{1}{R_1}-\frac{1}{R_2}\right) > 0$. This can occur in one of three ways:
(i) $\{R_1 < R_2\} \cup \{R_1, R_2 > 0\}$ (ii) $R_1 > 0, R_2 < 0$ (iii) $\{|R_1| > |R_2|\} \cup \{R_1, R_2 < 0\}$.
Hence the three lenses in Fig. (35-29a).
We have a diverging lens if the focal length is negative, which requires:
$\frac{1}{f} = (n-1)\left(\frac{1}{R_1}-\frac{1}{R_2}\right) < 0 \Rightarrow \left(\frac{1}{R_1}-\frac{1}{R_2}\right) < 0$. This can occur in one of three ways:
(i) $\{R_1 > R_2\} \cup \{R_1, R_2 > 0\}$ (ii) $R_1 > R_2 > 0$ (iii) $R_1 < 0, R_2 > 0$.
Hence the three lenses in Fig. (35-29b).

35-37: If you move to the mirror at 2.25 m/s, then your image moves toward the mirror at the same speed, but in the opposite direction. Therefore you see the image approaching YOU at 4.50 m/s, the sum of your speed and that of the image in the mirror.

35-38: The minimum length mirror for a man to see his full height h in, is h/2, as shown at right.

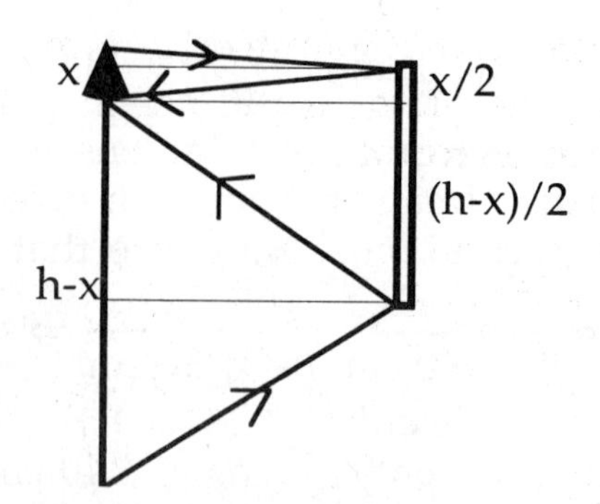

35-39: (a) $|m| = \frac{y'}{y} = \frac{350}{5.00} = 70.0 = \frac{s'}{s} \Rightarrow s = \frac{3.20 \text{ m}}{70.0} = 0.0457$ m is where the filament should be placed.

(b) $\frac{1}{s} + \frac{1}{s'} = \frac{2}{R} \Rightarrow \frac{2}{R} = \frac{1}{0.0457 \text{ m}} + \frac{1}{3.20 \text{ m}} \Rightarrow R = 0.0901 \text{ m}.$

35-40: (a) There are three image formed.

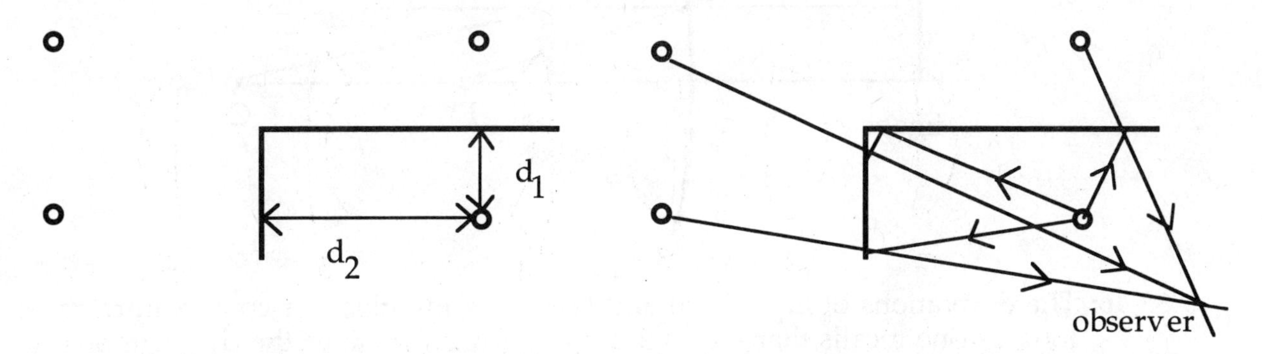

35-41: $|m| = 2.75 = \frac{s'}{s} = \frac{s + 5.00 \text{ m}}{s} \Rightarrow 1.75s = 5.00 \text{ m} \Rightarrow s = 2.86 \text{ m}.$

So the mirror is 7.86 m from the wall. Also:

$\frac{1}{s} + \frac{1}{s'} = \frac{2}{R} \Rightarrow \frac{2}{R} = \frac{1}{2.86 \text{ m}} + \frac{1}{7.86 \text{ m}} \Rightarrow R = 4.19 \text{ m}.$

35-42: $|m| = \frac{1}{3} = \frac{s'}{s} \Rightarrow s = 3s'$

$\Rightarrow \frac{1}{3s'} + \frac{1}{s'} = \frac{2}{R} \Rightarrow \frac{4}{3s'} = \frac{2}{R} \Rightarrow s' = \frac{2R}{3} \Rightarrow s = 2R.$

So the image is a distance 2R/3 from the mirror, while the object is a distance 2R from it.

35-43: (a) $\frac{1}{s} + \frac{1}{s'} = \frac{2}{R} \Rightarrow \frac{1}{8.0 \text{ cm}} + \frac{1}{s'} = \frac{2}{19.4 \text{ cm}} \Rightarrow s' = -46 \text{ cm}$, so the image is virtual.

(b) $m = -\frac{s'}{s} = -\frac{-46}{8.0} = 5.8$, so the image is erect, and its height is:

$y' = (5.8)y = (5.8)(5.0 \text{ mm}) = 29 \text{ mm}.$

(c) When the filament is 8 cm from the mirror, there is no place where a real image can be formed.

35-44: $\frac{1}{s} + \frac{1}{s'} = \frac{2}{R} \Rightarrow \frac{1}{11.0 \text{ m}} + \frac{1}{s'} = \frac{2}{-0.200 \text{ m}} \Rightarrow s' = -0.0991 \text{ m}.$

$y' = y\left(-\frac{s'}{s}\right) = (1.50 \text{ m})\left(-\frac{-0.0991}{11.0}\right) = 0.0135 \text{ m}$. So the height of the image is less than 1% of the true height of the car, and is less than the image would appear in a plane mirror at the same location.

35-45: (a) A real image is produced for virtual object positions between the focal point and vertex of the mirror. So for a 14 cm radius mirror, the virtual object positions must be between the vertex and 7.0 cm to the right of the mirror.

(b) The image orientation is erect, since $m = -\frac{s'}{s} = -\frac{s'}{-|s|} > 0$.

(c)

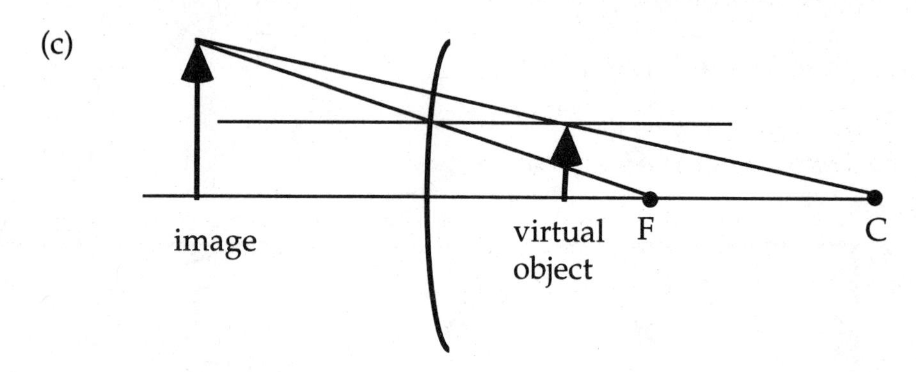

35-46: The derivations of Eqs. (35-6) and (35-7) are identical for convex mirrors, as long as one recalls that R and s' are negative. Consider the diagram below:

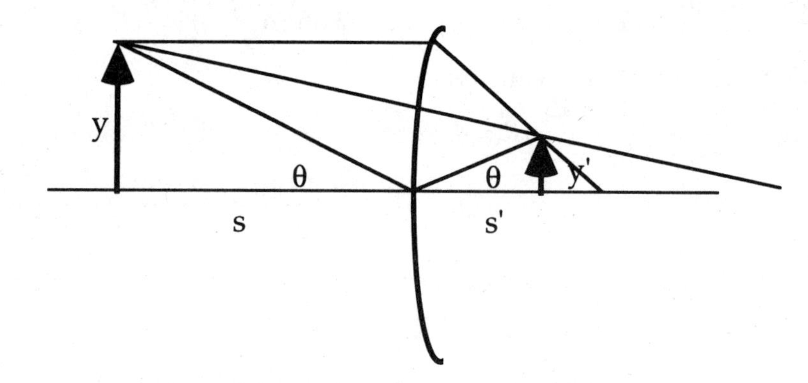

We have: $\frac{1}{\infty} + \frac{1}{s'} = \frac{2}{R} \Rightarrow s' = f = \frac{R}{2} \Rightarrow \frac{1}{s} + \frac{1}{s'} = \frac{2}{R} = \frac{1}{f}$

and $m = \frac{y'}{y} = -\frac{s'}{s}$, since s' is not on the outgoing side of the mirror.

35-47: $\frac{n_a}{s} + \frac{n_b}{s'} = \frac{n_b - n_a}{R}$, but $s = \infty, s' = 2R \Rightarrow \frac{1}{\infty} + \frac{n}{2R} = \frac{n-1}{R} \Rightarrow \frac{n}{2} = 1 \Rightarrow n = 2.00$.

35-48: $\frac{n_a}{s} + \frac{n_b}{s'} = 0 \Rightarrow \frac{n}{s} + \frac{1}{s'} = 0 \Rightarrow n = -\frac{s}{s'} = -\frac{2.00\text{ mm} + 0.70\text{ mm}}{-2.00\text{ mm}} = 1.35$.

35-49: (a) Reflection from the front face of the glass means that the image is just h below the glass surface, like a normal mirror.
(b) The reflection from the mirrored surface behind the glass will not be affected because of the intervening glass. The light travels through a distance

$2d$ of glass, so the path through the glass appears to be $\frac{2d}{n}$, and the image appears to be $h+\frac{2d}{n}$ behind the front surface of the glass.

(c) The distance between the two images is just $\frac{2d}{n}$.

35-50: $\frac{n_a}{s}+\frac{n_b}{s'}=0 \Rightarrow \frac{n}{20.0\text{ cm}}+\frac{1}{-14.0\text{ cm}}=0 \Rightarrow n=\frac{20.0}{14.0}=1.43.$

When viewed from the curved end of the rod:

$\frac{n_a}{s}+\frac{n_b}{s'}=\frac{n_b-n_a}{R} \Rightarrow \frac{n}{s}+\frac{1}{s'}=\frac{1-n}{R} \Rightarrow \frac{1.43}{20.0\text{ cm}}+\frac{1}{s'}=\frac{-0.43}{-8.00\text{ cm}} \Rightarrow s'=-56\text{ cm},$ so the image is 56 cm below the surface of the rod.

35-51: (a) From the diagram:

$\sin\theta=\frac{0.230}{R}=1.50\sin\theta'.$

But $\sin\theta'=\frac{r'}{R}\approx\frac{r}{R}=\frac{0.230}{R(1.50)}$

$\Rightarrow r=\frac{0.230\text{ cm}}{1.50}=0.153.$

So the diameter of the light hitting the surface is 2r = 0.306 cm.

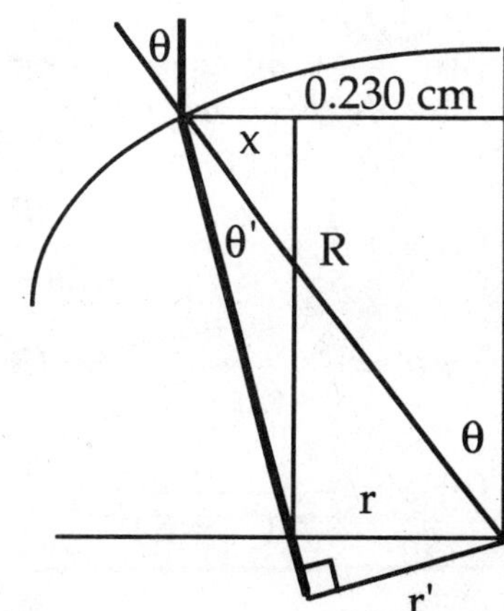

(b) There is no dependence on the radius of the glass sphere in the calculation above.

35-52: (a) $\frac{n_a}{s}+\frac{n_b}{s'}=\frac{n_b-n_a}{R} \Rightarrow \frac{n_a}{f}+\frac{n_b}{\infty}=\frac{n_b-n_a}{R}$ and $\frac{n_a}{\infty}+\frac{n_b}{f'}=\frac{n_b-n_a}{R}.$

$\Rightarrow \frac{n_a}{f}=\frac{n_b-n_a}{R}$ and $\frac{n_b}{f'}=\frac{n_b-n_a}{R} \Rightarrow \frac{n_a}{f}=\frac{n_b}{f'} \Rightarrow n_a=n_b\frac{f}{f'}.$

(b) $\frac{n_a}{s}+\frac{n_b}{s'}=\frac{n_b-n_a}{R} \Rightarrow \frac{n_b f}{sf'}+\frac{n_b}{s'}=\frac{n_b(1-f/f')}{R} \Rightarrow \frac{f}{s}+\frac{f'}{s'}=\frac{f'(1-f/f')}{R}=\frac{f'-f}{R}=1.$

(Note that the first two equations on the second line can be rewritten as $\frac{n_a}{n_b-n_a}=\frac{f}{R}$ and $\frac{n_b}{n_b-n_a}=\frac{f'}{R}$ so we can write $\frac{f'-f}{R}=1.$

35-53: (a) $\frac{1}{s}+\frac{1}{s'}=\frac{1}{f}$ and taking its derivative with respect to s we have:

$0=\frac{d}{ds}\left(\frac{1}{s}+\frac{1}{s'}-\frac{1}{f}\right)=-\frac{1}{s^2}-\frac{1}{s'^2}\frac{ds'}{ds} \Rightarrow \frac{ds'}{ds}=-\frac{s'^2}{s^2}=-m^2.$ But $\frac{ds'}{ds}=m' \Rightarrow m'=-m^2.$

(b) (i) Front face: $\frac{1}{s}+\frac{1}{s'}=\frac{2}{R} \Rightarrow \frac{1}{70.0\text{ cm}}+\frac{1}{s'}=\frac{2}{40.0\text{ cm}} \Rightarrow s'=28.0\text{ cm}.$

Rear face: $\frac{1}{s}+\frac{1}{s'}=\frac{2}{R} \Rightarrow \frac{1}{71.0\text{ cm}}+\frac{1}{s'}=\frac{2}{40.0\text{ cm}} \Rightarrow s'=27.8\text{ cm}.$

(ii) Front face: $m=-\frac{s'}{s}=-\frac{28.0}{70.0}=-0.400,\ m'=-m^2=-(-0.400)^2=-0.160.$

Rear face: $m=-\frac{s'}{s}=-\frac{27.8}{71.0}=-0.392,\ m'=-m^2=-(-0.392)^2=-0.158.$

(iii) So the front legs are magnified by 0.400, the back legs by 0.392, and the side legs by 0.159, the average of the front and back longitudinal magnifications.

35-54: $\frac{n_a}{s}+\frac{n_b}{s'}=\frac{n_b-n_a}{R}$ and taking its derivative with respect to s we have:

$$0=\frac{d}{ds}\left(\frac{n_a}{s}+\frac{n_b}{s'}-\frac{n_b-n_a}{R}\right)=-\frac{n_a}{s^2}-\frac{n_b}{s'^2}\frac{ds'}{ds} \Rightarrow \frac{ds'}{ds}=-\frac{s'^2}{s^2}\frac{n_a}{n_b}=-\left(\frac{s'^2}{s^2}\frac{n_a^2}{n_b^2}\right)\frac{n_b}{n_a}=-m^2\frac{n_b}{n_a}.$$

But $\frac{ds'}{ds}=m' \Rightarrow m'=-m^2\frac{n_b}{n_a}.$

35-55: (a) $\frac{1}{s}+\frac{1}{s'}=\frac{2}{R} \Rightarrow s'=\frac{sR}{2s-R} \Rightarrow v'=\frac{ds'}{dt}=\frac{ds'}{ds}\frac{ds}{dt}=v\left(\frac{R}{2s-R}-\frac{2sR}{(2s-R)^2}\right)$

$$\Rightarrow v'=-v\frac{R^2}{(2s-R)^2}=-(3.00\text{ m/s})\frac{(1.80\text{ m})^2}{(2(10.0\text{ m})-1.80\text{ m})^2}=-0.0205\text{ m/s}.$$

(b) $v'=-v\frac{R^2}{(2s-R)^2}=-(3.00\text{ m/s})\frac{(1.80\text{ m})^2}{(2(2.0\text{ m})-1.80\text{ m})^2}=-0.29\text{ m/s}.$

35-56: (a) The image from the left end acts as the object for the right end of the rod.

(b) $\frac{n_a}{s}+\frac{n_b}{s'}=\frac{n_b-n_a}{R} \Rightarrow \frac{1}{18.0\text{ cm}}+\frac{1.5}{s'}=\frac{0.50}{5.0\text{ cm}} \Rightarrow s'=33.8\text{ cm}.$

So the second object distance is $s_2=50.0\text{ cm}-33.8\text{ cm}=16.2\text{ cm}.$

Also: $m_1=-\frac{n_a s'}{n_b s}=-\frac{33.8}{(1.50)(18.0)}=-1.25.$

(c) The object is real and inverted.

(d) $\frac{n_a}{s_2}+\frac{n_b}{s_2'}=\frac{n_b-n_a}{R} \Rightarrow \frac{1.5}{16.2\text{ cm}}+\frac{1}{s_2'}=\frac{-0.50}{-10.0\text{ cm}} \Rightarrow s'=-23.6\text{ cm}.$

Also: $m_2=-\frac{n_a s'}{n_b s}=-\frac{(1.50)(-23.6)}{16.2}=2.18 \Rightarrow m=m_1 m_2=(-1.25)(2.18)=-2.75.$

So the final image is virtual, and inverted.

35-57: (a) $\frac{n_a}{s}+\frac{n_b}{s'}=\frac{n_b-n_a}{R} \Rightarrow \frac{1}{18.0\text{ cm}}+\frac{1.5}{s'}=\frac{0.50}{5.0\text{ cm}} \Rightarrow s'=33.8\text{ cm}.$

So the second object distance is $s_2=15.0\text{ cm}-33.8\text{ cm}=-18.8\text{ cm}.$

Also: $m_1=-\frac{n_a s'}{n_b s}=-\frac{33.8}{(1.50)(18.0)}=-1.25.$

(b) The object is real.

(c) $\frac{n_a}{s_2}+\frac{n_b}{s_2'}=\frac{n_b-n_a}{R} \Rightarrow \frac{1.5}{-18.8\text{ cm}}+\frac{1}{s_2'}=\frac{-0.50}{-10.0\text{ cm}} \Rightarrow s'=7.69\text{ cm}.$

Also: $m_2 = -\frac{n_a s'}{n_b s} = -\frac{(1.50)(7.69)}{-18.8} = 0.615 \Rightarrow m = m_1 m_2 = (-1.25)(0.615) = -0.769.$

(d) So the final image is real, and inverted.

(e) $y' = ym = (1.00 \text{ mm})(-0.769) = -0.769 \text{ mm}.$

35-58: For the water-benzene interface to get the apparent water depth:

$$\frac{n_a}{s} + \frac{n_b}{s'} = 0 \Rightarrow \frac{1.33}{5.00 \text{ cm}} + \frac{1.5}{s'} = 0 \Rightarrow s' = -5.64 \text{ cm}.$$

For the benzene-air interface, to get the total apparent distance to the bottom:

$$\frac{n_a}{s} + \frac{n_b}{s'} = 0 \Rightarrow \frac{1.50}{(5.64 \text{ cm} + 3.00 \text{ cm})} + \frac{1}{s'} = 0 \Rightarrow s' = -5.76 \text{ cm}.$$

35-59: $\frac{n_a}{s} + \frac{n_b}{s'} = \frac{n_b - n_a}{R} \Rightarrow \frac{1}{\infty} + \frac{1.50}{s'} = \frac{0.50}{3.00 \text{ cm}} \Rightarrow s' = 9.00 \text{ cm}$. So the object distance for the far side of the ball is $6.00 \text{ cm} - 9.00 \text{ cm} = -3.00 \text{ cm}$.

$\Rightarrow \frac{n_a}{s} + \frac{n_b}{s'} = \frac{n_b - n_a}{R} \Rightarrow \frac{1.50}{-3.00 \text{ cm}} + \frac{1}{s'} = \frac{-0.50}{-3.00 \text{ cm}} \Rightarrow s' = 1.50 \text{ cm}$, which is 4.50 cm from the center of the sphere.

35-60: (a) $\frac{n_a}{s} + \frac{n_b}{s'} = \frac{n_b - n_a}{R} \Rightarrow \frac{1}{16.0 \text{ cm}} + \frac{1.50}{s'} = \frac{0.50}{12.0 \text{ cm}} \Rightarrow s' = -72.0 \text{ cm}$. So the object distance for the far end of the rod is $40.0 \text{ cm} - (-72.0 \text{ cm}) = 112 \text{ cm}$.

$$\Rightarrow \frac{n_a}{s} + \frac{n_b}{s'} = \frac{n_b - n_a}{R} \Rightarrow \frac{1.50}{112 \text{ cm}} + \frac{1}{s'} = 0 \Rightarrow s' = -74.7 \text{ cm}.$$

(b) The magnification is the product of the two magnifications:

$$m_1 = -\frac{n_a s'}{n_b s} = -\frac{-72.0}{(1.50)(16.0)} = 3.00,\ m_2 = 1.00 \Rightarrow m = m_1 m_2 = 3.00.$$

35-61: $\frac{1}{s} + \frac{1}{s'} = \frac{1}{f} \Rightarrow \frac{1}{40.0 \text{ cm}} + \frac{1}{s'} = \frac{1}{20.0 \text{ cm}} \Rightarrow s' = 40.0 \text{ cm}.$

So the object distance for second lens is $26.0 \text{ cm} - (40.0 \text{ cm}) = -14.0 \text{ cm}$.

$$\Rightarrow \frac{1}{s} + \frac{1}{s'} = \frac{1}{f} \Rightarrow \frac{1}{-14.0 \text{ cm}} + \frac{1}{s'} = \frac{1}{20.0 \text{ cm}} \Rightarrow s' = 8.24 \text{ cm}.$$

So the object distance for third lens is $26.0 \text{ cm} - (8.24 \text{ cm}) = 17.8 \text{ cm}$.

$\Rightarrow \frac{1}{s} + \frac{1}{s'} = \frac{1}{f} \Rightarrow \frac{1}{17.8 \text{ cm}} + \frac{1}{s'} = \frac{1}{20.0 \text{ cm}} \Rightarrow s' = -159 \text{ cm}$, so the final image is virtual and 159 cm to the left of the third mirror, or equivalently 107 cm to the left of the first mirror.

35-62: (a) $s + s' = 16.0 \text{ cm}$ and $\frac{1}{s} + \frac{1}{s'} = \frac{1}{f} \Rightarrow \frac{1}{16.0 \text{ cm} - s'} + \frac{1}{s'} = \frac{1}{3.50 \text{ cm}}$

$$\Rightarrow (s')^2 - (16.0 \text{ cm})s' + 56.0 \text{ cm}^2 = 0 \Rightarrow s' = 10.8 \text{ cm},\ 5.17 \text{ cm} = s.$$

So the screen must either be 5.17 cm or 10.8 cm from the object.

(b) $s = 5.17$ cm: $m = -\frac{s'}{s} = -\frac{10.8}{5.17} = 2.09$.

$s = 10.8$ cm: $m = -\frac{s'}{s} = -\frac{5.17}{10.8} = 0.478$.

35-63: Parallel light coming in from the left is focused 9.0 cm from the left lens, which is 6.0 cm to the right of the second lens. Therefore:
$\frac{1}{s} + \frac{1}{s'} = \frac{1}{f} \Rightarrow \frac{1}{-6.00\text{ cm}} + \frac{1}{s'} = \frac{1}{9.00\text{ cm}} \Rightarrow s' = 3.60$ cm, to the right of the second lens, and this is where the first focal point of the eyepiece is located. The second focal point is obtained by sending in parallel light from the right, and the symmetry of the lens set-up enables us to immediately state that the second focal point is 3.60 cm to the left of the first lens.

35-64: (a) With two lenses of different focal length in contact, the image distance from the first lens becomes exactly minus the object distance for the second lens. So we have:

$$\frac{1}{s_1} + \frac{1}{s_1'} = \frac{1}{f_1} \Rightarrow \frac{1}{s_1'} = \frac{1}{f_1} - \frac{1}{s_1} \quad \text{and} \quad \frac{1}{s_2} + \frac{1}{s_2'} = \frac{1}{-s_1'} + \frac{1}{s_2'} = \left(\frac{1}{s_1} - \frac{1}{f_1}\right) + \frac{1}{s_2'} = \frac{1}{f_2}.$$

But overall for the lens system, $\frac{1}{s_1} + \frac{1}{s_2'} = \frac{1}{f} \Rightarrow \frac{1}{f} = \frac{1}{f_2} + \frac{1}{f_1}$.

(b) With water sitting in a meniscus lens, we have two lenses in contact. All we need in order to calculate the system's focal length is calculate the individual focal lengths, and then use the formula from part (a).

For the meniscus: $\frac{1}{f_m} = (n_b - n_a)\left(\frac{1}{R_1} - \frac{1}{R_2}\right) = (0.52)\left(\frac{1}{4.00\text{ cm}} - \frac{1}{8.00\text{ cm}}\right) = 0.065$.

For the water: $\frac{1}{f_w} = (n_b - n_a)\left(\frac{1}{R_1} - \frac{1}{R_2}\right) = (0.33)\left(\frac{1}{8.00\text{ cm}} - \frac{1}{\infty}\right) = 0.0413$.

$$\Rightarrow \frac{1}{f} = \frac{1}{f_2} + \frac{1}{f_1} = 0.106 \Rightarrow f = 9.41\text{ cm}.$$

35-65: $\frac{1}{s_1} + \frac{1}{s_1'} = \frac{1}{f_1} \Rightarrow \frac{1}{x - 15\text{ cm}} + \frac{1}{15\text{ cm}} = \frac{1}{f}$ and $\frac{1}{x - 13\text{ cm}} + \frac{1}{11\text{ cm}} = \frac{1}{f}$
$\Rightarrow x^2 - 28x + 113\text{ cm}^2 = 0 \Rightarrow x = 23.1\text{ cm},\ 4.86\text{ cm}$. But the object must be to the left of the lens, so s = 23.1 cm - 15 cm = 8.1 cm
The corresponding focal length is 5.28 cm.

35-66: (a) Starting with the two equations:
$\frac{n_a}{s_1} + \frac{n_b}{s_1'} = \frac{n_b - n_a}{R_1}$ and $\frac{n_b}{s_2} + \frac{n_c}{s_2'} = \frac{n_c - n_b}{R_2}$, and using $n_a = n_{liq} = n_c$, $n_b = n$, and $s_1' = -s_2$, we get: $\frac{n_{liq}}{s_1} + \frac{n}{s_1'} = \frac{n - n_{liq}}{R_1}$ and $\frac{n}{-s_1'} + \frac{n_{liq}}{s_2'} = \frac{n_{liq} - n}{R_2}$.

$$\Rightarrow \frac{1}{s_1} + \frac{1}{s_2'} = \frac{1}{s} + \frac{1}{s'} = \frac{1}{f'} = (n / n_{liq} - 1)\left(\frac{1}{R_1} - \frac{1}{R_2}\right).$$

(b) Comparing the equations for focal length in and out of air we have:

$$f(n-1) = f'(n/n_{liq} - 1) = f'\left(\frac{n - n_{liq}}{n_{liq}}\right) \Rightarrow f' = \left[\frac{n_{liq}(n-1)}{n - n_{liq}}\right] f.$$

35-67: The image formed by the converging lens is 25.0 cm from the converging lens, and becomes an image for the diverging lens at a position 12.5 cm to the right of the diverging lens. The total distance from the converging lens to the screen is 41.0 cm. So for the diverging lens, we have:

$$\frac{1}{s} + \frac{1}{s'} = \frac{1}{f} \Rightarrow \frac{1}{-12.5\text{ cm}} + \frac{1}{41.0\text{ cm} - 12.5\text{ cm}} = \frac{1}{f} \Rightarrow f = -22.3\text{ cm}.$$

35-68: We have images formed from both ends. From the first:

$$\frac{n_a}{s} + \frac{n_b}{s'} = \frac{n_b - n_a}{R} \Rightarrow \frac{1}{20.0\text{ cm}} + \frac{1.50}{s'} = \frac{0.50}{5.00\text{ cm}} \Rightarrow s' = 30.0\text{ cm}.$$

This image becomes the object for the second end:

$$\frac{n_a}{s} + \frac{n_b}{s'} = \frac{n_b - n_a}{R} \Rightarrow \frac{1.50}{d - 30.0\text{ cm}} + \frac{1}{52.0\text{ cm}} = \frac{-0.50}{-5.00\text{ cm}}$$

$$\Rightarrow d - 30.0\text{ cm} = 18.6\text{ cm} \Rightarrow d = 48.6\text{ cm}.$$

35-69: (a) Bouncing first off the convex mirror, then the concave mirror:

$$\frac{1}{s} + \frac{1}{s'} = \frac{2}{R} \Rightarrow \frac{1}{0.600\text{ m} - x} + \frac{1}{s'} = \frac{2}{-0.360\text{ m}} \Rightarrow \frac{1}{s'} = -5.56\text{ m}^{-1} - \frac{1}{x - 0.600\text{ m}}$$

$$\Rightarrow s' = \frac{x - 0.600\text{ m}}{-5.56\text{ m}^{-1}x + 4.33}.$$ But the object distance for the concave mirror is just

$$s = 0.600\text{ m} - s' = \frac{4.33x + 3.20\text{ m}}{5.56\text{ m}^{-1}x - 4.33}.$$

So for the concave mirror: $\frac{1}{s} + \frac{1}{s'} = \frac{2}{R} \Rightarrow \frac{5.56\text{ m}^{-1}x - 4.33}{4.33x + 3.20\text{ m}} + \frac{1}{x} = \frac{2}{0.360}$

$$\Rightarrow 18.5x^2 - 17.8x + 3.20 = 0 \Rightarrow x = 0.72\text{ m},\ 0.24\text{ m}.$$

But the object position must be between the mirrors, so the distance must be the smaller of the two above, 0.24 m, from the concave mirror.

(b) Now having the light bounce first from the concave mirror, and then the convex mirror, we have:

$$\frac{1}{s} + \frac{1}{s'} = \frac{2}{R} \Rightarrow \frac{1}{x} + \frac{1}{s'} = \frac{2}{0.360} \Rightarrow \frac{1}{s'} = 5.56\text{ m}^{-1} - \frac{1}{x} \Rightarrow s' = \frac{x}{5.56\text{m}^{-1}x - 1.00}.$$

But the object distance for the convex mirror is just

$$s = 0.600\text{ m} - s' = \frac{2.33x - 0.600\text{ m}}{5.56\text{ m}^{-1}x - 1}.$$

So for the convex mirror: $\frac{1}{s} + \frac{1}{s'} = \frac{2}{R} \Rightarrow \frac{5.56\text{ m}^{-1}x - 1}{2.33x - 0.600\text{ m}} + \frac{1}{x} = \frac{2}{-0.360}$

$$\Rightarrow 18.5x^2 - 2.00x - 0.600 = 0 \Rightarrow x = -0.13\text{ m},\ 0.24\text{ m}.$$

But the object position must be between the mirrors, so the distance must be 0.24 m from the concave mirror.

35-70: Light passing straight through the lens:

(a)

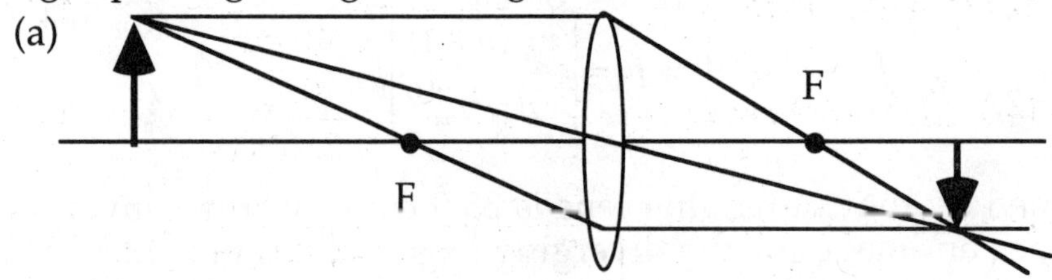

(b) $\frac{1}{s}+\frac{1}{s'}=\frac{1}{f}\Rightarrow\frac{1}{85.0\text{ cm}}+\frac{1}{s'}=\frac{1}{32.0\text{ cm}}\Rightarrow s'=51.3\text{ cm}$, to the right of the lens.

(c) The image is real.

(d) The image is inverted.

For light reflecting off the mirror, and then passing through the lens:

(a)

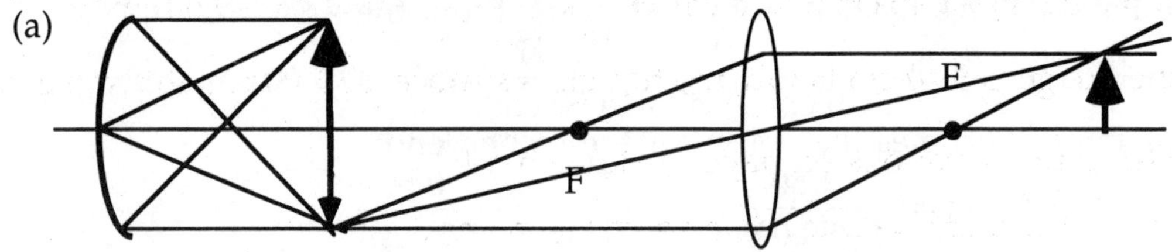

$\frac{1}{s}+\frac{1}{s'}=\frac{1}{f}\Rightarrow\frac{1}{20.0\text{ cm}}+\frac{1}{s'}=\frac{1}{10.0\text{ cm}}\Rightarrow s'=20.0\text{ cm}$, so the image from the mirror, which becomes the new object for the lens, is at the same location as the object. So the final image position is 51.3 cm to the right of the lens, as in the first case above.

(c) The image is real.

(d) The image is erect.

35-71: Entering the sphere: $\frac{n_a}{s}+\frac{n_b}{s'}=\frac{n_b-n_a}{R}\Rightarrow\frac{1}{3R}+\frac{4}{3s'}=\frac{1}{3R}\Rightarrow s'=\infty$.

$\frac{n_a}{s}+\frac{n_b}{s'}=\frac{n_b-n_a}{R}\Rightarrow\frac{1.33}{\infty}+\frac{1}{s'}=\frac{1}{3R}\Rightarrow s'=3R$.

So the final image is a distance 3R from the right-hand side of the sphere, or 4R to the right of the center of the globe.

35-72: (a) Treating each of the goblet surfaces as spherical surfaces, we have to pass, from left to right, through four interfaces. For the empty goblet:

$\frac{n_a}{s}+\frac{n_b}{s'}=\frac{n_b-n_a}{R}\Rightarrow\frac{1}{\infty}+\frac{1.50}{s_1'}=\frac{0.50}{4.00\text{ cm}}\Rightarrow s_1'=12\text{ cm}$

$\Rightarrow s_2=0.60\text{ cm}-12\text{ cm}=-11.4\text{ cm}\Rightarrow\frac{1.50}{-11.4\text{ cm}}+\frac{1}{s_2'}=\frac{-0.50}{3.40\text{ cm}}\Rightarrow s_2'=-64.6\text{ cm}$.

$\Rightarrow s_3=64.6\text{ cm}+6.80\text{ cm}=71.4\text{ cm}\Rightarrow\frac{1}{71.4\text{ cm}}+\frac{1.50}{s_3'}=\frac{0.50}{-3.40\text{ cm}}\Rightarrow s_3'=-9.31\text{ cm}$.

$\Rightarrow s_4=9.31\text{ cm}+0.60\text{ cm}=9.91\text{ cm}\Rightarrow\frac{1.50}{9.91\text{ cm}}+\frac{1}{s_4'}=\frac{-0.50}{-4.00\text{ cm}}\Rightarrow s_4'=-37.9\text{ cm}$.

So the image is 37.9 cm - 2(4.0 cm) = 29.9 cm to the left of the goblet.

(b) For the wine-filled goblet:

$\frac{n_a}{s}+\frac{n_b}{s'}=\frac{n_b-n_a}{R}\Rightarrow\frac{1}{\infty}+\frac{1.50}{s_1'}=\frac{0.50}{4.00\text{ cm}}\Rightarrow s_1'=12\text{ cm}$

$$\Rightarrow s_2 = 0.60\text{ cm} - 12\text{ cm} = -11.6\text{ cm} \Rightarrow \frac{1.50}{-11.4\text{ cm}} + \frac{1.37}{s_2'} = \frac{-0.13}{3.40\text{ cm}} \Rightarrow s_2' = 14.7\text{ cm}.$$

$$\Rightarrow s_3 = 6.80\text{ cm} - 14.7\text{ cm} = -7.9\text{ cm} \Rightarrow \frac{1.37}{-7.9\text{ cm}} + \frac{1.50}{s_3'} = \frac{0.13}{-3.40\text{ cm}} \Rightarrow s_3' = 11.1\text{ cm}.$$

$$\Rightarrow s_4 = 0.60\text{ cm} - 11.1\text{ cm} = -10.5\text{ cm} \Rightarrow \frac{1.50}{-10.5\text{ cm}} + \frac{1}{s_4'} = \frac{-0.50}{-4.00\text{ cm}} \Rightarrow s_4' = 3.73\text{ cm},$$

to the right of the goblet.

35-73: $\frac{n_a}{s} + \frac{n_b}{s'} = 0 \Rightarrow \frac{1}{7.00\text{ cm}} + \frac{1.50}{s_1'} = 0 \Rightarrow s_1' = -10.5\text{ cm}$

$$\Rightarrow s_2 = 3.00\text{ cm} + 10.5\text{ cm} = 13.5\text{ cm} \Rightarrow \frac{1.50}{13.5\text{ cm}} + \frac{1}{s_1'} = 0 \Rightarrow s_1' = -9.00\text{ cm}.$$

So the image is 9.00 cm below the top glass surface, or 1.00 cm above the page.

35-74: (a) Using the equations on page 1104, we have:

$$\frac{1}{f} = (n-1)\left(\frac{1}{R_1} - \frac{1}{R_2}\right) \Rightarrow \frac{1}{30\text{ cm}} = 0.50\left(\frac{2}{R}\right) \Rightarrow R = 30\text{ cm}.$$

And $\frac{n_a}{s} + \frac{n_b}{s'} = \frac{n_b - n_a}{R} \Rightarrow \frac{1}{90.0\text{ cm}} + \frac{1.50}{s_1'} = \frac{0.50}{30.0\text{ cm}} \Rightarrow s_1' = 270\text{ cm} = -s_2.$

$$\Rightarrow \frac{1.50}{-270\text{ cm}} + \frac{1.33}{s_2'} = \frac{-0.167}{-30.0\text{ cm}} \Rightarrow s_2' = 120\text{ cm}.$$

The mirror reflects the image back (since there is just 80 cm between the lens and mirror. So, the position of the image is 40 cm to the left of the mirror, or 40 cm to the right of the lens. So:

$\Rightarrow \frac{1.33}{40.0\text{ cm}} + \frac{1.50}{s_3'} = \frac{0.167}{30.0\text{ cm}} \Rightarrow s_3' = -54.0\text{ cm}$. Then leaving the glass back to air:

$\Rightarrow \frac{1.50}{54.0\text{ cm}} + \frac{1}{s_4'} = \frac{-0.500}{-30.0\text{ cm}} \Rightarrow s_4' = -90.0\text{ cm}$, to the right of the lens.

$$m = m_1 m_2 m_3 m_4 = \left(\frac{n_{a1}s_1'}{n_{b1}s_1}\right)\left(\frac{n_{a2}s_2'}{n_{b2}s_2}\right)\left(\frac{n_{a3}s_3'}{n_{b3}s_3}\right)\left(\frac{n_{a4}s_4'}{n_{b4}s_4}\right) = \left(\frac{270}{90}\right)\left(\frac{120}{-270}\right)\left(\frac{-54}{40}\right)\left(\frac{-90}{54}\right) = -3.00.$$

(Note all the indices of refraction cancel out.)

(b) The image is virtual. (c) The image is inverted.

(d) The final height is $y' = my = (-3.00)(3.00\text{ mm}) = 9.00\text{ mm}$.

35-75: At the first surface, $\frac{n_a}{s} + \frac{n_b}{s'} = 0 \Rightarrow s' = -\frac{n_b}{n_a}s = -\frac{1.52}{1.00}(-14.4\text{ cm}) = 22.9\text{ cm}.$

At the second surface,

$$s' = 14.7\text{ cm} - t = -\frac{n_b}{n_a}s = +\frac{1.00}{1.52}(21.9\text{ cm} - t) \Rightarrow 22.344 - 1.52t = 21.888 - t$$

$$\Rightarrow 0.52t = 0.456 \Rightarrow t = 0.88\text{ cm}.$$

(Note, as many significant figures as possible should be kept during the calculation, since numbers comparable in size are subtracted.)

35-76: The first image formed by the spherical mirror is the one where the light immediately strikes its surface, with out bouncing from the plane mirror.

$\frac{1}{s}+\frac{1}{s'}=\frac{1}{f}\Rightarrow\frac{1}{7.50\text{ cm}}+\frac{1}{s'}=\frac{1}{-18.0\text{ cm}}\Rightarrow s'=-5.29\text{ cm}$, and the image height:

$$y'=-\frac{s'}{s}y=-\frac{-5.29}{7.5}(0.300\text{ cm})=0.212\text{ cm}.$$

The second image is of the plane mirror image, located (15 cm + 7.5 cm) from the vertex of the spherical mirror. So:

$\frac{1}{s}+\frac{1}{s'}=\frac{1}{f}\Rightarrow\frac{1}{22.5\text{ cm}}+\frac{1}{s'}=\frac{1}{-18.0\text{ cm}}\Rightarrow s'=-10.0\text{ cm}$, and the image height:

$$y'=-\frac{s'}{s}y=-\frac{-10.0}{22.5}(0.300\text{ cm})=0.133\text{ cm}.$$

35-77: (a) The distance between image and object can be calculated by taking the derivative of the separation distance and minimizing it.

$$D=s+s' \quad \text{but} \quad s'=\frac{sf}{s-f}\Rightarrow D=s+\frac{sf}{s-f}=\frac{s^2}{s-f}$$

$$\Rightarrow\frac{dD}{ds}=\frac{d}{ds}\left(\frac{s^2}{s-f}\right)=\frac{2s}{s-f}-\frac{s^2}{(s-f)^2}=\frac{s^2-2sf}{(s-f)^2}=0$$

$\Rightarrow s^2-2sf=0\Rightarrow s(s-2f)=0\Rightarrow s=0,\,2f=s'$, so for a real image, the minimum separation between object and image is $4f$.

(b)

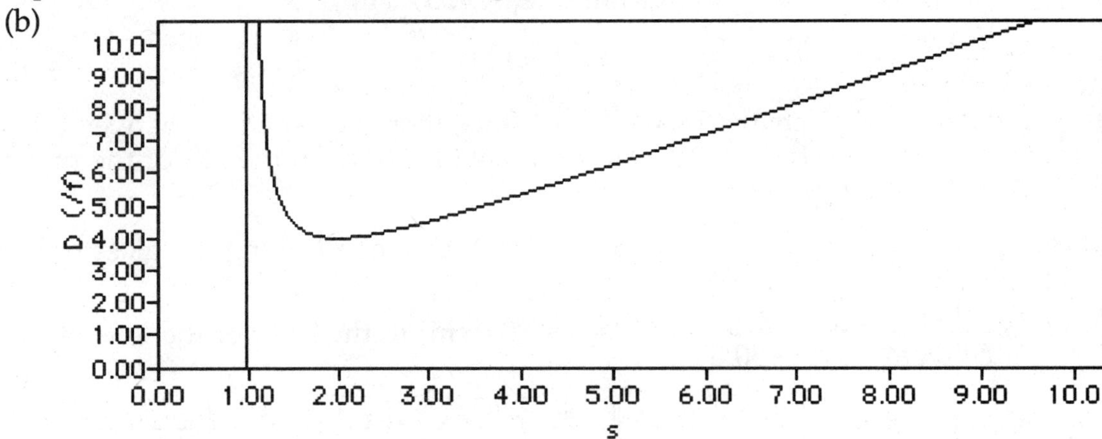

Note that the minimum does occur for $D=4f$.

35-78: (a) By the symmetry of image production, any image must be the same distance D as the object from the mirror intersection point. But if the images and the object are equal distances from the mirror intersection, they lie on a circle with radius equal to D.

(b) The center of the circle lies at the mirror intersection as discussed above.

(c)

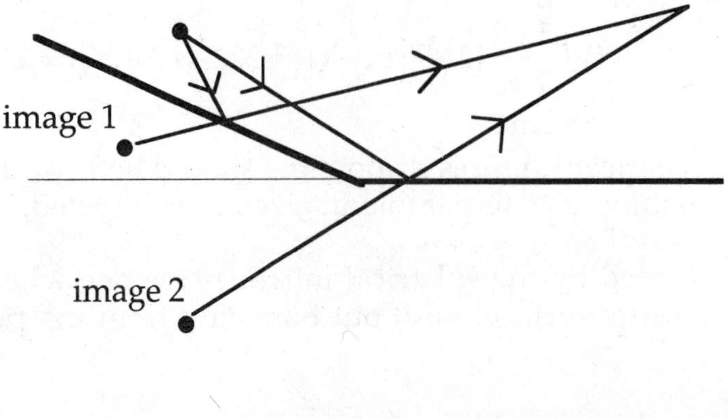

image 1

image 2

35-79:

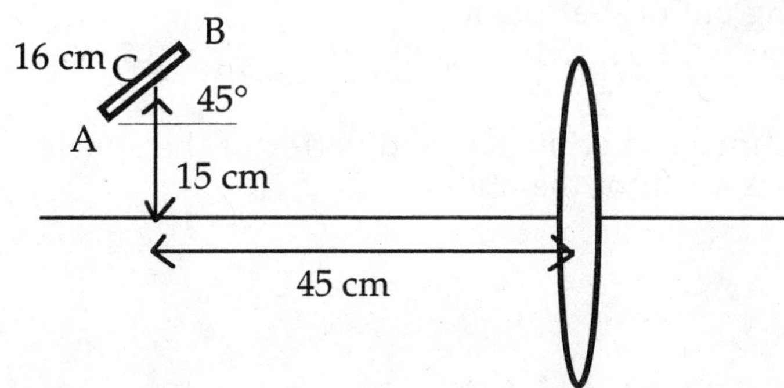

(a) For point C: $\frac{1}{s}+\frac{1}{s'}=\frac{1}{f}\Rightarrow\frac{1}{45.0\text{ cm}}+\frac{1}{s'}=\frac{1}{20.0\text{ cm}}\Rightarrow s'=36.0\text{ cm}.$

$y'=-\frac{s'}{s}y=-\frac{36.0}{45.0}(15.0\text{ cm})=-12.0\text{ cm}$, so the image of point C is 36.0 cm to the right of the lens, and 12.0 cm below the axis.

For point A: $s=45.0\text{ cm}+8.00\text{ cm}(\cos 45°)=50.7\text{ cm}$

$\frac{1}{s}+\frac{1}{s'}=\frac{1}{f}\Rightarrow\frac{1}{50.7\text{ cm}}+\frac{1}{s'}=\frac{1}{20.0\text{ cm}}\Rightarrow s'=33.0\text{ cm}.$

$y'=-\frac{s'}{s}y=-\frac{33.0}{45.0}(15.0\text{ cm}-8.00\text{ cm}(\sin 45°))=-6.10\text{ cm}$, so the image of point A is 33.0 cm to the right of the lens, and 6.10 cm below the axis.

For point B: $s=45.0\text{ cm}-8.00\text{ cm}(\cos 45°)=39.3\text{ cm}$

$\frac{1}{s}+\frac{1}{s'}=\frac{1}{f}\Rightarrow\frac{1}{39.3\text{ cm}}+\frac{1}{s'}=\frac{1}{20.0\text{ cm}}\Rightarrow s'=40.7\text{ cm}.$

$y'=-\frac{s'}{s}y=-\frac{40.7}{39.3}(15.0\text{ cm}+8.00\text{ cm}(\sin 45°))=-21.4\text{ cm}$, so the image of point B is 40.7 cm to the right of the lens, and 21.4 cm below the axis.

(b) The length of the pencil is the distance from point A to B:

$L=\sqrt{(x_A-x_B)^2+(y_A-y_B)^2}=\sqrt{(33.0\text{ cm}-40.7\text{ cm})^2+(6.10\text{ cm}-21.4\text{ cm})^2}$

$\Rightarrow L=17.1\text{ cm}.$

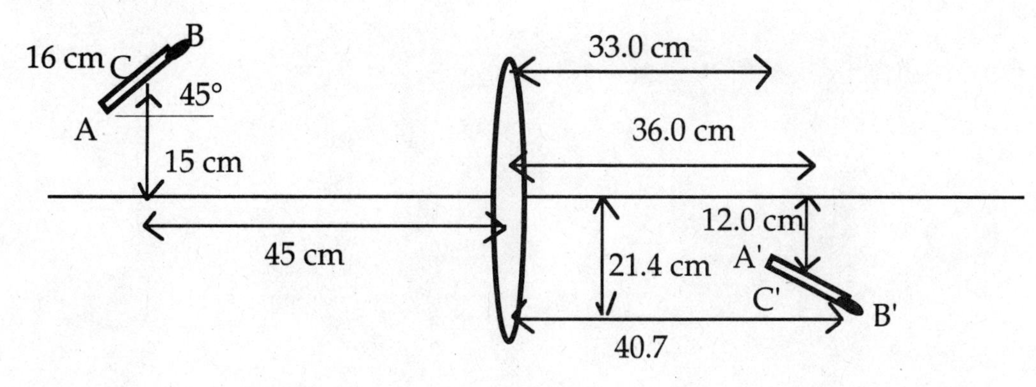

35-80: At the first interface:

$\frac{n_a}{s}+\frac{n_b}{s'}=\frac{n_b-n_a}{R}\Rightarrow\frac{1}{2R}+\frac{1.50}{s'}=\frac{0.50}{R}\Rightarrow s'=\infty.$

Next, reflection at the silvered hemisphere occurs, so:

$\frac{1}{s}+\frac{1}{s'}=\frac{2}{R}\Rightarrow\frac{1}{\infty}+\frac{1}{s'}=\frac{2}{R}\Rightarrow s'=\frac{R}{2}.$

Finally, passing out of the sphere:

$$\frac{n_a}{s}+\frac{n_b}{s'}=\frac{n_b-n_a}{R}\Rightarrow\frac{1.50}{3R/2}+\frac{1}{s'}=\frac{-0.50}{-R}\Rightarrow s'=-2R.$$

Thus the final image is at the silvered vertex of the sphere, opposite the entry vertex, which is 4R from the object.

Chapter 36: Optical Instruments

36-1: $\frac{1}{s}+\frac{1}{s'}=\frac{1}{f}\Rightarrow\frac{1}{s}+\frac{1}{14.6\text{ cm}}=\frac{1}{13.5\text{ cm}}\Rightarrow s=179\text{ cm}.$

36-2: $y'=-\frac{s'}{s}y=-\frac{f}{s}y=\frac{0.105\text{ m}}{3.86\times10^{8}\text{ m}}(3.48\times10^{6}\text{ m})=9.47\times10^{-4}\text{ m}=0.947\text{ mm}.$

36-3: (a) $|m|=\left|\frac{y'}{y}\right|=\left|\frac{s'}{s}\right|\Rightarrow s'=s\left|\frac{y'}{y}\right|=120\text{ m}\left(\frac{0.024\text{ m}}{100\text{ m}}\right)=0.0288\text{ m}=29\text{ mm}.$

$\Rightarrow\frac{1}{f}=\frac{1}{s}+\frac{1}{s'}=\frac{1}{1.2\times10^{5}\text{ mm}}+\frac{1}{29\text{ mm}}\Rightarrow f=29\text{ mm}$. So one should use the 28 mm lens.

(b) $s'=s\left|\frac{y'}{y}\right|=5.3\text{ m}\left(\frac{0.036\text{ m}}{0.90\text{ m}}\right)=0.212\text{ m}=212\text{ mm}.$

$\Rightarrow\frac{1}{f}=\frac{1}{s}+\frac{1}{s'}=\frac{1}{5.3\times10^{3}\text{ mm}}+\frac{1}{212\text{ mm}}\Rightarrow f=204\text{ mm}$. So one should use the 200 mm lens.

36-4: $\frac{1}{s}+\frac{1}{s'}=\frac{1}{f}\Rightarrow\frac{1}{4.75\text{ m}}+\frac{1}{s'}=\frac{1}{0.050\text{ m}}\Rightarrow s'=0.0505\text{ m}.$

$y'=-\frac{s'}{s}y=-\frac{0.0505}{4.75}1750\text{ mm}=19\text{ mm}$, so it fits on the $24\text{ mm}\times36\text{ mm}$ film.

36-5: (a) $f/2.8\Rightarrow2.8=\frac{f}{D}\Rightarrow D=\frac{f}{2.8}=\frac{85.0\text{ mm}}{2.8}=30.4\text{ mm}.$

(b) $f/5.6\Rightarrow D=\frac{f}{5.6}$, so the diameter is 0.5 times smaller, and the area is 0.25 times smaller. Therefore only a quarter of the light entered the aperture, and the film must be exposed four times as long for the correct exposure.

36-6: The aperture diameter determines the length of the exposure time required. Thus, the camera B exposure time must be one quarter of that of the camera A exposure, since its aperture is twice as large.

36-7: (a) $|m|=\frac{s'}{s}\approx\frac{f}{s}\Rightarrow|m|=\frac{28\text{ mm}}{200{,}000\text{ mm}}=1.4\times10^{-4}.$

(b) $|m|=\frac{s'}{s}\approx\frac{f}{s}\Rightarrow|m|=\frac{105\text{ mm}}{200{,}000\text{ mm}}=5.3\times10^{-4}.$

(c) $|m|=\frac{s'}{s}\approx\frac{f}{s}\Rightarrow|m|=\frac{300\text{ mm}}{200{,}000\text{ mm}}=1.5\times10^{-3}.$

36-8: (a) $s_1=\infty\Rightarrow s_1'=f_1=12\text{ cm}.$

(b) $s_2=4.0\text{ cm}-12\text{ cm}=-8\text{ cm}.$

(c) $\dfrac{1}{s}+\dfrac{1}{s'}=\dfrac{1}{f}\Rightarrow\dfrac{1}{-8\text{ cm}}+\dfrac{1}{s_2'}=\dfrac{1}{-12\text{ cm}}\Rightarrow s_2'=24\text{ cm}$, to the right.

(d) $s_1=\infty\Rightarrow s_1'=f_1=12\text{ cm}.$

$s_2=8.0\text{ cm}-12\text{ cm}=-4\text{ cm}.$

$$\frac{1}{s}+\frac{1}{s'}=\frac{1}{f}\Rightarrow\frac{1}{-4\text{ cm}}+\frac{1}{s_2'}=\frac{1}{-12\text{ cm}}\Rightarrow s_2'=6\text{ cm}.$$

36-9: (a) $\dfrac{1}{s}+\dfrac{1}{s'}=\dfrac{1}{f}\Rightarrow\dfrac{1}{s}+\dfrac{1}{5.00\text{ m}}=\dfrac{1}{0.100\text{ m}}\Rightarrow s=0.102\text{ m}=10.2\text{ cm}.$

(b) If the slide-lens distance was kept at 10 cm, then the image would be at infinity and could not be focused on a screen at any position.

36-10: (a) $\dfrac{1}{s}+\dfrac{1}{s'}=\dfrac{1}{f}\Rightarrow\dfrac{1}{s}+\dfrac{1}{8.00\text{ m}}=\dfrac{1}{0.120\text{ m}}\Rightarrow s=0.122\text{ m}=12.2\text{ cm}.$

(b) $|m|=\dfrac{s'}{s}=\dfrac{8.00}{0.122}=65.7\Rightarrow$ dimensions are $(24\text{ mm}\times36\text{ mm})m=(1.6\text{ m}\times2.4\text{ m})$.

36-11: (a) $f=\dfrac{1}{\text{power}}=\dfrac{1}{2.75\text{ m}^{-1}}=0.364\text{ m}=36.4\text{ cm}$. The near-point is normally at 25 cm: $\dfrac{1}{s}+\dfrac{1}{s'}=\dfrac{1}{f}\Rightarrow\dfrac{1}{25\text{ cm}}+\dfrac{1}{s'}=\dfrac{1}{36.4\text{ cm}}\Rightarrow s'=-80\text{ cm}$, in front of the eye.

(b) $f=\dfrac{1}{\text{power}}=\dfrac{1}{-0.830\text{ m}^{-1}}=-1.20\text{ m}=-120\text{ cm}$. The far point is ideally at infinity, so: $\dfrac{1}{s}+\dfrac{1}{s'}=\dfrac{1}{f}\Rightarrow\dfrac{1}{\infty}+\dfrac{1}{s'}=\dfrac{1}{-120\text{ cm}}\Rightarrow s'=-120\text{ cm}.$

36-12: $\dfrac{n_a}{s}+\dfrac{n_b}{s'}=\dfrac{n_b-n_a}{R}\Rightarrow\dfrac{1}{\infty}+\dfrac{1.40}{2.60\text{ cm}}=\dfrac{0.40}{R}\Rightarrow R=0.74\text{ cm}.$

36-13: $\dfrac{n_a}{s}+\dfrac{n_b}{s'}=\dfrac{n_b-n_a}{R}\Rightarrow\dfrac{1}{25.0\text{ cm}}+\dfrac{1.40}{2.60\text{ cm}}=\dfrac{0.40}{R}\Rightarrow R=0.69\text{ cm}.$

36-14: (a) $\dfrac{1}{f}=\dfrac{1}{s}+\dfrac{1}{s'}=\dfrac{1}{0.25\text{ m}}+\dfrac{1}{-0.700\text{ m}}\Rightarrow\text{power}=\dfrac{1}{f}=2.57\text{ diopters}.$

(b) $\dfrac{1}{f}=\dfrac{1}{s}+\dfrac{1}{s'}=\dfrac{1}{\infty}+\dfrac{1}{-0.700\text{ m}}\Rightarrow\text{power}=\dfrac{1}{f}=-1.43\text{ diopters}.$

36-15: $m=-\dfrac{s'}{s}=7.25\Rightarrow s'=-7.25s\Rightarrow\dfrac{1}{s}+\dfrac{1}{s'}=\dfrac{1}{s}+\dfrac{1}{-7.25s}=\dfrac{1}{3.00\text{ cm}}$

$\Rightarrow s=2.59\text{ cm},\ s'=-7.25s=-18.8\text{ cm}.$

36-16: (a) Angular magnification $M = \frac{25.0 \text{ cm}}{f} = \frac{25.0 \text{ cm}}{7.00 \text{ cm}} = 3.57.$

(b) $\frac{1}{s} + \frac{1}{s'} = \frac{1}{f} \Rightarrow \frac{1}{s} + \frac{1}{-25.0 \text{ cm}} = \frac{1}{7.00 \text{ cm}} \Rightarrow s = 5.47 \text{ cm}.$

36-17: (a) $\frac{1}{s} + \frac{1}{s'} = \frac{1}{f} \Rightarrow \frac{1}{s} + \frac{1}{-25.0 \text{ cm}} = \frac{1}{9.00 \text{ cm}} \Rightarrow s = 6.62 \text{ cm}.$

(b) $|m| = \frac{s'}{s} = \frac{25.0 \text{ cm}}{6.62 \text{ cm}} = 3.78 \Rightarrow y' = ym = (1.00 \text{ mm})(3.78) = 3.78 \text{ mm}$

36-18: Using the approximation $s_1 \approx f$, and then $|m_1| = \frac{s_1'}{f_1}$, we have:

$f = 16 \text{ mm}: s' = 120 \text{ mm} + 16 \text{ mm} = 136 \text{ mm}; s = 16 \text{ mm}$

$\Rightarrow |m_1| = \frac{s'}{s} = \frac{136 \text{ mm}}{16 \text{ mm}} = 8.5.$

$f = 4 \text{ mm}: s' = 120 \text{ mm} + 4 \text{ mm} = 124 \text{ mm}; s = 4 \text{ mm} \Rightarrow |m_1| = \frac{s'}{s} = \frac{124 \text{ mm}}{4 \text{ mm}} = 31.$

$f = 1.9 \text{ mm}: s' = 120 \text{ mm} + 1.9 \text{ mm} = 122 \text{ mm}; s = 1.9 \text{ mm}$

$\Rightarrow |m_1| = \frac{s'}{s} = \frac{122 \text{ mm}}{1.9 \text{ mm}} = 64.$

The eyepiece magnifies by either 5 or 10, so:

(a) The maximum magnification occurs for the 1.9 mm objective and 10x eyepiece: $\Rightarrow M = |m_1| m_e = (64)(10) = 640.$

(b) The minimum magnification occurs for the 16 mm objective and 5x eyepiece: $\Rightarrow M = |m_1| m_e = (8.5)(5) = 43.$

36-19: (a) The image from the objective is at the focal point of the eyepiece, so $s_1' = d_{oe} - f_2 = 22.6 \text{ cm} - 2.50 \text{ cm} = 20.1 \text{ cm}$

$\Rightarrow \frac{1}{s} + \frac{1}{s'} = \frac{1}{f} \Rightarrow \frac{1}{s} + \frac{1}{20.1 \text{ cm}} = \frac{1}{1.60 \text{ cm}} \Rightarrow s = 1.74 \text{ cm}.$

(b) $|m_1| = \frac{s'}{s} = \frac{20.1 \text{ cm}}{1.74 \text{ cm}} = 11.6.$

(c) The overall magnification is $|M| = |m_1| \frac{25.0 \text{ cm}}{f_2} = (11.6) \frac{25.0 \text{ cm}}{2.50 \text{ cm}} = 116.$

36-20: (a) $M = \frac{(250 \text{ mm}) s_1'}{f_1 f_2} = \frac{(250 \text{ mm})(180 \text{ mm} + 4.0 \text{ mm})}{(4.00 \text{ mm})(22.0 \text{ mm})} = 522.$

(b) $m = \frac{y'}{y} \Rightarrow y = \frac{y'}{m} = \frac{0.10 \text{ mm}}{522} = 1.9 \times 10^{-4} \text{ mm}.$

36-21: $\frac{f}{D} = 19.0 \Rightarrow f = (19.0)D = (19.0)(1.02 \text{ m}) = 19.4 \text{ m}.$

36-22: (a) $f = (f - \text{number})D = (8)(15 \text{ cm}) = 120 \text{ cm}.$

(b) $f = (f - \text{number})D = (6)(20 \text{ cm}) = 120 \text{ cm}.$

(c) $f = (f-\text{number})D = (4.5)(40\ \text{cm}) = 180\ \text{cm}.$
(d) The narrowest angle of view is from the telescope with the largest magnification, which comes from $|M| = \frac{f_1}{f_2} \propto f_1$, where f_1 is the objective's focal length. So the telescope of part (c) has the narrowest field of view. Conversely, the widest field of view comes from the telescope with the least magnification, which occurs for both (a) and (b).
The brightest image comes from the telescope which collects the most light (has the largest aperture) which is the telescope of part (c). The dimmest image comes from the telescope of part (a).

36-23: (a) $M = -\frac{f_1}{f_2} = -\frac{90.0\ \text{cm}}{30.0\ \text{cm}} = -3.00.$
(b) $\frac{1}{s} + \frac{1}{s'} = \frac{1}{f} \Rightarrow \frac{1}{2000\ \text{m}} + \frac{1}{s'} = \frac{1}{0.900\ \text{m}} \Rightarrow s' = 0.900\ \text{m}$, so the height of an image of a building is $|y'| = \frac{s'}{s}y = \frac{0.900}{2000}(80.0\ \text{m}) = 0.0360\ \text{m}.$
(c) $\theta' = M\theta = 3\arctan(80.0/2000) \approx 3(80.0)/(2000) = 0.120\ \text{rad}.$

36-24: $f_1 + f_2 = d_{ss'} \Rightarrow f_1 = d_{ss'} - f_2 = 2.20\ \text{m} - 0.0600\ \text{m} = 2.14\ \text{m}$
$\Rightarrow M = -\frac{f_1}{f_2} = -\frac{214}{6.00} = 35.7.$

36-25: $|y'| = y\frac{s'}{s} = y\frac{f}{s} = (3.48 \times 10^6\ \text{m})\frac{18\ \text{m}}{3.84 \times 10^8\ \text{m}} = 0.16\ \text{m} = 16\ \text{cm}.$

36-26: (a) $f_1 = \frac{R}{2} = 0.400\ \text{m} \Rightarrow d = f_1 + f_2 = 0.415\ \text{m}.$
(b) $|M| = \frac{f_1}{f_2} = \frac{0.400\ \text{m}}{0.0150\ \text{m}} = 26.7.$

36-27: $\frac{1}{s} + \frac{1}{s'} = \frac{1}{f} \Rightarrow \frac{1}{1.5\ \text{m} - 2.6\ \text{m}} + \frac{1}{1.5\ \text{m} + 0.25\ \text{m}} = \frac{1}{f} \Rightarrow f = -2.96\ \text{m} \Rightarrow R = 2f = -5.9\ \text{m}$
So the smaller mirror must be convex (negative focal length) and have a radius of curvature equal to 5.9 m.

36-28: We use the lens makers equation: $\frac{1}{s} + \frac{1}{s'} = \frac{1}{f} = (n-1)\left(\frac{1}{R_1} - \frac{1}{R_2}\right).$

violet: $\frac{1}{14.2\ \text{cm}} + \frac{1}{s'} = (1.537 - 1)\left(\frac{-1}{15.0\ \text{cm}} - \frac{1}{15.0\ \text{cm}}\right) \Rightarrow s' = -7.041\ \text{cm}$

red: $\frac{1}{14.2\ \text{cm}} + \frac{1}{s'} = (1.517 - 1)\left(\frac{-1}{15.0\ \text{cm}} - \frac{1}{15.0\ \text{cm}}\right) \Rightarrow s' = -7.176\ \text{cm}$

So the images are to the left of the lens, and the difference in their positions is: $\Delta s' = -7.041\ \text{cm} - (-7.176\ \text{cm}) = 0.135\ \text{cm}.$

36-29: We must consider both the apertures used and the length of the exposure.

For the apertures: $\frac{D_A}{D_B} = \frac{(5.6)^2}{(16)^2} = 0.1225.$

For the exposures: $\frac{t_A}{t_B} = \frac{1/125}{1/500} = 4.00.$

So the sensitivity: $\frac{A}{B} = \frac{t_B}{t_A} \cdot \frac{D_B}{D_A} = \frac{1}{(4.00)(0.1225)} = 2.04$, and film A is more sensitive than film B.

36-30: $\frac{1}{s} + \frac{1}{s'} = \frac{1}{f} \Rightarrow \frac{1}{7500 \text{ mm}} + \frac{1}{s'} = \frac{1}{50 \text{ mm}} \Rightarrow s' = 50.34 \text{ mm}.$

$\frac{1}{s} + \frac{1}{s'} = \frac{1}{f} \Rightarrow \frac{1}{750 \text{ mm}} + \frac{1}{s'} = \frac{1}{50 \text{ mm}} \Rightarrow s' = 53.57 \text{ mm}.$

$\Rightarrow \Delta s' = 53.57 \text{ mm} - (-50.34 \text{ mm}) = 3.2 \text{ mm}$, and the lens was moved away from the film to focus the second image.

36-31: (a) $m = -\frac{s'}{s} = \frac{y'}{y} = \frac{2}{3}\frac{(0.0360 \text{ m})}{(3.60 \text{ m})} \Rightarrow s' = (6.67 \times 10^{-3})s$

$\Rightarrow \frac{1}{s} + \frac{1}{s'} = \frac{1}{s} + \frac{1}{(6.67 \times 10^{-3})s} = \frac{1}{s}\left(1 + \frac{1}{6.67 \times 10^{-3}}\right) = \frac{1}{f} = \frac{1}{0.0500 \text{ m}} \Rightarrow s = 7.55 \text{ m}.$

(b) To just fill the frame, the magnification must be 0.01, so;

$\Rightarrow \frac{1}{s}\left(1 + \frac{1}{0.0100}\right) = \frac{1}{f} = \frac{1}{0.0500 \text{ m}} \Rightarrow s = 5.05 \text{ m}.$

36-32: $\frac{1}{s} + \frac{1}{s'} = \frac{1}{f} \Rightarrow \frac{1}{40{,}000 \text{ mm}} + \frac{1}{s'} = \frac{1}{50.0 \text{ mm}} \Rightarrow s' = 50.06 \text{ mm}.$

The resolution of 100 lines per millimeter means that the image line width is 0.0100 cm between lines. That is $y' = 0.0100 \text{ mm}$. But:

$\left|\frac{y'}{y}\right| = \left|\frac{s'}{s}\right| \Rightarrow y = |y'| \cdot \frac{s}{s'} = (0.0100 \text{ mm})\frac{40{,}000 \text{ mm}}{50.06 \text{ mm}} = 7.98 \text{ mm}$, which is the minimum separation between two lines 40.0 m away from the camera.

36-33: (a) From the diagram below, we see that $|m| = \frac{d}{W} = \frac{s'}{s} \Rightarrow \frac{1}{s'} = \frac{W}{sd}.$

$\Rightarrow \frac{1}{s} + \frac{1}{s'} = \frac{1}{s} + \frac{W}{sd} = \frac{1}{s}\left(1 + \frac{W}{d}\right) = \frac{d+W}{sd} = \frac{1}{f} \Rightarrow f = \frac{sd}{d+W}$. But when the object is much larger than the image we have the approximation:

$s' \approx f$ and $d + W \approx W \Rightarrow m = \frac{d}{W} \approx \frac{f}{s} \Rightarrow \tan\frac{\theta}{2} = \frac{W}{2} \cdot \frac{1}{s} = \frac{d}{2f} \Rightarrow \theta = 2\arctan\left(\frac{d}{2f}\right).$

(b) The film is 24 mm x 36 mm, so the diagonal length is just:

$d = \left(\sqrt{24^2 + 36^2}\right)\text{mm} = 43.3\text{ mm}$. So:

$f = 28\text{ mm}: \theta = 2\arctan\left(\dfrac{43.3\text{ mm}}{2(28\text{ mm})}\right) = 75°$.

$f = 105\text{ mm}: \theta = 2\arctan\left(\dfrac{43.3\text{ mm}}{2(105\text{ mm})}\right) = 23°$.

$f = 300\text{ mm}: \theta = 2\arctan\left(\dfrac{43.3\text{ mm}}{2(300\text{ mm})}\right) = 8.2°$.

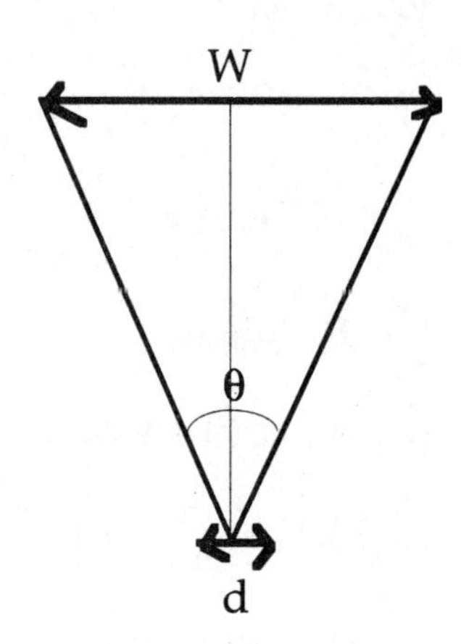

36-34: (a) $\dfrac{1}{s_1} + \dfrac{1}{s_1'} = \dfrac{1}{f_1}$ and $\dfrac{1}{s_2} + \dfrac{1}{s_2'} = \dfrac{1}{f_2}$, but $s_1' = -s_2$

$\Rightarrow \dfrac{1}{s_2} + \dfrac{1}{s_2'} = \dfrac{1}{-s_1'} + \dfrac{1}{s_2'} = -\left(\dfrac{1}{f_1} - \dfrac{1}{s_1}\right) + \dfrac{1}{s_2'} = \dfrac{1}{f_2} \Rightarrow \dfrac{1}{s_1} + \dfrac{1}{s_2'} = \dfrac{1}{f_1} + \dfrac{1}{f_2} = \dfrac{1}{f}$.

But the power is just one over the focal length, so $P_1 + P_2 = P$, and this is valid for all focal lengths.

(b) $P = P_1 + P_2 = \dfrac{1}{0.150\text{ m}} + \dfrac{1}{0.350\text{ m}} = 9.52$ diopters.

36-35: The near point is at infinity, so that is where the image must be found for any objects that are close. So:

$P = \dfrac{1}{f} = \dfrac{1}{s} + \dfrac{1}{s'} = \dfrac{1}{32\text{ cm}} + \dfrac{1}{-\infty} = \dfrac{1}{0.32\text{ m}} = 3.13$ diopters.

36-36: $\dfrac{n_a}{s} + \dfrac{n_b}{s'} = \dfrac{n_b - n_a}{R} \Rightarrow \dfrac{1}{30.0\text{ cm}} + \dfrac{1.40}{s'} = \dfrac{0.40}{0.75\text{ cm}} \Rightarrow s' = 2.8\text{ cm}$. This distance is greater than the normal eye, which has a cornea vertex to retina distance of about 2.6 cm.

36-37: (a) At age 10: $f_n = 7\text{ cm}$: $M = 2.5 = \dfrac{7\text{ cm}}{f} \Rightarrow f = 2.8\text{ cm}$.

(b) At age 30: $f_n = 14\text{ cm}$: $M = 2.5 = \dfrac{14\text{ cm}}{f} \Rightarrow f = 5.6\text{ cm}$.

(c) At age 60: $f_n = 200\text{ cm}$: $M = 2.5 = \dfrac{200\text{ cm}}{f} \Rightarrow f = 80\text{ cm}$.

(d) If the 2.8 cm focal length lens is used by the 60-year old, then

$M = \dfrac{200\text{ cm}}{f} = \dfrac{200\text{ cm}}{2.8\text{ cm}} = 71.4$.

(e) This does not mean that the older viewer sees a more magnified image. The object is over 28 times further away from the 60-year old, which is exactly the ratio needed to result in the magnification of 2.5 as seen by the 10-yr old.

36-38: (a) $\frac{1}{s}+\frac{1}{s'}=\frac{1}{f}\Rightarrow\frac{1}{s}+\frac{1}{-25\text{ cm}}=\frac{1}{f}\Rightarrow s=\frac{f(25\text{ cm})}{f+25\text{ cm}}$.

(b) Height $= y \Rightarrow \theta'=\arctan\left(\frac{y}{s}\right)=\arctan\left(\frac{y(f+25\text{ cm})}{f(25\text{ cm})}\right)\approx\frac{y(f+25\text{ cm})}{f(25\text{ cm})}$.

(c) $M=\frac{\theta'}{\theta}=\frac{y(f+25\text{ cm})}{f(25\text{ cm})}\cdot\frac{1}{y/25\text{ cm}}=\frac{f+25\text{ cm}}{f}$.

(d) If $f=10\text{ cm}\Rightarrow M=\frac{10\text{ cm}+25\text{ cm}}{10\text{ cm}}=3.5$. This is 1.4 times greater than the magnification obtained if the image is formed at infinity ($M_\infty=\frac{25\text{ cm}}{f}=2.5$).

(e) Having the first image form just within the focal length puts one in the situation described above, where it acts as a source which yields an enlarged virtual image. If the first image fell just outside the second focal point, then the image would be real and diminished.

36-39: (a) $\frac{1}{s_1}+\frac{1}{s_1'}=\frac{1}{f_1}\Rightarrow\frac{1}{s_1}+\frac{1}{14.0\text{ cm}}=\frac{1}{0.900\text{ cm}}\Rightarrow s_1=0.962\text{ cm}$.

$\frac{1}{s_2}+\frac{1}{s_2'}=\frac{1}{f_2}\Rightarrow\frac{1}{s_2}+\frac{1}{100\text{ cm}}=\frac{1}{6.00\text{ cm}}\Rightarrow s_2=6.38\text{ cm}$.

Also $m_1=-\frac{s_1'}{s_1}=-\frac{14.0\text{ cm}}{0.962\text{ cm}}=-14.6$ and $m_2=-\frac{s_2'}{s_2}=-\frac{100\text{ cm}}{6.38\text{ cm}}=-15.7$.

$\Rightarrow m_{total}=m_1m_2=(-14.6)(-15.7)=228$.

(b) $d=s_1'+s_2=14.0\text{ cm}+6.38\text{ cm}=20.4\text{ cm}$.

36-40: (a) From the figure, $u=\frac{y}{f_1}$ and $u'=\frac{y}{|f_2|}=-\frac{y}{f_2}$.

So the angular magnification is: $M=\frac{u'}{u}=-\frac{f_1}{f_2}$.

(b) $M=-\frac{f_1}{f_2}\Rightarrow f_2=-\frac{f_1}{M}=-\frac{90.0\text{ cm}}{3.00}=-30.0\text{ cm}$.

(c) The length of the telescope is $90.0\text{ cm}-30.0\text{ cm}=60.0\text{ cm}$, compared to the length of 120 cm for the telescope in **Ex. 36-23.**

36-41: (a) From Figure 36-22, we define $x=r_o-r_o'\Rightarrow r_o'=r_o-x=r_o-r_o\left(\frac{d}{f_1}\right)=\frac{r_o(f_1-d)}{f_1}$.

(b) $s_2=d-f_1\Rightarrow\frac{1}{d-f_1}+\frac{1}{s_2'}=\frac{1}{f_2}\Rightarrow\frac{1}{s_2'}=\frac{d-f_1-f_2}{f_2(d-f_1)}=\frac{|f_2|-f_1+d}{|f_2|(f_1-d)}\Rightarrow s_2'=\frac{|f_2|(f_1-d)}{|f_2|-f_1+d}$.

(c) $\frac{r_o'}{s_2'}=\frac{r_o}{f}\Rightarrow f=\frac{r_o}{r_o'}s_2'=\frac{f_1}{f_1-d}\cdot\frac{|f_2|(f_1-d)}{|f_2|-f_1+d}\Rightarrow f=\frac{f_1|f_2|}{|f_2|-f_1+d}$.

(d) $f_{\max}=\frac{f_1|f_2|}{|f_2|-f_1+d}=\frac{(10.0\text{ cm})(15.0\text{ cm})}{(15.0\text{ cm}-10.0\text{ cm}+0.0\text{ cm})}=30\text{ cm}$.

$$f_{\min} = \frac{f_1|f_2|}{|f_2| - f_1 + d} = \frac{(10.0 \text{ cm})(15.0 \text{ cm})}{(15.0 \text{ cm} - 10.0 \text{ cm} + 7.0 \text{ cm})} = 12.5 \text{ cm}.$$

If the effective focal length is 15 cm, then the separation is:

$$f = \frac{f_1|f_2|}{|f_2| - f_1 + d} \Rightarrow 15 \text{ cm} = \frac{(10.0 \text{ cm})(15.0 \text{ cm})}{(15.0 \text{ cm} - 10.0 \text{ cm} + d)}$$

$$\Rightarrow 15.0 \text{ cm} - 10.0 \text{ cm} + d = 10.0 \text{ cm} \Rightarrow d = 5.0 \text{ cm}.$$

36-42: First recall that $|M| = \frac{\theta'}{\theta}$, and that $\theta = \left|\frac{y_1'}{f_1}\right|$ and $\theta' = \left|\frac{y_2'}{s_2'}\right| \Rightarrow |M| = \left|\frac{y_2'}{s_2'} \cdot \frac{f_1}{y_1'}\right|$.

But since the image formed by the objective is used as the object for the eyepiece, $y_1' = y_2$. So $|M| = \left|\frac{y_2'}{s_2'} \cdot \frac{f_1}{y_2}\right| = \left|\frac{y_2'}{y_2} \cdot \frac{f_1}{s_2'}\right| = \left|\frac{s_2'}{s_2} \cdot \frac{f_1}{s_2'}\right| = \left|\frac{f_1}{s_2}\right|$.

Therefore, $s_2 = \frac{f_1}{|M|} = \frac{40.0 \text{ cm}}{24} = 1.67 \text{ cm}$, and this is just outside the eyepiece focal point.

Now the distance from the mirror vertex to the lens is $f_1 + s_2 = 41.7 \text{ cm}$, and so $\frac{1}{s_2} + \frac{1}{s_2'} = \frac{1}{f_2} \Rightarrow s_2' = \left(\frac{1}{1.50 \text{ cm}} - \frac{1}{1.667 \text{ cm}}\right)^{-1} = 15.0 \text{ cm}$. Thus we have a final image which is real and 15.0 cm from the eyepiece. (Take care to carry plenty of figures in the calculation because two close numbers are subtracted.)

36-43: (a) People with normal vision cannot focus on distant objects under water because the image is unable to be focused in a short enough distance to form on the retina. Equivalently, the radius of curvature of the normal eye is about five or six times too great for focusing at the retina to occur.

(b) When introducing glasses, let's first consider what happens at the eye:
$\frac{n_a}{s_2} + \frac{n_b}{s_2'} = \frac{n_b - n_a}{R} \Rightarrow \frac{1.33}{s_2} + \frac{1.40}{2.6 \text{ cm}} = \frac{0.07}{0.74 \text{ cm}} \Rightarrow s_2 = -3.00 \text{ cm}$. That is, the object for the cornea must be 3.00 cm behind the cornea. Now, assume the glasses are 2.00 cm in front of the eye, then:

$$s_1' = 2.00 \text{ cm} + s_2 = 5.00 \text{ cm} \Rightarrow \frac{1}{s_1} + \frac{1}{s_1'} = \frac{1}{f_1'} \Rightarrow \frac{1}{\infty} + \frac{1}{5.00 \text{ cm}} = \frac{1}{f_1'} \Rightarrow f_1' = 5.00 \text{ cm}.$$

This is the focal length in water, but to get it in air, we use the formula from **Pr. 35-66:** $f_1 = f_1'\left[\frac{n - n_{liq}}{n_{liq}(n-1)}\right] = (5.00 \text{ cm})\left[\frac{1.52 - 1.333}{1.333(1.52 - 1)}\right] = 1.34 \text{ cm}.$

Chapter 37: Interference

37-1: The brightest wavelengths are when constructive interference occurs:

$$d = m\lambda \Rightarrow \lambda = \frac{d}{m} \Rightarrow \lambda_3 = \frac{1800\text{ nm}}{3} = 600\text{ nm and } \lambda_4 = \frac{1800\text{ nm}}{4} = 450\text{ nm}.$$

37-2: Destructive interference occurs for:

$$\lambda = \frac{d}{m+1/2} \Rightarrow \lambda_3 = \frac{1800\text{ nm}}{3.5} = 514\text{ nm and } \lambda_4 = \frac{1800\text{ nm}}{4.5} = 400\text{ nm}.$$

37-3: Measuring with a ruler from both S_1 and S_2 to three different points in the antinodal line labeled m = 3, we find that the difference in path length is three times the wavelength of the wave, as measured from one crest to the next on the diagram.

37-4: (a) Along the y-axis, above and below the sources, the path length difference is equal to four wavelengths ($m = \pm 4$) but no other place in the plane has a greater path difference between the two sources, so the greatest value of $|m|$ is four.

(b) There are eight nodal curves ($-4 \le m \le +3$), the first between S_1 and the $m = +3$ antinodal line, and the last between the $m = -3$ antinodal line and S_2.

(c) The maximum value of the antinodal curves is given by the largest integer less than or equal to $\frac{d}{\lambda}$ and the minimum value is the smallest integer greater than or equal to $-\frac{d}{\lambda}$. These answers are obtained by requiring the antinodal lines to fit in between the sources.

37-5: (a) At S_1, $r_2 - r_1 = 4\lambda$, and this path difference stays the same all along the y-axis, so $m = +4$. At S_2, $r_2 - r_1 = -4\lambda$, and the path difference below this point, along the negative y-axis, stays the same, so $m = -4$.

(b)

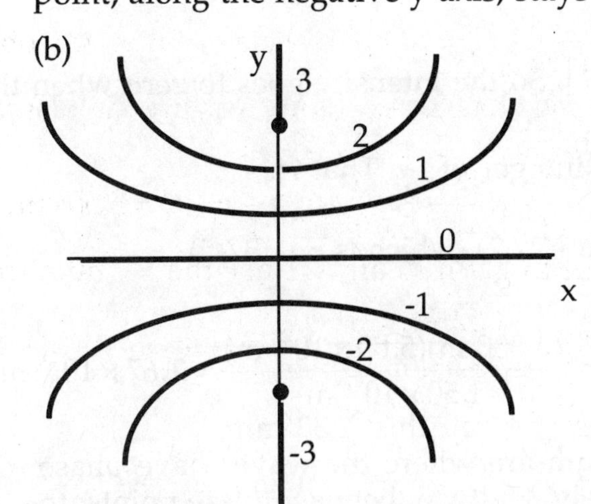

(c) The maximum and minimum m-values are determined by the largest integer less than or equal to $\frac{d}{\lambda}$.

(d) If $d = 5\frac{1}{2}\lambda \Rightarrow -5 \le m \le +5$, so there will be a total of 11 antinodes between the sources. (Another antinode cannot be squeezed in until the separation becomes six times the wavelength.)

37-6: (a) The path difference is 130 m, so for destructive interference:

$$\frac{\lambda}{2} = 130\,\text{m} \Rightarrow \lambda = 260\ \text{m}.$$

(b) The longest wavelength for constructive interference is $\lambda = 130$ m.

37-7: For constructive interference, we need $r_2 - r_1 - m\lambda \rightarrow (130\ \text{m} - x) - x - m\lambda$.

$$\Rightarrow x = 65\ \text{m} - \frac{m\lambda}{2} = 65\ \text{m} - \frac{mc}{2f} = 65\ \text{m} - \frac{m(3.00 \times 10^8\ \text{m/s})}{2(7.50 \times 10^6\ \text{Hz})} = 65\ \text{m} - m(20\ \text{m}).$$

$\Rightarrow x = 5$ m, 25 m, 45 m, 65 m, 85 m, 105 m, 125 m.

37-8: For bright fringes:

$$d = \frac{Rm\lambda}{y_m} = \frac{(1.20\ \text{m})(20)(5.89 \times 10^{-7}\ \text{m})}{0.0106\ \text{m}} = 1.33 \times 10^{-3}\ \text{m} = 1.33\ \text{mm}.$$

37-9: Recall $y_m = \dfrac{Rm\lambda}{d} \Rightarrow \Delta y_{23} = y_3 - y_2 = \dfrac{R\lambda(3-2)}{d} = \dfrac{(0.850\ \text{m})(6.00 \times 10^{-7}\ \text{m})}{3.00 \times 10^{-4}\ \text{m}}$

$\Rightarrow \Delta y_{23} = 1.70 \times 10^{-3}\ \text{m} = 1.70\ \text{mm}.$

37-10: (a) For the number of antinodes we have:

$\sin\theta = \dfrac{m\lambda}{d} = \dfrac{mc}{df} = \dfrac{m(3.00 \times 10^8\ \text{m/s})}{(15.0\ \text{m})(1.035 \times 10^8\ \text{Hz})} = 0.1932m$, so the maximum integer value is five, and the angles are $\pm 11.1°$, $\pm 22.7°$, $\pm 35.4°$, $\pm 50.6°$, $\pm 74.9°$.

(b) The nodes are given by $\sin\theta = \dfrac{(m + 1/2)\lambda}{d} = 0.1932(m + 1/2)$. So the angles are $\pm 5.5°$, $\pm 16.8°$, $\pm 28.9°$, $\pm 42.5°$, $\pm 60.3°$.

37-11: $\Delta y = \dfrac{R\lambda}{d} \Rightarrow \lambda = \dfrac{d\,\Delta y}{R} = \dfrac{(3.80 \times 10^{-4}\ \text{m})(2.27 \times 10^{-3}\ \text{m})}{2.50\ \text{m}} = 3.45 \times 10^{-7}\ \text{m}.$

37-12: From Eq. (37-14), $I = I_o \cos^2\left(\dfrac{\pi d}{\lambda}\sin\theta\right)$. So the intensity goes to zero when the cosine's argument becomes an odd integer of $\dfrac{\pi}{2}$. That is:

$\dfrac{\pi d}{\lambda}\sin\theta = (m + 1/2)\pi \Rightarrow d\sin\theta = \lambda(m + 1/2)$, which is Eq. (37-5).

37-13: (a) To the first maximum: $y_1 = \dfrac{R\lambda}{d} = \dfrac{(0.800\ \text{m})(5.00 \times 10^{-7}\ \text{m})}{1.50 \times 10^{-4}\ \text{m}} = 2.67 \times 10^{-3}\ \text{m}$. So the distance to the first minimum is one half this, 1.33 mm.

(b) The first maximum and minimum are where the waves have phase differences of zero and pi respectively. Halfway between these points, the phase difference between the waves is $\dfrac{\pi}{2}$. So:

$$I = I_o \cos^2\left(\frac{\phi}{2}\right) = I_o \cos^2\left(\frac{\pi}{4}\right) = \frac{I_o}{2} = 3.00 \times 10^{-6}\ \text{W/m}^2.$$

37-14: (a) The source separation is 120 m, and the wavelength of the wave is $\lambda = \frac{c}{f} = \frac{3.00 \times 10^8 \text{ m/s}}{8.20 \times 10^5 \text{ Hz}} = 366 \text{m}$. So there is only one antinode between the sources (m = 0), and it is a perpendicular bisector of the line connecting the sources.

(b) $I = I_o \cos^2\left(\frac{\phi}{2}\right) = I_o \cos^2\left(\frac{\pi d}{\lambda}\sin\theta\right) = I_o \cos^2\left(\frac{\pi(120 \text{ m})}{(366 \text{ m})}\sin\theta\right) = I_o \cos^2((1.03)\sin\theta)$

So, for $\theta = 30°,\ I = 0.757 I_o;\ \theta = 45°,\ I = 0.557 I_o;$
$\theta = 60°,\ I = 0.394 I_o;\ \theta = 90°,\ I = 0.265 I_o.$

37-15: (a) The distance from the central maximum to the first minimum is half the distance to the first maximum, so:

$y = \frac{R\lambda}{2d} = \frac{(0.600 \text{ m})(6.20 \times 10^{-7} \text{ m})}{2(3.00 \times 10^{-4} \text{ m})} = 6.20 \times 10^{-4} \text{ m}.$

(b) The intensity is half that of the maximum intensity when you are halfway to the first minimum, which is 3.10×10^{-4} m.

37-16: (a) $\lambda = \frac{c}{f} = \frac{3.00 \times 10^8 \text{ m/s}}{9.01 \times 10^7 \text{ Hz}} = 3.33 \text{m}$, and we have:

$\phi = \frac{2\pi}{\lambda}(r_1 - r_2) = \frac{2\pi}{3.33 \text{ m}}(2.2 \text{ m}) = 4.15 \text{ rad}.$

(b) $I = I_o \cos^2\left(\frac{\phi}{2}\right) = I_o \cos^2\left(\frac{4.15 \text{ rad}}{2}\right) = 0.233 I_o.$

37-17: In this situation, both reflected beams undergo half-cycle phase changes. So for the dark fringe spacing: $\Delta x = \frac{\lambda l}{2nh} = \frac{(5.00 \times 10^{-4} \text{ mm})(10 \text{ cm})}{2(1.50)(0.0020 \text{ cm})} = 0.83 \text{ mm}.$

37-18: (a) The number of wavelengths is given by the total extra distance traveled, divided by the wavelength, so the number is

$\frac{x}{\lambda} = \frac{2tn}{\lambda_o} = \frac{2(8.76 \times 10^{-6} \text{ m})(1.35)}{4.80 \times 10^{-7} \text{m}} = 36.5.$

(b) The phase difference for the two parts of the light is zero because the path difference is a half-integer multiple of the wavelength and the top surface reflection has a half-cycle phase shift, while the bottom surface does not.

37-19: $\Delta x = \frac{\lambda l}{2h} = \frac{\lambda}{2\tan\theta}$

$\Rightarrow \theta = \arctan\left(\frac{\lambda}{2\Delta x}\right) = \arctan\left(\frac{5.89 \times 10^{-7} \text{ m}}{2((1/1600) \text{ m})}\right) = 4.71 \times 10^{-4} \text{ rad} = 0.0270°.$

37-20: $\Delta x = \frac{\lambda l}{2h} = \frac{(5.20 \times 10^{-7} \text{ m})(8.00 \text{ cm})}{2(1.00 \times 10^{-4} \text{ m})} = 0.0208 \text{ cm}.$

So the number of fringes per centimeter is $\frac{1}{\Delta x} = 48.1 \text{ fringes/cm}.$

37-21: Both parts of the light undergo half-cycle phase shifts when they reflect, so for destructive interference $t = \frac{\lambda}{4} = \frac{\lambda_o}{4n} = \frac{5.00 \times 10^{-7}\ \text{m}}{4(1.42)} = 8.80 \times 10^{-8}\ \text{m} = 88.0\ \text{nm}.$

37-22: There is a half-cycle phase shift at both interfaces, so for destructive interference: $t = \frac{\lambda}{4} = \frac{\lambda_o}{4n} = \frac{620\ \text{nm}}{4(1.49)} = 104\ \text{nm}.$

37-23: (a) To have a strong reflection, constructive interference is desired. One part of the light undergoes a half-cycle phase shift, so:
$2d = \left(m + \frac{1}{2}\right)\frac{\lambda}{n} \Rightarrow \lambda = \frac{2dn}{\left(m+\frac{1}{2}\right)} = \frac{2(250\ \text{nm})(1.33)}{\left(m+\frac{1}{2}\right)} = \frac{665\ \text{nm}}{\left(m+\frac{1}{2}\right)}.$ For an integer value of zero, the wavelength is not visible (infrared) but for $m = 1$, the wavelength is 443 nm, which is blue.
(b) When the wall thickness is 340 nm, the first visible constructive interference occurs again for $m = 1$ and yields $\lambda = \frac{904\ \text{nm}}{\left(m+\frac{1}{2}\right)} = 603\ \text{nm}$, which is orange.

37-24: (a) Since there is a half-cycle phase shift at just one of the interfaces, the minimum thickness for constructive interference is:
$t = \frac{\lambda}{4} = \frac{\lambda_o}{4n} = \frac{600\ \text{nm}}{4(1.80)} = 83.3\ \text{nm}.$
(b) The next smallest thickness for constructive interference is with another half wavelength thickness added: $t = \frac{3\lambda}{4} = \frac{3\lambda_o}{4n} = \frac{3(600\ \text{nm})}{4(1.80)} = 250\ \text{nm}.$

37-25: The wavelength of the sound, assuming the speed of sound to be 344 m/s, is given by: $\lambda = \frac{v}{f} = \frac{344\ \text{m/s}}{3500\ \text{Hz}} = 0.0983\,\text{m} = 9.83\ \text{cm}$. Assuming the auditory canal acts like a non-reflective layer, the condition for maximum transmission is: $t = 2.5\ \text{cm} = \frac{\lambda}{4} \Rightarrow \lambda = 10\ \text{cm}$. This is very close to the wavelength of 3500 Hz sound waves, and hence the ear is sensitive to frequencies in this range.

37-26: A half-cycle phase change occurs, so for destructive interference
$t = \frac{\lambda}{2} = \frac{\lambda_o}{2n} = \frac{580\ \text{nm}}{2(1.33)} = 218\ \text{nm}.$

37-27: $x = \frac{m\lambda}{2} = \frac{1600(6.06 \times 10^{-7}\ \text{m})}{2} = 4.85 \times 10^{-4}\ \text{m}.$

37-28: (a) For Jan, the total shift was $\Delta x_1 = \frac{m\lambda_1}{2} = \frac{942(6.33 \times 10^{-7}\ \text{m})}{2} = 2.98 \times 10^{-4}\ \text{m}.$

For Linda, the total shift was $\Delta x_2 = \frac{m\lambda_2}{2} = \frac{942(5.89 \times 10^{-7}\ \text{m})}{2} = 2.77 \times 10^{-4}\ \text{m}.$

(b) The net displacement of the mirror is the difference of the above values: $\Delta x = \Delta x_1 - \Delta x_2 = 0.298 \text{ mm} - 0.277 \text{ mm} = 0.021 \text{ mm}$.

37-29: $E_{\min} = hf_{\min} = \dfrac{hc}{\lambda_{\max}} = \dfrac{(6.64\times10^{-34}\ \text{J}\cdot\text{s})(3.00\times10^{8}\ \text{m/s})}{7.00\times10^{-7}\text{m}} = 2.84\times10^{-19}\ \text{J}.$

$E_{\max} = hf_{\max} = \dfrac{hc}{\lambda_{\min}} = \dfrac{(6.64\times10^{-34}\ \text{J}\cdot\text{s})(3.00\times10^{8}\ \text{m/s})}{4.00\times10^{-7}\text{m}} = 4.97\times10^{-19}\ \text{J}.$

37-30: The energy emitted in 30 seconds is $E = Pt = (40{,}000\ \text{W})(30\ \text{s}) = 1.2\times10^{6}\ \text{J}$. So the number of photons emitted in that time, of a given frequency, is:

$$\#\text{photons} = \frac{E}{E_\gamma} = \frac{E}{hf} = \frac{1.2\times10^{6}\ \text{J}}{(6.64\times10^{-34}\ \text{J}\cdot\text{s})(9.93\times10^{7}\ \text{Hz})} = 1.8\times10^{31}.$$

37-31: (a) $F = \dfrac{\Delta p}{\Delta t} = \dfrac{\Delta E/c}{\Delta t} = \dfrac{1}{c}\dfrac{\Delta E}{\Delta t} = \dfrac{P}{c}.$

(b) $F = \dfrac{P}{c} = \dfrac{3.5\times10^{13}\ \text{W}}{3.00\times10^{8}\ \text{m/s}} = 1.2\times10^{5}\ \text{N}.$

37-32: The photon flux is $\Phi = \dfrac{1}{A}\dfrac{dN}{dt} = \dfrac{1}{A}\dfrac{dE/dt}{dE/dN} = \dfrac{1}{A}\dfrac{P}{hf} = \dfrac{I}{hf}$, where I is the intensity of the beam.

37-33: The momentum of the emitted photon must be the same magnitude as the recoiling iron nucleus, so:

$$p_{Fe} = mv = m\sqrt{\frac{2K}{m}} = \sqrt{2Km} = \sqrt{2(3.786\times10^{-22}\ \text{J})(9.454\times10^{-26}\text{kg})}$$

$\Rightarrow p_{Fe} = p_\gamma = 8.461\times10^{-24}\ \text{N}\cdot\text{s}.$

Also, $E = p_\gamma c = (8.461\times10^{-24}\ \text{N}\cdot\text{s})(3.00\times10^{8}\ \text{m/s}) = 2.54\times10^{-15}\ \text{J}$,

$\lambda = \dfrac{h}{p} = \dfrac{6.64\times10^{-34}\ \text{J}\cdot\text{s}}{8.461\times10^{-24}\ \text{N}\cdot\text{s}} = 7.84\times10^{-11}\ \text{m}$, and

$f = \dfrac{c}{\lambda} = \dfrac{3.00\times10^{8}\ \text{m/s}}{7.84\times10^{-11}\ \text{m}} = 3.83\times10^{18}\ \text{Hz}.$

37-34: (a) The de Broglie wavelength is $\lambda = \dfrac{h}{p} = \dfrac{h}{mv}$.

(b) A photon with the same wavelength as the particle has an energy equal to:

$E = hf = \dfrac{hc}{\lambda} = \dfrac{hc\,mv}{h} = mvc.$

(c) $E = mvc = (1.67\times10^{-27}\text{kg})(3.00\times10^{4}\ \text{m/s})(3.00\times10^{8}\ \text{m/s}) = 1.50\times10^{-14}\ \text{J}.$

In electron volts (dividing by $1.60\times10^{-19}\ \text{J/eV}$) the energy is $9.39\times10^{4}\ \text{eV}$.

37-35: (a) If the two sources are out of phase by one half-cycle, we must add an extra half a wavelength to the path difference equations Eq. (37–1) and Eq. (37–2).

This exactly changes one for the other, for $m \to m+\frac{1}{2}$ and $m+\frac{1}{2} \to m$, since m is any integer.
(b) If one source leads the other by a phase angle ϕ, the fraction of a cycle different is $\frac{\phi}{2\pi}$. Thus the path length difference for the two sources must be adjusted for both destructive and constructive interference by this amount. So for constructive interference: $r_1 - r_2 = (m + \phi/2\pi)\lambda$, and for destructive interference, $r_1 - r_2 = (m + 1/2 + \phi/2\pi)\lambda$.

37-36: For constructive interference: $d\sin\theta = m\lambda_1 \Rightarrow d\sin\theta = 3(656\ \text{nm}) = 1968\ \text{nm}$.
For destructive interference: $d\sin\theta = (m+\frac{1}{2})\lambda_2 \Rightarrow \lambda_2 = \dfrac{d\sin\theta}{m+\frac{1}{2}} = \dfrac{1968\ \text{nm}}{m+\frac{1}{2}}$.
So the possible wavelengths are $\lambda_2 = 562$ nm, for $m = 3$, and $\lambda_2 = 437$ nm, for $m = 4$.
Both d and θ drop out of the calculation since their combination is just the path difference, which is the same for both types of light.

37-37: Immersion in water just changes the wavelength of the light from **Ex. (37–9)**, so: $y = \dfrac{R\lambda}{dn} = \dfrac{y_{vacuum}}{n} = \dfrac{1.70\times10^{-3}\ \text{m}}{1.33} = 1.28\times10^{-3}\ \text{m} = 1.28\ \text{mm}$, using the solution from **Ex. (37–9)**.

37-38: $y = \dfrac{R\lambda}{d} \Rightarrow d = \dfrac{R\lambda}{y} = \dfrac{(2.50\ \text{m})(6.00\times10^{-7}\text{m})}{3.60\times10^{-3}\ \text{m}} = 4.17\times10^{-4}\ \text{m} = 0.417\ \text{mm}$.

37-39: (a) Hearing minimum intensity sound means that the path lengths from the individual speakers to you differ by a half-cycle, and are hence out of phase at that position.
(b) By moving the speakers toward you by 0.32 m, a maximum is heard, which means that the signals are back in phase. Therefore the wavelength of the signal is 0.64 m, and the frequency is $f = \dfrac{v}{\lambda} = \dfrac{340\ \text{m/s}}{0.64\ \text{m}} = 531\ \text{Hz}$.
(c) To reach the next maximum, one must move an additional distance of one wavelength, a distance of 0.64 m.

37-40: To find destructive interference, $d = r_2 - r_1 = \sqrt{(200\ \text{m})^2 + x^2} - x = (m+\frac{1}{2})\lambda$

$$\Rightarrow (200\ \text{m})^2 + x^2 = x^2 + \left[(m+\tfrac{1}{2})\lambda\right]^2 + 2x(m+\tfrac{1}{2})\lambda \Rightarrow x = \frac{20{,}000\ \text{m}^2}{(m+\frac{1}{2})\lambda} - \frac{1}{2}(m+\tfrac{1}{2})\lambda.$$

The wavelength is easily calculated by $\lambda = \dfrac{c}{f} = \dfrac{3.00\times10^8\ \text{m/s}}{5.00\times10^6\ \text{Hz}} = 60.0\ \text{m}$.
$\Rightarrow m = 0$: $x = 652$ m, and $m = 1$: $x = 177$ m, and $m = 2$: $x = 58.3$ m.

37-41: We need to find the positions of the first and second dark lines:

$$\theta_1 = \arcsin\left(\frac{\lambda}{2d}\right) = \arcsin\left(\frac{6.00\times10^{-7}\ \text{m}}{2(1.50\times10^{-6}\ \text{m})}\right) = 0.203$$

$$\Rightarrow y_1 = R\tan\theta_1 = (0.450\ \text{m})\tan(0.203) = 0.0919\ \text{m}.$$

Also $\theta_2 = \arcsin\left(\frac{3\lambda}{2d}\right) = \arcsin\left(\frac{3(6.00\times10^{-7}\text{ m})}{2(1.50\times10^{-6}\text{ m})}\right) = 0.644$

$\Rightarrow y_2 = R\tan\theta_2 = (0.450\text{ m})\tan(0.6435) = 0.3375\text{ m}$.

The fringe separation is then $\Delta y = y_2 - y_1 = 0.3375\text{ m} - 0.0919\text{ m} = 0.246\text{ m}$.

37-42: (a) The electric field is the sum of the two wave functions, and can be written:
$E_p(t) = E_1(t) + E_2(t) = E\cos(\omega t) + E\cos(\omega t + \phi) \Rightarrow E_p(t) = 2E\cos(\phi/2)\cos(\omega t + \phi/2)$.

(b) $E_p(t) = A\cos(\omega t + \phi/2)$, so comparing with part (a), we see that the amplitude of the wave (which is always positive) must be $A = 2E|\cos(\phi/2)|$.

(c) To have an interference maximum, $\frac{\phi}{2} = 2\pi m$. So, for example using $m = 1$, the relative phases are $E_1: \varphi = 0;\ E_2: \varphi = \phi = 4\pi;\ E_p: \varphi = \frac{\phi}{2} = 2\pi$, and all waves are in phase.

(d)To have an interference minimum, $\frac{\phi}{2} = \pi(m + \frac{1}{2})$. So, for example using $m = 0$, the relative phases are $E_1: \varphi = 0;\ E_2: \varphi = \phi = \pi;\ E_p: \varphi = \phi/2 = \pi/2$, and the resulting wave is out of phase by a quarter of a cycle from both of the original waves.

(e) The instantaneous magnitude of the Poynting vector is:
$|S| = \varepsilon_o c E_p^{\,2}(t) = \varepsilon_o c\left(4E^2\cos^2(\phi/2)\cos^2(\omega t + \phi/2)\right)$.

For a time average, $\cos^2(\omega t + \phi/2) = \frac{1}{2}$, so $|S_{av}| = 2\varepsilon_o c E^2\cos^2(\phi/2)$.

37-43: First we need to find the angles at which the intensity drops by one half from the value of the m-th bright fringe.

$$I = I_o\cos^2\left(\frac{\pi d}{\lambda}\sin\theta\right) = \frac{I_o}{2} \Rightarrow \frac{\pi d}{\lambda}\sin\theta \approx \frac{\pi d\theta_m}{\lambda} = (m + 1/2)\frac{\pi}{2}.$$

$\Rightarrow m = 0:\ \theta = \theta_m^- = \frac{\lambda}{4d};\quad m = 1:\ \theta = \theta_m^+ = \frac{3\lambda}{4d} \Rightarrow \Delta\theta_m = \frac{\lambda}{2d}$, so there is no dependence on the m-value of the fringe.

37-44: There is just one half-cycle phase change upon reflection, so for constructive interference: $2t = (m_1 + \frac{1}{2})\lambda_1 = (m_2 + \frac{1}{2})\lambda_2$. But the two different wavelengths differ by just one m-value, $m_2 = m_1 - 1$.

$$\Rightarrow (m_1 + \tfrac{1}{2})\lambda_1 = (m_1 - \tfrac{1}{2})\lambda_2 \Rightarrow m_1(\lambda_2 - \lambda_1) = \frac{\lambda_1 + \lambda_2}{2} \Rightarrow m_1 = \frac{\lambda_1 + \lambda_2}{2(\lambda_2 - \lambda_1)}$$

$$\Rightarrow m_1 = \frac{500.0\text{nm} + 590.9\text{nm}}{2(590.9\text{nm} - 500.0\text{nm})} = 6.00.$$

$$\Rightarrow 2t = (6 + \tfrac{1}{2})\frac{\lambda_1}{n} \Rightarrow t = \frac{13(500.0\text{ nm})}{2(1.52)} = 1069\text{ nm}.$$

37-45: (a) Intensified reflected light means we have constructive interference. There is one half-cycle phase shift, so:

$$2t = (m+\tfrac{1}{2})\frac{\lambda}{n} \Rightarrow \lambda = \frac{2tn}{(m+\frac{1}{2})} = \frac{2(425\,\text{nm})(1.50)}{(m+\frac{1}{2})} = \frac{1275\,\text{nm}}{(m+\frac{1}{2})}.$$

$\Rightarrow \lambda = 510\,\text{nm}$ $(m=2)$, is the only visible light wavelength that is intensified.

(b) Intensified transmitted light means we have destructive interference. There is still a one half-cycle phase shift, so:

$$2t = \frac{m\lambda}{n} \Rightarrow \lambda = \frac{2tn}{m} = \frac{2(425\,\text{nm})(1.50)}{m} = \frac{1275\,\text{nm}}{m}$$

$\Rightarrow \lambda = 638\,\text{nm}$ $(m=2)$ and $\lambda = 425\,\text{nm}$ $(m=3)$, are the only wavelengths of visible light that are intensified.

37-46: (a) There is one half-cycle phase shift, so for constructive interference:

$$2t = (m+\tfrac{1}{2})\frac{\lambda_o}{n} \Rightarrow \lambda = \frac{2tn}{(m+\frac{1}{2})} = \frac{2(430\,\text{nm})(1.40)}{(m+\frac{1}{2})} = \frac{1204\,\text{nm}}{(m+\frac{1}{2})}.$$ Therefore, we have constructive interference at $\lambda = 482\,\text{nm}$ $(m=2)$, which corresponds to blue.

(b) Beneath the water, looking for maximum intensity means that the reflected part of the wave at that wavelength must be weak, or have interfered destructively. So:

$$2t = \frac{m\lambda_o}{n} \Rightarrow \lambda_o = \frac{2tn}{m} = \frac{2(430\,\text{nm})(1.40)}{m} = \frac{1204\,\text{nm}}{m}.$$ Therefore the strongest transmitted wavelength (as measured in air) is $\lambda = 602\,\text{nm}$ $(m=2)$.

37-47: For maximum intensity, with a half-cycle phase shift,

$$2t = (m+\tfrac{1}{2})\lambda \text{ and } t = R - \sqrt{R^2 - r^2} \Rightarrow \frac{(2m+1)\lambda}{4} = R - \sqrt{R^2 - r^2}$$

$$\Rightarrow \sqrt{R^2 - r^2} = R - \frac{(2m+1)\lambda}{4} \Rightarrow R^2 - r^2 = R^2 + \left[\frac{(2m+1)\lambda}{4}\right]^2 - \frac{(2m+1)\lambda R}{2}$$

$$\Rightarrow r = \sqrt{\frac{(2m+1)\lambda R}{2} - \left[\frac{(2m+1)\lambda}{4}\right]^2} \Rightarrow r \approx \sqrt{\frac{(2m+1)\lambda R}{2}}, \text{ for } R >> \lambda.$$

The third bright ring is when $m = 2$:

$$r \approx \sqrt{\frac{(2(2)+1)(6.50\times10^{-7}\,\text{m})(1.15\,\text{m})}{2}} = 1.37\times10^{-3}\,\text{m} = 1.37\,\text{mm}.$$

So the diameter of the third bright ring is 2.73 mm.

37-48: As found in **Pr. (37–47)**, the radius of the m-th bright ring is in general:

$r \approx \sqrt{\frac{(2m+1)\lambda R}{2}}$, for $R >> \lambda$. Introducing a liquid between the lens and the plate just changes the wavelength from $\lambda \to \frac{\lambda}{n}$.

So: $$r(n) \approx \sqrt{\frac{(2m+1)\lambda R}{2n}} = \frac{r}{\sqrt{n}} = \frac{0.450\,\text{mm}}{\sqrt{1.30}} = 0.395\,\text{mm}.$$

37-49: (a) There is a half-cycle phase change at the glass, so for constructive interference:

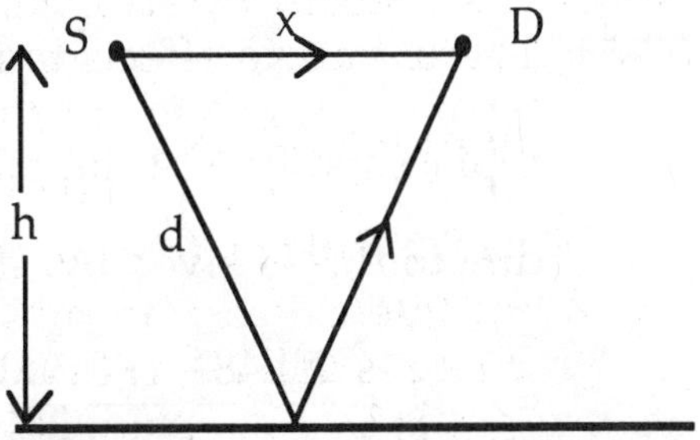

$$2d - x = 2\sqrt{h^2 + \left(\frac{x}{2}\right)^2} - x = (m + \tfrac{1}{2})\lambda$$

$$\Rightarrow \sqrt{x^2 + 4h^2} - x = (m + \tfrac{1}{2})\lambda.$$

Similarly for destructive interference:

$$\sqrt{x^2 + 4h^2} - x = m\lambda.$$

S x D h d

(b) The longest wavelength for constructive interference is when m = 0:

$$\lambda = \frac{\sqrt{x^2 + 4h^2} - x}{m + \frac{1}{2}} = \frac{\sqrt{(8.0 \text{ cm})^2 + 4(30.0 \text{ cm})^2} - 8.0 \text{ cm}}{1/2} = 105 \text{ cm}.$$

37-50: (a) At the water (or cytoplasm) to guanine interface, there is a half-cycle phase shift for the reflected light, but there is not one at the guanine to cytoplasm interface. Therefore there will always be one half-cycle phase difference between two neighboring reflected beams. For the guanine layers:

$$2t_g = (m + \tfrac{1}{2})\frac{\lambda}{n_g} \Rightarrow \lambda = \frac{2t_g n_g}{(m + \frac{1}{2})} = \frac{2(74 \text{nm})(1.80)}{(m + \frac{1}{2})} = \frac{266 \text{ nm}}{(m + \frac{1}{2})} \Rightarrow \lambda = 533 \text{ nm } (m = 0).$$

For the cytoplasm layers:

$$2t_c = (m + \tfrac{1}{2})\frac{\lambda}{n_c} \Rightarrow \lambda = \frac{2t_c n_c}{(m + \frac{1}{2})} = \frac{2(100 \text{nm})(1.333)}{(m + \frac{1}{2})} = \frac{267 \text{ nm}}{(m + \frac{1}{2})} \Rightarrow \lambda = 533 \text{ nm } (m = 0).$$

(b) By having many layers the reflection is strengthened, because at each interface some more of the transmitted light gets reflected back, increasing the total percentage reflected.

(c) At different angles, the path length in the layers change (always to a larger value than the normal incidence case). If the path length changes, then so do the wavelengths which will interfere constructively upon reflection.

37-51: (a) Adding glass over the top slit increases the effective path length from that slit to the screen. The interference pattern will therefore change, with the central maximum shifting downwards.

(b) Normally the phase shift is $\phi = \frac{2\pi d}{\lambda}\sin\theta$, but now there is an added shift from the glass, so the total phase shift is now

$$\phi = \frac{2\pi d}{\lambda}\sin\theta + \left(\frac{2\pi Ln}{\lambda} - \frac{2\pi L}{\lambda}\right) = \frac{2\pi d}{\lambda}\sin\theta + \frac{2\pi L(n-1)}{\lambda} = \frac{2\pi}{\lambda}(d\sin\theta + L(n-1)).$$

So the intensity becomes $I = I_o\cos^2\frac{\phi}{2} = I_o\cos^2\left(\frac{\pi}{\lambda}(d\sin\theta + L(n-1))\right)$.

(c) The maxima occur at $\frac{\pi}{\lambda}(d\sin\theta + L(n-1)) = m\pi \Rightarrow d\sin\theta = m\lambda - L(n-1)$.

37-52: The passage of fringes indicates an effective change in path length, since the wavelength of the light is getting shorter as more gas enters the tube.

$$\Delta m = \frac{L}{\lambda / n} - \frac{L}{\lambda} = \frac{L}{\lambda}(n-1) \Rightarrow (n-1) = \frac{\Delta m \lambda}{L}.$$

So here: $(n-1) = \frac{52(5.89 \times 10^{-7} \text{ m})}{0.0800 \text{m}} = 3.83 \times 10^{-4}$.

37-53: There are two effects to be considered: first, the expansion of the rod, and second, the change in the rod's refractive index. The extra length of rod replaces a little of the air so that the change in the number of wavelengths due to this is given by: $\Delta N_1 = \frac{2n_{glass}\Delta L}{\lambda_o} - \frac{2n_{air}\Delta L}{\lambda_o} = \frac{2(n_{glass}-1)L_o\alpha\Delta T}{\lambda_o}$.

$$\Rightarrow \Delta N_1 = \frac{2(1.48-1)(0.300\text{ m})(5.00\times10^{-6}/\text{C}^\circ)(5.00\text{ C}^\circ)}{5.89\times10^{-7}\text{ m}} = 1.22.$$

The change in the number of wavelengths due to the change in refractive index of the rod is:

$$\Delta N_2 = \frac{2\Delta n_{glass}L_o}{\lambda_o} = \frac{2(2.50\times10^{-5}/\text{C}^\circ)(5.00\text{ C}^\circ/\text{min})(1.00\text{ min})(0.0300\text{ m})}{5.89\times10^{-7}\text{m}} = 12.73.$$

So the total change in the number of wavelengths as the rod expands is $\Delta N = 12.73 + 1.22 = 14.0$ fringes/minute.

37-54: (a) Since we can approximate the angles of incidence on the prism as being small, Snell's Law tells us that an incident angle of θ on the flat side of the prism enters the prism at an angle of θ/n, where n is the index of refraction of the prism. Similarly on leaving the prism, the in-going angle is $\theta/n - A$ from the normal, and the outgoing, relative to the prism, is $n(\theta/n - A)$. So the beam leaving the prism is at an angle of $\theta' = n(\theta/n - A) + A$ from the optical axis. So $\theta - \theta' = (n-1)A$.

At the plane of the source S_o, we can calculate the height of one image above the source: $\frac{d}{2} = \tan(\theta-\theta')a \approx (\theta-\theta')a = (n-1)Aa \Rightarrow d = 2aA(n-1)$.

(b) To find the spacing of fringes on a screen, we use:

$$\Delta y = \frac{R\lambda}{d} = \frac{R\lambda}{2aA(n-1)} = \frac{(2.00\text{ m}+0.200\text{ m})(5.00\times10^{-7}\text{ m})}{2(0.200\text{ m})(3.50\times10^{-3}\text{ rad})(1.50 - 1.00)} = 1.57\times10^{-3}\text{ m}.$$

Chapter 38: Diffraction

38-1: The angle to the first dark fringe is simply:

$$\theta = \arctan\left(\frac{y_{\text{max}}}{2}\frac{1}{x}\right) = \arctan\left(\frac{0.027\text{ m}}{2}\frac{1}{6.0\text{ m}}\right) = 0.129°.$$

38-2: (a) $y_1 = \frac{x\lambda}{a} \Rightarrow a = \frac{x\lambda}{y_1} = \frac{(4.00\text{m})(5.00\times10^{-7}\text{ m})}{4.00\times10^{-3}\text{ m}} = 5.00\times10^{-4}\text{ m}.$

(b) $a = \frac{x\lambda}{y_1} = \frac{(4.00\text{m})(5.00\times10^{-5}\text{ m})}{4.00\times10^{-3}\text{ m}} = 5.00\times10^{-2}\text{ m} = 5.00\text{ cm}.$

(c) $a = \frac{x\lambda}{y_1} = \frac{(4.00\text{m})(5.00\times10^{-10}\text{ m})}{4.00\times10^{-3}\text{ m}} = 5.00\times10^{-7}\text{ m}.$

38-3: $y_1 = \frac{x\lambda}{a} \Rightarrow \lambda = \frac{y_1 a}{x} = \frac{(1.25\times10^{-3}\text{ m})(8.00\times10^{-4}\text{ m})}{3.00\text{ m}} = 3.33\times10^{-7}\text{ m}.$

38-4: $y_1 = \frac{x\lambda}{a} \Rightarrow a = \frac{x\lambda}{y_1} = \frac{(0.400\text{m})(5.46\times10^{-7}\text{ m})}{9.75\times10^{-3}\text{ m}} = 2.24\times10^{-5}\text{ m}.$

38-5: (a) $y_1 = \frac{x\lambda}{a} = \frac{(4.00\text{ m})(6.33\times10^{-7}\text{ m})}{2.50\times10^{-4}\text{ m}} = 0.0101\text{ m}$. So the width of the brightest fringe is twice this distance to the first minimum, 0.0203 m.

(b) $y_2 = \frac{2x\lambda}{a} = \frac{2(4.00\text{ m})(6.33\times10^{-7}\text{ m})}{2.50\times10^{-4}\text{ m}} = 0.0203\text{ m}$. So the width of the first bright fringe on the side of the central maximum, is the distance from y_2 to y_1, which is 0.0101 m.

38-6: $D = 2y_1 = \frac{2x\lambda}{a} = \frac{2(3.00\text{ m})(5.89\times10^{-7}\text{ m})}{8.50\times10^{-4}\text{ m}} = 4.16\times10^{-3}\text{ m}.$

38-7: The angle to the first minimum is $\theta = \arcsin\left(\frac{\lambda}{a}\right) = \arcsin\left(\frac{8.00\text{ cm}}{10.0\text{ cm}}\right) = 0.927\text{ rad}$.
So the distance from the central maximum to the first minimum is just $y_1 = x\tan\theta = (50.0\text{ cm})\tan(0.927\text{ rad}) = \pm 66.7\text{cm}$.

38-8: The total intensity is given by drawing an arc of a circle that has length E_0 and finding the length of the cord which connects the starting and ending points of the curve. So graphically we can find the electric field at a point by examining the geometry as shown below for three cases.

38-8 cont:

(a) $\beta = \dfrac{2\pi a}{\lambda}\sin\theta = \dfrac{2\pi a}{\lambda}\cdot\dfrac{\lambda}{2a} = \pi$. From the diagram, $\pi\dfrac{E_p}{2} = E_o \Rightarrow E_p = \dfrac{2}{\pi}E_o$.

So the intensity is just:

$$I = \left(\frac{2}{\pi}\right)^2 I_o = \frac{4I_o}{\pi^2}.$$

This agrees with Eq. (38–5).

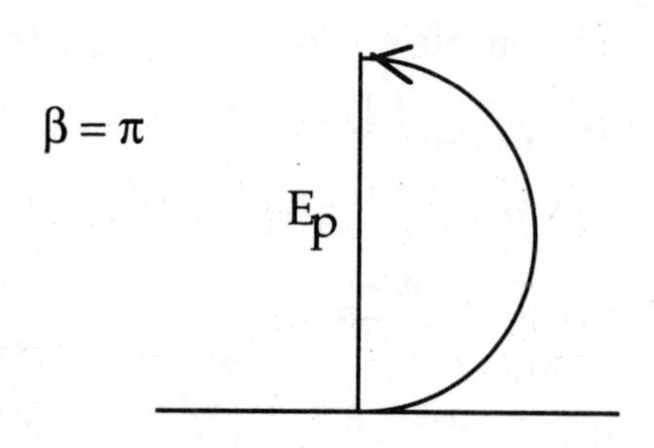

(b) $\beta = \dfrac{2\pi a}{\lambda}\sin\theta = \dfrac{2\pi a}{\lambda}\cdot\dfrac{\lambda}{a} = 2\pi$. From the diagram, it is clear that the total amplitude is zero, as is the intensity. This also agrees with Eq. (38–5).

(c) $\beta = \dfrac{2\pi a}{\lambda}\sin\theta = \dfrac{2\pi a}{\lambda}\cdot\dfrac{3\lambda}{2a} = 3\pi$. From the diagram, $3\pi\dfrac{E_p}{2} = E_o \Rightarrow E_p = \dfrac{2}{3\pi}E_o$.

So the intensity, is just:

$$I = \left(\frac{2}{3\pi}\right)^2 I_o = \frac{4}{9\pi^2}I_o.$$

This agrees with Eq. (38–5).

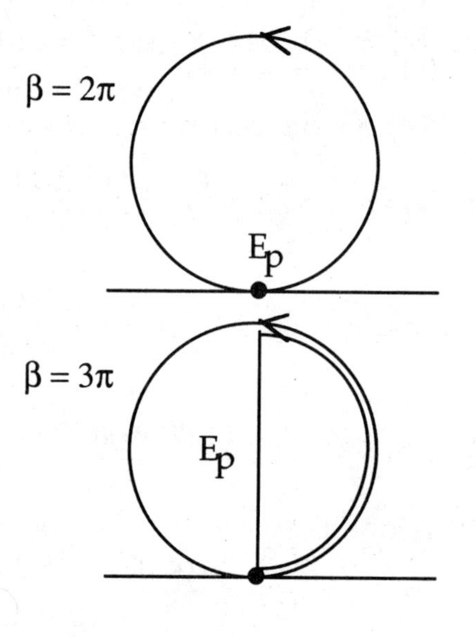

38-9: (a) $\beta = \dfrac{2\pi a}{\lambda}\sin\theta \Rightarrow \lambda = \dfrac{2\pi a}{\beta}\sin\theta = \dfrac{2\pi(1.25\times10^{-4}\text{ m})}{48.0\text{ rad}}\sin 5.35° = 1.53\times10^{-6}\text{ m}.$

(b) $I = I_o\left(\dfrac{\sin(\beta/2)}{\beta/2}\right)^2 = I_o\left(\dfrac{\sin(48.0/2)}{48.0/2}\right)^2 = (1.42\times10^{-3})I_o.$

38-10: $\beta = \dfrac{2\pi a}{\lambda}\sin\theta \approx \dfrac{2\pi a}{\lambda}\cdot\dfrac{y}{x} = \dfrac{2\pi(4.50\times10^{-4}\text{m})}{(6.20\times10^{-7}\text{ m})(3.00\text{ m})}y = (1520\text{ m}^{-1})y.$

(a) $y = 1.00\times10^{-3}$ m: $\dfrac{\beta}{2} = \dfrac{(1520\text{ m}^{-1})(1.00\times10^{-3}\text{ m})}{2} = 0.760.$

$\Rightarrow I = I_o\left(\dfrac{\sin(\beta/2)}{\beta/2}\right)^2 = I_o\left(\dfrac{\sin(0.760)}{0.760}\right)^2 = 0.822\,I_o$

(b) $y = 3.00\times10^{-3}$ m: $\dfrac{\beta}{2} = \dfrac{(1520\text{ m}^{-1})(3.00\times10^{-3}\text{ m})}{2} = 2.28.$

$\Rightarrow I = I_o\left(\dfrac{\sin(\beta/2)}{\beta/2}\right)^2 = I_o\left(\dfrac{\sin(2.28)}{2.28}\right)^2 = 0.111\,I_o.$

(c) $y = 5.00\times10^{-3}$ m: $\dfrac{\beta}{2} = \dfrac{(1520\text{ m}^{-1})(5.00\times10^{-3}\text{ m})}{2} = 3.80.$

$\Rightarrow I = I_o\left(\dfrac{\sin(\beta/2)}{\beta/2}\right)^2 = I_o\left(\dfrac{\sin(3.80)}{3.80}\right)^2 = 0.0259\,I_o.$

38-11: (a) $y_1 = \frac{x\lambda}{a} = \frac{(4.00\text{ m})(5.90\times10^{-7}\text{ m})}{2.00\times10^{-4}\text{ m}} = 0.0118\text{ m}.$

(b) $\beta = \frac{2\pi a}{\lambda}\sin\theta \approx \frac{2\pi a}{\lambda}\cdot\frac{y_1/2}{x} = \frac{2\pi a}{\lambda}\cdot\frac{x\lambda}{2ax} = \pi.$

$\Rightarrow I = I_o\left(\frac{\sin(\beta/2)}{\beta/2}\right)^2 = (5.00\times10^{-6}\text{ W/m}^2)\left(\frac{\sin(\pi/2)}{\pi/2}\right)^2 = 2.03\times10^{-6}\text{ W/m}^2.$

38-12: (a) $\theta = 0$: $\beta = \frac{2\pi a}{\lambda}\sin 0° = 0.$

(b) At the second minimum from the center $\beta = \frac{2\pi a}{\lambda}\sin\theta = \frac{2\pi a}{\lambda}\cdot\frac{2\lambda}{a} = 4\pi.$

(c) $\beta = \frac{2\pi a}{\lambda}\sin\theta = \frac{2\pi(2.50\times10^{-4}\text{ m})}{5.00\times10^{-7}\text{ m}}\sin 7.0° = 383\text{ rad}.$

38-13: $\beta = \frac{2\pi a}{\lambda}\sin\theta \Rightarrow \lambda = \frac{2\pi a}{\beta}\sin\theta = \frac{2\pi(4.00\times10^{-4}\text{ m})}{\pi/2}\sin 3.2° = 8.9\times10^{-5}\text{ m}.$

38-14: By examining the diagram, we see that every fourth slit cancels each other.

38-15: With four slits there must be four vectors in each phasor diagram, with the orientation of each successive one determined by the relative phase shifts. So:

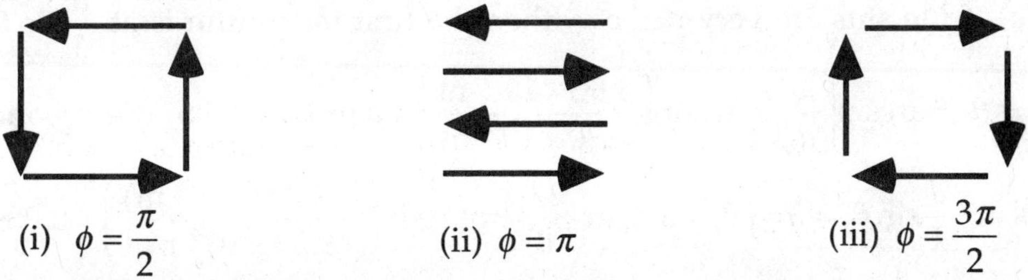

(i) $\phi = \frac{\pi}{2}$ (ii) $\phi = \pi$ (iii) $\phi = \frac{3\pi}{2}$

We see that destructive interference occurs from adjacent slits in case (ii) and from alternate slits in cases (i) and (iii).

38-16: (a) Ignoring diffraction, the first five maxima will occur as given by:

$d\sin\theta = m\lambda \Rightarrow \theta = \arcsin\left(\frac{m\lambda}{d}\right) = \arcsin\left(\frac{m\lambda}{4a}\right)$, for $m = 1,2,3,4,5.$

(b) $\beta = \frac{2\pi a}{\lambda}\sin\theta = \frac{2\pi a}{\lambda}\cdot\frac{m\lambda}{d} = \frac{m\pi}{2}$, and $\phi = \frac{2\pi d}{\lambda}\sin\theta = \frac{2\pi d}{\lambda}\cdot\frac{m\lambda}{d} = 2\pi m.$

So including diffraction, the intensity:

$$I = I_o\cos^2\frac{\phi}{2}\left(\frac{\sin(\beta/2)}{\beta/2}\right)^2 = I_o\cos^2\frac{2\pi m}{2}\left(\frac{\sin(m\pi/4)}{m\pi/4}\right)^2 = I_o\left(\frac{\sin(m\pi/4)}{m\pi/4}\right)^2.$$

So for $m=1$: $I_1=\left(\frac{\sin(\pi/4)}{\pi/4}\right)^2 I_o=0.811I_o$; $m=2$: $I_2=\left(\frac{\sin(\pi/2)}{\pi/2}\right)^2 I_o=0.405I_o$;

$m=3$: $I_3=\left(\frac{\sin(3\pi/4)}{3\pi/4}\right)^2 I_o=0.0901I_o$; $m=4$: $I_4=\left(\frac{\sin(\pi)}{\pi}\right)^2 I_o=0$;

$m=5$: $I_5=\left(\frac{\sin(5\pi/4)}{5\pi/4}\right)^2 I_o=0.0324I_o$.

38-17: (a) If $\frac{d}{a}=3$, then there are 5 fringes: $m=0,\pm1,\pm2$.

(b) On one side of the central maximum, within the first diffraction maximum there are two fringes, $m=1,2$.

38-18: Diffraction dark fringes occur for $\sin\theta=\frac{m_d\lambda}{a}$, and interference maxima occur for $\sin\theta=\frac{m_i\lambda}{d}$. Setting them equal to each other yields a missing bright spot whenever the destructive interference matches the bright spots. That is: $\frac{m_i\lambda}{d}=\frac{m_d\lambda}{a}\Rightarrow m_i=\frac{d}{a}m_d=3m_d$. That is, the missing parts of the pattern occur for $m_i=3,6,9...=3m$, for $m=$ integers..

38-19: (a) If the slits are very narrow, then the first maximum is at $\frac{d}{\lambda}\sin\theta_1=1$.

$\Rightarrow\theta_1=\arcsin\left(\frac{\lambda}{d}\right)=\arcsin\left(\frac{4.90\times10^{-7}\text{ m}}{6.30\times10^{-4}\text{ m}}\right)=0.0446°$. Also, the second maximum is at $\frac{d}{\lambda}\sin\theta_2=2\Rightarrow\theta_2=\arcsin\left(\frac{2\lambda}{d}\right)=\arcsin\left(\frac{2(4.90\times10^{-7}\text{ m})}{6.30\times10^{-4}\text{ m}}\right)=0.0891°$.

(b) $I_1=I_o\left(\frac{\sin(\pi a\sin\theta_1/\lambda)}{\pi a\sin\theta_1/\lambda}\right)^2=I_o\left(\frac{\sin(\pi a/d)}{\pi a/d}\right)^2$

$$\Rightarrow I_1=I_o\left(\frac{\sin(\pi(4.20\times10^{-4}\text{ m})/(6.30\times10^{-4}\text{ m}))}{\pi(4.20\times10^{-4}\text{ m})/(6.30\times10^{-4}\text{ m})}\right)^2=0.170I_o.$$

And $I_2=I_o\left(\frac{\sin(\pi a\sin\theta_2/\lambda)}{\pi a\sin\theta_2/\lambda}\right)^2=I_o\left(\frac{\sin(2\pi a/d)}{2\pi a/d}\right)^2$

$$\Rightarrow I_2=I_o\left(\frac{\sin(2\pi(4.20\times10^{-4}\text{ m})/(6.30\times10^{-4}\text{ m}))}{2\pi(4.20\times10^{-4}\text{ m})/(6.30\times10^{-4}\text{ m})}\right)^2=0.0427I_o.$$

38-20: We will use $I=I_o\cos^2\frac{\phi}{2}\left(\frac{\sin(\beta/2)}{\beta/2}\right)^2$, and must calculate the phases ϕ and β.

Using $\beta/2=\frac{\pi a}{\lambda}\sin\theta$, and $\phi=\frac{2\pi d}{\lambda}\sin\theta$, we have:

(a) $\theta = 1.25\times10^{-4}$ rad : $\beta/2 = 0.314$, and $\phi = 1.57 \Rightarrow I = 0.484\,I_o$.
(b) $\theta = 2.50\times10^{-4}$ rad: $\beta/2 = 0.628$, and $\phi = 3.14 \Rightarrow I = 0$.
(c) $\theta = 3.00\times10^{-4}$ rad : $\beta/2 = 0.754$, and $\phi = 3.77 \Rightarrow I = (0.0787)\,I_o$

38-21: 400 slits / mm $\Rightarrow d = \dfrac{1}{4.00\times10^5\ \text{m}^{-1}} = 2.50\times10^{-6}$ m. Then:

$$d\sin\theta = m\lambda \Rightarrow \theta = \arcsin\left(\frac{m\lambda}{d}\right) = \arcsin\left(\frac{m(6.00\times10^{-7}\ \text{m})}{2.50\times10^{-6}\ \text{m}}\right) = \arcsin((0.240)m)$$

$\Rightarrow m = 1$: $\theta = 13.9°$; $m = 2$: $\theta = 28.7°$; $m = 3$: $\theta = 46.1°$.

38-22: 6000 slits / cm $\Rightarrow d = \dfrac{1}{6.00\times10^5\ \text{m}^{-1}} = 1.67\times10^{-6}$ m.

(a) $d\sin\theta = m\lambda \Rightarrow \lambda = \dfrac{d\sin\theta}{m} = \dfrac{(1.67\times10^{-6}\ \text{m})\sin 12.8°}{1} = 3.69\times10^{-7}$ m.

(b) $m = 2$: $\theta = \arcsin\left(\dfrac{m\lambda}{d}\right) = \arcsin\left(\dfrac{2(3.69\times10^{-7}\ \text{m})}{1.67\times10^{-6}\ \text{m}}\right) = 26.2°$.

38-23: 4000 slits / cm $\Rightarrow d = \dfrac{1}{4.00\times10^5\ \text{m}^{-1}} = 2.50\times10^{-6}$ m. So for the α-hydrogen line, we have: $\theta = \arcsin\left(\dfrac{m\lambda}{d}\right) = \arcsin\left(\dfrac{m(6.56\times10^{-7}\ \text{m})}{2.50\times10^{-6}\ \text{m}}\right) = \arcsin((0.262)m)$.

$\Rightarrow m = 1$: $\theta_1 = 15.2°$; $m = 2$: $\theta = 31.6°$. And for the δ-hydrogen line, the angle's given by: $\theta = \arcsin\left(\dfrac{m\lambda}{d}\right) = \arcsin\left(\dfrac{m(4.10\times10^{-7}\ \text{m})}{2.50\times10^{-6}\ \text{m}}\right) = \arcsin((0.164)m)$.

$\Rightarrow m = 1$: $\theta_1 = 9.44°$; $m = 2$: $\theta = 19.1°$; $\Rightarrow \Delta\theta_1 = 5.8°$, $\Delta\theta_2 = 12.5°$.

38-24: 350 slits/mm $\Rightarrow d = \dfrac{1}{3.50\times10^5\ \text{m}^{-1}} = 2.86\times10^{-6}$ m, and $d\sin\theta = m\lambda$.

$$\Rightarrow m = 1:\ \theta_{400} = \arcsin\left(\frac{\lambda}{d}\right) = \arcsin\left(\frac{4.00\times10^{-7}\ \text{m}}{2.86\times10^{-6}\ \text{m}}\right) = 8.05°.$$

$$\theta_{700} = \arcsin\left(\frac{\lambda}{d}\right) = \arcsin\left(\frac{7.00\times10^{-7}\ \text{m}}{2.86\times10^{-6}\ \text{m}}\right) = 14.18°.$$

$\Rightarrow \Delta\theta_1 = 14.18° - 8.05° = 6.13°$.

$$\Rightarrow m = 3:\ \theta_{400} = \arcsin\left(\frac{3\lambda}{d}\right) = \arcsin\left(\frac{3(4.00\times10^{-7}\ \text{m})}{2.86\times10^{-6}\ \text{m}}\right) = 24.8°.$$

$$\theta_{700} = \arcsin\left(\frac{3\lambda}{d}\right) = \arcsin\left(\frac{3(7.00\times10^{-7}\ \text{m})}{2.86\times10^{-6}\ \text{m}}\right) = 47.3°.$$

$\Rightarrow \Delta\theta_1 = 47.3° - 24.8° = 22.5°$.

38-25: $\theta = \arcsin\left(\frac{m\lambda}{d}\right) = \arcsin\left(\frac{m(6.328\times10^{-7}\text{ m})}{1.60\times10^{-6}\text{ m}}\right) = \arcsin((0.396)m)$.
$\Rightarrow m=1\text{: } \theta_1 = 23.3°;\ m=2\text{: } \theta = 52.3°$. All other m-values lead to angles greater than 90°.

38-26: $R = \frac{\lambda}{\Delta\lambda} = Nm \Rightarrow N = \frac{\lambda}{m\Delta\lambda} = \frac{6.5645\times10^{-7}\text{ m}}{2(6.5645\times10^{-7}\text{ m} - 6.5627\times10^{-7}\text{ m})} = 1820$ slits.

38-27: For x-ray diffraction, $2d\sin\theta = m\lambda \Rightarrow d = \frac{m\lambda}{2\sin\theta}$

$\Rightarrow d = \frac{2(8.20\times10^{-11}\text{ m})}{2\sin 23.5°} = 2.06\times10^{-10}\text{ m}$.

38-28: For the first order maximum in Bragg reflection:
$2d\sin\theta = m\lambda \Rightarrow \lambda = \frac{2d\sin\theta}{m} = \frac{2(4.00\times10^{-10}\text{ m})\sin 37.2°}{1} = 4.84\times10^{-10}\text{ m}$.

38-29: $\sin\theta_1 = 1.22\frac{\lambda}{D} \Rightarrow D = \frac{1.22\lambda}{\sin\theta_1} = 1.22\lambda\frac{h}{W} = 1.22(0.040\text{ m})\frac{9.0\times10^{5}\text{ m}}{4.2\times10^{4}\text{ m}}$
$\Rightarrow D = 1.05\text{ m}$.

38-30: $\sin\theta_1 = 1.22\frac{\lambda}{D} \Rightarrow \lambda = \frac{D\sin\theta_1}{1.22} \approx \frac{D\theta_1}{1.22} = \frac{(8.00\times10^{6}\text{ m})(1.00\times10^{-8})}{1.22}$
$\Rightarrow \lambda = 0.0656\text{ m} = 6.56\text{ cm}$.

38-31: $\sin\theta_1 = 1.22\frac{\lambda}{D} = 1.22\frac{4.8\times10^{-7}\text{ m}}{8.4\times10^{-6}\text{ m}} = 7.0\times10^{-2}$. The screen is 4.0 m away, so the diameter of the Airy ring is given from simple trigonometry:
$D = 2y = 2x\tan\theta \approx 2x\sin\theta = 2(4.0\text{m})(7.0\times10^{-2}) = 56\text{ cm}$.

38-32: The image is 15.0 cm from the lens, and from the diagram and Rayleigh's criteria, the diameter of the circles is twice the "height" as given by:
$D = 2|y'| = \frac{2s'}{s}y = \frac{2fy}{s} = \frac{2(0.150\text{ m})(6.00\times10^{-3}\text{ m})}{40.0\text{ m}} = 4.50\times10^{-5}\text{ m} = 0.0450\text{ mm}$.

38-33: $\sin\theta_1 = 1.22\frac{\lambda}{D} \Rightarrow D = \frac{1.22\lambda}{\sin\theta_1} \approx 1.22\lambda\frac{R}{W} = 1.22(5.0\times10^{-7}\text{ m})\frac{1.49\times10^{11}\text{ m}}{3.0\times10^{4}\text{ m}} = 3.05\text{m}$.

38-34: $\sin\theta_1 = 1.22\frac{\lambda}{D} = \frac{y}{s} \Rightarrow s = \frac{yD}{1.22\lambda} = \frac{(6.00\times10^{-3}\text{ m})(0.0520\text{ m})}{1.22(5.50\times10^{-7}\text{ m})} = 465\text{m}$.

38-35: The best resolution is 0.3 arcseconds, which is about $8.33\times10^{-5}°$.
(a) $D = \frac{1.22\lambda}{\sin\theta_1} = \frac{1.22(5.5\times10^{-7}\text{ m})}{\sin(8.33\times10^{-5}°)} = 0.46\text{m} \approx 0.5\text{m}$.

(b) The Keck telescope is able to gather more light than the Hale telescope, and hence it can detect fainter objects. However, its larger size does not allow it to have greater resolution – atmospheric conditions limit the resolution.

38-36: $D = \dfrac{1.22\lambda}{\sin\theta_1} = \dfrac{1.22(5.5\times10^{-7}\text{ m})}{\sin(1/60)^\circ} = 2.31\times10^{-3}\text{ m} = 2.3\text{ mm}.$

38-37: (a) $I = I_o/2 \Rightarrow \dfrac{1}{\sqrt{2}} = \dfrac{\sin(\pi a\sin\theta/\lambda)}{\pi a\sin\theta/\lambda} \equiv \dfrac{\sin x}{x} = 0.7071$. Solving for x through trial and error, and remembering to use radians throughout, one finds $x = 1.39$ rad and $\beta = 2x = 2.78$ rad. Also, $\Delta\theta = |\theta_+ - \theta_-| = 2\theta_+$, and

$$\beta = \frac{2\pi a}{\lambda}\sin\theta \Rightarrow \sin\theta_+ = \frac{\lambda\beta}{2\pi a} = \frac{\lambda}{a}\left(\frac{2.78\text{ rad}}{2\pi\text{ rad}}\right) = 0.442\frac{\lambda}{a}.$$

(i) $\dfrac{a}{\lambda} = 2 \Rightarrow \sin\theta_+ = 0.221 \Rightarrow \theta_+ = 0.223\text{ rad} \Rightarrow \Delta\theta = 0.446\text{ rad}.$

(ii) $\dfrac{a}{\lambda} = 5 \Rightarrow \sin\theta_+ = 0.0885 \Rightarrow \theta_+ = 0.0886\text{ rad} \Rightarrow \Delta\theta = 0.177\text{ rad}.$

(iii) $\dfrac{a}{\lambda} = 10 \Rightarrow \sin\theta_+ = 0.0442 \Rightarrow \theta_+ = 0.0443\text{ rad} \Rightarrow \Delta\theta = 0.0885\text{ rad}.$

(b) For the first minimum, $\sin\theta_o = \dfrac{\lambda}{a}$.

(i) $\dfrac{a}{\lambda} = 2 \Rightarrow \theta_o = \arcsin\left(\dfrac{1}{2}\right) = 0.524\text{ rad} \Rightarrow 2\theta_o = 1.05\text{ rad}.$

(ii) $\dfrac{a}{\lambda} = 5 \Rightarrow \theta_o = \arcsin\left(\dfrac{1}{5}\right) = 0.201\text{ rad} \Rightarrow 2\theta_o = 0.402\text{ rad}.$

(iii) $\dfrac{a}{\lambda} = 10 \Rightarrow \theta_o = \arcsin\left(\dfrac{1}{10}\right) = 0.100\text{ rad} \Rightarrow 2\theta_o = 0.200\text{ rad}.$

Both methods show the central width getting smaller as the slit width *a* is increased.

38-38: If the apparatus of **Ex. (38–6)** is placed in water, then all that changes is the wavelength $\lambda \to \lambda' = \dfrac{\lambda}{n}$.

So: $D' = 2y_1' = \dfrac{2x\lambda'}{a} = \dfrac{2x\lambda}{an} = \dfrac{D}{n} = \dfrac{4.16\times10^{-3}\text{ m}}{1.33} = 3.13\times10^{-3}\text{ m} = 3.13\text{ mm}.$

38-39: (a) $y_1 = \dfrac{x\lambda}{a} = \dfrac{(1.40\text{ m})(6.00\times10^{-7}\text{ m})}{3.00\times10^{-4}\text{ m}} = 2.80\times10^{-3}\text{ m}.$

(b) $\dfrac{\sin(\pi a\sin\theta/\lambda)}{\pi a\sin\theta/\lambda} = \dfrac{1}{\sqrt{2}} \Rightarrow \dfrac{\pi a\sin\theta}{\lambda} = 1.39$

$$\Rightarrow \sin\theta = \frac{(1.39)(6.00\times10^{-7}\text{m})}{\pi(3.00\times10^{-4}\text{ m})} = 8.85\times10^{-4}$$

$$\Rightarrow y = x\tan\theta \approx x\sin\theta = (1.40\text{ m})(8.85\times10^{-4}) = 1.24\times10^{-3}\text{ m} = 1.24\text{ mm}.$$

38-40: (a) $I = I_o\left(\frac{\sin\gamma}{\gamma}\right)^2$. The maximum intensity occurs when the derivative of the intensity function with respect to γ is zero.

$$\frac{dI}{d\gamma} = I_o\frac{d}{d\gamma}\left(\frac{\sin\gamma}{\gamma}\right)^2 = 2\left(\frac{\sin\gamma}{\gamma}\right)\left(\frac{\cos\gamma}{\gamma} - \frac{\sin\gamma}{\gamma^2}\right) = 0 \Rightarrow \frac{\cos\gamma}{\gamma} = \frac{\sin\gamma}{\gamma^2} \Rightarrow \gamma\cos\gamma = \sin\gamma$$

$\Rightarrow \gamma = \tan\gamma$.

(b)

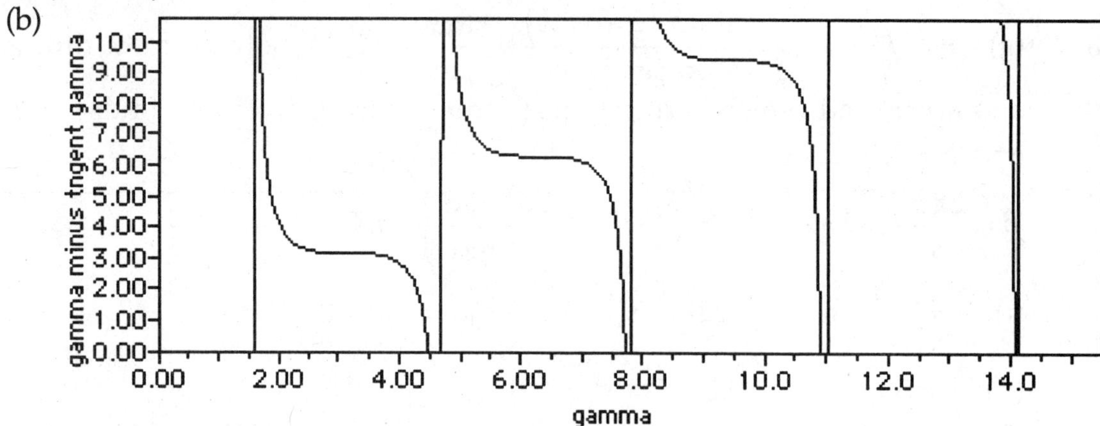

The graph above is a plot of $f(\gamma) = \gamma - \tan\gamma$. So when it equal zero, one has an intensity maximum. Getting estimates from the graph, and then using trial and error to narrow in on the value, we find that the three smallest γ-values are γ= 4.49 rad, 7.73 rad, and 10.9 rad.

38-41: The phase shift for adjacent slits is $\phi = \frac{2\pi d}{\lambda}\sin\theta \approx \frac{2\pi d\theta}{\lambda} \Rightarrow \theta = \frac{\phi\lambda}{2\pi d}$.

So, with the principle maxima at phase shift values of $\phi = 2\pi m$, and $(N-1)$ minima between the maxima, the phase shift between the minima adjacent to the maximum, and the maximum itself, must be $\pm\frac{2\pi}{N}$.

Therefore total phase shifts of these minima are $2\pi m \pm \frac{2\pi}{N}$.

Hence the angle at which they are found, and the angular width, will be:

$$\theta_\pm = \frac{\lambda}{2\pi d}\left(2\pi m \pm \frac{2\pi}{N}\right) = \frac{m\lambda}{d} \pm \frac{\lambda}{dN} \Rightarrow \Delta\theta_\pm = \frac{2\lambda}{dN}.$$

38-42: (a) $E_P{}^2 = E_{P_x}{}^2 + E_{P_y}{}^2$. So, from the diagram at right, we have:

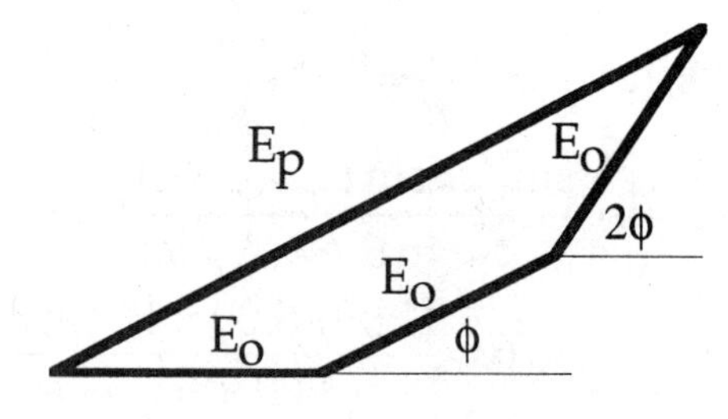

$$\frac{E_P{}^2}{E_o{}^2} = (1+\cos\phi+\cos 2\phi)^2 + (\sin\phi + \sin 2\phi)^2$$

$$= \left(2\cos^2\phi + \cos\phi\right)^2 + (\sin\phi + 2\sin\phi\cos\phi)^2$$

$$= \left(\cos^2\phi + \sin^2\phi\right)(1+2\cos\phi)^2$$

$$\Rightarrow \frac{E_P{}^2}{E_o{}^2} = (1+2\cos\phi)^2 \Rightarrow E_P = E_o(1+2\cos\phi).$$

(b) $\phi = \dfrac{2\pi d}{\lambda}\sin\theta \Rightarrow I_P = I_o\left(1 + 2\cos\left(\dfrac{2\pi d \sin\theta}{\lambda}\right)\right)^2$. This is graphed below:

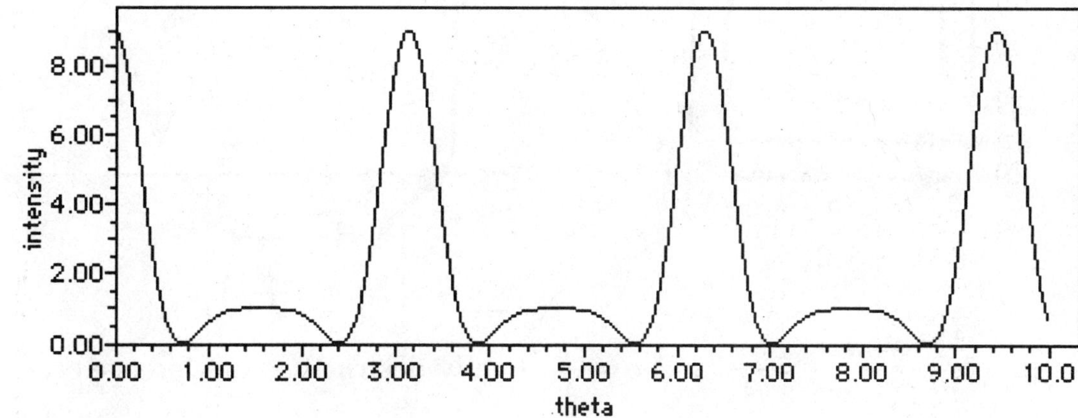

(c) (i) At $\theta = 0$, $I_P = I_o(1 + 2\cos(0°))^2 = 9I_o$.

(ii) The principle maximum is when $I_o \dfrac{2\pi d \sin\theta}{\lambda} = 2\pi m \Rightarrow d\sin\theta = m\lambda$.

(iii) & (iv) The minima occur at $2\cos\left(\dfrac{2\pi d\sin\theta}{\lambda}\right) = -1 \Rightarrow \dfrac{2\pi d \sin\theta}{\lambda} = \dfrac{2\pi m}{3}$

$\Rightarrow d\sin\theta = \dfrac{m\lambda}{3}$, with m not divisible by 3. Thus there are two minima between every principal maximum.

(v) The secondary maxima occur when $\cos\left(\dfrac{2\pi d\sin\theta}{\lambda}\right) = -1 \Rightarrow I_P = I_o = \dfrac{I_{max}}{9}$.

Also $\dfrac{2\pi d\sin\theta}{\lambda} = m\pi \Rightarrow d\sin\theta = \dfrac{m\lambda}{2}$.

All of these findings agree with the N-slit statements in Section 38–5.

(d) Below are phasor diagrams for specific phase shifts.

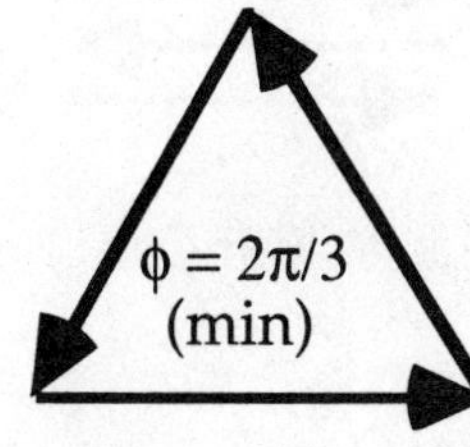

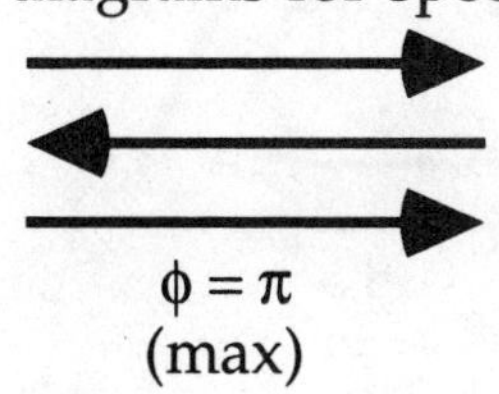

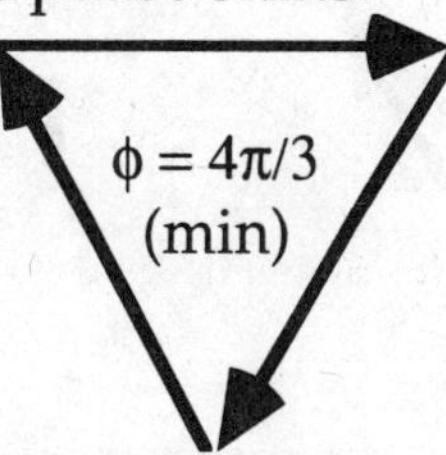

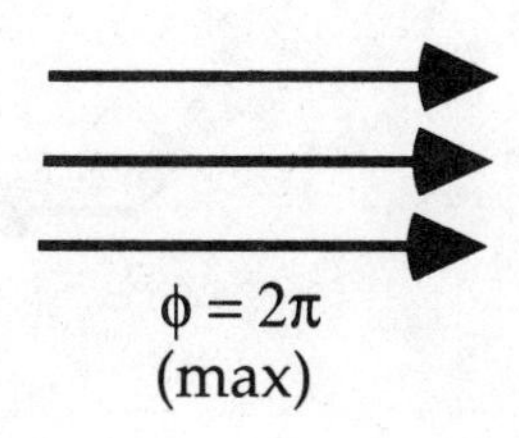

38-43: (a) For eight slits, the phasor diagrams must have eight vectors:

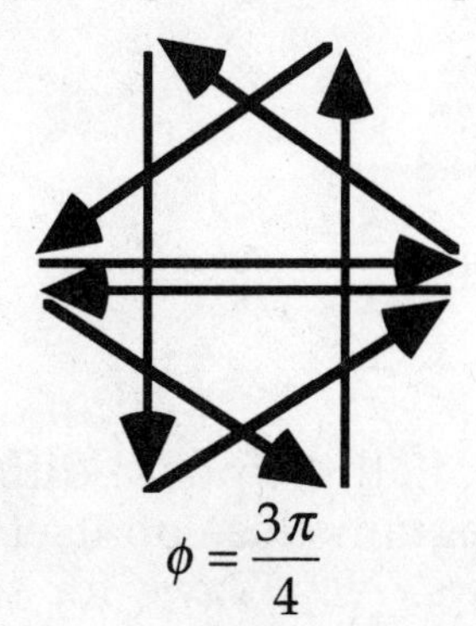

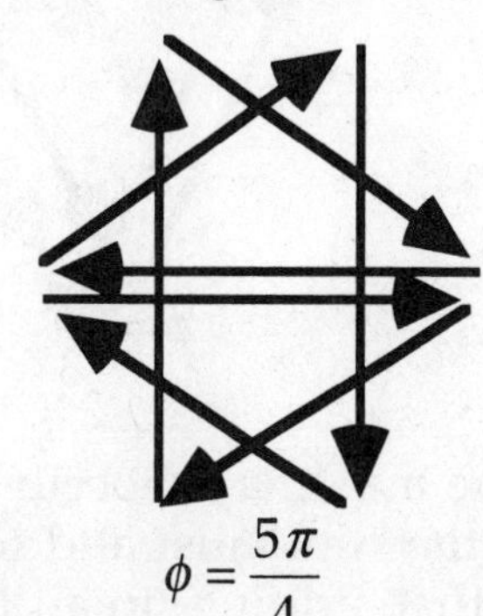

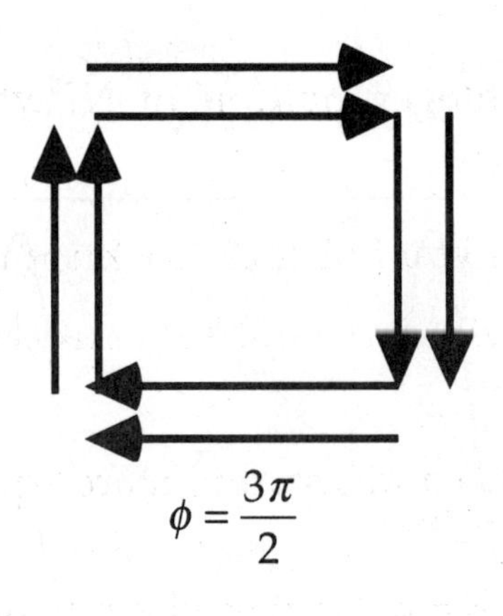

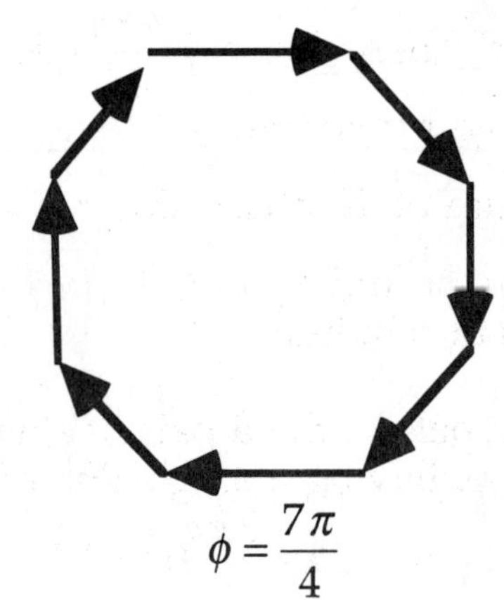

(b) For $\phi = \frac{3\pi}{4}$, $\phi = \frac{5\pi}{4}$, and $\phi = \frac{7\pi}{4}$, totally destructive interference occurs between slits four apart. For $\phi = \frac{3\pi}{2}$, totally destructive interference occurs with every second slit.

38-44: For six slits, the phasor diagrams must have six vectors.
(a) Zero phase difference between adjacent slits means that the total amplitude is $6E$, and the intensity is $36I$.

(b) If the phase difference is 2π, then we have the same phasor diagram as above, and equal amplitude, $6E$, and intensity, $36I$.
(c) There is an interference minimum whenever the phasor diagrams close on themselves, such as in the five cases below.

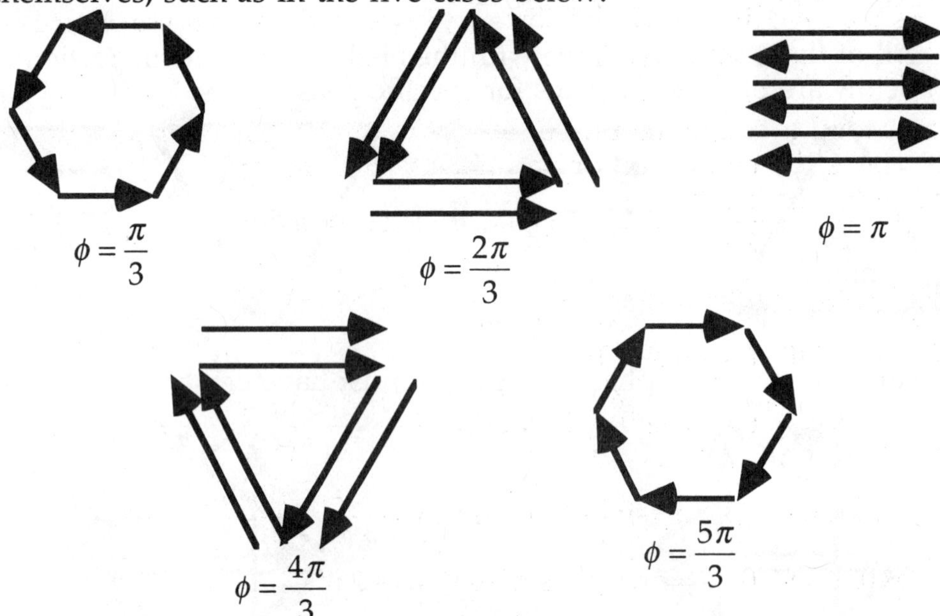

38-45: (a) For the maxima to occur for N slits, the sum of all the phase differences between the slits must add to zero (the phasor diagram close on itself). This requires that, adding up all the relative phase shifts, $N\phi = 2\pi m$, for some integer m. Therefore $\phi = \frac{2\pi m}{N}$, for m not an integer multiple of N, which would give a maximum.

(b) The sum of N phase shifts $\phi = \frac{2\pi m}{N}$ brings you full circle back to the maximum, so only the N-1 previous phases yield minima between each pair of principal maxima.

38-46: As shown below, the a pair of slits whose width and separation are equal is the same as having a single slit, of twice the width.

a

d = a

a

$\phi = \frac{2\pi d}{\lambda}\sin\theta$, so $\beta = \frac{2\pi a}{\lambda}\sin\theta = \phi$. So then the intensity is

$$I = I_o \cos^2(\beta/2)\left(\frac{\sin^2(\beta/2)}{(\beta/2)^2}\right) = I_o \frac{(2\sin(\beta/2)\cos(\beta/2))^2}{\beta^2}$$

$$\Rightarrow I = I_o \frac{\sin^2\beta}{\beta^2} = I_o \frac{\sin^2(\beta'/2)}{(\beta'/2)^2}, \text{ where } \beta' = \frac{2\pi(2a)}{\lambda}\sin\theta, \text{ which}$$

is Eq. (38–5) with double the slit width.

38-47: For 7000 slits / cm $\Rightarrow d = \frac{1}{7.00\times10^5\ \text{m}^{-1}} = 1.43\times10^{-6}\ \text{m}$.

$$d\sin\theta = m\lambda \Rightarrow \lambda = \frac{d\sin\theta}{m} \Rightarrow \lambda_{4\max} = \frac{d}{4} = \frac{1.43\times10^{-6}\ \text{m}}{4} = 3.57\times10^{-7}\text{m}.$$

38-48: (a) As the rays first reach the slits there is already a phase difference between adjacent slits of $\frac{2\pi d\sin\theta'}{\lambda}$.

This, added to the usual phase difference introduced after passing through the slits yields the condition for an intensity maximum:

$$\frac{2\pi d\sin\theta}{\lambda} + \frac{2\pi d\sin\theta'}{\lambda} = 2\pi m \Rightarrow d(\sin\theta + \sin\theta') = m\lambda.$$

(b) 600 slits / mm $\Rightarrow d = \frac{1}{6.00\times10^5\ \text{m}^{-1}} = 1.67\times10^{-6}\ \text{m}$.

$\theta' = 0°$: $m = 0$: $\theta = \arcsin(0) = 0$.

$$m = 1:\ \theta = \arcsin\left(\frac{\lambda}{d}\right) = \arcsin\left(\frac{4.00\times10^{-7}\ \text{m}}{1.67\times10^{-6}\ \text{m}}\right) = 13.9°.$$

$$m = -1:\ \theta = \arcsin\left(-\frac{\lambda}{d}\right) = \arcsin\left(-\frac{4.00\times10^{-7}\ \text{m}}{1.67\times10^{-6}\ \text{m}}\right) = -13.9°.$$

$\theta' = 20.0°$: $m = 0$: $\theta = \arcsin(-\sin 20.0°) = -20.0°$.

$$m = 1:\ \theta = \arcsin\left(\frac{4.00\times10^{-7}\ \text{m}}{1.67\times10^{-6}\ \text{m}} - \sin 20.0°\right) = -5.88°.$$

$$m = -1:\ \theta = \arcsin\left(-\frac{4.00\times10^{-7}\ \text{m}}{1.67\times10^{-6}\ \text{m}} - \sin 20.0°\right) = -35.6°.$$

38-49: For 800 slits / mm $\Rightarrow d = \dfrac{1}{8.00\times 10^{5}\ \text{m}^{-1}} = 1.25\times 10^{-6}\ \text{m}$.

We need $d\sin\theta = m\lambda \Rightarrow \sin\theta = \dfrac{m\lambda}{d} \le 1$, if the whole spectrum is to be seen.

$\lambda_1 = 4.00\times 10^{-7}\text{m}:\ m=1:\ \dfrac{\lambda_1}{d} = 0.32;\ m=2:\ \dfrac{2\lambda_1}{d} = 0.64$.

$\lambda_1 = 7.00\times 10^{-7}\text{m}:\ m=1:\ \dfrac{\lambda_2}{d} = 0.56;\ m=2:\ \dfrac{2\lambda_2}{d} = 1.12$. So the second order does not contain the violet end of the spectrum, and therefore only the first order diffraction pattern contains all colors of the spectrum.

38-50: (a) $2d\sin\theta = m\lambda \Rightarrow \theta = \arcsin\left(\dfrac{m\lambda}{2d}\right) = \arcsin\left(m\dfrac{0.0950\ \text{nm}}{2(0.282\ \text{nm})}\right) = \arcsin(0.1684\ m)$.
So for $m=1:\ \theta = 9.7°$, $m=2:\ \theta = 19.7°$, $m=3:\ \theta = 30.4°$, $m=4:\ \theta = 42.4°$, and $m=5:\ \theta = 57.4°$. No larger m values yield answers.

(b) If the separation $d = \dfrac{a}{\sqrt{2}}$, then $\theta = \arcsin\left(\dfrac{\sqrt{2}m\lambda}{2a}\right) = \arcsin\left(0.1684\sqrt{2}\,m\right)$.

So for $m=1:\ \theta = 13.8°$, $m=2:\ \theta = 28.4°$, $m=3:\ \theta = 45.6°$, and $m=4:\ \theta = 72.3°$. No larger m values yield answers.

38-51: To resolve two objects, according to Rayleigh's criterion, one must be located at the first minimum of the other. In this case, knowing the equation for the angle to the first minimum, and also the objects' separation and distance away, the sine of the angle subtended by them is easily calculated to be:
$\sin\theta = \dfrac{\lambda}{a} = \dfrac{\Delta x}{R} \Rightarrow R = \dfrac{a\Delta x}{\lambda} = \dfrac{(4.00\times 10^{-4}\ \text{m})(3.00\ )}{5.00\times 10^{-7}\ \text{m}} = 2400\ \text{m} = 2.40\ \text{km}$.

38-52: $\sin\theta = 1.22\dfrac{\lambda}{D} \Rightarrow \theta = \arcsin\left(1.22\dfrac{\lambda}{D}\right)$. So for

(a) Mt. Palomar: $\theta = \arcsin\left(1.22\dfrac{(5.00\times 10^{-7}\ \text{m})}{(5.08\ \text{m})}\right) = (6.88\times 10^{-6})°$.

(b) Arecibo: $\theta = \arcsin\left(1.22\dfrac{(0.210\ \text{m})}{(305\ \text{m})}\right) = 0.0481°$.

38-53: Diffraction limited seeing and Rayleigh's criterion tell us:
$\sin\theta = 1.22\dfrac{\lambda}{D} = \dfrac{(1.22)(5.50\times 10^{-7}\ \text{m})}{(4.00\times 10^{-3}\ \text{m})} = 1.68\times 10^{-4}\ \text{m}$. But now the height of the astronaut can be calculated from the angle (above) and the object separation (55 m). We have:
$\dfrac{\Delta x}{h} = \tan\theta \Rightarrow h = \dfrac{\Delta x}{\tan\theta} \approx \dfrac{\Delta x}{\sin\theta} = \dfrac{55.0\ \text{m}}{1.68\times 10^{-4}} = 3.28\times 10^{5}\ \text{m} = 328\ \text{km}$.

38-54: (a) From the segment dy', the fraction of the amplitude of E_o that gets through is $E_o\left(\dfrac{dy'}{a}\right) \Rightarrow dE = E_o\left(\dfrac{dy'}{a}\right)\sin(\omega t - kx)$.

(b) The path difference between each little piece is $y'\sin\theta \Rightarrow kx = k(D - y'\sin\theta)$
$\Rightarrow dE = \frac{E_o dy'}{a}\sin(\omega t - k(D - y'\sin\theta))$. This can be rewritten as
$$dE = \frac{E_o dy'}{a}(\sin(\omega t - kD)\cos(ky'\sin\theta) + \sin(ky'\sin\theta)\cos(\omega t - kD)).$$
(c) So the total amplitude is given by the integral over the slit of the above.
$$\Rightarrow E = \int_{-a/2}^{a/2} dE = \frac{E_o}{a}\int_{-a/2}^{a/2} dy'(\sin(\omega t - kD)\cos(ky'\sin\theta) + \sin(ky'\sin\theta)\cos(\omega t - kD)).$$
But the second term integrates to zero, so we have:
$$E = \frac{E_o}{a}\sin(\omega t - kD)\int_{-a/2}^{a/2} dy'(\cos(ky'\sin\theta)) = E_o\sin(\omega t - kD)\left[\left(\frac{\sin(ky'\sin\theta)}{ka\sin\theta/2}\right)\right]_{-a/2}^{a/2}$$
$$\Rightarrow E = E_o\sin(\omega t - kD)\left(\frac{\sin(ka(\sin\theta)/2)}{ka(\sin\theta)/2}\right) = E_o\sin(\omega t - kD)\left(\frac{\sin(\pi a(\sin\theta)/\lambda)}{\pi a(\sin\theta)/\lambda}\right).$$
At $\theta = 0, \frac{\sin[...]}{[...]} = 1 \Rightarrow E = E_o\sin(\omega t - kD)$.

(d) Since $I = E^2 \Rightarrow I = I_o\left(\frac{\sin(ka(\sin\theta)/2)}{ka(\sin\theta)/2}\right)^2 = I_o\left(\frac{\sin(\beta/2)}{\beta/2}\right)^2$, where we have used $I_o = E_o^{\,2}\sin^2(\omega t - kx)$.

Chapter 39: Relativity

39-1: If O'sees simultaneous flashes then O will see the A(A') flash first since O would believe that the A' flash must have travelled longer to reach O', and hence started first.

39-2: $\gamma \Delta t = 0.310\ \text{ms}$.

39-3: (a) $\Delta t = \dfrac{\Delta t_o}{\sqrt{1-u^2/c^2}} \Rightarrow 1-\dfrac{u^2}{c^2} = \left(\dfrac{\Delta t_o}{\Delta t}\right)^2$

$$\Rightarrow u = c\sqrt{1-\left(\frac{\Delta t_o}{\Delta t}\right)^2} = c\sqrt{1-\left(\frac{2.2}{19}\right)^2}$$

$\therefore u = 0.993c$

(b) $\Delta x = u\Delta t = (0.993)(3.00\times 10^8\ \text{m/s})(19\times 10^{-6}\ \text{s})$

$= 5700\ \text{m}$.

39-4: (a) $\gamma\tau = 1.2\times 10^{-7}$ s. (b) $u\gamma\tau = 37$ m.

39-5: $\Delta t_o = \sqrt{1-u^2/c^2}\,\Delta t = \sqrt{1-\left(\dfrac{6.95\times 10^6\ \text{m/s}}{3.00\times 10^8\ \text{m/s}}\right)^2}\,(1\ \text{yr})$

$\Rightarrow \Delta t_o = (0.9997)\ \text{yr}$

$\Rightarrow (\Delta t - \Delta t_o) = (2.7\times 10^{-4})\ \text{yr} = 8500\ \text{s} = 2.4\ \text{hrs}$.

The least time elapses on the rocket's clock.

39-6: a) The frame in which the source (the searchlight) is stationary is the spacecraft's frame, so 4.50×10^{-5} s is the proper time. b) To three figures, $u = c$. Solving Eq. (39-7) for u/c in terms of γ,

$$\frac{u}{c} = \sqrt{1-(1/\gamma)^2} \approx 1-\frac{1}{2\gamma^2}.$$

Using $1/\gamma = \Delta t_0/\Delta t = 4.50\times 10^{-5}\ \text{s}/3.25\times 10^{-3}\ \text{s}$ gives $u/c = 0.999904$.

39-7: $\sqrt{1-u^2/c^2} = (1-u^2/c^2)^{1/2} \approx 1-\dfrac{u^2}{2c^2}+\ldots$

$$\Rightarrow (\Delta t - \Delta t_o) = (1-\sqrt{1-u^2/c^2})(\Delta t) = \frac{u^2}{2c^2}\Delta t = \frac{(400\ \text{m/s})^2(5\ \text{hrs})\cdot(360)}{2(3.00\times 10^8\ \text{m/s})^2}$$

$\Rightarrow (\Delta t - \Delta t_o) = 1.6\times 10^{-8}\ \text{s}$.

The clock on the plane shows the shorter elapsed time.

39-8: $\gamma = 1.120$ (a) The time may be found from Eq. (39-6) with Δt_0 being the time measured by the teenage pilot and $\Delta t = 2.75\times10^8\ \text{m}/0.450\ c$, so

$$\Delta t_0 = \frac{\Delta t}{\gamma} = \frac{2.75\times10^8\ \text{m}}{\gamma(0.450\,c)} = 1.82\ \text{s}.$$

b) During the time found in part (b), which is the proper time in the racer's frame, the freighter has moved a distance $(1.82\ \text{s})(0.450c) = 2.46\times10^8\ \text{m}$.

c) $\Delta t_0 = 1.82\text{s}/\gamma = 1.63\text{s}$.

39-9: (a) $l_o = 1200\text{m}$

$$\Rightarrow l = l_o\sqrt{1-\frac{u^2}{c^2}} = l_o(1200\text{m})\sqrt{1-\frac{(8.00\times10^7\ \text{m/s})^2}{(3.00\times10^8\ \text{m/s})^2}}$$

$$= (1200\text{m})(0.964) = 1160\text{m}$$

(b) $\Delta t_o = \dfrac{l_o}{u} = \dfrac{1200\text{m}}{8.00\times10^7\ \text{m/s}} = 1.50\times10^{-5}\text{s}$

(c) $\Delta t = \dfrac{l}{u} = \dfrac{1160\text{m}}{8.00\times10^7\ \text{m/s}} = 1.45\times10^{-5}\text{s}.$

39-10: $\gamma = 1/0.914$, so

$$u = c\sqrt{1-(1/\gamma)^2} = 0.405c = 1.21\times10^8\ \text{m/s}.$$

39-11: $l = l_o\sqrt{1-\dfrac{u^2}{c^2}} \Rightarrow l_o = \dfrac{l}{\sqrt{1-u^2/c^2}}$

$$\Rightarrow l_o = \frac{86.5\text{m}}{\sqrt{1-\left(\frac{0.800\,c}{c}\right)^2}} = 144\text{m}.$$

39-12: (a) $t = \dfrac{9.50\times10^4\ \text{m}}{0.99860c} = 3.17\times10^{-4}\ \text{s}.$ (b) $h' = \dfrac{h}{\gamma} = 5.03\ \text{km}.$

(c) $\dfrac{h'}{0.99860c} = 1.68\times10^{-5}\ \text{s}$, and $\dfrac{t}{\gamma} = 1.68\times10^{-5}\ \text{s}.$

39-13: $\gamma = \dfrac{1}{\sqrt{1-(u/c)^2}} = \dfrac{1}{\sqrt{1-(0.9954)^2}} = \dfrac{1}{0.09581} = 10.44.$

(a) $l = \dfrac{l_o}{\gamma} = \dfrac{45\text{km}}{10.44} = 4.31\text{km}$

(b) In muon's frame: $d = u\Delta t = (0.9954c)(2.2\times10^{-6}\text{s}) = 0.66\text{km}.$

$$\Rightarrow \% = \frac{d}{h} = \frac{0.66}{4.31} = 0.15 = 15\%$$

(c) In earth's frame: $\Delta t = \Delta t_o \gamma = (2.2\times10^{6}\,\text{s})(10.44) = 2.3\times10^{-5}\,\text{s}$

$\Rightarrow d' = u\Delta t = (0.9954c)(2.3\times10^{-5}\,\text{s}) = 6.9\,\text{km}$

$\Rightarrow \% = \dfrac{d'}{h'} = \dfrac{6.9\,\text{km}}{45.0\,\text{km}} = 15\%.$

39-14: Multiplying the last equation of (39-22) by u and adding to the first to eliminate t gives

$$x' + ut' = \gamma x\left(1 - \frac{u^2}{c^2}\right) = \frac{1}{\gamma}x,$$

and multiplying the first by $\frac{u}{c^2}$ and adding to the last to eliminate x gives

$$t' + \frac{u}{c^2}x' = \gamma t\left(1 - \frac{u^2}{c^2}\right) = \frac{1}{\gamma}t,$$

so $x=\gamma(x'+ut')$ and $t=\gamma(t'+ux'/c^2)$, which is indeed the same as Eq. (39-22) with the primed coordinates replacing the unprimed, and a change of sign of u.

39-15: Eq 39-19: $x' = \dfrac{x-ut}{\sqrt{1-u^2/c^2}}$ Eq 39-20: $x' = -ut' + x\sqrt{1-u^2/c^2}$

Equate: $(x-ut)\gamma = -ut' + \dfrac{x}{\gamma}$

$$\Rightarrow t' = \left(-\frac{x\gamma}{u} + t\gamma + \frac{x}{u\gamma}\right) = t\gamma + \frac{x}{u}\left(\frac{1}{\gamma} - \gamma\right)$$

$$\frac{1}{\gamma} - \gamma = \sqrt{1-(u/c)^2} - \frac{1}{\sqrt{1-(u/c)^2}} = \frac{1-(u/c)^2-1}{\sqrt{1-(u/c)^2}} = \frac{-u^2/c^2}{\sqrt{1-(u/c)^2}} = \gamma u^2/c^2$$

$$\Rightarrow t' = t\gamma + \frac{xu\gamma}{c^2} = \frac{t - ux/c^2}{\sqrt{1-(u/c)^2}}$$

39-16: Starting from Eq. (39-23),

$$v' = \frac{v-u}{1-uv/c^2}$$

$$v'(1-uv/c^2) = v - u$$

$$v' + u = v + v'uv/c^2$$

$$= v(1 + uv'/c^2)$$

from which Eq. (39-24) follows. This is the same as switching the primed and unprimed coordinates and changing the sign of u.

39-17: (a) $v = \dfrac{v'+u}{1+uv'/c^2} = \dfrac{0.600c + 0.500c}{1+(0.600)(0.500)} = 0.846c$

(b) $v = \dfrac{v'+u}{1+uv'/c^2} = \dfrac{0.900c + 0.500c}{(0.900c)} = 0.966c$

(c) $v = \dfrac{v'+u}{1+uv'/c^2} = \dfrac{0.900c + 0.500c}{(0.900c)} = 0.997c.$

39-18: $\gamma = 1.667,\ \gamma = 5/3$ if $u = (4/5)c$. (a) In Mavis' frame $x' = 0$ and $t = 2.50$ s, so from the result of Exercise 39-14 or Example 39-7, $x = \gamma ut' = 1.00 \times 10^9\,\text{m},\quad t = \gamma t' = 4.17\,\text{s}$. (b) The 2.50-s interval in Mavis' frame is the proper time Δt_0 in Eq. (39-6), so $\Delta t = \gamma \Delta t_0 = 4.17\,\text{s}$, as in part (a). (c) $(4.17\,\text{s})(0.800c) = 1.00 \times 10^9\,\text{m}$, which is the distance x found in part (a).

39-19: $v' = \dfrac{v-u}{1-uv/c^2} \Rightarrow v' - \dfrac{uvv'}{c^2} = v - u$

$\Rightarrow u\left(1 - \dfrac{vv'}{c^2}\right) = v - v' \Rightarrow u = \dfrac{v-v'}{(1-vv'/c^2)}$

$\Rightarrow u = \dfrac{0.290c - 0.840c}{(1-(0.290)(0.840))} = -0.727c$

$\Rightarrow$ moving opposite the rocket, ie. away from earth.

39-20: (a) In Eq. (39-24), $u = 0.600c,\ v' = 0.800c$ and so $v = 0.946c$. (b) $\dfrac{\Delta x}{v} = 14.1\,\text{s}$.

39-21: $v' = \dfrac{v-u}{1-\dfrac{uv}{c^2}} = \dfrac{0.700c - 0.900c}{1-(0.7)(0.9)} = -0.541c$

$\Rightarrow$ cruiser in moving toward the fighter at $0.541c$.

39-22: In the frame of one of the particles, u and v' in Eq. (39-24) are both $0.630c$, and $v = 0.902c = 2.71 \times 10^8\ \text{m/s}$.

39-23: $v = \dfrac{v'+u}{1+\dfrac{uv'}{c^2}} = \dfrac{-0.900c + 0.700c}{1+(-0.900)(0.700)} = -0.541c$

39-24: If u is assumed to be positive, the spacetime diagram of Fig. (39-16) may be used to see that in Stanley's frame $t_1 < t_2$. Using a spacetime diagram for the frame with Mavis at rest, Stanley is moving with a relative speed that is negative with respect to a system in which $x_2' > x_1'$, and that diagram again shows that $t_1 < t_2$.

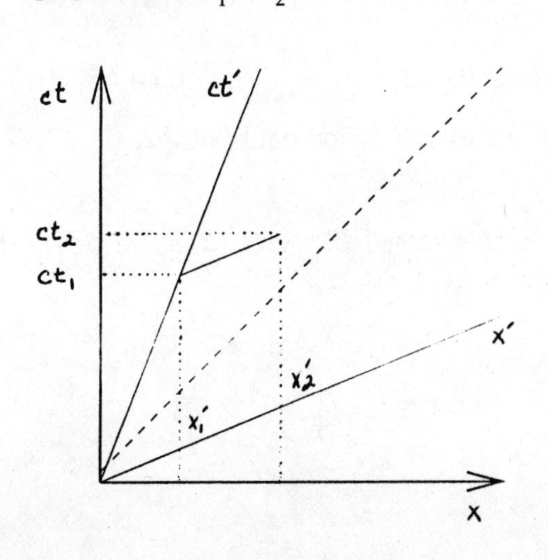

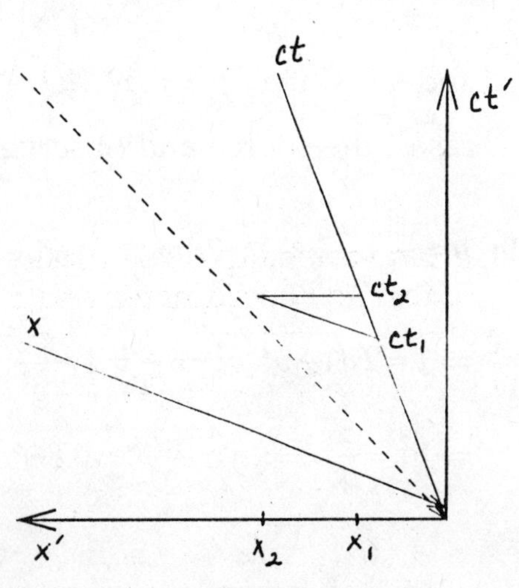

39-25:

(a) Mavis sees event 1 as occuring before event 2, as seen on the spacetime diagram at right, just as Stanley does.
(b) Mavis does not see the events occuring at the same place, and sees event 2 occuring closer to the x origin than event 1.

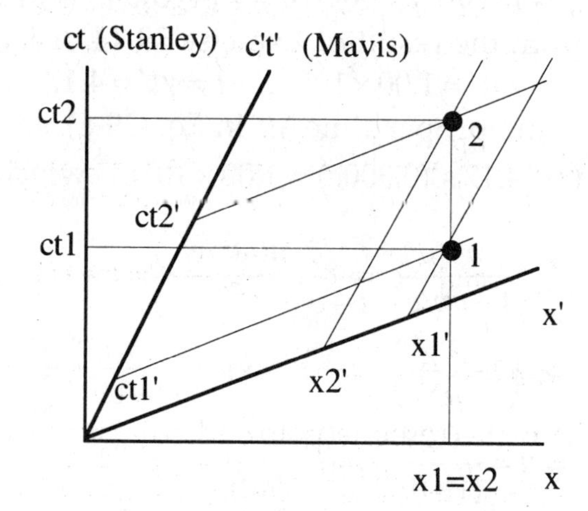

(a) Mavis sees event 1 before event 2.
(b) Don't occur at the same place.

39-26: Using $u = -0.600c = -(3/5)c$ in Eq. (39-26) gives

$$f = \sqrt{\frac{1-(3/5)}{1+(3/5)}}f_0 = \sqrt{\frac{2/5}{8/5}}f_0 = f_0/2.$$

39-27: $f = \sqrt{\frac{c+u}{c-u}}f_o \Rightarrow (c-u)f^2 = (c+u)f_o^{\,2}$

$$\Rightarrow u = \frac{c(f^2 - f_o^{\,2})}{f_o^{\,2} + f^2} = \frac{c((f/f_o)^2 - 1)}{(f/f_o)^2 + 1} = \frac{c((\lambda_o/\lambda)^2 - 1)}{((\lambda_o/\lambda)^2 + 1)}$$

$$\therefore u = c\frac{((675/525)^2 - 1)}{((675/525)^2 + 1)} = 0.246c = 7.38 \times 10^7 \text{ m/s}.$$

39-28: Solving Eq. (39-26) for u/c,

$$\frac{u}{c} = \frac{1-(f/f_0)^2}{1+(f/f_0)^2},$$

and so (a) if $f/f_0 = 0.99, (u/c) = 0.0100$, and (b) if $f/f_0 = \frac{1}{3}, (u/c) = 4/5$. In both cases, the source and observer are moving away from each other.

39-29: $p = \frac{mv}{\sqrt{1-v^2/c^2}} = 2mv$

$$\Rightarrow 1 = 2\sqrt{1-v^2/c^2} \Rightarrow \frac{1}{4} = 1 - \frac{v^2}{c^2}$$

$$\Rightarrow v^2 = \frac{3}{4}c^2 \Rightarrow v = \frac{\sqrt{3}}{2}c = 0.866c.$$

39-30: (a) $\gamma = 1.01$, so $(v/c) = 0.140$ and $v = 4.21 \times 10^8$ m/s. (b) The relativistic expression is *always* larger in magnitude than the non-relativistic expression.

39-31: (a) $F = \dfrac{dp}{dt} = \dfrac{d}{dt}\left(\dfrac{mv}{\sqrt{1-v^2/c^2}}\right) = \dfrac{ma}{\sqrt{1-v^2/c^2}} - \dfrac{\frac{1}{2}mv \cdot \frac{1}{2}(2va/c^2)}{(1-v^2/c^2)^{3/2}}$

$$= ma\left(\frac{(1-v^2/c^2)+v^2/c^2}{(1-v^2/c^2)^{3/2}}\right) = ma/(1-v^2/c^2)^{3/2}$$

$$\Rightarrow a = \frac{F}{m}(1-\frac{v^2}{c^2})^{3/2}$$

(b) If the force is perpendicular to velocity then denominator is constant

$$\Rightarrow F = \frac{dp}{dt} = \frac{m\,dv/dt}{\sqrt{1-v^2/c^2}} \Rightarrow a = \frac{F}{m}\sqrt{1-v^2/c^2}$$

39-32: The force is found from Eq. (39-33) or Eq. (39-34). (a) Indistinguishable from $F = ma = 0.145\text{ N}$. (b) $\gamma^3 ma = 1.75\text{ N}$. (c) $\gamma^3 ma = 51.7\text{ N}$. (d) $\gamma ma =$ 0.145 N, 0.333 N, 1.03 N.

39-33: $K = (\gamma - 1)mc^2 \approx \dfrac{1}{2}mv^2 + \dfrac{3}{8}\dfrac{mv^4}{c^2} + \ldots$

if $K - K_o = 1.01\dfrac{1}{2}mv^2 \Rightarrow \dfrac{3}{8}\dfrac{v^4}{c^2} = \dfrac{0.01}{2}v^2$

$$\Rightarrow \frac{300}{4}v^2 = c^2 \Rightarrow v = \sqrt{\frac{4}{300}}c = 0.115c = 3.5 \times 10^7\text{ m/s}$$

39-34: (a) $(\gamma_f - 1)mc^2 = (3.22 \times 10^{-3})mc^2$. (b) $(\gamma_f - \gamma_i)mc^2 = 2.73mc^2$. (c) The result of part (b) is far larger than that of part (a).

39-35: (a) $K = \dfrac{mc^2}{\sqrt{1-v^2/c^2}} - mc^2 = mc^2$

$$\Rightarrow \frac{1}{\sqrt{1-v^2/c^2}} = 2 \Rightarrow \frac{1}{4} = 1 - \frac{v^2}{c^2} \Rightarrow v = \sqrt{\frac{3}{4}} = 0.866c$$

(b) $K = 6mc^2 \Rightarrow \dfrac{1}{\sqrt{1-v^2/c^2}} = 7 \Rightarrow \dfrac{1}{49} = 1 - \dfrac{v^2}{c^2} \Rightarrow v = \sqrt{\dfrac{48}{49}}c = 0.990c.$

39-36: $E = 2mc^2 = 2(9.11 \times 10^{-31}\text{ kg})(3.00 \times 10^8\text{ m/s})^2 = 1.64 \times 10^{-13}\text{ J} = 1.02\text{ MeV}.$

39-37: (a) $v = 8\times10^{7}\,\text{m/s} \Rightarrow \gamma = \dfrac{1}{\sqrt{1-v^2/c^2}} = 1.0376$

$m = m_e \qquad K_o = \dfrac{1}{2}mv^2 = 2.92\times10^{-15}\,\text{J}$

$K = (\gamma-1)mc^2 = 3.08\times10^{-15}\,\text{J} \therefore \dfrac{K}{K_o} = 1.05.$

(b) $v = 2.85\times10^{8}\,\text{m/s} \therefore \gamma = 3.203$

$K_o = \dfrac{1}{2}mv^2 = 3.70\times10^{-14}\,\text{J}$

$K = (\gamma-1)mc^2 = 1.81\times10^{-13}\,\text{J} \qquad K/K_o = 4.88.$

39-38: $(2.40\times10^{-28}\,\text{kg})(3.00\times10^{8}\,\text{m/s})^2 = 2.16\times10^{-11}\,\text{J} = 135\,\text{MeV}.$

39-39: (a) $E = 0.380\,\text{MeV} = 3.80\times10^{5}\,\text{eV}$

(b) $E = K + mc^2 = 3.80\times10^{5}\,\text{eV} + \dfrac{(9.11\times10^{-31}\,\text{kg})(3.00\times10^{8}\,\text{m/s})^2}{1.6\times10^{-19}\,\text{J/eV}}$

$= 8.92\times10^{5}\,\text{eV}$

(c) $E = \dfrac{mc^2}{\sqrt{1-v^2/c^2}} \Rightarrow v = c\sqrt{1-\left(\dfrac{mc^2}{E}\right)^2}$

$= c\sqrt{1-\left(\dfrac{5.11\times10^{5}\,\text{eV}}{8.92\times10^{5}\,\text{eV}}\right)^2} = 2.46\times10^{8}\,\text{m/s}$

(d) Nonrel: $K = \dfrac{1}{2}mv^2 \Rightarrow v = \sqrt{\dfrac{2K}{m}} = \sqrt{\dfrac{2(3.8\times10^{5}\,\text{eV})(1.6\times10^{-19}\,\text{J/eV})}{9.11\times10^{-31}\,\text{kg}}}$

$= 3.65\times10^{8}\,\text{m/s}.$

39-40: (a) The fraction of the initial mass that becomes energy is

$1-\dfrac{(4.0015\,\text{u})}{2(2.0136\,\text{u})} = 6.382\times10^{-3}$, and so the energy released per kilogram is

$(6.382\times10^{-3})(1.00\,\text{kg})(3.00\times10^{8}\,\text{m/s})^2 = 5.74\times10^{14}\,\text{J}.$

(b) $\dfrac{1.0\times10^{19}\,\text{J}}{5.74\times10^{14}\,\text{J/kg}} = 1.7\times10^{4}\,\text{kg}.$

39-41: $(m = 3.32\times10^{-27}\,\text{kg}, \quad p = 8.25\times10^{-11}\,\text{kg}\cdot\text{m/s})$

(a) $E = \sqrt{(mc^2)^2 + (pc)^2}$

$= 3.88\times10^{-10}\,\text{J}$

(b) $K = E - mc^2 = 3.88\times10^{-10} - (3.32\times10^{-27}\,\text{kg})c^2 = 8.9\times10^{-11}\,\text{J}$

(c) $\dfrac{K}{mc^2} = \dfrac{8.9\times10^{-11}\,\text{J}}{(3.32\times10^{-27}\,\text{kg})c^2} = 0.30$

39-42: $E=\left(m^2c^4+p^2c^2\right)^{1/2}=mc^2\left(1+\left(\frac{p}{mc}\right)^2\right)^{1/2}$

$$\approx mc^2\left(1+\frac{1}{2}\frac{p^2}{m^2c^2}\right)=mc^2+\frac{p^2}{2m},$$

the sum of the rest mass energy and the classical kinetic energy.

39-43: Need $a=b \Rightarrow l_o=a, l=b$

$$\therefore \frac{l}{l_o}=\frac{b}{a}=\frac{b}{1.50b}=\sqrt{1-u^2/c^2}$$

$$\Rightarrow u=c\sqrt{1-\left(\frac{b}{a}\right)^2}=c\sqrt{1-\left(\frac{1}{1.5}\right)^2}=0.745\text{c}.$$

$$=2.24\times10^8\,\text{m/s}.$$

39-44: The change in the astronaut's biological age is Δt_0 in Eq. (39-6), and Δt is the distance to the star as measured from earth, divided by the speed. Combining, the astronaut's biological age is

$$19\text{ yr}+\frac{26.5\text{ yr } c}{\gamma u}=19\text{ yr}+\frac{26.5\text{ yr}}{\gamma(0.9930)}=22.2\text{ yr}.$$

39-45: (a) $d=c\Delta t\Rightarrow\Delta t=\frac{d}{c}=\frac{2200\,\text{m}}{3\times10^8\,\text{m/s}}=7.33\times10^{-6}\,\text{s}$

But $\frac{\Delta t_o}{\Delta t}=\sqrt{1-\frac{u^2}{c^2}}\Rightarrow\left(\frac{u}{c}\right)^2=1-\left(\frac{\Delta t_o}{\Delta t}\right)^2$

$$\Rightarrow(1-\Delta)^2=1-\left(\frac{\Delta t_o}{\Delta t}\right)^2$$

$$\Rightarrow 1-2\Delta\approx 1-\left(\frac{\Delta t_o}{\Delta t}\right)^2\Rightarrow\Delta=\frac{1}{2}\left(\frac{\Delta t_o}{\Delta t}\right)^2=\frac{1}{2}\left(\frac{2.6\times10^{-8}}{7.32\times10^{-6}}\right)$$

$$\therefore\Delta=6.3\times10^{-6}.$$

(b) $E=\gamma mc^2=\left(\frac{\Delta t}{\Delta t_o}\right)\text{mc}^2=\left(\frac{7.33\times10^{-6}}{2.6\times10^{-8}}\right)139.6\,\text{MeV}$

$$\Rightarrow E=3.9\times10^4\text{ MeV}=39\text{ GeV}$$

39-46: One dimension of the cube appears contracted by a factor of $\frac{1}{\gamma}$, so the volume in S' is $a^3/\gamma=a^3\sqrt{1-(u/c)^2}$.

39-47: Heat in $Q=mL_f=(5.00\,\text{kg})(3.34\times10^5\,\text{J/kg})$

$$=1.67\times10^6\,\text{J}$$

$$\Rightarrow\Delta m=\frac{Q}{c^2}=\frac{(1.67\times10^6\,\text{J})}{(3.00\times10^8\,\text{m/s})^2}=1.86\times10^{-11}\,\text{kg}.$$

39-48: (a) $v = \frac{p}{m} = \frac{(E/c)}{m} = \frac{E}{mc}$, where the atom and the photon have the same magnitude of momentum, E/c. (b) $v = \frac{E}{mc} = \ll c$, so $E \ll mc^2$.

39-49: (a) $E = \gamma mc^2$ and $\gamma = 10 = \frac{1}{\sqrt{1-(v/c)^2}} \Rightarrow \frac{v}{c} = \sqrt{\frac{\gamma^2 - 1}{\gamma^2}}$

$$\Rightarrow \frac{v}{c} = c\sqrt{\frac{99}{100}} = 0.995$$

(b) $(pc)^2 = m^2v^2\gamma^2c^2$

$$E^2 = m^2c^4\left(\left(\frac{v}{c}\right)^2\gamma^2 + 1\right)$$

$$\Rightarrow \frac{E^2 - (pc)^2}{E^2} = \frac{1}{1+\gamma^2\left(\frac{v}{c}\right)^2} = \frac{1}{1+\left(10/(0.995)\right)^2} = 0.01 = 1\%$$

39-50: (a) $(9.00\ \text{kg})(1.00\times10^{-4})(3.00\times10^{8}\ \text{m/s})^2 = 8.10\times10^{13}\ \text{J}$.

(b) $(\Delta E/\Delta t) = (8.10\times10^{13}\ \text{J})/(4.00\times10^{-6}\ \text{s}) = 2.03\times10^{19}\ \text{W}$.

(c) $M = \frac{\Delta E}{gh} = \frac{(8.10\times10^{13}\ \text{J})}{(9.80\ \text{m/s}^2)(1.00\times10^{3}\ \text{m})} = 8.27\times10^{9}\ \text{kg}$.

39-51: $x'^2 = c^2t'^2$

$$\Rightarrow (x-ut)^2\gamma^2 = c^2\gamma^2(t-ux/c^2)^2$$

$$\Rightarrow x - ut = c(t - ux/c^2)$$

$$\Rightarrow x(1+\frac{u}{c}) = \frac{1}{c}x(u+c) = t(u+c) \Rightarrow x = ct$$

$$\Rightarrow x^2 = c^2t^2.$$

39-52: a) From Eq. (39-38),

$$K - \frac{1}{2}mv^2 = \frac{3}{8}m\frac{v^4}{c^2} = \frac{3}{8}(80.0\text{kg})\frac{(3.00\times10^4\ \text{m/s})^4}{(3.00\times10^8\ \text{m/s})^2} = 270\ \text{J}.$$

(b) $\frac{(3/8)mv^4/c^2}{(1/2)mv^2} = \frac{3}{4}\left(\frac{v}{c}\right)^2 = 7.50\times10^{-9}$.

39-53: Speed in glass $v = \frac{c}{n} = \frac{c}{1.62} = 1.85\times10^8\text{m/s}$

$$\gamma = \frac{1}{\sqrt{1-v^2/c^2}} = 1.271$$

$$\Rightarrow K = (\gamma - 1)mc^2 = (0.271)(0.511\text{MeV}) = 0.138\,\text{MeV} = 1.38\times10^5\text{eV}$$

39-54: (a) 80.0m/s is non-relativistic, and $K=\frac{1}{2}mv^2=186\text{ J}$.

(b) $(\gamma-1)mc^2=1.31\times10^{15}\text{ J}$.

(c) In Eq. (39-24),

$v'=2.20\times10^8\text{ m/s},\ u=-1.80\times10^8\text{ m/s}$, and so $v=7.14\times10^7\text{ m/s}$.

(d) $\dfrac{20.0\text{m}}{\gamma}=13.6\text{m}$.

(e) $\dfrac{20.0\text{ m}}{2.20\times10^8\text{ m/s}}=9.09\times10^{-8}\text{ s}$.

(f) $t'=\dfrac{t}{\gamma}=6.18\times10^{-8}\text{ s}$, or $t'=\dfrac{13.6\text{m}}{2.20\times10^8\text{ m/s}}=6.18\times10^{-8}\text{ s}$.

39-55: Longer wavelength (redshift) implies recession

using the result of 39-27: $u=c\dfrac{(\lambda_o/\lambda)^2-1}{(\lambda_o/\lambda)^2+1}$

$$\Rightarrow u=c\left[\frac{(121.6/346.2)^2-1}{(121.6/346.2)^2+1}\right]=-0.780\text{c}=-2.341\times10^8\text{m/s}$$

($\because\Rightarrow$recession)

39-56: The car had better be moving non-relativistically, so the Doppler shift formula (Eq. (39-26)) becomes $f=f_0(1-(u/c))$. This is the frequency with which the radar waves strike the car, and the car reradiates at f. The receiver detects a frequency $f(1-(u/c))=f_0(1-(u/c))^2\approx f_0(1-2(u/c))$, so $\Delta f=2f_0(u/c)$ and the fractional frequency shift is $\dfrac{\Delta f}{f_0}=2(u/c)$. In this case,

$$\frac{\Delta f}{f_0}=2(35.8\text{ m/s})/(3.00\times10^8\text{ m/s})=2.39\times10^{-7}.$$

39-57: $a=\dfrac{dv}{dt}=\dfrac{F}{m}(1-v^2/c^2)^{3/2}$

$$\Rightarrow\int_0^v\frac{dv}{(1-(v^2/c^2))^{3/2}}=\frac{F}{m}\int_0^t dt=\frac{F}{m}t$$

$$\Rightarrow c\int_0^{v/c}\frac{dx}{(1-x^2)^{3/2}}=c\left.\frac{x}{\sqrt{1-x^2}}\right|_o^{v/c}=\frac{v}{\sqrt{1-(v/c)^2}}=\frac{F}{m}t$$

$$\Rightarrow v^2=\left(\frac{Ft}{m}\right)^2\left(1-\left(\frac{v}{c}\right)^2\right)=\left(\frac{Ft}{m}\right)^2-v^2\left(\frac{Ft}{mc}\right)^2$$

$$\Rightarrow v=\frac{Ft/m}{\sqrt{1+(Ft/mc)^2}}.$$

So as $t\to\infty,\ v\to c$.

39-58: Setting $x = 0$ in Eq. (39-22), the first equation becomes $x' = -\gamma ut$ and the last, upon multiplication by c, becomes $ct' = \gamma ct$. Squaring and subtracting gives

$$c^2t'^2 - x'^2 = \gamma^2\left(c^2t^2 - u^2t^2\right) = c^2t^2,$$

or $x' = c\sqrt{t'^2 - t^2} = 5.33\times10^8\ \text{m}$.

39-59: (a) since the two triangles are similar:
$H = A\gamma = mc^2\gamma = E$
(b) $O = \sqrt{H^2 - A^2} = \sqrt{E^2 - (mc^2)^2} = pc$
(c) $K = E - mc^2$
The kinetic energy can be obtained by the difference between the hypoteneuse and adjacent edge lengths.

39-60: (a) As in the hint, both the sender and the receiver measure the same distance. However, in our frame, the ship has moved between emission of succesive wavefronts, and we can use the time $T=1/f$ as the proper time, with the result that $f = \gamma f_0 > f_0$. (b) Approaching: 4.90 $f_0 = 17.1\ \text{MHz}, f - f_0$ $= 13.6\ \text{MHz}$. Moving away: $0.204f_0 = 0.714\text{MHz}, f - f_0 = -2.79\text{MHz}$.
(c) $\gamma f_0 = 2.55f_0 = 8.93\ \text{MHz},\ f - f_0 = 5.43\text{MHz}$. The frequency and frequency shift are the respective averages of those found in part (b). To see this, note that

$$\frac{1}{2}\left[\sqrt{\frac{c+u}{c-u}} + \sqrt{\frac{c-u}{c+u}}\right] = \frac{1}{2}\left[\frac{c+u}{\sqrt{c^2-u^2}} + \frac{c-u}{\sqrt{c^2-u^2}}\right] = \gamma.$$

39-61:

The crux of this problem is the question of simultaneity. To be "in the barn at one time" for the runner is different than for a stationary observer in the barn.
The diagram at right shows the rod fitting into the barn at time t = 0, according to the stationary observer.

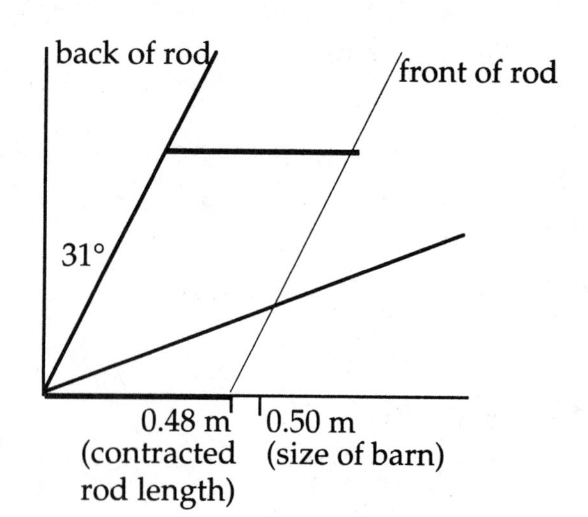

The diagram at right is in the runner's frame of reference.
The front of the rod enters the barn at time t_1 and leaves the back of the barn at time t_2.
However, the back of the rod does not enter the front of the barn until the later time t_3.

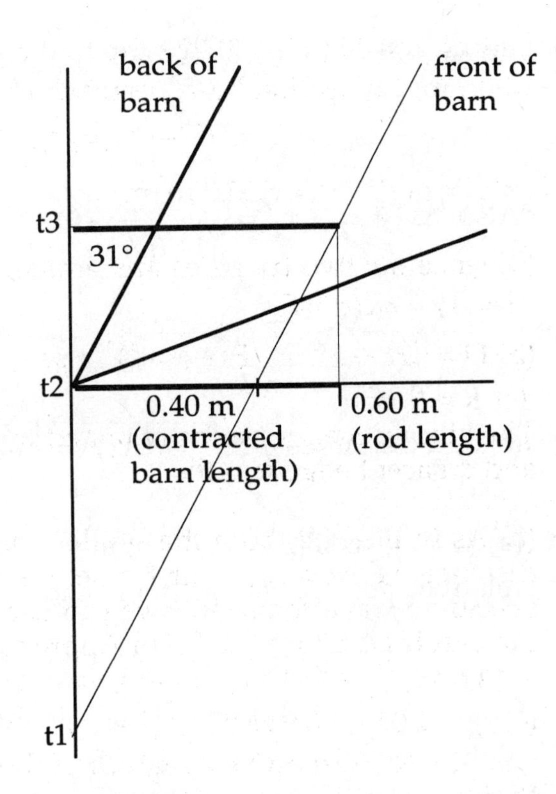

39-62: In Eq. (39-24), $u = V, v' = (c/n)$, and so

$$v = \frac{(c/n)+V}{1+\frac{cV}{nc^2}} = \frac{(c/n)+V}{1+(V/nc)}.$$

For V non-relativistic, this is

$$\begin{aligned} v &\approx ((c/n)+V)(1-(V/nc)) \\ &= (c/n)+V-(V/n^2)-(V^2/nc) \\ &\approx \frac{c}{n}+\left(1-\frac{1}{n^2}\right)V \end{aligned}$$

so $k = \left(1-\frac{1}{n^2}\right)$. For water, $n = 1.333$ and $k = 0.437$.

39-63: (a) Want $\Delta t' = t_2{}' - t_1{}'$

$x_1{}' = (x_1 - ut_1)\gamma = x_2{}' = (x_2 - ut_2)\gamma$

$\Rightarrow u = \dfrac{x_2 - x_1}{t_2 - t_1} = \dfrac{\Delta x}{\Delta t}$

And $\Delta t' = \gamma\left(t_2 - t_1 - u\dfrac{(x_2 - x_1)}{c^2}\right)$. Since $u = \Delta x/\Delta t$,

$$\Delta t' = \gamma\left(\Delta t - \frac{u\Delta x}{c^2}\right) = \gamma\left(\Delta t - \frac{\Delta x^2}{c^2 \Delta t}\right) \Rightarrow \Delta t' = \Delta t\sqrt{1-\left(\tfrac{\Delta x}{\Delta t}\right)^2/c^2} = \sqrt{\Delta t^2 - \frac{\Delta x^2}{c^2}}$$

There's no physical solution for $\Delta t < \dfrac{\Delta x}{c} \Rightarrow \Delta x \geq c\Delta t$

(b) Simultaneously $\Rightarrow \Delta t' = 0 \quad \therefore t_1 - \frac{ux_1}{c^2} = t_2 - \frac{ux_2}{c^2}$

$$\Rightarrow \Delta t = \frac{u}{c^2}\Delta x \quad \Rightarrow u = \frac{c^2 \Delta t}{\Delta x}$$

Also $\Delta x' = x_2{}' - x_1{}' = \frac{1}{\sqrt{1-(u/c)^2}}(\Delta x - u\Delta t) = \frac{1}{\sqrt{1-c^2\Delta t^2/\Delta x^2}}\left(\Delta x - c^2\frac{\Delta t^2}{\Delta x}\right)$

$$= \Delta x\sqrt{1-\frac{c^2\Delta t^2}{\Delta x^2}}$$

$$\Rightarrow \Delta x' = \sqrt{\Delta x^2 - c^2\Delta t^2}$$

(c) part (b): $\Delta t = \frac{1}{c}\sqrt{(\Delta x')^2 - (\Delta x)^2} = \frac{1}{c}\sqrt{(2.5\text{ m})^2 - (1.25\text{ m})^2} = 7.22\times10^{-9}\text{s}$

39-64: (a) $(100\text{s})(0.600)(3.00\times10^8\text{ m/s}) = 1.80\times10^{10}\text{ m}$. (b) In Stanley's frame, the relative speed of the tachyons and the ship is 3.40*c*, and so the time $t_2 = 100\text{s} + \frac{1.80\times10^{10}\text{ m}}{3.4c} = 118\text{s}$. At t_2 Stanley measures that Mavis is a distance from him of $(118\text{ s})(0.600)(3.00\times10^8\text{m/s}) = 2.12\times10^{10}\text{m}$. (c) From Eq. (39-24), with $v' = -4.00c$ and $u = 0.600c$, $v = +2.43c$, with the plus sign indicating a direction in the same direction as Mavis' motion (that is, away from Stanley). (d) As the result of part (c) suggests, Stanley would see the tachyons moving toward Mavis, and hence t_3 is the time they would have left Stanley in order to reach Mavis at the distance found in part (b), or $118\text{ s} - \frac{2.12\times10^{10}\text{ m}}{2.43c} = 89\text{s}$, and so Stanley is in trouble before he knows what hit him.

39-65: (a) $K = (\gamma - 1)mc^2 \qquad E = \gamma mc^2$

$$E = K + mc^2 = \sqrt{(mc^2)^2 + (pc)^2}$$

$$\Rightarrow \gamma = 1 + \left(\frac{m_\pi c^2}{m_p c^2}\right) = 1 + \frac{139.6\text{MeV}}{938\text{MeV}}$$

$$\gamma = 1.149 = \frac{1}{\sqrt{1-\left(\frac{v'}{c}\right)^2}} \Rightarrow v' = c\sqrt{\frac{\gamma^2-1}{\gamma^2}} = 0.492\text{c}$$

lab: $v' = \frac{v'+u}{1+uv'/c^2} = \frac{2v'}{1+v'^2/c^2} = 0.793\text{c} \Rightarrow \gamma = 1.64$

$$\Rightarrow \text{K} = (\gamma - 1)\text{mc}^2 = (0.640)(938)\text{MeV} = 600\text{MeV}.$$

(b) $\frac{K}{\text{rest mass}} = \frac{600\text{MeV}}{2(139.6)} = 2.15$

39-66: For any function $f = f(x,t)$ and $x = x(x',t'), t = t(x',t')$, let $F(x',t') = f(x(x',t'),t(x',t'))$ and use the standard (but mathematically improper) notation $F(x',t') = f(x',t')$. The chain rule is then

$$\frac{\partial f(x',t')}{\partial x} = \frac{\partial f(x,t)}{\partial x'}\frac{\partial x'}{\partial x} + \frac{\partial f(x',t')}{\partial t'}\frac{\partial t'}{\partial x},$$
$$\frac{\partial f(x',t')}{\partial t} = \frac{\partial f(x,t)}{\partial x'}\frac{\partial x'}{\partial t} + \frac{\partial f(x',t')}{\partial t'}\frac{\partial t'}{\partial t}.$$

In this solution, the explicit dependence of the functions on the sets of dependent variables is suppressed, and the above relations are then

$$\frac{\partial f}{\partial x} = \frac{\partial f}{\partial x'}\frac{\partial x'}{\partial x} + \frac{\partial f}{\partial t'}\frac{\partial t'}{\partial x}, \quad \frac{\partial f}{\partial t} = \frac{\partial f}{\partial x'}\frac{\partial x'}{\partial t} + \frac{\partial f}{\partial t'}\frac{\partial t'}{\partial t}.$$

(a) $\frac{\partial x'}{\partial x} = 1, \frac{\partial x'}{\partial t} = -v, \frac{\partial t'}{\partial x} = 0$ and $\frac{\partial t'}{\partial t} = 1$. Then, $\frac{\partial y}{\partial x} = \frac{\partial y}{\partial x'}$ and $\frac{\partial^2 y}{\partial x^2} = \frac{\partial^2 y}{\partial x'^2}$. For the time derivative, $\frac{\partial f}{\partial t} = -v\frac{\partial y}{\partial x'} + \frac{\partial y}{\partial t'}$. To find the second time derivative, the chain rule must be applied to both terms; that is,

$$\frac{\partial}{\partial t}\frac{\partial y}{\partial x'} = -v\frac{\partial^2 y}{\partial x'^2} + \frac{\partial^2 y}{\partial t' \partial x'},$$
$$\frac{\partial}{\partial t}\frac{\partial y}{\partial t'} = -v\frac{\partial^2 y}{\partial x' \partial t'} + \frac{\partial^2 y}{\partial t'^2}.$$

Using these in $\frac{\partial^2 y}{\partial t^2}$, collecting terms and equating the mixed partial derivatives gives

$$\frac{\partial^2 y}{\partial t^2} = v^2\frac{\partial^2 y}{\partial x'^2} - 2v\frac{\partial^2 y}{\partial x' \partial t'} + \frac{\partial^2 y}{\partial t'^2}$$

and using this and the above expression for $\frac{\partial^2 y}{\partial x'^2}$ gives the result.

(b) For the Lorentz transformation, $\frac{\partial x'}{\partial x} = \gamma, \frac{\partial x'}{\partial t} = \gamma v, \frac{\partial t'}{\partial x} = -\gamma v / c^2$ and $\frac{\partial t'}{\partial t} = \gamma$. The first partials are then

$$\frac{\partial y}{\partial x} = \gamma\frac{\partial y}{\partial x'} - \gamma\frac{v}{c^2}\frac{\partial y}{\partial t'}, \quad \frac{\partial y}{\partial t} = -\gamma v\frac{\partial y}{\partial x'} + \gamma\frac{\partial y}{\partial t'}$$

and the second partials are (again equating the mixed partials)

$$\frac{\partial^2 y}{\partial x^2} = \gamma^2\frac{\partial^2 y}{\partial x'^2} + \gamma^2\frac{v^2}{c^4}\frac{\partial^2 y}{\partial t'^2} - 2\gamma^2\frac{v}{c^2}\frac{\partial^2 y}{\partial x' \partial t'}$$
$$\frac{\partial^2 y}{\partial t^2} = \gamma^2 v^2\frac{\partial^2 y}{\partial x'^2} + \gamma^2\frac{\partial^2 y}{\partial t'^2} - 2\gamma^2 v\frac{\partial^2 y}{\partial x' \partial t'}.$$

Substituting into the wave equation and combining terms (note that the mixed partials cancel),

$$\frac{\partial^2 y}{\partial x^2}-\frac{1}{c^2}\frac{\partial^2 y}{\partial t^2}=\gamma^2\left(1-\frac{v^2}{c^2}\right)\frac{\partial^2 y}{\partial x'^2}+\gamma^2\left(\frac{v^2}{c^4}-\frac{1}{c^2}\right)\frac{\partial^2 y}{\partial t'^2}$$

$$=\frac{\partial^2 y}{\partial x'^2}-\frac{1}{c^2}\frac{\partial^2 y}{\partial t'^2}$$

$$=0.$$

39-67: (a) $a'=\dfrac{dv}{dt'}$ $\qquad dt'=\gamma(dt-udx/c^2)$

$$dv'=\frac{dv}{(1-uv/c^2)}+\frac{v-u}{(1-uv/c^2)^2}\frac{u}{c^2}dv$$

$$\frac{dv'}{dv}=\frac{1}{1-uv/c^2}+\frac{v-u}{(1-uv/c^2)^2}\cdot\left(\frac{u}{c^2}\right)$$

$$\therefore dv'=dv\left(\frac{1}{1-uv/c^2}+\frac{(v-u)u/c^2}{(1-uv/c^2)^2}\right)=dv\left(\frac{1-u^2/c^2}{(1-uv/c^2)^2}\right)$$

$$\therefore a'=\frac{dv\dfrac{(1-u^2/c^2)}{(1-uv/c^2)^2}}{\gamma dt-u\gamma dx/c^2}=\frac{dv}{dt}\cdot\frac{(1-u^2/c^2)}{(1-uv/c^2)^2}\cdot\frac{1}{\gamma(1-uv/c^2)}$$

$$=a(1-u^2/c^2)^{3/2}(1-uv/c^2)^{-3}$$

(b) Changing frames from $S'\rightarrow S$ just involves changing $a\rightarrow a', v\rightarrow -v'$

$$\Rightarrow a=a'(1-u^2/c^2)^{3/2}\left(1+\frac{uv'}{c^2}\right)^{-3}.$$

39-68: (a) The speed v' is measured relative to the rocket, and so for the rocket and its occupant, $v'=0$. The acceleration as seen in the rocket is given to be $a'=g$, and so the acceleration as measured on the earth is

$$a=\frac{du}{dt}=g\left(1-\frac{u^2}{c^2}\right)^{3/2}.$$

(b) With $v_1=0$ when $t=0$,

$$dt=\frac{1}{g}\frac{du}{(1-u^2/c^2)^{3/2}}$$

$$\int_0^{t_1}dt=\frac{1}{g}\int_0^{v_1}\frac{du}{(1-u^2/c^2)^{3/2}}$$

$$t_1=\frac{v_1}{g\sqrt{1-v_1^2/c^2}}.$$

(c) $dt'=\gamma dt=dt/\sqrt{1-u^2/c^2}$, so the relation in part (b) between dt and du, expressed in terms of dt' and du, is

$$dt'=\gamma dt=\frac{1}{\sqrt{1-u^2/c^2}}\frac{du}{g(1-u^2/c^2)^{3/2}}=\frac{1}{g}\frac{du}{(1-u^2/c^2)^2}.$$

Integrating as above (perhaps using the substitution $z = u/c$) gives

$$t_1' = \frac{c}{g}\,\text{arctanh}\left(\frac{v_1}{c}\right).$$

For those who wish to avoid inverse hyperbolic functions, the above integral may done by the method of partial fractions;

$$g\,dt' = \frac{du}{(1+u/c)(1-u/c)} = \frac{1}{2}\left[\frac{du}{1+u/c} + \frac{du}{1-uc}\right],$$

which integrates to

$$t_1' = \frac{c}{2g}\ln\left(\frac{c+v_1}{c-v_1}\right).$$

(d) Solving the expression from part (c) for v_1 in terms of $t_1, (v_1/c) = \tanh(gt_1'/c)$, so that $\sqrt{1-(v_1/c)^2} = 1/\cosh(gt_1'/c)$, using the appropriate indentities for hyperbolic functions. Using this in the expression found in part (b),

$$t_1 = \frac{c}{g}\frac{\tanh(gt_1'/c)}{1/\cosh(gt_1'/c)} = \frac{c}{g}\sinh(gt_1'/c),$$

which may be rearranged slightly as

$$\frac{gt_1}{c} = \sinh\left(\frac{gt_1'}{c}\right).$$

If hyperbolic functions are not used , v_1 in terms of t_1' is found to be

$$\frac{v_1}{c} = \frac{e^{gt_1'/c} - e^{-gt_1'/c}}{e^{gt_1'/c} + e^{-gt_1'/c}},$$

which is the same as $\tanh(gt_1'/c)$. Inserting this expression into the result of part (b) gives, after much algebra,

$$t_1 = \frac{c}{2g}\left(e^{gt_1'/c} - e^{-gt_1'/c}\right),$$

which is equivalent to the expression found using hyperbolic functions.
(e) After the first acceleration period (of 5 years by Stella's clock), the elapsed time on earth is

$$t_1' = \frac{c}{g}\sinh(gt_1'/c) = 2.65\times10^9\ \text{s} = 84.0\ \text{yr}.$$

The elapsed time will be the same for each of the four parts of the voyage, so when Stella has returned, Terra has aged 336 yr and the year is 2436. (Keeping more precision than is given in the problem gives February 7 of that year.)

39-69: (a) $f_o = 4.5679\times10^{14}\ \text{Hz}$ $\quad f_+ = 4.5685\times10^{14}\ \text{Hz}$ $\quad f_- = 4.5661\times10^{14}\ \text{Hz}$

$$\left.\begin{aligned} f_+ &= \sqrt{\frac{c+(u+v)}{c-(u+v)}}\cdot f_o \\ f_- &= \sqrt{\frac{c+(u-v)}{c-(u-v)}}\cdot f_o \end{aligned}\right\} \Rightarrow f_+^2(c-(u+v)) = f_o^2(c+(u+v))$$

$$\Rightarrow (u+v) = \frac{(f_+/f_o)^2 - 1}{(f_+/f_o)^2 + 1} c$$

$\Rightarrow u + v = 39402\,\text{m/s} \quad u - v = -118200\,\text{m/s}$

$\Rightarrow u = -39400\,\text{m/s} \Rightarrow$ moving away.

(b) $v = 78820\,\text{m/s} \qquad T = 90.0\,\text{days}$.

$\Rightarrow 2\pi R = vt \Rightarrow R = \dfrac{(78820\ \text{m/s})(90)(24)(3600\ \text{s})}{2\pi} = 9.5\times 10^{10}\,\text{m} \approx 0.65$ earth-sun distance.

Also: $\dfrac{(Gm^2)}{(2R)^2} = \dfrac{mv^2}{R} \Rightarrow m = \dfrac{4Rv^2}{G} = 3.6\times 10^{31}\,\text{kg} \approx 18\,\text{m}_{\text{sun}}$.

Chapter 40: Photons, Electrons, and Atoms

40-1: (a) $E = hf \Rightarrow f = \dfrac{E}{h} = \dfrac{(3.25\times10^{6}\text{eV})(1.60\times10^{-19}\text{J/eV})}{6.63\times10^{-34}\text{J}\cdot\text{s}} = 7.86\times10^{20}\text{Hz}$

(b) $\lambda = \dfrac{c}{f} = \dfrac{3.00\times10^{8}\text{m/s}}{7.86\times10^{20}\text{Hz}} = 3.81\times10^{-13}\text{m}.$

(c) λ is of the same magnitude as a nuclear radius.

40-2: $\dfrac{dN}{dt} = \dfrac{(dE/dt)}{(dE/dN)} = \dfrac{P}{hf} = \dfrac{P\lambda}{hc} = 1.78\times10^{20}/\text{s}.$

40-3: $f = \dfrac{c}{\lambda} = \dfrac{3.00\times10^{8}\text{m/s}}{6.10\times10^{-7}\text{m}} = 4.92\times10^{14}\text{Hz}$

$p = \dfrac{h}{\lambda} = \dfrac{6.63\times10^{-34}\text{J}\cdot\text{s}}{6.10\times10^{-7}\text{m}} = 1.09\times10^{-27}\text{kg}\cdot\text{m/s}$

$E = pc = (1.09\times10^{-27}\text{kg}\cdot\text{m/s})(3.00\times10^{8}\text{m/s}) = 3.26\times10^{-19}\text{J}$

$= 2.04\text{ eV}.$

40-4: (a) $Pt = (0.600\text{W})(20.0\times10^{-3}\text{s}) = 0.0120\text{J} = 7.49\times10^{16}\text{eV}.$

(b) $hf = \dfrac{hc}{\lambda} = 3.05\times10^{-19}\text{J} = 1.90\text{ eV}.$

(c) $\dfrac{Pt}{hf} = 3.94\times10^{16}.$

40-5: (a) $E = pc = (4.25\times10^{-27}\text{kg}\cdot\text{m/s})(3.00\times10^{8}\text{m/s})$

$= 1.27\times10^{-18}\text{J} = \dfrac{1.27\times10^{-18}\text{J}}{1.60\times10^{-19}\text{J/eV}} = 7.93\text{ eV}$

(b) $p = \dfrac{h}{\lambda} \Rightarrow \lambda = \dfrac{h}{p} = \dfrac{(6.63\times10^{34}\text{J}\cdot\text{s})}{(4.25\times10^{-27}\text{kg}\cdot\text{m/s})} = 1.56\times10^{-7}\text{m}.$

This is ultraviolet radiation.

40-6: (a) The threshold frequency is found by setting $V = 0$ in Eq.(40-4), $f_0 = \phi/h$.

(b) $\phi = hf_0 = \dfrac{hc}{\lambda} = 7.21\text{ eV}.$

40-7: $\dfrac{1}{2}mv_{\max}^2 = hf - \phi = (6.63\times10^{-34}\text{J}\cdot\text{s})(3.4\times10^{15}\text{Hz}) - (5.1\text{eV})(1.60\times10^{-19}\text{J/eV})$

$= 1.44\times10^{-18}\text{J}$

$\Rightarrow v_{\max} = \sqrt{\dfrac{2(1.44\times10^{-18}\text{J})}{9.11\times10^{-31}\text{kg}}} = 1.8\times10^{6}\text{m/s}$

40-8: $E = hf - \phi = hc\left(\frac{1}{\lambda} - \frac{1}{\lambda_0}\right) = 1.97\,\text{eV}.$

40-9: (a) The work function $\phi = hf - eV_o = \frac{hc}{\lambda} - eV_o$

$$\Rightarrow \phi = \frac{(6.63\times 10^{-34}\,\text{J}\cdot\text{s})(3.00\times 10^{8}\,\text{m/s})}{2.54\times 10^{-7}\,\text{m}} - (1.60\times 10^{-19}\,\text{C})(0.181\,\text{V})$$
$$= 7.53\times 10^{-19}\,\text{J}.$$

The threshold frequency implies $\phi = hf_{th} \Rightarrow \frac{hc}{\lambda_{th}} \Rightarrow \lambda_{th} = \frac{hc}{\phi}$

$$\Rightarrow \lambda_{th} = \frac{(6.63\times 10^{-34}\,\text{J}\cdot\text{s})(3.00\times 10^{8}\,\text{m/s})}{7.53\times 10^{-19}\,\text{J}} = 2.64\times 10^{-7}\,\text{m}$$

(b) $\phi = 7.53\times 10^{-19}\,\text{J} = 4.70\,\text{eV}$, as found in part (a), and this is the value from Table 40-1.

40-10: (a) From Eq. (40-4),

$$V = \frac{1}{e}\left(\frac{hc}{\lambda} - \phi\right) = \frac{(4.136\times 10^{-15}\,\text{eV}\cdot\text{s})(3.00\times 10^{8}\,\text{m/s})}{(280\times 10^{-9}\,\text{m})} - 2.3\,\text{V} = 2.1\,\text{V}.$$

(b) The stopping potential, multiplied by the electron charge, is the maximum kinetic energy, 2.1 eV.

(c) $v = \sqrt{\frac{2K}{m}} = \sqrt{\frac{2eV}{m}} = \sqrt{\frac{2(1.602\times 10^{-19}\,\text{C})(2.1\,\text{V})}{(9.11\times 10^{-31}\,\text{kg})}} = 8.7\times 10^{5}\,\text{m/s}.$

40-11: (a) $\frac{1}{\lambda} = R\left(\frac{1}{2^2} - \frac{1}{n^2}\right)$

$$\Rightarrow n = 4: \quad \frac{1}{\lambda} = R\left(\frac{1}{4} - \frac{1}{16}\right) = \frac{3}{16}R$$

$$\Rightarrow \lambda = \frac{16}{3R} = \frac{16}{3(1.10\times 10^{7})}\,\text{m} = 4.86\times 10^{-7}\,\text{m} = 486\,\text{nm}$$

$$\Rightarrow f = \frac{c}{\lambda} = \frac{3.00\times 10^{8}\,\text{m/s}}{4.86\times 10^{-7}\,\text{m}} = 6.17\times 10^{14}\,\text{Hz}$$

(b) $\lambda = 4.86\times 10^{-7}\,\text{m}$, as found in part (a).

40-12: Lyman: largest is $n = 2, \lambda = \frac{(4/3)}{R} = \frac{(4/3)}{(1.097\times 10^{7}\,\text{m}^{-1})} = 122\,\text{nm}$, in the ultraviolet. Smallest is $n = \infty, \lambda = \frac{1}{R} = 91.2\,\text{nm}$, also ultraviolet. Paschen: largest is $n = 4, \lambda = \frac{(144/7)}{R} = 1875\,\text{nm}$, in the infrared. Smallest is $n = \infty, \lambda = \frac{9}{R} = 820\,\text{nm}$, also infrared.

40-13: $\Delta E = \frac{hc}{\lambda} \Rightarrow \lambda = \frac{hc}{\Delta E} = \frac{(6.63\times10^{-34}\,\text{J}\cdot\text{s})(3.00\times10^{8}\,\text{m/s})}{(3.40\,\text{eV})(1.60\times10^{-19}\,\text{J/eV})}$

$= 3.65\times10^{-7}\,\text{m} = 365\,\text{nm}$

40-14: (a) $-E_1 = 20$ eV. (b) The 3-2, 3-1, and 2-1 transitions are possible, with energies 5 eV, 15 eV and 10 eV. (c) There is no energy level 8 eV above the ground state energy, so the photon will not be absorbed. (d) The work function must be more than 5 eV, but not larger than 8 eV.

40-15: (a) $E_\gamma = \frac{hc}{\lambda} = \frac{(6.63\times10^{-34}\,\text{J}\cdot\text{s})(3.00\times10^{8}\,\text{m/s})}{7.35\times10^{-7}\,\text{m}} = 2.71\times10^{-19}\,\text{J}$

$= 1.69$ eV

So the internal energy of the atom increases by 1.69 eV to $E = -8.92\,\text{eV} + 1.69\,\text{eV} = -7.23\,\text{eV}$

(b) $E_\gamma = \frac{hc}{\lambda} = \frac{(6.63\times10^{-34}\,\text{J}\cdot\text{s})(3.00\times10^{8}\,\text{m/s})}{3.60\times10^{-7}\,\text{m}} = 5.53\times10^{-19}\,\text{J}$

$= 3.45$ eV

So the final internal energy of the atom decreases to $E = -1.35\,\text{eV} - 3.45\,\text{eV} = -4.80\,\text{eV}$.

40-16: (a) Equating initial kinetic energy and final potential energy and solving for the separation radius r,

$$r = \frac{1}{4\pi\varepsilon_0}\frac{(79e)(2e)}{K}$$

$$= (8.988\times10^{9}\,\text{N}\cdot\text{m}^2/\text{C}^2)\frac{(158)(1.602\times10^{-19}\,\text{C})}{(4.78\times10^{6}\,\text{J/C})} = 4.76\times10^{-14}\,\text{m}.$$

(b) The above result may be substituted into Coulomb's law, or, the relation between the magnitude of the force and the magnitude of the potential energy in a Coulombic field is $F = \frac{K}{r} = 16.1\text{N}$.

40-17: (a) $U = \frac{q_1 q_2}{4\pi\varepsilon_0 r} = \frac{(2e)(79e)}{4\pi\varepsilon_0 r} = \frac{158(1.60\times10^{-19}\,\text{C})^2}{4\pi\varepsilon_0(5.60\times10^{-14}\,\text{m})}$

$\Rightarrow U = 6.49\times10^{-13}\,\text{J} = 4.06\times10^{6}\,\text{eV} = 4.06\,\text{MeV}$

(b) $K_1 + U_1 = K_2 + U_2 \Rightarrow K_1 = U_2 = 6.49\times10^{-13}\,\text{J} = 4.06\,\text{MeV}$

(c) $K = \frac{1}{2}mv^2 \Rightarrow v = \sqrt{\frac{2K}{m}} = \sqrt{\frac{2(6.49\times10^{-13}\,\text{J})}{6.64\times10^{-27}\,\text{kg}}} = 1.40\times10^{7}\,\text{m/s}$

40-18: (a), (b) For either atom, the magnitude of the angular momentum is $\frac{h}{2\pi} = 1.05\times10^{-34}\,\text{kg}\cdot\text{m}^2/\text{s}$.

40-19: (a) $v_n = \frac{1}{\varepsilon_0}\frac{e^2}{2nh}$: $n=1 \Rightarrow v_1 = \frac{(1.60\times10^{-19}\,\text{C})^2}{\varepsilon_0 2(6.63\times10^{-34}\,\text{J}\cdot\text{s})} = 2.18\times10^6\,\text{m/s}$

$$h = 2 \Rightarrow v_2 = \frac{v_1}{2} = 1.09\times10^6\,\text{m/s}$$

$$h = 3 \Rightarrow v_3 = \frac{v_1}{3} = 7.27\times10^5\,\text{m/s}$$

(b) Orbital period $= \frac{2\pi r_n}{v_n} = \frac{2\varepsilon_0 n^2h^2/me^2}{1/\varepsilon_0\cdot e^2/2nh} = \frac{4\varepsilon_0^2 n^3h^3}{me^4}$

$$n = 1 \Rightarrow T_1 = \frac{4\varepsilon_0^2(6.63\times10^{-34}\,\text{J}\cdot\text{s})^3}{(9.11\times10^{-31}\,\text{kg})(1.60\times10^{-19}\,\text{C})^4} = 1.53\times10^{-16}\,\text{s}$$

$$n = 2:\quad T_2 = T_1(2)^3 = 1.22\times10^{-15}\,\text{s}$$

$$n = 3:\quad T_3 = T_1(3)^3 = 4.13\times10^{-15}\,\text{s}$$

(c) number of orbits $= \frac{1.0\times10^{-8}\,\text{s}}{1.22\times10^{-15}\,\text{s}} = 8.2\times10^6.$

40-20: (a) Using the values from Appendix F, keeping eight significant figures, gives $R = 1.0973731\times10^7\,\text{m}^{-1}$. (Note: on some standard calculators, intermediate values in the calculation may have exponents that exceed 100 in magnitude. If this is the case, the numbers must be manipulated in a different order.) (b) Using the eight-figure value for R gives $E = \frac{hc}{\lambda} = hcR = 2.1798741\times10^{-18}\text{J}$ $= 13.605670$ eV. (c) Using the value for the proton mass as given in Appendix F gives $R = 1.0967758\times10^7\,\text{m}^{-1}$, which agrees with the five-figure valued obtained using $m_r = 0.99946\,m$. (d) $E = 2.17868751\times10^{-18}\text{J} = 13.598292\,\text{eV}$.

40-21: (a) Following the derivation for the hydrogen atom we see that for Li^{2+} all we need do is replace e^2 by $3e^2$. Then $E_n(\text{Li}^{2+}) = -\frac{1}{\varepsilon_0^2}\frac{m(3e^2)^2}{8n^2h^2} = 9E_n(\text{H})$

$$\Rightarrow E_n(\text{Li}^{2+}) = 9\left(\frac{-13.60\,\text{eV}}{n^2}\right)$$

So for the ground state, $E_1(\text{Li}^{2+}) = -122$ eV

(b) The ionization energy is the energy difference between the $n\to\infty$ and n=1 levels. So it is just 122 eV for Li^{2+}, which is 9 times that of hydrogen.

(c) $\frac{1}{\lambda} = \frac{m(3e^2)^2}{8\varepsilon_0^2h^3c}\left(\frac{1}{n_1^2}-\frac{1}{n_2^2}\right) = (9.87\times10^7\,\text{m}^{-1})\left(\frac{1}{n_1^2}-\frac{1}{n_2^2}\right)$

So for n = 2 to n = 1, $\frac{1}{\lambda} = (9.87\times10^7\,\text{m}^{-1})\left(1-\frac{1}{4}\right) = 7.40\times10^7\,\text{m}$

$\Rightarrow \lambda = 1.35\times10^{-8}\,\text{m}$. This is 9 times shorter than that from the hydrogen atom.

(d) $r_n = (\text{Li}^{2+}) = \frac{\varepsilon_0 n^2h^2}{\pi m(3e^2)} = \frac{1}{3}r_n(\text{H})$

40-22: $\frac{1}{\lambda}=R\left(1-\frac{1}{9}\right)$, so $\lambda=103\,\text{nm}$ and $f=\frac{c}{\lambda}=2.93\times10^{15}\,\text{Hz}$.

40-23: $E_\gamma=\frac{hc}{\lambda}=\frac{(6.63\times10^{-34}\,\text{J}\cdot\text{s})(3.00\times10^{8}\,\text{m/s})}{6.33\times10^{-7}\,\text{m}}=3.14\times10^{-19}\,\text{J}.$

Total energy in 1 second from the laser $E=Pt=2.50\times10^{-3}\,\text{J}$.

So the number of photons emitted per second is $\frac{E}{E_\gamma}=\frac{2.50\times10^{-3}\,\text{J}}{3.14\times10^{-19}\,\text{J}}=7.96\times10^{15}$.

40-24: $20.66\,\text{eV}-18.70\,\text{eV}=1.96\,\text{eV}=3.14\times10^{-19}\,\text{J}$, and $\lambda=\frac{c}{f}=\frac{hc}{E}=632\,\text{nm}$, in good agreement.

40-25: $\frac{n_{5s}}{n_{3p}}=e^{-(E_{5s}-E_{3p})/kT}$

But $E_{5s}=20.66\,\text{eV}=3.31\times10^{-18}\,\text{J}$

$E_{3p}=18.70\,\text{eV}=2.99\times10^{-18}\,\text{J}$

$\Rightarrow\Delta E=3.14\times10^{-19}\,\text{J}$

$\Rightarrow\frac{n_{5s}}{n_{3p}}=e^{-(3.14\times10^{-19}\,\text{J})/(1.38\times10^{-23}\,\text{J/K})(300\,\text{K})}=1.2\times10^{-33}$

40-26: $\frac{1}{100}=0.010=e^{-\Delta E/kT}$, so

$$T=\frac{\Delta E}{k\ln(100)}=\frac{(3/4)hcR}{k\ln(100)}=2.57\times10^{4}\,\text{K}.$$

40-27: An electron's energy after being accelerated by a voltage V is just $E=qV$. The most energetic photon able to be produced by the electron is just: $\lambda=\frac{hc}{E}=\frac{hc}{qV}$

$$\Rightarrow\lambda=\frac{(6.63\times10^{-34}\,\text{J}\cdot\text{s})(3.00\times10^{8}\,\text{m/s})}{(1.60\times10^{-19}\,\text{C})(1.7\times10^{4}\,\text{V})}=7.31\times10^{-11}\,\text{m}$$

$$=0.0731\,\text{nm}.$$

40-28: (a) $\frac{hc}{e\lambda}=1.03\times10^{4}\,\text{V}$. b) The shortest wavelength would correspond to the maximum electron energy, eV, and so $\lambda=\frac{hc}{eV}=0.0310\,\text{nm}$.

40-29: $eV_{AC}=hf_{\max}=\frac{hc}{\lambda_{\min}}$

$$\Rightarrow\lambda_{\min}=\frac{hc}{eV_{AC}}=\frac{(6.63\times10^{-34}\,\text{J}\cdot\text{s})(3.00\times10^{8}\,\text{m/s})}{(1.60\times10^{-19}\,\text{C})(5000\,\text{V})}=2.49\times10^{-10}\,\text{m}$$

This is the same answer as would be obtained if electrons of this energy were used. Electron beams are much more easily produced and accelerated than proton beams.

40-30: (a) $\lambda = \frac{hc}{E} = 0.0827\,\text{nm}$. b) From Eq. (40-21),

$\Delta\lambda = (h/mc)(1-\cos\phi) = (2.426\,\text{pm})(1-\cos 60.0°) = 1.21\,\text{pm}$, and $\lambda' = 0.0839\,\text{nm}$.

(b) $E = \frac{hc}{\lambda} = 14.8\,\text{keV}$.

40-31: The derivation of Eq. (40-21) is explicitly shown in equations (40-22) through (40-25) with the final substitution of $p' = h/\lambda'$ and $p = h/\lambda$ yielding

$$\lambda' - \lambda = \frac{h}{mc}(1-\cos\phi).$$

40-32: From Eq. (40-21), $\cos\phi = 1 - \frac{\Delta\lambda}{(h/mc)}$, and so a) $\cos\phi = 1 - \frac{0.0027\,\text{nm}}{0.002426\,\text{nm}} = -0.113$, and $\phi = 96.5°$. (b) The photon is undeflected, $\cos\phi = 1$ and $\phi = 0$.

40-33:
$$\lambda' - \lambda = \frac{h}{mc}(1-\cos\phi) \Rightarrow \lambda'_{\text{max}} = \lambda + \frac{h}{mc}(1-(-1)) = \lambda + \frac{2h}{mc}$$

$$\Rightarrow \lambda'_{\text{max}} = 8.40\times10^{-11}\,\text{m} + 2(2.426\times10^{-12}\,\text{m}) = 8.89\times10^{-11}\,\text{m}$$

40-34: From Eq.(40-28), (a) $\lambda_{\text{m}} = \frac{2.898\times10^{-3}\,\text{m}\cdot\text{K}}{3.00\,\text{K}} = 0.966\,\text{mm}$, and $f = \frac{c}{\lambda_{\text{m}}} = 3.10\times10^{11}\,\text{Hz}$. Note that a more precise value of the Wien displacement law constant has been used. (b) A factor of 100 increase in the temperature lowers λ_{m} by a factor of 100 to 9.66 μm and raises the frequency by the same factor, to $3.10\times10^{13}\,\text{Hz}$. (c) Similarly, λ_{m} = 966 nm and $f = 3.10\times10^{14}\,\text{Hz}$.

40-35: Eq. (40-30): $I(\lambda) = \frac{2\pi hc^2}{\lambda^5\left(e^{hc/\lambda kT}-1\right)}$

but $e^x = 1 + x + \frac{x^2}{2} + \ldots \approx 1 + x \quad \text{for } x \ll 1$

$$\Rightarrow I(\lambda) \approx \frac{2\pi hc^2}{\lambda^5(hc/\lambda kT)} = \frac{2\pi ckT}{\lambda^4} = \text{Eq. (40-29)}$$

which is Rayleigh's distribution.

40-36: From Eq. (40-29), $T = \frac{2.898\times10^{-3}\,\text{m}\cdot\text{K}}{550\times10^{-9}\,\text{m}} = 5.27\times10^{3}\,\text{K}$.

40-37: (a) To find the maximum in the Plauck distribution:

$$\frac{dI}{d\lambda}=\frac{d}{d\lambda}\left(\frac{2\pi hc^2}{\lambda^5\left(e^{\alpha/\lambda}-1\right)}\right)=0=-5\frac{\left(2\pi hc^2\right)}{\lambda^5\left(e^{\alpha/\lambda}-1\right)}-\frac{2\pi hc^2\left(-\alpha/\lambda^2\right)}{\lambda^5\left(e^{\alpha/\lambda}-1\right)^2}$$

$$\Rightarrow -5\left(e^{\alpha/\lambda}-1\right)\lambda=\alpha$$

$$\Rightarrow -5e^{\alpha/\lambda}+5=\alpha/\lambda$$

$$\Rightarrow \text{Solve } 5-x=5e^{x} \text{ where } x=\frac{\alpha}{\lambda}=\frac{hc}{\lambda kT}.$$

Its root is 4.965, so $\frac{\alpha}{\lambda}=4.965$

$$\Rightarrow \lambda=\frac{hc}{(4.965)kT}.$$

(b) $\lambda_m T=\frac{hc}{(4.965)k}=\frac{\left(6.63\times10^{-34}\,\text{J}\cdot\text{s}\right)\left(3.00\times10^{8}\,\text{m/s}\right)}{(4.965)\left(1.38\times10^{-23}\,\text{J/K}\right)}=2.90\times10^{-3}\,\text{m}\cdot\text{K}$

40-38: Combining equations (40-26) and (40-28),

$$\lambda_{\text{m}}=\frac{2.898\times10^{-3}\,\text{m}\cdot\text{K}}{(I/\sigma)^{1/4}}$$

$$=\frac{\left(2.898\times10^{-3}\,\text{m}\cdot\text{K}\right)}{\left(5.93\times10^{3}\,\text{W/m}^2/5.67\times10^{-8}\,\text{W/m}^2\cdot\text{K}^4\right)^{1/4}}$$

$$=5.10\times10^{-6}\,\text{m}=5.10\,\mu\text{m}.$$

40-39: $\lambda_m=\frac{2.90\times10^{-3}\,\text{m}\cdot\text{K}}{T}=\frac{2.90\times10^{-3}\,\text{m}\cdot\text{K}}{2.726\,\text{K}}=1.06\times10^{-3}\,\text{m}$

$$=1.06\,\text{mm}$$

This is in the microwave part of the electromagnetic spectrum.

40-40: (a) As in Example 40-11, using four-place values for the physical constants,

$$\frac{hc}{\lambda kT}=95.80,$$

from which

$$\frac{I(\lambda)\Delta\lambda}{\sigma T^4}=6.44\times10^{-38}.$$

(b) With T = 2000 K and the same values for λ and $\Delta\lambda$,

$$\frac{hc}{kT}=14.37$$

and so

$$\frac{I(\lambda)\Delta\lambda}{\sigma T^4}=7.54\times10^{-6}.$$

(c) With $T=6000\,\text{K}$, $\frac{hc}{\lambda T}=4.790$ and

$$\frac{I(\lambda)\Delta\lambda}{\sigma T^4}=1.36\times10^{-3}.$$

(d) For these temperatures, the intensity varies strongly with temperature, although for even higher temperatures the intensity in this wavelength interval would decrease. From the Wein displacement law, the temperature that has the peak of the corresponding distribution in this wavelength interval is 5800 K (see Example 40-10), close to that used in part (c).

40-41: Given a source of spontaneous emission photons we can imagine we have a uniform source of photons over a long period of time (any one direction as likely as any other for emission). If a certain number of photons pass out through an area A, a distance D from the source, then at a distance 2D, those photons are spread out over an area $(2)^2 = 4$ times the original area A (because of how surface areas of sphere increase). Thus the number of photons per unit area DECREASE as the inverse square of the distance from the source.

40-42: (a) See Problem 40-2: $\frac{dN}{dt} = \frac{(dE/dt)}{(dE/dN)} = \frac{P}{hf} = \frac{P\lambda}{hc} = 4.5 \times 10^{19}/\text{s}.$

(b) The area would be $\frac{(dN/dt)}{((dN/dt)/A)} = 4.53 \times 10^4\ \text{m}^2$, and the radius of a sphere with this area is $r = \sqrt{A/4\pi} = 60\,\text{m}$.

40-43: (a) Energy to dissociate on AgBr molecule is just

$$E = \frac{E(\text{mole})}{1\ \text{mole}} = \frac{1.00 \times 10^5\,\text{J}}{6.02 \times 10^{23}} = \frac{1.66 \times 10^{-19}\,\text{J}}{(1.60 \times 10^{-19}\,\text{J/eV})} = 1.04\ \text{eV}$$

(b) $\lambda = \frac{hc}{E} = \frac{(6.63 \times 10^{-34}\,\text{J}\cdot\text{s})(3.00 \times 10^8\,\text{m/s})}{1.66 \times 10^{-19}\,\text{J}} = 1.20 \times 10^{-6}\,\text{m}$

(c) $f = \frac{c}{\lambda} = \frac{3.00 \times 10^8\,\text{m/s}}{1.20 \times 10^{-6}\,\text{m}} = 2.51 \times 10^{14}\,\text{Hz}$

(d) $E = hf = (6.63 \times 10^{-34}\,\text{J}\cdot\text{s})(9.99 \times 10^7\,\text{Hz}) = \frac{6.62 \times 10^{-26}\,\text{J}}{1.60 \times 10^{-19}\,\text{J/eV}} = 4.14 \times 10^{-7}\,\text{eV}$

(e) Even though a 50 kW radio station emitts huge numbers of photons, each individual photon has insufficient energy to dissociate the AgBr molecule, while a finetly's visible photons do have enough energy.

40-44: $\lambda = \frac{hc}{E} = \frac{hc}{\frac{3}{2}kT} = \frac{2}{3}\frac{hc}{kT} = 3.20 \times 10^{-5}\,\text{m} = 32.0\,\mu\text{m}.$

40-45: Recall $eV_0 = \frac{hc}{\lambda} - \phi$

$$\Rightarrow e(V_{02} - V_{01}) = \frac{hc}{\lambda_2} - \frac{hc}{\lambda_1} \Rightarrow \Delta V_0 = \frac{hc}{e}\left(\frac{1}{\lambda_2} - \frac{1}{\lambda_1}\right)$$

$$\Rightarrow \Delta V_0 = \frac{(6.63 \times 10^{-34}\,\text{J}\cdot\text{s})(3.00 \times 10^8\,\text{m/s})}{(1.60 \times 10^{-11}\,\text{C})}\left(\frac{1}{3.10 \times 10^{-7}\,\text{m}} - \frac{1}{3.80 \times 10^{-7}\,\text{m}}\right)$$

$= 0.739$ V

So the change in the stopping potential is an increase of 0.739 V.

40-46: (a) $\lambda_0 = \dfrac{hc}{E}$, and the wavelengths are: cesium: 590 nm, copper: 264 nm, potassium: 539 nm, zinc: 288 nm. (b) The wavelengths of copper and zinc are in the infrared, and visible light is not energetic enough to overcome the threshold energy of these metals.

40-47: (a) Plot: Below is the graph of frequency versus stopping potential.

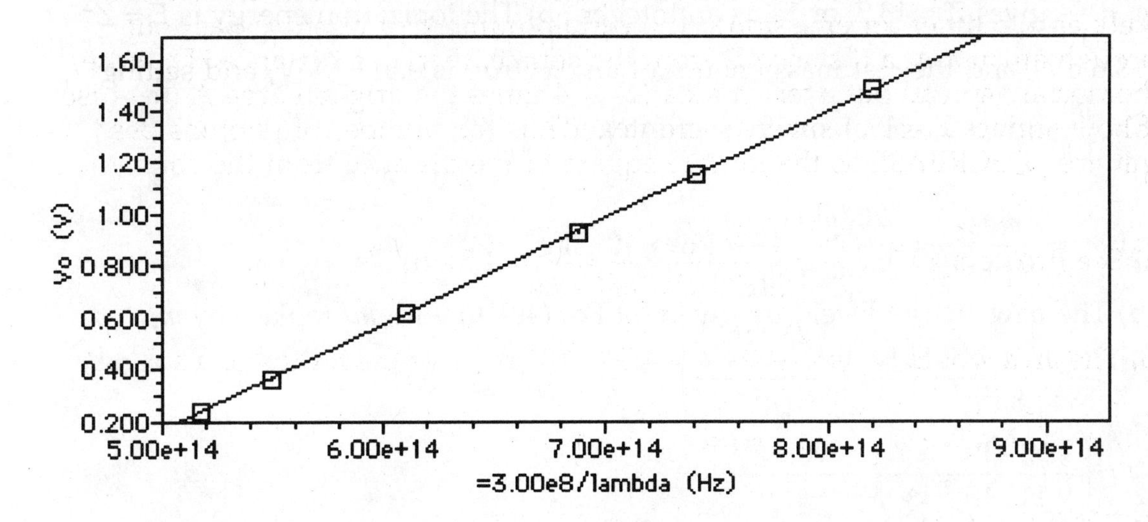

Threshold frequency is when the stopping potential is zero.

$$\Rightarrow f_{th} = \frac{1.89}{4.11\times10^{-15}}\,\text{Hz} = 4.60\times10^{14}\,\text{Hz}$$

(b) Threshold wavelength is $\lambda_{th} = \dfrac{c}{f_{th}} = \dfrac{(3.00\times10^{8}\,\text{m/s})}{4.60\times10^{14}\,\text{Hz}} = 6.52\times10^{-7}\,\text{m}$

(c) The work function is just $hf_{th} = (6.63\times10^{-34}\,\text{J}\cdot\text{s})(4.60\times10^{14}\,\text{Hz}) = 3.05\times10^{-19}\,\text{J}$ $= 1.91$ eV.

(d) The slope of the graph is $m = \dfrac{V_0}{f} = \dfrac{h}{e} \Rightarrow h = me$

$$= (4.11\times10^{-15}\,\text{V/Hz})(1.60\times10^{-19}\,\text{C})$$

$$= 6.58\times10^{-34}\,\text{J}\cdot\text{s}$$

40-48: (a) The levels are –1.0 eV, –3.0 eV, –7.0 eV and –10.0 eV. b) The atom is initially in the –1.0 eV-state, and –1.0 eV – (–3.0 eV) = 2.0 eV, –1.0 eV – (–7.0 eV) = 6.0 eV, –1.0 eV – (–10.0 eV) = 9.0 eV, –3.0 eV – (–7.0 eV) = 4.0 eV, –3.0 eV – (–10.0 eV) = 7.0 eV and –7.0 eV – (–10.0 eV) = 3.0 eV.

40-49: The maximum energy available to be deposited in the atom is

$$E_\gamma = \frac{hc}{\lambda} = \frac{(6.63\times10^{-34}\,\text{J}\cdot\text{s})(3.00\times10^{8}\,\text{m/s})}{5.34\times10^{-8}\,\text{m}}$$

$$= \frac{3.72 \times 10^{-18}\,\text{J}}{1.60 \times 10^{-19}\,\text{J}} = 23.25\,\text{eV}$$

To ionize an electron, from the ground state, 13.60 eV is needed, so the remaining energy is $K = 23.25\,\text{eV} - 13.60\,\text{eV} = 9.7\,\text{eV}$

40-50: From Eq. (40-14), the speed in the ground state is $v_1 = Z(2.19 \times 10^6\,\text{m/s})$. Setting $v_1 = \frac{c}{10}$ gives Z = 13.7, or 14 as an integer. b) The ionization energy is $E = Z^2$ (13.6 eV), and the rest mass energy of an electron is 0.511 MeV, and setting $E = \frac{mc^2}{100}$ gives Z = 19.4, or 19 as an integer.

40-51: (a) $m_r = \frac{m_1 m_2}{m_1 + m_2} = \frac{207 m_e m_p}{207 m_e + m_p} = 1.69 \times 10^{-28}\,\text{kg}$.

(b) The new energy levels are given by Eq. (40-16) with m_e replace by m_r.

$$E_n = -\frac{1}{\epsilon_0^2}\frac{m_r e^4}{8n^2h^2} = \left(\frac{m_r}{m_e}\right)\left(\frac{-13.60\,\text{eV}}{n^2}\right)$$

$$= \left(\frac{1.69 \times 10^{-28}\,\text{kg}}{9.11 \times 10^{-31}\,\text{kg}}\right)\left(\frac{-13.06\,\text{eV}}{n^2}\right)$$

$$= \frac{-2.53 \times 10^3\,\text{eV}}{n^2}$$

$$\Rightarrow E_1 = -2.53\,\text{keV}$$

(c) $$\frac{1}{\lambda} = \frac{1}{hc}(E_1 - E_2) = \frac{(-2.53 \times 10^3\,\text{eV}/4 - (-2.53 \times 10^3\,\text{eV}))(1.60 \times 10^{-19}\,\text{J/eV})}{(6.63 \times 10^{-34}\,\text{J}\cdot\text{s})(3.00 \times 10^8\,\text{m/s})}$$

$$= 1.53 \times 10^9\,\text{m}^{-1}$$

$$\Rightarrow \lambda = 6.55 \times 10^{-10}\,\text{m}.$$

40-52: (a) $m_{\text{r,d}} = m_\text{e}\frac{m_\text{d}}{m_\text{e} + m_\text{d}} = m_\text{e}(1 + (m_\text{e}/m_\text{d}))^{-1}$, and insertion of the numerical values gives $m_{\text{r,d}} = 0.999728\, m_\text{e}$. b) Let $\lambda_0 = \frac{(4/3)}{R}$, $\lambda' = \frac{(4/3)}{R'}$, where R' is the Rydberg constant evaluated with $m = m_{\text{r,p}} = 0.999456$, so $R' = \frac{m_{\text{r,p}}}{m_\text{e}}R$, and $\lambda'' = \frac{(4/3)}{R''}$, $R'' = \frac{m_{\text{r,d}}}{m_\text{e}}R$. Then

$$\Delta\lambda = \frac{4}{3}\frac{1}{R}\left(\frac{R}{R'} - \frac{R}{R''}\right) = \lambda_0\left(\frac{m_\text{e}}{m_{\text{r,p}}} - \frac{m_\text{e}}{m_{\text{r,d}}}\right) = 0.033\,\text{nm}.$$

40-53: (a) The H_β line is emitted by an electron in the n = 4 energy level, $E_4 = \frac{-13.60\,\text{eV}}{(4)^2} = -0.85\,\text{eV}$. The ground state energy is $E_1 = -13.60$ V, so one must add at least 13.60 eV – 0.85 eV = 12.75 eV if the H_β line is to be emitted.

(b) The possible emitted photons are $4 \to 3$ $4 \to 2$ $4 \to 1$ $3 \to 2$ $3 \to 1$ $2 \to 1$, with the wavelengths given by $\frac{1}{\lambda} = \frac{1}{hc} = (E_i - E_f) = \frac{+13.60\,\text{eV}}{hc}\left(\frac{1}{n_f^2} - \frac{1}{n_i^2}\right)$. This yields the wavelengths of 1880 nm, 486 nm, 97.2 nm, 656 nm, 103 nm, and 122 nm for the respective photon above.

40-54: $\frac{3}{2}kT = \Delta E$, so $T = \frac{2}{3}\frac{\Delta E}{k} = 7.89 \times 10^4\,\text{K}.$

40-55: The average kinetic energy of a gas atom is $\frac{3}{2}kT$.

So $\frac{3}{2}kT = 4.0\,\text{eV} \Rightarrow T = \frac{2(4.0\,\text{eV})(1.60 \times 10^{-19}\,\text{J/eV})}{3(1.38 \times 10^{-23}\,\text{J/K})} = 3.1 \times 10^4\,\text{K}$. If the metal temperature was this high, the electron's kinetic energy would have them escaping from the metal. But this temperature is far above the melting point of any metals.

40-56: (a) In terms of the satellite's mass M, orbital radius R and orbital period T,

$$n = \frac{2\pi}{h}L = \frac{2\pi}{h}MR^2\left(\frac{2\pi}{T}\right) = \frac{4\pi^2 MR^2}{hT}.$$

Using the given numerical values, $n = 1.08 \times 10^{46}$. (b) The angular momentum of the satellite in terms of its orbital speed V, mass and radius is $L = MVR$, so $V^2 = (L/MR)^2$, and its angular acceleration is $\frac{V^2}{R} = \frac{L^2}{M^2R^3}$.

Newton's law of gravitation can then be expressed as

$$\frac{GM_{\text{earth}}}{R^2} = \frac{L^2}{M^2R^3}, \quad \text{or} \quad R = \frac{L^2}{GM_{\text{earth}}M^2}.$$

If $L = nh/2\pi$,

$$R = n^2\left(\frac{h^2}{4\pi^2 GM_{\text{earth}}M^2}\right) = kn^2.$$

(c) $\Delta R = 2kn\Delta n$, and for the next orbit, $\Delta n = 1$, and $\Delta R = n\left(h^2/2\pi^2 GM_{\text{earth}}M^2\right)$. Insertion of numerical values from Appendix F and using n from part (a) gives $\Delta R = 1.5 \times 10^{-39}\,\text{m}$, which is (d) not observable. (e) The quantum and classical orbit do correspond, either would be correct, but only the classical calculation is userful.

40-57: (a) Quantization of angular momentum implies $L = mvr = n\frac{h}{2\pi}$

$$\Rightarrow v = \frac{nh}{2\pi mr}.$$

But $F = +Dr = \frac{mv^2}{r} \Rightarrow r^2 = \frac{mv^2}{D} = \frac{n^2h^2}{4\pi^2 mr^2 D}$

$$\Rightarrow r_n = \left(\frac{n^2h^2}{4\pi^2 mD}\right)^{1/4}.$$

(b) The energy $E = K + U = \frac{1}{2}mv^2 + \frac{1}{2}Dr^2$, since $F = -Dr$ is completely analagous to $F = -kx \Rightarrow U = \frac{1}{2}kx^2$. So $E = \frac{1}{2}m \cdot \left(\frac{D}{m}r^2\right) + \frac{1}{2}Dr^2 = Dr^2$

$$\Rightarrow E_n = D \cdot \frac{nh}{2\pi\sqrt{mD}} = \sqrt{\frac{D}{m}} \cdot \frac{nh}{2\pi}.$$

(c) Photon energies $E_\gamma = E_i - E_f = (n_i - n_f)\sqrt{\frac{D}{m}}\frac{h}{2\pi}$.

$\Rightarrow E_\gamma = n\sqrt{\frac{D}{m}}\frac{h}{2\pi}$ where n= integers > 0.

(d) This could describe a charged mass attached to a spring, being spun in a circle.

40-58: (a) IV (1–1.0 %) = 743 W. (b) $\frac{dT}{dt} = (dE/dt)/(dE/dT)$, and $\frac{dE}{dT} = mc$, so

$$\frac{dT}{dt} = \frac{(743 \text{ W})}{(0.300 \text{ kg})(147 \text{J/kg} \cdot \text{K})} = 16.8 \text{ K/s}.$$

(c) A high melting point is essential.

40-59: The transition energy equals the sum of the recoiling atom's kinetic energy and the photon's energy

$$E_{tr} = E_k + E_{\gamma'}$$

$$\Rightarrow E_{\gamma'} = \frac{hc}{\lambda'} = E_{tr} - E_k \Rightarrow \lambda' = \frac{hc}{E_{tr} - E_k}.$$

If the recoil is neglected $\lambda = \frac{hc}{E_{tr}}$

$$\Rightarrow \Delta\lambda = hc\left(\frac{1}{E_{tr} - E_k} - \frac{1}{E_{tr}}\right) = \frac{hc}{E_{tr}}\left(\frac{1}{1 - E_k/E_{tr}} - 1\right)$$

$$\approx \frac{hc}{E_{tr}}\left(\left(1 + \frac{E_k}{E_{tr}} - \dots\right) - 1\right)$$

$$\Rightarrow \Delta\lambda = hc\left(\frac{E_k}{E_{tr}^2}\right) = hc\frac{E_k}{(hc/\lambda)^2} = \left(\frac{E_k}{hc}\right)\lambda^2$$

Conservation of momentum, assuming atom initially at rest, yields:

$$P_\gamma = \frac{h}{\lambda} = P_k \Rightarrow E_k = \frac{P_k^2}{2m} = \frac{h^2}{2m\lambda^2}$$

$$\Rightarrow \Delta\lambda = \left(\frac{h^2}{2mhc\lambda^2}\right)\lambda^2 = \frac{h}{2mc}$$

For the hydrogen atom: $\Delta\lambda = \frac{(6.63 \times 10^{-34} \text{ J} \cdot \text{s})}{2(1.67 \times 10^{-27} \text{ kg})(3.00 \times 10^8 \text{ m/s})} = 6.6 \times 10^{-16} \text{ m}$

40-60: (a) $\Delta\lambda = 2\lambda_c = 0.049$ nm, so $\lambda' = 0.1449$ nm.b) $\Delta E = hc\left(\frac{1}{\lambda} - \frac{1}{\lambda'}\right) = 4.80\times10^{-17}$ J

(b) $\Delta E = hc\left(\frac{1}{\lambda} - \frac{1}{\lambda'}\right) = 4.80\times10^{-17}\ \text{J} = 300$ eV.

This will be the kinetic energy of the electron. (c) The kinetic energy is far less than the rest mass energy, so a non-relativistic calculation is adequate; $v = \sqrt{2K/m} = 1.03\times10^7$ m/s.

40-61: (a) Largest wavelength shift:

$$\Delta\lambda = \frac{h}{mc}(1-(-1)) = \frac{2(6.63\times10^{-34}\,\text{J}\cdot\text{s})}{(9.11\times10^{-31}\,\text{kg})(3.00\times10^8\,\text{m/s})} = 4.85\times10^{-12}\,\text{m}$$

(b) We want $\Delta\lambda = \lambda' - \lambda = 2\lambda - \lambda = \lambda = \frac{h}{mc}(1-\cos\phi)$

Smallest energy implies largest λ so $\cos\phi = -1$

$$\Rightarrow \lambda = \frac{2h}{mc}$$

$$\Rightarrow E = \frac{hc}{\lambda} = \frac{mc^2}{2} = \frac{5.11\times10^5\,\text{eV}}{2} = 2.56\times10^5\,\text{eV}.$$

40-62: (a) At $\cos\phi = -1$ ($\phi = 180°$), the wavelength shift is $2\lambda_c = 4.85$ pm.
(b) The Compton wavelength for a proton is less than λ_c by a factor of the ratio of the masses, and the maximum wavelength change for scattering from a proton is $(4.85\times10^{-12}\,\text{m})\frac{9.11\times10^{-31}\,\text{kg}}{1.673\times10^{-27}\,\text{kg}} = 2.64\times10^{-15}\,\text{m}$.

40-63: (a) $\Delta E = h(f_2 - f_1) = hc\left(\frac{1}{\lambda_2} - \frac{1}{\lambda_1}\right)$

$$= (6.63\times10^{-34}\,\text{J}\cdot\text{s})(3.00\times10^8\,\text{m/s})\left(\frac{1}{1.28\times10^{-10}\,\text{m}} - \frac{1}{1.20\times10^{-10}\,\text{m}}\right)$$

$$= -1.04\times10^{-16}\,\text{J}$$

The kinetic energy of the electron is $1.04\times10^{-16}\,\text{J} = 647\text{eV}$
(b) If all the energy of the electron is lost in the emission of a photon, then

$$\lambda = \frac{hc}{E} = \frac{(6.63\times10^{-34}\,\text{J}\cdot\text{s})(3.00\times10^8\,\text{m/s})}{1.04\times10^{-16}\,\text{J}}$$

$$= 1.91\times10^{-9}\,\text{m} = 1.91\text{nm}.$$

40-64: (a) $\Delta\lambda_1 = (h/mc)(1-\cos\theta_1)$, $\Delta\lambda_2 = (h/mc)(1-\cos\theta_2)$, and so the overall wavelength shift is $\Delta\lambda = (h/mc)(2-\cos\theta_1 - \cos\theta_2)$. (b) In general, the total shift produced by two successive scatterings of $\theta/2$ is less than that produced by a single scattering by θ. To see this, use the identity

$$1-\cos\phi = 2(1-\cos^2(\theta/2)) = 2(1-\cos(\phi/2))(1+\cos(\theta/2))$$

along with the fact that $\cos(\theta/2) \geq 0$ *for* $0 \leq \theta \leq 180°$. (c) $(h/mc)2(1-\cos 30.0°) = 0.268(h/mc)$. (d) $(h/mc)(1-\cos 60°) = 0.500(h/mc)$, which is indeed greater than the shift found in part (c).

40-65: For Compton scattering, $\lambda' - \lambda = \dfrac{h}{mc}(1 - \cos\phi)$

$$\lambda' - \lambda = hc\left(\frac{1}{E'} - \frac{1}{E}\right) \Rightarrow (1 - \cos\phi) = mc^2\left(\frac{1}{E'} - \frac{1}{E}\right)$$

$$\Rightarrow \cos\phi = 1 - \left(5.11\times10^5\,\text{eV}\right)\left(\frac{1}{7.50\times10^3\,\text{eV}} - \frac{1}{2.00\times10^6\,\text{eV}}\right)$$

$$= 1 - 0.426 = 0.574$$

$$\Rightarrow \phi = 55.0^\circ$$

40-66: $I = \sigma T^4$, $P = IA$, and $\Delta E = Pt$; combining,

$$t = \frac{\Delta E}{A\sigma T^4} = \frac{(400\,\text{J})}{\left(2.00\times10^{-6}\,\text{m}^2\right)\left(5.67\times10^{-8}\,\text{W/m}^2\cdot\text{K}^4\right)(373.15\,\text{K})^4} = 1.82\times10^5\,\text{s},$$

about fifty hours.

40-67: (a) $I(\lambda) = \dfrac{2\pi hc^2}{\lambda^5\left(e^{hc/\lambda kT} - 1\right)}$ but $\lambda = \dfrac{c}{f}$

$$\Rightarrow I(f) = \frac{2\pi hc^2}{(c/f)^5\left(e^{hf/kT} - 1\right)} = \frac{2\pi hf^5}{c^3\left(e^{hf/kT} - 1\right)}$$

(b) $\int_0^\infty I(\lambda)\,d\lambda = \int_\infty^0 I(f)\,df\left(\dfrac{-c}{f^2}\right)$

$$= \int_0^\infty \frac{2\pi hf^3\,df}{c^2\left(e^{hf/kT} - 1\right)} = \frac{2\pi(kT)^4}{c^2h^3}\int_0^\infty \frac{x^3}{e^x - 1}dx$$

$$= \frac{2\pi(kT)^4}{c^2h^3}\frac{1}{240}(2\pi)^4 = \frac{(2\pi)^5(kT)^4}{240h^3c^2} = \frac{2\pi^5k^4T^4}{15c^2h^3}$$

(c) The expression $\dfrac{2\pi^5k^4T^4}{15h^3c^2} = \sigma$ as shown in Eq (40-34). Plugging in the values for the constants we get $\sigma = 5.67\times10^{-8}\,\text{W/m}^2\cdot\text{K}^4$

40-68: (a) The final energy of the photon is $E' = \dfrac{hc}{\lambda'}$, and $E = E' + K$, where K is the kinetic energy of the electron after the collision. Then,

$$\lambda = \frac{hc}{E' + K} = \frac{hc}{(hc/\lambda') + K} = \lambda'\left(1 + (\lambda' K/hc)\right)^{-1}.$$

Using $K = mc^2(\gamma - 1)$ (the relativistic expression must be used for three-figure accuracy), $\lambda = 0.00546$ nm. (b) $\phi = \arccos\left(1 - \dfrac{\Delta\lambda}{(h/mc)}\right)$, and using the above value and $(h/mc) = 2.426\times10^{-12}$ m gives $\phi = 38.8^\circ$.

40-69: (a) The period was found in Ex (40-19b): $T = \dfrac{4\varepsilon_0^2 n^3 h^3}{me^4}$ and frequency is just

$$f = \frac{1}{T} = \frac{me^4}{4\varepsilon_0^2 n^3 h^3}.$$

(b) Eq. (40-6) tells us that $f = \frac{1}{h}(E_2 - E_1)$. So $f = \dfrac{me^4}{8\varepsilon_0^2 h^3}\left(\dfrac{1}{n_2^2} - \dfrac{1}{n_1^2}\right)$ (from Eq. (40-16))

If $n_2 = n$ and $n_1 = n + 1$, then $\dfrac{1}{n_2^2} - \dfrac{1}{n_1^2} = \dfrac{1}{n^2} - \dfrac{1}{(n+1)^2}$

$$= \frac{1}{n^2}\left(1 - \frac{1}{(1+1/n)^2}\right) \approx \frac{1}{n^2}\left(1 - \left(1 - \frac{2}{n} + \ldots\right)\right)$$

$$= \frac{2}{n^3} \quad \text{for large } n.$$

$$\Rightarrow f \approx \frac{me^4}{4\varepsilon_0^2 n^3 h^3}.$$

40-70: Each photon has momentum $p = \dfrac{h}{\lambda}$, and if the rate at which the photons strike the surface is (dN/dt), the force on the surface is $(h/\lambda)(dN/dt)$, and the pressure is $(h/\lambda)(dN/dt)/A$. The intensity is

$$I = (dN/dt)(E)/A = (dN/dt)(hc/\lambda)/A,$$

and comparison of the two expressions gives the pressure as (I/c).

40-71: Momentum:

$$\vec{p} + \vec{P} = \vec{p}\,' + \vec{P}' \Rightarrow p - P = -p' - P'$$

$$\Rightarrow p' = P - (p + P')$$

energy: $pc + E = p'c + E'$

$$= p'c + \sqrt{(P'c)^2 + (mc^2)^2}$$

$$\Rightarrow (pc - p'c + E)^2 = (P'c)^2 + (mc^2)^2$$

$$= (Pc)^2 + ((p+p')c)^2 - 2P(p+p')c^2 + (mc^2)^2$$

$$(pc - p'c)^2 + E^2 = E^2 + (pc + p'c)^2 - 2(Pc^2)(p+p') + 2Ec(p-p')$$

$$-4pp'c^2 + 2Ec(p-p') + 2(Pc^2)(p+p') = 0$$

$$\Rightarrow p'(Pc^2 - 2pc^2 - Ec) = p(-Ec - Pc^2)$$

$$\Rightarrow p' = p\frac{Ec + Pc^2}{2pc^2 + Ec - Pc^2} = p\frac{E + Pc}{2pc + (E - Pc)}$$

$$\Rightarrow \lambda' = \lambda\left(\frac{2hc/\lambda + (E - Pc)}{E + Pc}\right) = \lambda\left(\frac{E - Pc}{E + Pc}\right) + \frac{2hc}{E + Pc}$$

$$\Rightarrow \lambda' = \frac{(\lambda(E - Pc) + 2hc)}{E + Pc}$$

If $E \gg mc^2$, $Pc = \sqrt{E^2 - (mc^2)^2} = E\sqrt{1 - \left(\frac{mc^2}{E}\right)^2}$

$$\approx E\left(1 - \frac{1}{2}\left(\frac{mc^2}{E}\right)^2 + \dots\right)$$

$$\Rightarrow E - Pc \approx \frac{1}{2}\frac{(mc^2)^2}{E}$$

$$\Rightarrow \lambda' \approx \frac{\lambda(mc^2)^2}{2E(2E)} + \frac{hc}{E} = \frac{hc}{E}\left(1 + \frac{m^2c^4\lambda}{4hcE}\right)$$

(b) If $\lambda = 6.328 \times 10^{-7}\,\text{m}$, $E = 2.0 \times 10^{10}\,\text{eV} = 3.2 \times 10^{-9}\,\text{J}$

$$\Rightarrow \lambda' \approx (6.22 \times 10^{-17}\,\text{m})(1 + 1.671)$$

$$= 1.66 \times 10^{-16}\,\text{m}$$

(c) These photons are gamma rays.

Chapter 41: The Wave Nature of Particles

41-1: (a) $\lambda_e = \dfrac{h}{m_e v} = \dfrac{6.63\times10^{-34}\,\mathrm{J\cdot s}}{(9.11\times10^{-31}\,\mathrm{kg})(6.80\times10^{6}\,\mathrm{m/s})} = 1.07\times10^{-10}\,\mathrm{m}$

(b) $\lambda_p = \dfrac{m_e}{m_p}\lambda_e = \left(\dfrac{9.11\times10^{-31}\,\mathrm{kg}}{1.67\times10^{-27}\,\mathrm{kg}}\right)1.07\times10^{-10}\,\mathrm{m} = 5.84\times10^{-14}\,\mathrm{m}$

41-2: (a) $E = \dfrac{hc}{\lambda} = \dfrac{(4.136\times10^{-15}\,\mathrm{eV\cdot s})\,(3.00\times10^{8}\,\mathrm{m/s})}{(0.20\times10^{-9}\,\mathrm{m})} = 6.2\,\mathrm{keV}.$

(b) $K = \dfrac{p^2}{2m} = \dfrac{(h/\lambda)^2}{2m} = \dfrac{((6.626\times10^{-34}\,\mathrm{J\cdot s})/(0.20\times10^{-9}\,\mathrm{m}))^2}{2(9.11\times10^{-31}\,\mathrm{kg})} = 6.0\times10^{-18}\,\mathrm{J} = 37\,\mathrm{eV}.$

Note that the kinetic energy found this way is much smaller than the rest energy, so the nonrelavisitic approximation is appropriate.

(c) $K = \dfrac{p^2}{2m} = \dfrac{(h/\lambda)^2}{2m} = \dfrac{((6.626\times10^{-34}\,\mathrm{J\cdot s})/(0.20\times10^{-9}\,\mathrm{m}))^2}{2(6.64\times10^{-27}\,\mathrm{kg})} = 8.3\times10^{-22}\,\mathrm{J} =$ $5.2\,\mathrm{meV}$. Again, the nonrelavisitic approximation is appropriate.

41-3: (a) $E_{blue} = \dfrac{hc}{\lambda_{\mathrm{blue}}} = \dfrac{(6.63\times10^{-34}\,\mathrm{J\cdot s})(3.00\times10^{8}\,\mathrm{m/s})}{(4.00\times10^{-7}\,\mathrm{m})}$

$= \dfrac{4.97\times10^{-19}\,\mathrm{J}}{1.60\times10^{-19}\,\mathrm{J/eV}} = 3.10\,\mathrm{eV}$

$E_{red} = \dfrac{hc}{\lambda_{red}} = \dfrac{(6.63\times10^{-34}\,\mathrm{J\cdot s})(3.00\times10^{8}\,\mathrm{m/s})}{(7.00\times10^{-7}\,\mathrm{m})}$

$= \dfrac{2.84\times10^{-19}\,\mathrm{J}}{1.60\times10^{-19}\,\mathrm{J/eV}} = 1.77\,\mathrm{eV}$

So the proton energies of the visible spectrum are from about 1.8 eV to 3.1 eV.

(b) Electron, with this energy:

$\lambda = \dfrac{h}{p} = \dfrac{h}{\sqrt{2mE}} \Rightarrow \lambda_1 = \dfrac{(6.63\times10^{-34}\,\mathrm{J\cdot s})}{\sqrt{2(9.11\times10^{-31}\,\mathrm{kg})(4.97\times10^{-19}\,\mathrm{J})}} = 9.2\times10^{-10}\,\mathrm{m}$

$\lambda_2 = \dfrac{(6.63\times10^{-34}\,\mathrm{J\cdot s})}{\sqrt{2(9.11\times10^{-31}\,\mathrm{kg})(2.84\times10^{-19}\,\mathrm{J})}} = 7.0\times10^{-10}\,\mathrm{m}$

41-4: $\lambda = \dfrac{h}{p} = \dfrac{h}{\sqrt{2mE}}$

$= \dfrac{(6.626\times10^{-34}\,\mathrm{J\cdot s})}{\sqrt{2(6.64\times10^{-27}\,\mathrm{kg})(4.78\times10^{6}\,\mathrm{eV})(1.602\times10^{-19}\,\mathrm{J/eV})}} = 6.57\times10^{-15}\,\mathrm{m}.$

41-5: $\lambda = \dfrac{h}{p} = \dfrac{h}{mv} = \dfrac{6.63\times10^{-34}\,\mathrm{J\cdot s}}{(2000\,\mathrm{kg})(24.0\,\mathrm{m/s})} = 1.38\times10^{-38}\,\mathrm{m}$

We should not expect the car to exhibit wavelike properties.

41-6: Combining Equations 39-39 and 39-40 gives

$$p = mc\sqrt{\gamma^2 - 1}.$$

(a) $\lambda = \frac{h}{p} = (h/mc)/\sqrt{\gamma^2 - 1} = 4.43 \times 10^{-12}$ m. (The incorrect nonrelativistic calculation gives 5.05×10^{-12} m.)

(b) $(h/mc)/\sqrt{\gamma^2 - 1} = 7.07 \times 10^{-13}$ m.

41-7: (a) In the Bohr model $mv\,r_n = \frac{nh}{2\pi}$. The de Broglie wavelength is $\lambda = \frac{h}{p} = \frac{h}{mv} = \frac{2\pi\, r_n}{n}$

for n=1: $r_1 = a_0 = 5.29 \times 10^{-11}\,\text{m} \Rightarrow \lambda_1 = 2\pi(5.29 \times 10^{-11}\text{ m}) = 3.32 \times 10^{-10}$ m

This equals the orbit circumference.

(b) n=4: $r_4 = (4)^2 a_0 = 16a_0 \Rightarrow \lambda_4 = \frac{2\pi(16a_0)}{4} = 4\lambda,$

$\Rightarrow \lambda_4 = 1.33 \times 10^{-9}\,\text{m}$

The de Broglie wavelength is a quarter of the circumference of the orbit, $2\pi\, r_4$.

41-8: (a) For a nonrelativistic particle, $K = \frac{p^2}{2m}$, so

$$\lambda = \frac{h}{p} = \frac{h}{\sqrt{2Km}}.$$

(b) $(6.626 \times 10^{-34}\text{ J}\cdot\text{s})/\sqrt{2(200\text{ eV})(1.602 \times 10^{-19}\text{ J/eV})(9.11 \times 10^{-31}\text{ kg})}$
$= 8.67 \times 10^{-11}$ m.

41-9: Surface scattering implies $d\sin\theta = m\lambda$

If m=1: $\theta = \arcsin[\lambda/d]$

But $\lambda = \frac{h}{p} = \frac{h}{\sqrt{2mE}} = \frac{6.63 \times 10^{-34}\,\text{J}\cdot\text{s}}{\sqrt{2(6.64 \times 10^{-27}\,\text{kg})(1200\,\text{eV})(1.60 \times 10^{-19}\,\text{J/eV})}}$

$\Rightarrow \lambda = 4.15 \times 10^{-13}\,\text{m}.$

So $\theta = \arcsin\left[\frac{4.15 \times 10^{-13}\,\text{m}}{6.34 \times 10^{-11}\,\text{m}}\right] = 0.375°$

41-10: Solving Eq. (41-3) for d and expressing the wavelength in terms of the kinetic energy, $\lambda = (h/p) = (h/\sqrt{2mK})$, and so

$$d = \frac{h}{\sin\theta\sqrt{2mK}}.$$

Note that m is the electron mass, not the diffraction order. Substitution of numerical values gives $d = 5.86 \times 10^{-11}$ m.

41-11: From Ex. (41-9), for m=1, $\lambda = d\sin\theta = \dfrac{h}{\sqrt{2mE}}$

$$\Rightarrow E = \frac{h^2}{2md^2\sin^2\theta} = \frac{(6.63\times10^{-34}\,\text{J}\cdot\text{s})^2}{2(1.675\times10^{-27}\,\text{kg})(9.60\times10^{-11}\,\text{m})^2\sin^2(38.2^\circ)}$$

$$\Rightarrow E = 3.72\times10^{-20}\,\text{J} = 0.233\,\text{eV}.$$

41-12: (a) $\Delta v_y = (\Delta p_y / m) > \dfrac{h}{2\pi m \Delta y}$

$$= \frac{(6.626\times10^{-34}\ \text{J}\cdot\text{s})}{2\pi(1.673\times10^{-27}\ \text{kg})(5.0\times10^{-12}\ \text{m})} = 1.3\times10^4\ \text{m/s}.$$

(b) $\Delta z > \dfrac{h}{2\pi\Delta p_z} = \dfrac{h}{2\pi m\Delta v_z} = 2.89\times10^{-4}\ \text{m}.$

41-13: Heisenberg's Uncertainly Principle tells us that:

$$\Delta x\Delta p_x \geq \frac{h}{2\pi}$$

Here $\Delta x\Delta p_x = (1.4\times10^{-10}\,\text{m})(5.0\times10^{-25}\,\text{kg}\cdot\text{m/s}) = 7.0\times10^{-35}\,\text{J}\cdot\text{s}$

But $\dfrac{h}{2\pi} = 1.05\times10^{-34}\,\text{J}\cdot\text{s}$

Therefore $\Delta x\Delta p_x < \dfrac{h}{2\pi}$ so the claim *is not valid.*

41-14: (a) $(\Delta x)(m\Delta v_x) \geq h/2\pi$, and setting $\Delta v_x = (0.010)v_x$ and the product of the uncertainties equal to $h/2\pi$ (for the minimum uncertainty) gives $v_x = h/(2\pi m(0.010)\Delta x) = 39\ \text{m/s}$. (b) Repeating with the proton mass gives 21 mm/s.

41-15: To find a particle's lifetime we need to know the uncertainty in its energy.

$$\Delta E = (\Delta m)c^2 = (0.028)(5)(1.67\times10^{-27}\,\text{kg})(3.00\times10^8\,\text{m/s})^2$$

$$= 2.10\times10^{-11}\,\text{J}$$

$$\Rightarrow \Delta t \approx \frac{h}{2\pi\Delta E} = \frac{6.63\times10^{-34}\,\text{J}\cdot\text{s}}{2\pi(2.10\times10^{-11}\,\text{J})} = 5.0\times10^{-24}\,\text{s}$$

41-16: $\Delta E > \dfrac{h}{2\pi\Delta t} = 2.9\times10^{-32}\ \text{J} = 1.8\times10^{-13}\ \text{eV}.$

41-17: (a) We recall $\lambda = \dfrac{h}{p} = \dfrac{h}{\sqrt{2mE}}$ But the energy of a particle accelerated through a potential is just $E = q\Delta V$

$$\Rightarrow \lambda_e = \frac{h}{\sqrt{2mq\Delta V}} = \frac{(6.63\times10^{-34}\,\text{J}\cdot\text{s})}{\sqrt{2(9.11\times10^{-31}\,\text{kg})(1.60\times10^{-19}\,\text{C})(700\,\text{V})}}$$

$$\Rightarrow \lambda_e = 4.64\times10^{-11}\,\text{m}$$

(b) For a proton, all that changes is the mass, so

$$\lambda_p = \sqrt{\frac{m_e}{m_p}}\lambda_e = \sqrt{\frac{9.11\times10^{-31}\,\text{kg}}{1.67\times10^{-27}\,\text{kg}}}\cdot(4.64\times10^{-11}\,\text{m})$$

$$\Rightarrow \lambda_e = 1.08\times10^{-12}\,\text{m}.$$

41-18: (a) $eV = K = \frac{p^2}{2m} = \frac{(h/\lambda)^2}{2m}$, so $V = \frac{(h/\lambda)^2}{2me} = 940$ V. b) The voltage is reduced by the ratio of the particle masses, $(940\,\text{V})\frac{9.11\times10^{-31}\,\text{kg}}{1.673\times10^{-27}\,\text{kg}} = 0.512\,\text{V}$.

41-19: $f(x,y) = \left(\frac{x-iy}{x+iy}\right)$ and $f^*(x,y) = \left(\frac{x+iy}{x-iy}\right)$

$$\Rightarrow |f|^2 = ff^* = \left(\frac{x-iy}{x+iy}\right)\cdot\left(\frac{x+iy}{x-iy}\right) = 1.$$

41-20: $\Psi^* = \psi^*\sin\omega t$, so

$$|\Psi|^2 = |\Psi^*\Psi| = \psi^*\psi\sin^2\omega t = |\psi|^2\sin^2\omega t.$$

$|\Psi|^2$ is not time-independent, so Ψ is not the wavefunction for a stationary state.

41-21: (a) $\psi(x) = A\sin kx$. The probability density is $|\psi|^2 = A^2\sin^2 kx$, and this is greatest when $\sin^2 kx = 1 \Rightarrow kx = \frac{n\pi}{2}$, $n = 1, 3, 5...$

$$\Rightarrow x = \frac{n\pi}{2k} = \frac{n\pi}{2(2\pi/\lambda)} = \frac{n\lambda}{4}, n = 1, 3, 5...$$

(b) The probability is zero when $|\psi|^2 = 0$, which requires

$$\sin^2 kx = 0 \Rightarrow kx = n\pi \Rightarrow x = \frac{n\pi}{k} = \frac{n\lambda}{2}, n = 0, 1, 2...$$

41-22: (a) The uncertainty in the particle postion is proportional to the width of $\psi(x)$, and is inversely proportional to $\sqrt{\alpha}$. This can be seen by either plotting the function for different values of α, finding the expectation value $\langle x^2\rangle = \int\psi^2x^2dx$ for the normalized wave function or by finding the full width at half-maximum. The particle's uncertainty in position decreases with increasing α. The dependence of the expectation value $\langle x^2\rangle$ on α may be found by considering

$$\langle x^2\rangle = \frac{\int_{-\infty}^{\infty} x^2e^{-2\alpha x^2}dx}{\int_{-\infty}^{\infty} e^{-2\alpha x^2}dx}$$

$$= -\frac{1}{2}\frac{\partial}{\partial\alpha}\ln\left[\int_{-\infty}^{\infty} e^{-2\alpha x^2}dx\right]$$

$$= -\frac{1}{2}\frac{\partial}{\partial\alpha}\ln\left[\frac{1}{\sqrt{2\alpha}}\int_{-\infty}^{\infty} e^{-u^2}du\right]$$

$$= \frac{1}{4\alpha},$$

where the substitution $u = \sqrt{\alpha}x$ has been made. (b) Since the uncertainty in position decreases, the uncertainty in momentum must increase.

41-23: The de Broglie wavelength of the blood cell is

$$\lambda = \frac{h}{mv} = \frac{(6.63\times10^{-34}\,\mathrm{J\cdot s})}{(1.00\times10^{-14}\,\mathrm{kg})(7.00\times10^{-3}\,\mathrm{m/s})} = 9.47\times10^{-18}\,\mathrm{m}$$

We need not be concerned about wave behavior.

41-24: (a) $\lambda = \frac{h}{\sqrt{2mK}} = 1.73\times10^{-10}$ m.

(b) $\frac{R}{v} = \frac{R}{\sqrt{2E/m}} = 4.77\times10^{-7}$ s.

(c) The width ω is $\omega = 2R\frac{\lambda}{a}$, and $\omega = \Delta v_y t = \Delta p_y t/m$, where t is the time found in part (b). Combining the expressions for ω, $\Delta p_y = \frac{2m\lambda R}{at} = 1.10\times10^{-28}$ kg$\cdot$m/s.

(d) $\Delta y = \frac{h}{2\pi\Delta p_y} = 0.95\,\mu$m, which is the same order of magnitude.

41-25: (a) Recall $\lambda = \frac{h}{p} = \frac{h}{\sqrt{2mE}} = \frac{h}{\sqrt{2mq\Delta V}}$ from Ex. (41-17).

So for an electron: $\lambda = \frac{6.63\times10^{-34}\,\mathrm{J\cdot s}}{\sqrt{2(9.11\times10^{-31}\,\mathrm{kg})(1.60\times10^{-19}\,\mathrm{C})(85.0\,V)}}$

$\Rightarrow \lambda = 1.33\times10^{-10}\,\mathrm{m}$

(b) For an alpha particle:

$$\lambda = \frac{6.63\times10^{-34}\,\mathrm{J\cdot s}}{\sqrt{2(6.64\times10^{-27}\,\mathrm{kg})2(1.60\times10^{-19}\,\mathrm{C})(85.0\,\mathrm{V})}} = 1.10\times10^{-12}\,\mathrm{m}$$

41-26: The wavelength is $\lambda = \frac{h}{\sqrt{(E/c)^2 - (mc)^2}}$. Using the given numbers, and converting from GeV to joules, a) 4.35×10^{-16} m, and b) 4.20×10^{-13} m.

41-27: Using the relativistic energy formula:

$$E^2 = (pc)^2 + (mc^2)^2 \Rightarrow p = \frac{1}{c}\sqrt{E^2 - (mc^2)^2}$$

The actual energy $E = K + mc^2 = 1.02\,\mathrm{MeV} + 0.511\,\mathrm{MeV} = 1.53\,\mathrm{MeV}$

$\Rightarrow E = (1.53\times10^6\,\mathrm{eV})(1.60\times10^{-19}\,\mathrm{J/eV}) = 2.45\times10^{-13}\,\mathrm{J}$

So $\lambda = \frac{h}{p} = \frac{hc}{\sqrt{E^2 - (mc^2)^2}}$,

$$\Rightarrow \lambda = \frac{(6.63 \times 10^{-34}\,\text{J}\cdot\text{s})(3.00 \times 10^{8}\,\text{m/s})}{\sqrt{(2.45 \times 10^{-13}\,\text{J})^2 - [(5.11 \times 10^{5}\,\text{eV})(1.60 \times 10^{-19}\,\text{J/eV})]^2}}$$

$\lambda = 8.61 \times 10^{-13}\,\text{m}$

41-28: $\Delta p \sim \frac{h}{2\pi a_0} = \frac{(6.626 \times 10^{-34}\ \text{J}\cdot\text{s})}{2\pi(0.5292 \times 10^{-10}\ \text{m})} = 2.0 \times 10^{-24}\ \text{kg}\cdot\text{m/s}$, which is comparable to $mv_1 = 2.0 \times 10^{-24}\ \text{kg}\cdot\text{m/s}$.

41-29: $\Delta x = 0.40\lambda = 0.40\frac{h}{p}$.

But $\Delta x \Delta p_x \geq \frac{h}{2\pi} \Rightarrow \Delta p_x(\text{min}) = \frac{h}{2\pi\Delta x} = \frac{p}{2\pi(0.4)} = 0.40p$

41-30: (a) $\frac{(6.626 \times 10^{-34}\ \text{J}\cdot\text{s})}{2\pi(5.0 \times 10^{-15}\ \text{m})} = 2.1 \times 10^{-20}\ \text{kg}\cdot\text{m/s}$.

(b) $K = \sqrt{(pc)^2 + (mc^2)^2} - mc^2 = 1.3 \times 10^{-13}\ \text{J} = 0.82\ \text{MeV}$.

(c) The result of part (b), about $1\,\text{MeV} = 1 \times 10^{6}\ \text{eV}$, many orders of magnitude larger than the potential energy of an electron in a hydrogen atom.

41-31: (a) $\Delta p(\text{min}) = \frac{h}{2\pi\Delta x} = \frac{6.63 \times 10^{-34}\,\text{J}\cdot\text{s}}{2\pi(5.0 \times 10^{-15}\,\text{m})} = 2.1 \times 10^{-20}\,\text{kg}\cdot\text{m/s}$

(b) $E = \sqrt{(pc)^2 + (mc^2)^2}$

$$= \sqrt{[(2.1 \times 10^{-20}\,\text{kg}\cdot\text{m/s})(3.0 \times 10^{8}\,\text{m/s})]^2 + [(9.11 \times 10^{-31}\,\text{kg})(3.0 \times 10^{8}\ \text{m/s})^2]^2}$$

$$= 6.3 \times 10^{-12}\,\text{J} = 39.5\,\text{MeV}$$

(c) The coulomb potential energy is $U = \frac{q_1 q_2}{4\pi\varepsilon_0 V}$

$$\Rightarrow U = \frac{(1.60 \times 10^{-19}\,\text{C})^2}{4\pi\varepsilon_0(5.0 \times 10^{-15}\,\text{m})} = 4.60 \times 10^{-14}\,\text{J} = 0.29\,\text{MeV}$$

Hence there is not enough energy to "hold" the electron in the nucleus.

41-32: (a) Take the direction of the electron beam to be the x-direction and the direction of motion perpendicular to the beam to be the y-direction. Then, the uncertainty Δr in the position of the point where the electrons strike the screen is

$$\Delta r = \Delta v_y t = \frac{\Delta p_y}{m}\frac{x}{v_x}$$

$$= \frac{h}{2\pi m \Delta y}\frac{x}{\sqrt{2K/m}}$$

$$= 9.56 \times 10^{-10}\ \text{m},$$

which is (b) far too small to affect the clarity of the picture.

41-33: $\Delta E = \dfrac{h}{2\pi\Delta t} = \dfrac{(6.63\times10^{-34}\,\text{J}\cdot\text{s})}{2\pi(8.4\times10^{-17}\,\text{s})} = 1.26\times10^{-18}\,\text{J}$

But $\Delta E = (\Delta m)c^2 \Rightarrow \Delta m = \dfrac{\Delta E}{c^2} = \dfrac{1.26\times10^{-18}\,\text{J}}{(3.0\times10^{8}\,\text{m/s})^2} = 1.4\times10^{-35}\,\text{kg}$

$\Rightarrow \dfrac{\Delta m}{m} = \dfrac{1.4\times10^{-35}\,\text{kg}}{264(9.11\times10^{-31}\,\text{kg})} = 5.8\times10^{-8}$

41-34: (a) 3.38 eV, with a wavelength of $\lambda = \dfrac{hc}{E} = 3.67\times10^{-7}\ \text{m} = 367\,\text{nm}$.

(b) $\dfrac{h}{2\pi\Delta t} = 5.3\times10^{-29}\ \text{J} = 3.3\times10^{-10}\ \text{eV}$.

(c) $\lambda E = hc$, so $(\Delta\lambda)E + \lambda\Delta E = 0$, and $|\Delta E/E| = |\Delta\lambda/\lambda|$, so $\Delta\lambda = \lambda|\Delta E/E| = 3.6\times10^{-17}\ \text{m}$.

41-35: (a) $\lambda = \dfrac{h}{mv} \Rightarrow v = \dfrac{h}{m\lambda} = \dfrac{6.63\times10^{-34}\,\text{J}\cdot\text{s}}{(60\,\text{kg})(1.0\text{m})} = 1.1\times10^{-35}\ \text{m/s}$

(b) $t = \dfrac{d}{v} = \dfrac{0.80\,\text{m}}{1.1\times10^{-35}\,\text{m/s}} = 7.2\times10^{34}\,\text{s} = 2.3\times10^{27}\ \text{years}$.

Therefore, we will not notice diffraction effects while passing through doorways.

41-36: $\sin\theta' = \dfrac{\lambda'}{\lambda}\sin\theta$, and $\lambda' = (h/p) = (h/\sqrt{2mE})$, and so

$$\theta' = \arcsin\left(\frac{h}{\lambda\sqrt{2mE}}\sin\theta\right) = 18.4°.$$

41-37: $\Delta E = \dfrac{h'}{2\pi\Delta t} = \dfrac{6.63\times10^{-22}\,\text{J}\cdot\text{s}}{2\pi(4.50\times10^{-3}\text{s})} = \dfrac{2.34\times10^{-20}\,\text{J}}{1.60\times10^{-19}\,\text{J/eV}}$

$\Rightarrow \Delta E = 0.147\,\text{eV}$.

41-38: (a) Using the given approximation,

$E = \frac{1}{2}\left((h/x)^2/m + kx^2\right), (dE/dx) = kx - (h^2/mx^3)$, and the minimum energy occurs when $kx = (h^2/mx^3)$, or $x^2 = \dfrac{h}{\sqrt{mk}}$. The minimum energy is then $h\sqrt{k/m}$.

(b) They are the same.

41-39: (a) $U = A|x|$ but $F = -\dfrac{dU}{dx}$

For $x > 0$, $|x| = x \Rightarrow F = -A$.

For $x < 0$, $|x| = -x \Rightarrow F = A$

So $F(x) = -\dfrac{A|x|}{x}$ for $x \neq 0$.

(b) From Pr. (41-38), $E = K + U = \dfrac{p^2}{2m} + A|x|$, and $px \approx h$

$$\Rightarrow E = \frac{h^2}{2mx^2} + A|x|$$

For $x > 0$; $E = \dfrac{h^2}{2mx^2} + Ax$. The minimum energy occurs

$$\text{when } \frac{dE}{dx} = 0$$

$$\Rightarrow \frac{dE}{dx} = 0 = -\frac{h^2}{mx^3} + A \Rightarrow \quad x' = \left(\frac{h^2}{mA}\right)^{1/3}$$

$$\text{So } E_{\min} = \frac{h^2}{2m(h^2/mA)^{2/3}} + A\left(\frac{h^2}{mA}\right)^{1/3} = \frac{3}{2}\left(\frac{h^2A^2}{m}\right)^{1/3}$$

41-40: For this wave function, $\Psi^* = \psi_1^* e^{i\omega_1 t} + \psi_2^* e^{i\omega_2 t}$, so

$$\Psi^2 = \Psi^*\Psi$$

$$= \left(\psi_1^* e^{i\omega_1 t} + \psi_2^* e^{i\omega_2 t}\right)\left(\psi_1 e^{-i\omega_1 t} + \psi_2 e^{-i\omega_2 t}\right)$$

$$= \psi_1^*\psi_1 + \psi_2^*\psi_2 + \psi_1^*\psi_2 e^{i(\omega_1 - \omega_2)t} + \psi_2^*\psi_1 e^{i(\omega_2 - \omega_1)t}.$$

The frequencies ω_1 and ω_2 are given as not being the same, so $|\Psi|^2$ is not time-independent, and Ψ is not the wave function for a stationary state.

41-41: (a) The ball is in a cube of volume 1000 cm^3 to start with, and hence has an uncertainty of 10 cm in any direction. $\Delta x = 0.10\,\text{m}$. (The x- direction in the horizontal, side-to-side direction.)

Now $\Delta p_x = \dfrac{h}{2\pi\Delta x} = \dfrac{0.0663\,\text{J}\cdot\text{s}}{2\pi(0.10\,\text{m})} = 0.11\,\text{kg}\cdot\text{m/s}$.

(b) The time of flight is $t = \dfrac{20\,\text{m}}{5.0\,\text{m/s}} = 4.0\,\text{s}$.

So the uncertainty in the x-direction at the catcher is

$$\Delta x = (\Delta v)t = \left(\frac{\Delta p}{m}\right)t = \left(\frac{0.11\,\text{kg}\cdot\text{m/s}}{0.30\,\text{kg}}\right)(4.0\,\text{s})$$

$$\Rightarrow \Delta x = 1.4\,\text{m}.$$

41-42: (a) $|\psi|^2 = A^2x^2e^{-2(\alpha x^2 + \beta y^2 + \gamma z^2)}$. To save some algebra, let $u = x^2$, so that

$|\psi|^2 = ue^{-2\alpha u}f(y,z)$, and $\dfrac{\partial}{\partial u}|\psi|^2 = (1 - 2\alpha u)|\psi|^2$; the maximum occurs at

$u_0 = \dfrac{1}{2\alpha}$, $x_0 = \pm\dfrac{1}{\sqrt{2\alpha}}$.

(b) ψ vanishes at $x = 0$, so the probability of finding the particle in the $x = 0$ plane is zero. The wavefunction vanishes for $x = \pm\infty$.

41-43: (a) $B(k) = e^{-\alpha^2 k^2}$ $\quad B(0) = B_{max} = 1$

$$B(k_h) = \frac{1}{2} = e^{-\alpha^2 k_h^2} \Rightarrow \ln(1/2) = -\alpha^2 k_h^2$$

$$\Rightarrow k_h = \frac{1}{\alpha}\sqrt{\ln(2)} = w_k.$$

Using tables: (b) $\psi(x) = \int_0^{\infty} e^{-\alpha^2 k^2} \cos kx dk = \frac{\sqrt{\pi}}{2\alpha}(e^{-x^2/4\alpha^2})$.

$\psi(x)$ is a maximum when $x = 0$.

(c) $\psi(x_h) = \frac{\sqrt{\pi}}{4\alpha}$ when $e^{-x_h{}^2/4\alpha^2} = \frac{1}{2} \Rightarrow \frac{-x_h{}^2}{4\alpha^2} = \ln\left(\frac{1}{2}\right)$

$$\Rightarrow x_h = 2\alpha\sqrt{\ln 2} = w_x$$

(d) $w_p w_x = \left(\frac{h w_k}{2\pi}\right) w_x = \frac{h}{2\pi}\left(\frac{1}{\alpha}\sqrt{\ln 2}\right)\left(2\alpha\sqrt{\ln 2}\right) = \frac{h}{2\pi}(2\ln 2) = \frac{h \ln 2}{\pi}$.

41-44: (a) For a standing wave, $n\lambda = 2L$, and

$$E_n = \frac{p^2}{2m} = \frac{(h/\lambda)^2}{2m} = \frac{n^2 h^2}{8mL^2}.$$

(b) With $L = a_0 = 0.5292 \times 10^{-10}$ m, $E_1 = 2.15 \times 10^{-17}$ J $= 134$ eV.

41-45: Time of flight of the marble, from free-fall kinematic equation is just $t = \sqrt{\frac{2y}{g}} = \sqrt{\frac{2(20.0\,\text{m})}{9.81\,\text{m/s}^2}} = 2.02\,\text{s}$

$$\Delta x_f = \Delta x_i + (\Delta v_x)t = \Delta x_i + \left(\frac{\Delta p_x}{m}\right)t = \frac{h\,t}{2\pi \Delta x_i m} + \Delta x_i$$

$$\frac{d(\Delta x_f)}{d(\Delta x_i)} = 0 = \frac{-ht}{2\pi m(\Delta x_i)^2} + 1 \Rightarrow \Delta x_i(\text{min}) = \sqrt{\left(\frac{ht}{2\pi m}\right)}$$

$$\Rightarrow \Delta x_f(\text{min}) = \sqrt{\frac{ht}{2\pi m}} + \sqrt{\frac{ht}{2\pi m}} = \sqrt{\frac{2ht}{\pi m}}$$

$$\Delta x_f(\text{min}) = \sqrt{\frac{2(6.63 \times 10^{-34}\,\text{J}\cdot\text{s})(2.02\,\text{s})}{\pi(0.0300\,\text{kg})}} = 1.7 \times 10^{-16}\,\text{m}$$

Chapter 42: Quantum Mechanics

42-1: (a) $E_n = \dfrac{n^2h^2}{8mL^2} \Rightarrow E_1 = \dfrac{h^2}{8mL^2} = \dfrac{(6.63\times10^{-34}\,\text{J}\cdot\text{s})^2}{8(0.20\text{kg})(1.5\text{m})^2}$

$\Rightarrow E_1 = 1.22\times10^{-67}\,\text{J}$

(b) $E = \dfrac{1}{2}mv^2 \Rightarrow v = \sqrt{\dfrac{2E}{m}} = \sqrt{\dfrac{2(1.2x10^{-67}\,\text{J})}{0.20\text{kg}}} = 1.1\times10^{-33}\,\text{m/s}$

$\Rightarrow t = \dfrac{d}{v} = \dfrac{1.5\text{m}}{1.1\times10^{-33}\,\text{m/s}} = 1.4\times10^{33}\,\text{s}$

(c) $E_2 - E_1 = \dfrac{h^2}{8mL^2}(4-1) = \dfrac{3h^2}{8mL^2} = 3(1.22\times10^{-67}\,\text{J})$

$= 3.7\times10^{-67}\,\text{J}.$

42.2: (a) The third excited state is $n = 4$, so

$$\Delta E = (4^2-1)\frac{h^2}{8mL^2}$$

$$= \frac{15(6.626\times10^{-34}\,\text{J}\cdot\text{s})^2}{8(9.11\times10^{-31}\,\text{kg})(0.125\times10^{-9}\,\text{m})^2} = 5.78\times10^{-17}\,\text{J} = 361\,\text{eV}.$$

(b) $\lambda = \dfrac{hc}{\Delta E} = \dfrac{(6.63\times10^{-34}\,\text{J}\cdot\text{s})(3.0\times10^8\,\text{m/s})}{5.78\times10^{-17}\,\text{J}}$

$\lambda = 3.44\,\text{nm}$

42-3: $E_2 - E_1 = \dfrac{h^2}{8mL^2}(4-1) = \dfrac{3h^2}{8mL^2}$

$\Rightarrow L = h\sqrt{\dfrac{3}{8m(E_2-E_1)}}$

$= (6.63\times10^{-34}\,\text{J}\cdot\text{s})\sqrt{\dfrac{3}{8(9.11\times10^{-31}\,\text{kg})(4.0\,\text{eV})(1.60\times10^{-19}\,\text{J/eV})}}$

$\Rightarrow L = 5.3\times10^{-10}\,\text{m}$

42-4: From Eq. (42-5),

$$L = \frac{h}{\sqrt{8mE_1}}$$

$$= \frac{(6.626\times10^{-34}\,\text{J}\cdot\text{s})}{\sqrt{8(1.673\times10^{-27}\,\text{kg})(5.0\times10^6\,\text{MeV})(1.602\times10^{-19}\,\text{J/eV})}}$$

$= 6.4\times10^{-15}\,\text{m}.$

42-5: Recall $\lambda = \dfrac{h}{p} = \dfrac{h}{\sqrt{2mE}}$

(a) $E_1 = \dfrac{h^2}{8mL^2} \Rightarrow \lambda_1 = \dfrac{h}{\sqrt{2mh^2/8mL^2}} = 2L = 2(5.0\times10^{-10}\,\text{m}) = 1.0\times10^{-9}\,\text{m}$

$$p_1 = \frac{h}{\lambda_1} = \frac{(6.63 \times 10^{-34}\,\text{J.s})}{1.0 \times 10^{-9}\,\text{m}} = 6.63 \times 10^{-25}\,\text{kg} \cdot \text{m/s}$$

(b) $E_2 = \dfrac{4h^2}{8mL^2} \Rightarrow \lambda_2 = L = 5.0 \times 10^{-10}\,\text{m}$

$$p_2 = \frac{h}{\lambda_2} = 2p_1 = 1.3 \times 10^{-24}\,\text{kg} \cdot \text{m/s}$$

(c) $E_3 = \dfrac{9h^2}{8mL^2} \Rightarrow \lambda_3 = \dfrac{2}{3}L = 3.3 \times 10^{-10}\,\text{m}$

$$p_3 = 3p_1 = 2.0 \times 10^{-24}\,\text{kg} \cdot \text{m/s}$$

42-6: (a) From Eq. (42-9), the wave function for $n = 1$ vanishes only at $x = 0$ and $x = L$ in the range $0 \le x \le L$. (b) In the range for x, the sine term is a maximum only at the middle of the box, $x = L/2$. (c) The answers to parts (a) and (b) are consistent with the figure.

42-7: The first excited state or $(n = 2)$ wavefunction

is $\psi_2(x) = \sqrt{\dfrac{2}{L}} \sin\left(\dfrac{2\pi x}{L}\right)$

So $|\psi_2(x)|^2 = \dfrac{2}{L}\sin^2\left(\dfrac{2\pi x}{L}\right)$

(a) If the probability amplitude is zero, then $\sin^2\left(\dfrac{2\pi x}{L}\right) = 0$

$$\Rightarrow \frac{2\pi x}{L} = m\pi \Rightarrow x = \frac{Lm}{2}, m = 0, 1, 2...$$

So probability is zero for $x = 0, \dfrac{L}{2}, L$.

(b) The probability is largest if $\sin\left(\dfrac{2\pi x}{L}\right) = \pm 1$.

$$\Rightarrow \frac{2\pi x}{L} = (2m+1)\frac{\pi}{2} \Rightarrow x = (2m+1)\frac{L}{4}$$

So probability is largest for $x = \dfrac{L}{4}$ and $\dfrac{3L}{4}$.

(c) These answers are consistent with the zeros and maxima of Fig 42-4.

42-8: $\dfrac{d^2\psi}{dx^2} = -k^2\psi$, and for ψ to be a solution of Eq. (42-12), $k^2 = E\dfrac{8\pi^2 m}{h^2} = E\dfrac{2m}{\hbar^2}$.

(b) The wave function must vanish at the rigid walls; the given function will vanish at $x = 0$ for any k, but to vanish at $x = L$, $kL = n\pi$ for integer n.

42-9: (a) Eq. (42-13): $-\dfrac{h^2}{8\pi^2 m} \cdot \dfrac{d^2\psi}{dx^2} = E\psi$.

$$\frac{d^2}{dx^2}\psi = \frac{d^2}{dx^2}(A\cos kx) = \frac{d}{dx}(-Ak\sin kx) = -Ak^2\cos kx$$

$$\Rightarrow \frac{Ak^2h^2}{8\pi^2 m}\cos kx = EA\cos kx$$

$$\Rightarrow E = \frac{k^2h^2}{8\pi^2 m} \Rightarrow k = \sqrt{\frac{8\pi^2 mE}{h^2}} = \frac{\sqrt{2mE}}{\hbar}$$

(b) This is not an acceptable wave function for a box with rigid walls since we need $\psi(0) = \psi(L) = 0$, but this $\psi(x)$ has maxima there. It doesn't satisfy the boundary condition.

42-10:

$$-\frac{\hbar^2}{2m}\frac{d^2\psi'}{dx^2} + U\psi' = -\frac{\hbar^2}{2m}C\frac{d^2\psi}{dx^2} + UC\psi$$

$$= C\left[-\frac{\hbar^2}{2m}\frac{d^2\psi}{dx^2} + U\psi\right]$$

$$= CE\psi = EC\psi = E\psi',$$

and so ψ' is a solution to Eq. (42-17) with the same energy.

42-11: Eq. (42-17): $\frac{-\hbar^2}{2m}\frac{d^2\psi}{dx^2} + U\psi = E\psi$

Let $\psi = A\psi_1 + B\psi_2$

$$\Rightarrow \frac{-\hbar^2}{2m}\frac{d^2}{dx^2}(A\psi_1 + B\psi_2) + U(A\psi_1 + B\psi_2) = E(A\psi_1 + B\psi_2)$$

$$\Rightarrow A\left(-\frac{\hbar^2}{2m}\frac{d^2\psi_1}{dx^2} + U\psi_1 - E\psi_1\right) + B\left(-\frac{\hbar^2}{2m}\frac{d^2\psi_2}{dx^2} + U\psi_2 - E\psi_2\right) = 0$$

But each of ψ_1 and ψ_2 satisfy Schröedinger's Equation separately so the equation still holds true, for any A or B.

42-12:

$$-\frac{\hbar^2}{2m}\frac{d^2\psi}{dx^2} + U\psi = BE_1\psi_1 + CE_2\psi_2.$$

If ψ were a solution with energy E, then

$$BE_1\psi_1 + CE_2\psi_2 = BE\psi_1 + CE\psi_2 \text{ or}$$

$$B(E_1 - E)\psi_1 = C(E - E_2)\psi_2.$$

This would mean that ψ_1 is a constant multiple of ψ_2, and from the result of Exercise 42-10, ψ_1 and ψ_2 would be wavefunctions with the same energy. However, $E_1 \neq E_2$, so this is not possible, and ψ cannot be a solution to Eq. (42-17).

42-13: Eq. (42-20): $\psi = A\sin\frac{\sqrt{2mE}}{\hbar}x + B\cos\frac{\sqrt{2mE}}{\hbar}x$

$$\frac{d^2\psi}{dx^2} = -A\left(\frac{2mE}{\hbar^2}\right)\sin\frac{\sqrt{2mE}}{\hbar}x - B\left(\frac{2mE}{\hbar^2}\right)\cos\frac{\sqrt{2mE}}{\hbar}x$$

$$= \frac{-2mE}{\hbar^2}(\psi) = Eq.(42-19).$$

42-14:

$$\frac{d\psi}{dx} = \kappa(Ce^{\kappa x} - De^{-\kappa x}),$$

$$\frac{d^2\psi}{dx^2} = \kappa^2(Ce^{\kappa x} + De^{-\kappa x}) = \kappa^2\psi$$

for all constants C and D. Hence ψ is a solution to Eq. (42-17) for

$$-\frac{\hbar^2}{2m}\kappa^2 + U_0 = E, \text{ or } \kappa = [2m(U_0 - E)]^{1/2}/\hbar,$$

and κ is real for $E < U_0$.

42-15: (a) Eq. (42-17): $\frac{-\hbar^2}{2m}\frac{d^2\psi}{dx^2} + U\psi = E\psi$

Left hand side: $\frac{-\hbar^2}{2m}\frac{d^2}{dx^2}(A\sin kx) + U_0 A\sin kx$

$$= \frac{\hbar^2 k^2}{2m}A\sin kx + U_0 A\sin kx$$

$$= \left(\frac{\hbar^2 k^2}{2m} + U_0\right)\psi.$$

But $\frac{\hbar^2 k^2}{2m} + U_0 > U_0 > E$ for constant k. But $\frac{\hbar^2 k^2}{2m} + U_0$ should equal E $\Rightarrow$ no solution

(b) If $E > U_0$, then $\frac{\hbar^2 k^2}{2m} + U_0 = E$ is consistent and so $\psi = A\sin kx$ is a solution of Eq (42-17) for this case.

42-16: From Example 42-5 and Fig. (42-8), $U_0 - E_1 = 5.375E_\infty$ and the maximum wavelength of the photon would be

$$\lambda = \frac{hc}{U_0 - E_1} = \frac{hc}{(5.375)(h^2/8mL^2)} = \frac{8mL^2c}{(5.375)h} = 3.5 \times 10^{-7}\text{ m}.$$

42-17: From Fig. 42-8, $U_0 = 6E_\infty$, $E_1 = 0.625E_\infty$, $E_3 = 5.09E_\infty$ and

$$E_\infty = \frac{\pi^2\hbar^2}{2mL^2} = \frac{\pi^2(1.054 \times 10^{-34}\text{ J}\cdot\text{s.})^2}{2(1.67 \times 10^{-27}\text{ kg})(8.0 \times 10^{-15}\text{ m})^2}$$

$\Rightarrow E_\infty = 5.12 \times 10^{-13}\text{ J}.$

The transition energy is $E_3 - E_1 = (5.09 - 0.625)(5.12 \times 10^{-13}\text{ J})$

$$= 2.29 \times 10^{-12}\text{ J}$$

The wavelength of the photon absorbed is then

$$\lambda = \frac{hc}{E_3 - E_1} = \frac{(6.63 \times 10^{-34}\text{ J}\cdot\text{s.})(3.00 \times 10^8\text{ m/s})}{2.29 \times 10^{-12}\text{ J}} = 8.7 \times 10^{-14}\text{ m}$$

42-18: Using the expression as given in Eq. (42-24), the probabilities are a) 2.7×10^{-3}, b) 1.4×10^{-4} and c) 1.6×10^{-6}.

42-19: (a) Probability of tunneling is $T = Ge^{-2\kappa L}$

where $G = 16\dfrac{E}{U_0}\left(1-\dfrac{E}{U_0}\right) = 16.\left(\dfrac{46}{55}\right)\left(1-\dfrac{46}{55}\right) = 2.19$

and $\kappa = \dfrac{\sqrt{2m(U_0 - E)}}{\hbar} = \dfrac{\sqrt{2(9.11\times10^{-31}\text{kg})(55\text{ eV} - 46\text{ eV})(1.60\times10^{-19}\text{J/eV})}}{1.054\times10^{-34}\text{J}\cdot\text{s.}}$

$= 1.54\times10^{10}\text{ m}^{-1}$

So $T_e = 2.19e^{-2(1.54\times10^{10}\text{ m}^{-1})(2.0\times10^{-10}\text{ m})} = 2.19e^{-6.15} = 0.0047$

(b) For a proton, $\kappa' = \sqrt{\dfrac{m_p}{m_e}}\,\kappa = \sqrt{\dfrac{1.67\times10^{-27}}{9.11\times10^{-31}}}\cdot\kappa$

$\Rightarrow \kappa' = 6.59\times10^{11}\text{m}^{-1}$

$\Rightarrow T = 2.19\,e^{-2(6.59\times10^{11}\text{ m}^{-1})(2.0\times10^{-10}\text{ m})} = 2.19e^{-264}$

$\Rightarrow T \approx 10^{-114}.$

42-20: (a) In Eq. (42-24), $\kappa = 4.374\times10^{14}\text{m}^{-1}$, $2\kappa L = 1.7495, A = 0.6144$, and so $T = 0.11$. b) $\kappa = 1.383\times10^{15}\text{m}^{-1}$, $2\kappa L = 5.532$, $A = 3.84$, and $T = 0.015$.

42-21: The ground state energy of a simple harmonic oscillator is

$E_0 = \dfrac{1}{2}\hbar\omega = \dfrac{1}{2}\hbar\sqrt{\dfrac{k'}{m}} = \dfrac{(1.034\times10^{-34}\text{J}\cdot\text{s})}{2}\sqrt{\dfrac{130N/\text{m}}{0.200\text{ kg}}}$

$\Rightarrow E_0 = 1.34\times10^{-33}\text{J} = 8.40\times10^{-15}\text{eV}$

Also $E_{n+1} - E_n = \hbar\omega = 2E_0 = 1.68\times10^{-14}\text{eV}$

$= 2.69\times10^{-33}\text{ J.}$

Such smaller energies are unimportant for the motion of the block

42-22: Let $\sqrt{mk'}/2\hbar = \delta$, and so $\dfrac{d\psi}{dx} = -2x\delta\psi$ and $\dfrac{d^2\psi}{dx^2} = (4x^2\delta^2 - 2\delta)\psi$, and ψ is a solution of Eq. (42-25) if $E = \dfrac{\hbar^2}{m}\delta = \dfrac{1}{2}\hbar\sqrt{k'/m} = \dfrac{1}{2}\hbar\omega.$

42-23: The photon's energy is $E_\gamma = \dfrac{hc}{\lambda}$.

The transition energy is $\Delta E = E_1 - E_0 = \hbar\omega\left(\dfrac{3}{2}-\dfrac{1}{2}\right) = \hbar\omega = \hbar\sqrt{\dfrac{k'}{m}}$

$\Rightarrow \dfrac{2\pi\hbar c}{\lambda} = \hbar\sqrt{\dfrac{k'}{m}} \Rightarrow k' = \dfrac{4\pi^2c^2m}{\lambda^2} = \dfrac{4\pi^2(3.00\times10^9\text{ m/s})^2(4.7\times10^{-26}\text{ kg})}{(3.40\times10^{-4}\text{ m})^2}$

$\Rightarrow k' = 1.4\text{ N/m.}$

42-24: (a) $\dfrac{|\psi(A)|^2}{|\psi(0)|^2} = \exp\left(-\dfrac{\sqrt{mk'}}{\hbar}A^2\right) = \exp\left(-\sqrt{mk'}\,\dfrac{\omega}{k'}\right) = e^{-1} = 0.368.$

This is more or less what is shown in Fig. (42-19).

(b) $\dfrac{|\psi(2A)|^2}{|\psi(0)|^2} = \exp\left(-\dfrac{\sqrt{mk'}}{\hbar}(2A)^2\right) = \exp\left(-\sqrt{mk'}\,4\dfrac{\omega}{k'}\right) = e^{-4} = 1.83\times10^{-2}$.
The figure cannot be read this precisely, but the qualitative decrease in amplitude with distance is clear.

42-25: For an excited level of the harmonic oscillator

$$E_n = \left(n+\frac{1}{2}\right)\hbar\omega = \frac{1}{2}k'A^2$$

$$\Rightarrow A = \sqrt{\frac{(2n+1)\hbar\omega}{k'}}$$ This is the uncertainty in the position.

Also $E_n = \left(n+\frac{1}{2}\right)\hbar\omega = \frac{1}{2}mv_{\text{max}}^2$

$$\Rightarrow v_{\text{max}} = \sqrt{\frac{(2n+1)\hbar\omega}{m}}$$

$$\Rightarrow \Delta x\Delta p = A\cdot(mv_{\text{max}}) = \sqrt{\frac{(2n+1)\hbar\omega}{k'}}\cdot\sqrt{(2n+1)\hbar\omega m}$$

$$= (2n+1)\,\hbar\omega\sqrt{\frac{m}{k'}} = (2n+1)\hbar$$

So $\Delta x\Delta p = (2n+1)\hbar$, which agrees for the ground state $(n=0)$ with $\Delta x\Delta p = \hbar$.
The uncertainty is seen to increase with n.

42-26: (a) $R_n = \dfrac{(n+1)^2 - n^2}{n^2} = \dfrac{2n+1}{n^2} = \dfrac{2}{n} + \dfrac{1}{n^2}$.
This is never larger than it is for $n=1$, and $R_1 = 3$. (b) R approaches zero; in the classical limit, there is no quantization, and the spacing of successive levels is vanishingly small compared to the energy levels.

42-27: The transition energy $\Delta E = E_2 - E_1 = \dfrac{h^2}{8mL^2}(4-1)$

$$\Rightarrow \lambda = \frac{hc}{\Delta E} = \frac{8mcL^2}{3h} = \frac{8(9.11\times10^{-31}\,\text{kg})(3.00\times10^8\,\text{m/s})(3.52\times10^{-9}\,\text{m})^2}{3(6.63\times10^{-34}\,\text{J}\cdot\text{s})}$$

$$\Rightarrow \lambda = 1.36\times10^{-5}\,\text{m}.$$

42-28: Using the normalized wave function $\psi_1 = \sqrt{2/L}\,\sin(\pi x/L)$, the probabilities $|\psi|^2 dx$ are a) $(2/L)\,\sin^2(\pi/4)\,dx = dx/L$, b) $(2/L)\,\sin^2(\pi/2)\,dx = 2\,dx/L$ and c) $(2/L)\sin^2(3\pi/4) = dx/L$.

42-29: $\psi_2(x) = \sqrt{\dfrac{2}{L}}\cdot\sin\left(\dfrac{2\pi x}{L}\right)$

$$\Rightarrow |\psi_2|^2 dx = \frac{2}{L}\sin^2\left(\frac{2\pi x}{L}\right)dx$$

(a) $x = \frac{L}{4} : |\psi_2|^2\, dx = \frac{2}{L}\sin^2\left(\frac{\pi}{2}\right) dx = \frac{2dx}{L}$

(b) $x = \frac{L}{2} : |\psi_2|^2\, dx = \frac{2}{L}\sin^2(\pi)\, dx = 0$

(c) $x = \frac{3L}{4} : |\psi_2|^2\, dx = \frac{2}{L}\sin^2\left(\frac{3\pi}{2}\right) dx = \frac{2\,dx}{L}$

42-30: (a) $\frac{2}{L}\int_0^{L/4} \sin^2 \frac{\pi x}{L}\, dx = \frac{2}{L}\int_0^{L/4} \frac{1}{2}\left(1 - \cos\frac{2\pi x}{L}\right) dx$

$$= \frac{1}{L}\left(x - \frac{L}{2\pi}\sin\frac{2\pi x}{L}\right)_0^{L/4}$$

$$= \frac{1}{4} - \frac{1}{2\pi},$$

about 0.0908.
(b) Repeating with limits of $L/4$ and $L/2$ gives

$$\frac{1}{L}\left(x - \frac{L}{2\pi}\sin\frac{2\pi x}{L}\right)_{L/4}^{L/2} = \frac{1}{4} + \frac{1}{2\pi},$$

about 0.0409. (c) The particle is much more likely to be nearer the middle of the box than the edge. (d) The results sum to exactly 1/2, which means that the particle is as likely to be between $x = 0$ and $x = L/2$ as it is to be between $x = L/2$ and $x = L$. (e) These results are represented in Fig. (42-4b).

42-31: (a) $P = \int_{L/4}^{3L/4} |\psi_1|^2\, dx = \int_{L/4}^{3L/4} \frac{2}{L}\sin^2\left(\frac{\pi x}{L}\right) dx$

Let $z = \frac{\pi x}{L} \Rightarrow dz = dx\frac{\pi}{L}$

$$\Rightarrow P = \frac{2}{\pi}\int_{\pi/4}^{3\pi/4} \sin^2 z\, dz = \frac{1}{\pi}\left[z - \frac{1}{2}\sin 2z\right]_{\pi/4}^{3\pi/4}$$

$$= \frac{1}{2} + \frac{1}{\pi}$$

(b) $P = \int_{L/4}^{3L/4} |\psi_2|^2\, dx = \int_{L/4}^{3L/4} \frac{2}{L}\sin^2\left(\frac{2\pi x}{L}\right) dx$

$$\Rightarrow P = \frac{1}{\pi}\int_{\pi/4}^{3\pi/2} \sin^2 z\, dz = \frac{1}{2\pi}\left[z - \frac{1}{2}\sin 2z\right]_{\pi/2}^{3\pi/2}$$

$$= \frac{1}{2}$$

(c) This is consistent with Fig. 42-4(b) since more of ψ_1 is between $x = \frac{L}{4}$ and $\frac{3L}{4}$ than ψ_2, and the proportions appear correct.

42-32: (a) $\psi = 0$, so $\dfrac{d\psi}{dx} = 0$. (b) $\dfrac{d\psi}{dx} = \dfrac{\pi}{L}\sqrt{\dfrac{2}{L}}\cos(\pi x / L)$, and as $x \to L, \dfrac{d\psi}{dx} \to -\sqrt{2\pi^2 / L^3}$. (c) Clearly not. In Eq. (42-17), any point where $U(x)$ is singular necessitates a discontinuity in $\dfrac{d\psi}{dx}$.

42-33: (a) $\psi_1 = \sqrt{\dfrac{2}{L}}\sin\left(\dfrac{\pi x}{L}\right)$

$$\Rightarrow \frac{d\psi_1}{dx} = \sqrt{\frac{2}{L}}\cdot\frac{\pi}{L}\cos\left(\frac{\pi x}{L}\right) = \sqrt{\frac{2\pi^2}{L^3}}\left(1 - \frac{1}{2}\left(\frac{\pi x}{L}\right)^2 + \ldots\right)$$

$$\Rightarrow \frac{d\psi_1}{dx} \approx \sqrt{\frac{2\pi^2}{L^3}} \text{ for } x \approx 0$$

(b) $\psi_2 = \sqrt{\dfrac{2}{L}}\sin\dfrac{2\pi x}{L} \Rightarrow \dfrac{d\psi_2}{dx} = \sqrt{\dfrac{2}{L}}\cdot\dfrac{2\pi}{L}\cdot\cos\dfrac{2\pi x}{L}$

$$\Rightarrow \frac{d\psi_2}{dx} \approx \sqrt{\frac{8\pi^2}{L^3}} \text{ for } x \approx 0.$$

(c) For $x \approx L, \cos\left(\dfrac{\pi x}{L}\right) \approx \cos\pi = -1$

$$\Rightarrow \frac{d\psi_1}{dx} \approx -\sqrt{\frac{2\pi^2}{L^3}} \text{ for } x \approx L$$

(d) $\dfrac{d\psi_2}{dx} = \sqrt{\dfrac{8\pi^2}{L^3}}\cos\left(\dfrac{2\pi x}{L}\right) \approx \sqrt{\dfrac{8\pi^2}{L^3}}$ for $x \approx L$, since $\cos 2\pi = 1$

(e) The slope of the wavefunction is greatest for $\psi_2 (n = 2)$ close to the walls of the box, so shown in Fig. 42-4.

42-34: (a) The time-dependent equation, with the separated form for $\Psi(x,t)$ as given becomes

$$i\hbar\psi(-i\omega) = \left(-\frac{\hbar^2}{2m}\frac{d^2\psi}{dx^2} + U(x)\psi\right).$$

Since ψ is a solution of the time-independent solution with energy E, the term in parenthesis is $E\psi$, and so $\omega\hbar = E$, and $\omega = (E/\hbar)$. b) $(e^{-i\omega\hbar t})^*(e^{-i\omega\hbar t}) = e^0 = 1$, so $|\Psi|^2 = |\psi|^2$.

42-35: (a) We set the solutions for inside and outside the well equal to each other at the well boundaries, $x = 0$ and L.

$x = 0: A\sin(0) + B = C \Rightarrow B = C$, since we must have $D = 0$ for $x < 0$

$x = L: A\sin\dfrac{\sqrt{2mE}\,L}{\hbar} + B\cos\dfrac{\sqrt{2mE}\,L}{\hbar} = +De^{-\kappa L}$ since $C = 0$ for $x > L$

$$\Rightarrow \frac{A\sin kL + B\cos kL = De^{-\kappa L}}{\text{where } k = \dfrac{\sqrt{2mE}}{\hbar}}$$

(b) Requiring continuous boundary condition yields

$x = 0: \dfrac{d\psi}{dx} = kA\cos(k\cdot 0) - kB\sin(k\cdot 0) = kA = \kappa Ce^{K\cdot 0}$

$\Rightarrow kA = \kappa C$

$x = L:\ kA\cos kL - kB\sin kL = -\kappa De^{-\kappa L}.$

42-36: For a wave function ψ with wavenumber k, $\dfrac{d^2\psi}{dx^2} = -k^2\psi$, and using this in Eq. (42-17) with $U = U_0 > E$ gives $k^2 = 2m(E - U_0)/\hbar^2$, or $k = i\sqrt{2m(U_0 - E)}/\hbar = i\kappa$.

42-37: (a) $T = \left[1 + \dfrac{(U_0 \sinh\kappa L)^2}{4E(U_0 - E)}\right]^{-1}$

If $\kappa L >> 1$, then $\sinh\kappa L \approx \dfrac{1}{2}e^{\kappa L}$

$$\Rightarrow T \approx \left[1 + \frac{U_0^2 e^{2\kappa L}}{16E(U_0 - E)}\right]^{-1} = \left[\frac{U_0^2 e^{2\kappa L}}{16E(U_0 - E)}\right]^{-1}$$

$$\Rightarrow T \approx Ge^{-2\kappa L} \text{ where } G = 16\frac{E}{U_0}\left(1 - \frac{E}{U_0}\right)$$

(b) As $E \to U_0$, $\kappa \to 0 \Rightarrow \sinh\kappa L \to \kappa L$

$$\Rightarrow T \approx \left[1 + \frac{U_0^2\kappa^2 L^2}{4E(U_0 - E)}\right]^{-1} \approx \left[\frac{1 + 2U_0^2 L^2 m}{4E\hbar^2}\right]^{-1}$$

since $\kappa^2 = \dfrac{2m(U_0 - E)}{\hbar^2}$

But $U_0 \approx E \Rightarrow \dfrac{U_0{}^2}{E} = E$

$$\Rightarrow T \approx \left[1 + \left(\frac{2mE}{\hbar^2}\right)\left(\frac{L}{2}\right)^2\right]^{-1} = \left[1 + \left(\frac{kL}{2}\right)^2\right]^{-1}$$

since $k^2 = \dfrac{2mE}{\hbar^2}$

42-38: In Eq. (42-24), $G = 3.36$, so $\kappa L = 4.060$; $\kappa = 8.873 \times 10^9$ /m, and so $L = 4.58 \times 10^{-10}$ m $= 0.46$ nm.

42-39: Eq. (42-25): $\dfrac{-\hbar^2}{2m}\dfrac{d^2\psi}{dx^2} + \dfrac{1}{2}k'x^2\psi = E\psi$

Now, $\dfrac{d}{dx^2}(Cxe^{-m\omega x^2/2\hbar}) = \dfrac{d}{dx}\left[Ce^{-m\omega x^2/2\hbar} - \dfrac{Cm\omega 2x^2 e^{-m\omega x^2/2\hbar}}{2\hbar}\right]$

$$= \left(Ce^{-m\omega x^2/2\hbar}\right)\left[\frac{-2xm\omega}{2\hbar} - \frac{4m\omega x}{2\hbar} + \frac{4m^2\omega^2 x^3}{4\hbar^2}\right]$$

$$=\left(Cxe^{-m\omega x^2/2\hbar}\right)\left[\frac{-3m\omega}{\hbar}+\left(\frac{m\omega}{\hbar}\right)^2 x^2\right]$$

$$\Rightarrow \text{Eq. (42-25)}: \ \psi\left[\frac{+3\omega\hbar}{2}-\frac{m\omega^2}{2}x^2+\frac{1}{2}k'x^2\right]=\psi E$$

but $\omega^2=\dfrac{k'}{m}\Rightarrow E=\dfrac{3\hbar\omega}{2}$.

42-40: With $u(x)=x^2$, $a=m\omega/\hbar$, $|\psi|^2$ is of the form $C^*Cu(x)e^{-au(x)}$, which has a maximum at $au=1$, or $\dfrac{m\omega}{\hbar}x^2=1$, $x=\pm\sqrt{\hbar/m\omega}$. b) ψ and hence $|\psi|^2$ vanish at $x=0$, and as $x\to\pm\infty$, $|\psi|^2\to 0$. The result of part (a) is $\pm A/\sqrt{3}$, found from $n=1$ in Eq. (42-30), and this is consistent with Fig. (42-19). The figure also shows a minimum at $x=0$ and a rapidly decreasing $|\psi|^2$ as $x\to\pm\infty$.

42-41: The angular frequency $\omega=\dfrac{2\pi}{T}=\dfrac{2\pi}{0.500s}=12.6\text{ rad/s}$.
Ground state harmonic oscillator energy is given by

$$E_0=\frac{1}{2}\hbar\omega=\frac{1}{2}(1.054\times10^{-34}\text{J}\cdot\text{s})(12.6\text{ rad/s})=6.62\times10^{-34}\text{J})$$

$$=\frac{6.62\times10^{-34}\text{ J}}{1.60\times10^{-19}\text{ J/eV}}=4.14\times10^{-15}\text{ eV}$$

$$\Delta E=E_{n+1}-E_n=\hbar\omega\left(\left(n+\frac{3}{2}\right)-\left(n+\frac{1}{2}\right)\right)=\hbar\omega$$

$$\Rightarrow \Delta E=2E_0=1.32\times10^{-33}\text{ J}=8.28\times10^{-15}\text{ eV}$$

These values are too small to be detected.

42-42: $E=\dfrac{1}{2}mv^2=(n+(1/2))\hbar\omega=(n+(1/2))\,hf$, and solving for n,

$$n=\frac{\frac{1}{2}mv^2}{hf}-\frac{1}{2}=\frac{(1/2)(0.025\text{ kg})(0.40\text{ m/s})^2}{(6.626\times10^{-34}\text{J}\cdot\text{s})(2.00\text{Hz})}-\frac{1}{2}=1.5\times10^{30}.$$

42-43: (a) Eq. (42-32): $\dfrac{-\hbar^2}{2m}\left(\dfrac{\partial^2\psi}{\partial x^2}+\dfrac{\partial^2\psi}{\partial y^2}+\dfrac{\partial^2\psi}{\partial z^2}\right)+U\psi=E\psi$.

ψ_{n_x},ψ_{n_y} and ψ_{n_z} are all solutions to the 1-D Schröeding equation, so

$$\frac{-\hbar^2}{2m}\frac{d^2\psi_{n_x}}{dx^2}+\frac{1}{2}k'x^2\psi_{n_x}=E_{n_x}\psi_{n_x}.$$

and similarly for ψ_{n_y} and ψ_{n_z}.
Now if $\psi=\psi_{n_x}(x)\psi_{n_y}(y)\psi_{n_z}(z)$ then

$$\frac{\partial^2\psi}{\partial x^2}=\left(\frac{d^2\psi_{n_x}}{dx}\right)\psi_{n_y}\psi_{n_z},\quad \frac{\partial^2\psi}{\partial y^2}=\left(\frac{d^2\psi_{n_y}}{dy^2}\right)\psi_{n_x}\psi_{n_z}$$

and $\dfrac{\partial^2\psi}{\partial z^2} = \left(\dfrac{d^2\psi_{n_z}}{dz^2}\right)\psi_{n_x}\psi_{n_y}.$

Therefore: $\dfrac{-\hbar^2}{2m}\left(\dfrac{\partial^2\psi}{\partial x^2}+\dfrac{\partial^2\psi}{\partial y^2}+\dfrac{\partial^2\psi}{\partial z^2}\right)+\dfrac{1}{2}k'(x^2+y^2+z^2)\psi$

$$=\left[\frac{-\hbar^2}{2m}\left(\frac{d^2\psi_{n_x}}{dx^2}\right)+\frac{1}{2}k'x^2\psi_{n_x}\right]\psi_{n_y}\psi_{n_z}.$$

$$+\left[\frac{-\hbar^2}{2m}\left(\frac{d^2\psi_{n_y}}{dy^2}\right)+\frac{1}{2}k'y^2\psi_{n_y}\right]\psi_{n_x}\psi_{n_z}$$

$$+\left[\frac{-\hbar^2}{2m}\left(\frac{d^2\psi_{n_z}}{dz^2}\right)+\frac{1}{2}k'z^2\psi_{n_z}\right]\psi_{n_x}\psi_{n_y}$$

$$=\left[E_{n_x}+E_{n_y}+E_{n_z}\right]\psi_{n_x}\psi_{n_y}\psi_{n_z}$$

$$=\left[E_{n_x}+E_{n_y}+E_{n_z}\right]\psi = E_{n_xn_yn_z}\psi$$

$$\Rightarrow E_{n_xn_yn_z}=\left[\left(n_x+\frac{1}{2}\right)+\left(n_y+\frac{1}{2}\right)+\left(n_z+\frac{1}{2}\right)\right]\hbar\omega$$

$$\Rightarrow E_{n_xn_yn_z}=\left[n_x+n_y+n_z+\frac{3}{2}\right]\hbar\omega.$$

(b) Ground state energy $E_{000}=\dfrac{3}{2}\hbar\omega$

First excited state energy $E_{100}=E_{001}=E_{010}=\dfrac{5}{2}\hbar\omega.$

(c) As seen in (b) there is just one set of quantum number (0, 0, 0) for the ground state and three possibilities (1, 0, 0), (0, 1, 0) and (0, 0, 1) for the first excited state.

42-44: Let $\omega_1=\sqrt{k_1'/m}, \omega_2=\sqrt{k_2'/m}, \psi_{n_x}(x)$ be a solution to Eq. (42-25) with $E_{n_x}=\left(n_x+\dfrac{1}{2}\right)\hbar\omega_1$, $\psi_{n_x}(y)$ be a similar solution, and $\psi_{n_z}(z)$ be a solution of Eq. (42-25) but with z as the independent variable instead of x, and energy $E_{n_z}=\left(n_z+\dfrac{1}{2}\right)\omega_2$. (a) As in Problem 42-43, look for a solution of the form $\psi(x,y,z)=\psi_{n_x}(x)\psi_{n_y}(y)\psi_{n_z}(z)$.

Then,

$$-\frac{\hbar^2}{2m}\frac{\partial^2\psi}{\partial x^2}=\left(E_{n_x}-\frac{1}{2}k_1'x^2\right)\psi$$

with similar relations for $\dfrac{\partial^2\psi}{\partial y^2}$ and $\dfrac{\partial^2\psi}{\partial z^2}$. Adding,

$$-\frac{\hbar^2}{2m}\left(\frac{\partial^2\psi}{\partial x^2}+\frac{\partial^2\psi}{\partial y^2}+\frac{\partial^2\psi}{\partial z^2}\right)=\left(E_{n_x}+E_{n_y}+E_{n_z}-\frac{1}{2}k_1'x^2-\frac{1}{2}k_1'y^2-\frac{1}{2}k_2'z^2\right)\psi$$

$$=\left(E_{n_x}+E_{n_y}+E_{n_z}-U\right)\psi$$

$$=(E-U)\psi$$

where the energy E is

$$E=E_{n_x}+E_{n_y}+E_{n_z}=\hbar\left[(n_x+n_y+1)\,\omega_1^2+\left(n_z+\frac{1}{2}\right)\omega_2^2\right],$$

n_x, n_y and n_z all nonnegative integers. (b) The ground level corresponds to $n_x=n_y=n_z=0$, and $E=\hbar(\omega_1^2+\omega_2^2/2)$. The first excited level corresponds to $n_x=n_y=0$ and $n_z=1$, as $\omega_2^2<\omega_1^2$, and $E=\hbar(\omega_1^2+(3/2)\omega_2^2)$. There is only one set of quantum numbers for both the ground state and the first excited state.

42-45: (a) $\psi(x)=A\sin kx$ and $\psi(-L/2)=0=\psi(+L/2)$

$$\Rightarrow 0=A\sin\left(\frac{+kL}{2}\right)\Rightarrow\frac{+kL}{2}=n\pi\Rightarrow k=\frac{2n\pi}{L}=\frac{2\pi}{\lambda}$$

$$\Rightarrow\lambda=\frac{L}{n}\Rightarrow p_n=\frac{h}{\lambda n}=\frac{nh}{L}$$

$$\Rightarrow E_n=\frac{p_n^{\;2}}{2m}=\frac{n^2h^2}{2mL^2},\ \text{where } n=1,2..$$

(b) $\psi(x)=A\cos kx$ and $\psi(-L/2)=0=\psi(+L/2)$

$$\Rightarrow 0=A\cos\left(\frac{kL}{2}\right)\Rightarrow\frac{kL}{2}=(2n+1)\frac{\pi}{2}\Rightarrow k=\frac{(2n+1)\pi}{L}=\frac{2\pi}{\lambda}$$

$$\Rightarrow\lambda=\frac{2L}{(2n+1)}\Rightarrow p_n=\frac{(2n+1)h}{2L}$$

$$\Rightarrow E_n=\frac{(2n+1)^2h^2}{8mL^2}\qquad n=0,1,2\ldots$$

(c) The combination of all the energies in parts (a) and (b) is the same energy levels as given in Eq. (42-5), where

$$E_n=\frac{n^2h^2}{8mL^2}$$

(d) Part (a)'s wave functions are odd, and part (b)'s are even.

42-46: (a) As with the particle in a box, $\psi(x)=A\sin kx$, where A is a constant and $k^2=2mE/\hbar^2$. Unlike the particle in a box, however, k and hence E do not have simple forms. (b) For $x>L$, the wave function must have the form of Eq. (42-22). For the wave function to remain finite as $x\to\infty$, $C=0$. The constant $\kappa^2=2m(U_0-E)/\hbar$, as in Eq. (42-21) and Eq. (42-22). (c) At $x=L$, $A\sin kL=De^{-\kappa L}$ and $kA\cos kL=-\kappa De^{-\kappa L}$. Dividing the second of these by the first gives

$$k\cot kL=-\kappa,$$

a transcendental equation that must be solved numerically for different values of the length L and the ratio E/U_0.

Chapter 43: Atomic Structure

43-1: $U = \frac{1}{4\pi\varepsilon_0} \cdot \frac{q_1 q_2}{r} = \frac{-1}{4\pi\varepsilon_0} \frac{(1.60\times10^{-19}\ \text{C})^2}{1.0\times10^{-10}\ \text{m}} = -2.3\times10^{-18}\ \text{J}$

$= \frac{-2.3\times10^{-18}\ \text{J}}{1.60\times10^{-19}\ \text{J/eV}} = -14.4\ \text{eV}.$

43-2: The (l, m_l) combinations are (0, 0), (1, 0), (1, ±1), (2, 0), (2, ±1), (2, ±2), (3, 0), (3, ±1), (3, ±2), (3, ±3), (4, 0), (4, ±1), (4, ±2), (4, ±3), and (4, ±4), a total of 25.
(b) Each state has the same energy (n is the same), $-\frac{13.60\ \text{eV}}{25} = -0.544\ \text{eV}.$

43-3: $L = \sqrt{l(l+1)}\,\hbar \Rightarrow l(l+1) = \left(\frac{L}{\hbar}\right)^2 = \left(\frac{3.653\times10^{-34}\,\text{J}\cdot\text{s}}{1.054\times10^{-34}\,\text{J}\cdot\text{s}}\right)$

$\Rightarrow l(l+1) = 12.0 \Rightarrow l = 3$

43-4: (a) $m_{l_{max}} = 2$, so $L_{z_{max}} = 2\hbar$. (b) $\sqrt{l(l+1)}\hbar = \sqrt{6}\hbar = 2.45\hbar$. (c) The angle is $\arccos\left(\frac{L_z}{L}\right) = \arccos\left(\frac{m_l}{\sqrt{6}}\right)$, and the angles are, for $m_l = -2$ to $m_l = 2$, 144.7°, 114.1°, 90.0°, 65.9°, 35.3°. The angle corresponding to $m_l = l$ will always be larger for larger l.

43-5: $L = \sqrt{l(l+1)}\,\hbar$. The maximum orbital quantum number $l = n-1$. So if:

$n = 1 \qquad l = 0 \qquad \Rightarrow \quad L = 0$

$n = 10 \qquad l = 9 \qquad \Rightarrow \quad L = \sqrt{9(10)}\,\hbar = 9.49\,\hbar$

$n = 100 \qquad l = 99 \qquad \Rightarrow \quad L = \sqrt{99(100)}\,\hbar = 99.5\,\hbar.$

The maximum angular momentum value gets closer to the Bohr model value the larger the value of n.

43-6: $e^{im_l\phi} = \cos(m_l\phi) + i\sin(m_l\phi)$, and to be periodic with period 2π, $m_l 2\pi$ must be an integer multiple of 2π, so m_l must be an integer.

43-7: $P(a) = \int_0^a |\psi_{1s}|^2 dV = \int_0^a \frac{1}{\pi a^3} e^{-2r/a} (4\pi r^2 dr)$

$\Rightarrow P(a) = \frac{4}{a^3}\int_0^a r^2 e^{-2r/a} dr = \frac{4}{a^3}\left[\left(\frac{-ar^2}{2} - \frac{a^2 r}{2} - \frac{a^3}{4}\right) e^{-2r/a}\right]_0^a$

$= \frac{4}{a^3}\left[\left(\frac{-a^3}{2} - \frac{a^3}{2} - \frac{a^3}{4}\right) e^{-2} + \frac{a^3}{4} e^0\right]$

$\Rightarrow P(a) = 1 - 5e^{-2}.$

43-8: (a) As in Example 43-3, the probability is

$$P = \int_0^{a/2} |\psi_{1s}|^2 4\pi r^2\, dr = \frac{4}{a^3}\left[\left(-\frac{ar^2}{2} - \frac{a^2 r}{2} - \frac{a^3}{4}\right)e^{-2r/a}\right]_0^{a/2}$$

$$= 1 - \frac{5e^{-1}}{2} = 0.0803.$$

(b) The difference in the probabilities is

$$(1 - 5e^{-2}) - (1 - (5/2)e^{-1}) = (5/2)(e^{-1} - 2e^{-2}) = 0.243.$$

43-9: (a) $|\psi|^2 = \psi * \psi = |R(r)|^2 |\Theta(\theta)|^2 (Ae^{-im_l\phi})(Ae^{+im_l\phi})$

$= A^2 |R(r)|^2 |\Theta(\theta)|^2$, which is independent of ϕ

(b) $\int_0^{2\pi} |\Phi(\theta)|^2 d\phi = A^2 \int_0^{2\pi} d\phi = 2\pi A^2 = 1$

$$\Rightarrow A = \frac{1}{\sqrt{2\pi}}.$$

43-10: The energy will be proportional to the reduced mass.

(a) 13.60 eV/0.99946 = 13.61 eV. (Use of more precise numerical values gives 13.6056981 eV.)

(b) 13.60 eV/2 = 6.80 eV

(c) (13.60 eV)(186) = 2530 eV.

43-11: (a) $m_r = m : a_1 = \frac{\varepsilon_0 h^2}{\pi m e^2} = \frac{\varepsilon_0 (6.63\times10^{-34}\ \mathrm{J\cdot s})^2}{\pi(9.11\times10^{-31}\ \mathrm{kg})(1.60\times10^{-19}\ \mathrm{C})^2}$

$\Rightarrow a_1 = 5.29\times10^{-11}$ m

(b) $m_r = \frac{m}{2} \Rightarrow a_2 = 2a_1 = 1.06\times10^{-10}$ m.

(c) $m_r = 186\ \mathrm{m} \Rightarrow a_3 = \frac{1}{186} a_1 = 2.85\times10^{-13}$ m.

43-12: (a) $\frac{e}{2m} n\hbar = \frac{e}{m}\hbar = 2\mu_B = 1.85\times10^{-23}\ \mathrm{A\cdot m^2}$,

(b) $\mu B = (1.855\times10^{-23}\ \mathrm{A\cdot m^2})(2.00\ \mathrm{T}) = 3.71\times10^{-23}\ \mathrm{J} = 2.32\times10^{-4}$ eV.

43-13: $3p \Rightarrow n = 3, l = 1 \quad U = m_l \mu_B B \Rightarrow B = \frac{U}{m_l \mu_B}$.

$$B = \frac{(3.25\times10^{-5}\ \mathrm{eV})}{1(5.788\times10^{-5}\ \mathrm{eV/T})} = 0.562\ \mathrm{T}.$$

43-14: (a) $\Delta E = \mu_B B = 4.63\times10^{-5}$ eV. b) $m_l = -2$, the lowest possible value of m_l.

43-15: (a) *f*-state $\Rightarrow l = 3 \Rightarrow$ # of states is $(2l+1) = 7$ $(m_l = 0, \pm1, \pm2, \pm3)$.

(b) $\Delta U = \mu_B B = (5.788\times10^{-5}\ \mathrm{eV/T})(1.25\ \mathrm{T}) = 7.24\times10^{-5}$ eV

$= 1.16\times10^{-23}$ J.

(c) $\Delta U_{-3+3} = 6\mu_B B = 4.34\times10^{-4}\ \mathrm{eV} = 6.95\times10^{-23}$ J

43-16: (a) $\lambda = \dfrac{hc}{\Delta E} = \dfrac{(4.136\times10^{-15}\ \text{eV}\cdot\text{s})(3.00\times10^{8}\text{m/s})}{(5.9\times10^{-6}\ \text{eV})} = 21.0\ \text{cm}$, a short radio wave.
(b) As in Example 43-7, the effective field is $B = \Delta E/2\mu_{\text{B}} = 5.1\times10^{-2}\ \text{T}$, far smaller than that found in the example.

43-17: (a) Classically $L = I\omega$, and $I = \dfrac{2}{5}mR^2$ for a uniform sphere.

$$\Rightarrow L = \frac{2}{5}m\omega R^2 = \sqrt{\frac{3}{4}}\hbar$$

$$\Rightarrow \omega = \frac{5\hbar}{2mR^2}\sqrt{\frac{3}{4}} = \frac{5(1.054\times10^{-34}\ \text{J}\cdot\text{s})}{2(9.11\times10^{-31}\ \text{kg})(1.0\times10^{-17}\ \text{m})^2}\sqrt{\frac{3}{4}}$$

$$\Rightarrow \omega = 2.5\times10^{30}\ \text{rad/s}.$$

(b) $v = r\omega = (1.0\times10^{-17}\ \text{m})(2.5\times10^{30}\ \text{rad/s}) = 2.5\times10^{13}\ \text{m/s}$ since this is faster than the speed of light this model is invalid.

43-18: (a) See Example 43-6; $\mu_{\text{B}}B = 7.88\times10^{-24}\ \text{J} = 4.92\times10^{-5}\ \text{eV}$. (b) For $n = 1, l = 0$ and there is no orbital angular momentum. If $n \neq 1$, there may or may not be an orbital angular momentum interaction, depending on l.

43-19: $U = -\mu_z B = (2.00232)\dfrac{e}{2m}S_z B$

$$\Rightarrow U = (2.00232)\frac{e}{2m}(m_s\hbar)B$$

$$\Rightarrow U = (2.00232)\mu_{\text{B}}m_s B \quad \text{where } \mu_{\text{B}}\frac{e\hbar}{2m}.$$

So the energy difference is $\Delta U = (2.00232)\mu_{\text{B}}B\left(\dfrac{1}{2} - \left(-\dfrac{1}{2}\right)\right)$

$$\Rightarrow \Delta U = (2.00232)(5.788\times10^{-5}\ \text{eV/T})(0.730\ \text{T})$$
$$= 8.46\times10^{-5}\ \text{eV}.$$

And the lower energy level is $m_s = -\dfrac{1}{2}$.

43-20: The allowed (l, j) combinations are $\left(0, \dfrac{1}{2}\right), \left(1, \dfrac{1}{2}\right), \left(1, \dfrac{3}{2}\right), \left(2, \dfrac{3}{2}\right)$ and $\left(2, \dfrac{5}{2}\right)$.

43-21: j quantum numbers are either $l + \dfrac{1}{2}$ or $l - \dfrac{1}{2}$. So if $j = 5/2$ and $7/2$, then $l = 3$. The letter used to describe $l = 3$ is "f".

43-22: However the number of electrons is obtained, the results must be consistent with Table (43-43-3); adding two more electrons to the zinc configuration gives $1s^2 2s^2 2p^6 3s^2 3p^6 4s^2 3d^{10} 4p^2$.

43-23: The ten lowest energy levels for electron are in the $n=1$ and $n=2$ shells.

$n=1,\ l=0,\ m_l=0,\ m_s=\pm\frac{1}{2}:2$ states.

$n=2,\ l=0,\ m_l=0,\ m_s=\pm\frac{1}{2}:2$ states.

$n=2,\ l=1,\ m_l=0,\pm1,\ m_s=\pm\frac{1}{2}:6$ states.

43-24: For the outer electrons, there are more inner electrons to screen the nucleus.

43-25: Using Eq. (43-27) for the ionization energy: $E_n=\frac{-Z_{\text{eff}}^{\ 2}}{n^2}(13.6\text{ eV})$. The 5*s* electron sees $Z_{\text{eff}}=2.771$ and $n=5$

$\Rightarrow E_5=\frac{-(2.771)^2}{5^2}(13.6\text{ eV})=4.18\text{ eV}.$

43-26: (a) $E_2=-\frac{13.6\text{ eV}}{4}Z_{\text{eff}}^2$, so $Z_{\text{eff}}=1.26$. (b) Similarly, $Z_{\text{eff}}=2.26$. (c) Z_{eff} becomes larger going down the columns in the periodic table.

43-27: (a) Again using $E_n=\frac{-Z_{\text{eff}}^{\ 2}}{n^2}(13.6\text{ eV})$, the outmost electron of the Be^+ L shell ($n=2$) sees the inner two electron shield two proton so $Z_{\text{eff}}=2$.

$\Rightarrow E_2=\frac{-2^2}{2^2}(13.6\text{ eV})=-13.6\text{ eV}.$

(b) For Ca^+, outer shell has $n=4$, so $E_4=\frac{-2^2}{4^2}(13.6\text{ eV})=-3.4\text{ eV}.$

43-28: For the 4*s* state, $E=-4.339\text{ eV}$ and $Z_{\text{eff}}=4\sqrt{(-4.339)/(-13.6)}=2.26$. Similarly, $Z_{\text{eff}}=1.79$ for the 4*p* state and 1.05 for the 4*d* state. The electrons in the states with higher l tend to be further away from the filled subshells and the screening is more complete.

43-29: (a) Nitrogen is the seventh element ($Z=7$). N^{2+} has two electron removed, so there are 5 remaining electron $\Rightarrow$ electron configuration is $1s^2 2s^2 2p$.

(b) $E=\frac{-Z_{\text{eff}}^{\ 2}}{n^2}(13.6\text{ eV})=\frac{-(7-4)^2}{2^2}(13.6\text{ eV})=-30.6\text{ eV}$

(c) Phosphorous is the fifteenth element ($Z=15$). P^{2+} has 13 electrons, so the electron configuration is $1s^2 2s^2 2p^6 3s^2 3p$.

(d) The least tightly held electron: $E=-\frac{(15-12)^2}{3^2}(13.6\text{ eV})=-13.6\text{ eV}.$

43-30: Using Eq. (43-30),

$$Z\approx 1+\sqrt{\frac{14.0\times10^3\text{ eV}}{10.2\text{ eV}}}=38.0,$$

which corresponds to the element strontium (Sr).

43-31: (a) $Z = 16$: $f = (2.48\times10^{15}\ \text{Hz})(15)^2 = 5.58\times10^{17}\ \text{Hz}$

$E = hf = 3.70\times10^{-16}\ \text{J} = 2.31\ \text{keV}$

$$\lambda = \frac{c}{f} = \frac{3.00\times10^{8}\ \text{m/s}}{5.58\times10^{17}\ \text{Hz}} = 5.38\times10^{-10}\ \text{m}.$$

(b) $Z = 24$: $f = 1.31\times10^{18}\ \text{Hz}$

$E = 5.43\ \text{keV}$

$\lambda = 2.29\times10^{-10}\ \text{m}$

(c) $Z = 41$: $f = 3.97\times10^{18}\ \text{Hz},\ E = 16.4\ \text{keV},\ \lambda = 7.56\times10^{-11}\ \text{m}$

43-32: (a) For large values of n, the inner electrons will completely shield the nucleus, so $Z_{\text{eff}} = 1$ and the ionization energy would be $\frac{13.60\ \text{eV}}{n^2}$.

(b) $\frac{13.60\ \text{eV}}{290^2} = 1.62\times10^{-4}\ \text{eV},\ r_{290} = (290)^2 a_0 = 4.45\times10^{-6}\ \text{m}$. (c) Similarly for $n = 732,\ -E_n = 2.54\times10^{-5}\ \text{eV},\ r_n = 2.83\times10^{-5}\ \text{m}$.

43-33: (a) $\theta_L = \arccos\left(\frac{L_z}{L}\right)$. This is smaller for L_z and L as large as possible. Thus $l = n-1$ and $m_l = l = n-1$

$$\Rightarrow L_z = m_l\,\hbar = (n-1)\hbar \text{ and } L = \sqrt{l(l+1)}\,\hbar = \sqrt{(n-1)n}\,\hbar$$

$$\Rightarrow \theta_L = \arccos\left(\frac{n-1}{\sqrt{n(n-1)}}\right).$$

(b) The largest angle implies $l = n-1,\ m_l = -l = -(n-1)$

$$\Rightarrow \theta_L = \arccos\left[\frac{-(n-1)}{\sqrt{n(n-1)}}\right]$$

$$= \arccos\left[-\sqrt{(1-1/n)}\right]$$

43-34: (a)

$$L_x^2 + L_y^2 = L^2 - L_z^2 = l(l+1)\hbar^2 - m_l^2\hbar^2$$

so $\sqrt{L_x^2 + L_y^2} = \sqrt{l(l+1) - m_l^2}\,\hbar$. (b) This is the magnitude of the component of angular momentum perpendicular to the z-axis. (c) The maximum value is $\sqrt{l(l+1)}\,\hbar = L$, when $m_l = 0$. That is, if the electron is known to have no z-component of angular momentum, the angular momentum must be perpendicular to the z-axis. (d) If $m_l = l$, $\sqrt{L_x^2 + L_y^2}\,\hbar = \sqrt{l}\,\hbar$; the perpendicular component of angular momentum will not vanish if $l \neq 0$.

43-35: (a) $E_{1s} = -\frac{1}{(4\pi\varepsilon_0)^2}\frac{me^4}{2\hbar^2}$ and $U(r) = \frac{-1}{4\pi\varepsilon_0}\frac{e^2}{r}$

If $E_{1s} = U(r)$, then $\frac{1}{(4\pi\varepsilon_0)^2}\frac{me^2}{2\hbar^2} = \frac{+1}{(4\pi\varepsilon_0)}\frac{e^2}{r}$

$$\Rightarrow r = \frac{4\pi\varepsilon_0 2\hbar^2}{me^2} = 2a$$

(b) $P(r>2a)=\int_{2a}^{\infty}|\psi_{1s}|^2 dV = 4\pi\int_{2a}^{\infty}|\psi_{1s}|^2 r^2 dr$ and ψ is $=\frac{1}{\sqrt{\pi a^3}}e^{-r/a}$

$$\Rightarrow P(r>2a)=\frac{4}{a^3}\int_{2a}^{\infty} r^2 e^{-2r/a}dr$$

$$\Rightarrow P(r>2a)=-\frac{4}{a^3}\left[e^{-2r/a}\left(\frac{ar^2}{2}+\frac{a^2r}{2}+\frac{a^3}{4}\right)\right]_{2a}^{\infty}$$

$$\Rightarrow P(r>2a)=\frac{4}{a^3}e^{-4}\left(2a^3+a^3+\frac{a^3}{4}\right)$$

$$=13e^{-4}=0.238$$

43-36: See Example 43-3; $r^2|\psi|^2 = Cr^2e^{-2r/a}$, $\frac{dr^2|\psi|^2}{dr}=Ce^{-2r/a}(2r-(2r^2/a))$, and for a maximum, $r=a$, the distance of the electron from the nucleus in the Bohr model.

43-37: (a) If normalized, then $\int_0^{\infty}|\psi_{2s}|^2 dV = 4\pi\int_0^{\infty}|\psi_{2s}|^2 r^2 dr = 1$

$$\Rightarrow I=\int_0^{\infty}\frac{4\pi}{32\pi a^3}r^2\left(2-\frac{r}{a}\right)^2 e^{-r/a}dr$$

$$=\frac{1}{8a^3}\int_0^{\infty}\left(4r^2-\frac{4r^3}{a}+\frac{r^4}{a^2}\right)e^{-r/a}dr$$

But recall $\int_0^{\infty}x^n e^{-\alpha x}dx=\frac{n!}{\alpha^{n+1}}$.

So $I=\frac{1}{8a^3}\left[4(2)a^3-\frac{4}{a}(6)a^4+\frac{1}{a^2}(24)a^5\right]$

$\Rightarrow I=\frac{1}{8a^3}\left[8a^3-24a^3+24a^3\right]=1$ and ψ_{2s} is normalized.

(b) We carry out the same calculation as part (a) except now the upper limit on the integral is $4a$, not infinity.

So $I=\frac{1}{8a^3}\int_0^{4a}\left(4r^2-4\frac{r^3}{a}+\frac{r^4}{a^2}\right)e^{-r/a}dr$

Now the necessary integral formulas are:

$$\int r^2e^{-r/a}dr=-e^{-r/a}(r^2a+2ra^2+2a^3)$$

$$\int r^3e^{-r/a}dr=-e^{-r/a}(r^3a+3r^2a^2+6ra^3+6a^3)$$

$$\int r^4e^{-r/a}dr=-e^{-r/a}(r^4a+4r^3a^2+12r^2a^3+24ra^4+24a^5)$$

All the integrals are evaluated at the limits $r=0$ and $4a$. After carefully plugging in the limits and collecting like terms we have:

$$I=\frac{1}{8a^3}\cdot a^3[(8-24+24)+e^{-4}(-104+568-824)]$$

$$\Rightarrow I=\frac{1}{8}(8-360e^{-4})=0.176=\text{Prob}(r<4a)$$

43-38: (a) Since the given $\psi(r)$ is real, $r^2|\psi|^2 = r^2\psi^2$. The probability density will be an extreme when

$$\frac{d}{dr}(r^2\psi^2) = 2\left(r\psi^2 + r^2\psi\frac{d\psi}{dr}\right) = 2r\psi\left(\psi + r\frac{d\psi}{dr}\right) = 0.$$

This occurs at $r = 0$, a minimum, and when $\psi = 0$, also a minimum. A maximum must correspond to $\psi + r\frac{d\psi}{dr} = 0$. Within a multiplicative constant,

$$\psi(r) = (2 - r/a)e^{-r/2a}, \quad \frac{d\psi}{dr} = -\frac{1}{a}(2 - r/2a)e^{-r/2a},$$

and the condition for a maximum is

$$(2 - r/a) = (r/a)(2 - r/2a), \quad \text{or} \quad r^2 - 6ra + 4a^2 = 0.$$

The solutions to the quadratic are $r = a(3 \pm \sqrt{5})$. The ratio of the probability densities at these radii is 3.68, with the larger density at $r = a(3 + \sqrt{5})$. b) $\psi = 0$ at $r = 2a$. Parts (a) and (b) are consistent with Fig. (43-2); note the two relative maxima, one on each side of the minimum of zero at $r = 2a$.

43-39: (a) The photon energy equals the atom's transition energy. The hydrogen atom decays from $n = 2$ to $n = 1$, so:

$$\Delta E = -13.60 \text{ eV}\left(\frac{1}{(2)^2} - \frac{1}{(1)^2}\right) = (10.2 \text{ eV})(1.60 \times 10^{-19} \text{ J/eV})$$

$$= 1.63 \times 10^{-18} \text{ J}$$

$$\Rightarrow \lambda = \frac{hc}{\Delta E} = \frac{(6.63 \times 10^{-34} \text{ J}\cdot\text{s})(3.00 \times 10^8 \text{ m/s})}{1.63 \times 10^{-18} \text{ J}} = 1.22 \times 10^{-7} \text{ m}$$

(b) The change in an energy level due to an external magnetic field is just $U = m_l \mu_B B$. The ground state has $m_l = 0$, and it is not shifted. The $n = 2$ state has $m_l = -1$, so it is shifted by

$$U = (-1)(9.274 \times 10^{-24} \text{ J/T})(1.75 \text{ T})$$

$$= -1.62 \times 10^{-23} \text{ J}$$

Problem 41-34(c) tells us that $\frac{\Delta\lambda}{\lambda} = \frac{\Delta E}{E}$

$$\Rightarrow \Delta\lambda = \lambda\frac{\Delta E}{E} = (1.22 \times 10^{-7} \text{ m})\left(\frac{1.62 \times 10^{-23} \text{ J}}{1.63 \times 10^{-18} \text{ J}}\right) = 1.2 \times 10^{-12} \text{ m}$$

Since the $n = 2$ level is lowered in energy (brought closes to the $n = 1$ level) the change in energy is less, and the photon wavelength increases due to the magnetic field.

43-40: (a) $Z^2(-13.6 \text{ eV}) = -1.10 \text{ keV}$. (b) The negative of the result of part (a), 1.10 keV. (c) The radius of the ground state orbit is inversely proportional to the nuclear charge, and $\frac{a}{Z} = 5.88 \times 10^{-12}$ m. (d) $\lambda = \frac{hc}{\Delta E} = \frac{(4/3)hc}{E_0}$, where E_0 is the energy found in part (b), and $\lambda = 1.50$ nm.

43-41: Decay from $3d$ to $2p$ state in hydrogen means that $n = 3 \rightarrow n = 2$ and $m_l = \pm 2, \pm 1, 0 \rightarrow m_l = \pm 1, 0$. However selection rules limit the possibilities for decay. The emitted photon carries off one unit of angular momentum so l must change by 1 and hence m_l must change by 0 or ± 1. The shift in the transition energy from the zero field value is just

$$U = (m_{l_3} - m_{l_2})\mu_B B = \frac{e\hbar B}{2m}(m_{l_3} - m_{l_2})$$

where m_{l_3} is the $3d$ m_l value and m_{l_2} is the $2p$ m_l value. Thus there are only three different energy shifts. They and the transitions that have them, labeled by m are:

$$\frac{e\hbar B}{2m} : 2 \rightarrow 1, \quad 1 \rightarrow 0, \quad 0 \rightarrow -1$$

$$0 : 1 \rightarrow 1, \quad 0 \rightarrow 0, \quad -1 \rightarrow -1$$

$$-\frac{e\hbar B}{2m} : 0 \rightarrow 1, \quad -1 \rightarrow 0, \quad -2 \rightarrow -1$$

43-42: (a) See Problem 43-18: For $m_l = 2$, 2.55×10^{-4} eV. For $m_l = 1$, 1.27×10^{-4} eV.

(b) See Problem 41-34; $|\Delta\lambda| = \lambda_0 \dfrac{|\Delta E|}{E_0}$, where $E_0 = (13.6 \text{ eV})((1/4) - (1/9))$ and

$\lambda_0 = \left(\dfrac{36}{5}\right)\dfrac{1}{R} = 6.563 \times 10^{-7}$ m. Then, $|\Delta\lambda| = 4.42 \times 10^{-11}$ m $= 0.0442$ nm. The wavelength corresponds to a larger energy change, and so the wavelength is smaller.

43-43: From section 40-7: $\dfrac{n_1}{n_0} = e^{-(E_1 - E_0)/kT}$ We need to know the difference in energy between the $m_s = +\dfrac{1}{2}$ and $m_s = -\dfrac{1}{2}$ states. $U = -\mu_z B = 2.00232\, \mu_B m_s B$

So $U_{\frac{1}{2}} - U_{-\frac{1}{2}} = 2.00232\, \mu_B B$

$$\Rightarrow \frac{n_{1/2}}{n_{-1/2}} = e^{-(2.00232)\mu_B B/kT}$$

$$= e^{-(2.00232)(9.274\times 10^{-24}\ \text{J/T})B/(1.381\times 10^{-23}\ \text{J/K})(300\ \text{K})}$$

$$= e^{-(4.482\times 10^{-3}\ \text{T}^{-1})B}$$

(a) $B = 5.00 \times 10^{-5}$ T $\Rightarrow \dfrac{n_{1/2}}{n_{-1/2}} = 0.9999998.$

(b) $B = 0.500$ T $\Rightarrow \dfrac{n_{1/2}}{n_{-1/2}} = 0.9978.$

(c) $B = 5.00$ T $\Rightarrow \dfrac{n_{1/2}}{n_{-1/2}} = 0.978.$

43-44: $\dfrac{e\hbar}{m}B = \dfrac{hc}{\lambda}$, or $B = \dfrac{2\pi cm}{\lambda e} = 0.397$ T.

43-45: (a) To calculate the total number of states for the nth principle quantum number shall we must multiply all the possibilities. The spin states multiply everything by 2. The maximum l value is $(n-1)$, and each l value has $(2l+1\ m_l)$ values.

So the total number of states is

$$N = 2\sum_{l=0}^{n-1}(2l+1) - 2\sum_{l=0}^{n-1}1 + 4\sum_{l=0}^{n-1}l$$

$$= 2n + \frac{4(n-1)(n)}{2} = 2n + 2n^2 - 2n$$

$$N = 2n^2$$

(b) The $n = 5$ shell (O-shell) has 50 states.

43-46: The effective field is that which gives rise to the observed difference in the energy level transition,

$$B = \frac{\Delta E}{\mu_B} = \frac{hc}{\mu_B}\left(\frac{\lambda_1 - \lambda_2}{\lambda_1 \lambda_2}\right) = \frac{2\pi mc}{e}\left(\frac{\lambda_1 - \lambda_2}{\lambda_1 \lambda_2}\right).$$

Substitution of numerical values gives $B = 3.64 \times 10^{-3}$ T, much smaller than that for sodium.

43-47: (a) The minimum wavelength means the largest transition energy. The highest shell is $n = 4$, and the transition is to $n = 1$. An approximation is to ignore the energy of the $n = 4$ level compared to $n = 1$ (there's a factor of 16 difference). Then $\Delta E = E_1 = (Z-1)^2(13.6 \text{ eV})$. For calcium, $Z = 20$

$$\Rightarrow \Delta E = 4.9 \times 10^3 \text{ eV} = 7.9 \times 10^{-16} \text{ J}$$

$$\Rightarrow \lambda = \frac{hc}{\Delta E} = \frac{(6.63 \times 10^{-34} \text{ J}\cdot\text{s})(3.00 \times 10^8 \text{ m/s})}{7.9 \times 10^{-16} \text{ J}} = 2.5 \times 10^{-10} \text{ m}$$

For the longest wavelength, we need the smallest transition energy, so this is the $n = 2 \rightarrow n = 1$ transition (K_α). So we use Moseley's Law:

$$f = (2.48 \times 10^{15} \text{ Hz})(20-1)^2 = 9.0 \times 10^{17} \text{ Hz}$$

$$\Rightarrow \lambda = \frac{c}{f} = 3.4 \times 10^{-10} \text{ m}.$$

(b) The niobium, $z = 41$, the minimum wavelength is

$$\lambda = \frac{hc}{(z-1)^2(13.6 \text{ eV})} = \frac{(6.63 \times 10^{-34} \text{ J}\cdot\text{s})(3.00 \times 10^8 \text{ m/s})}{(40)^2(13.6 \text{ eV})(1.60 \times 10^{-19} \text{ J/eV})}$$

$$\Rightarrow \lambda = 5.7 \times 10^{-11} \text{ m}$$

The maximum wavelength is $\lambda = \frac{c}{f} = \frac{(13.00 \times 10^8 \text{ m/s})}{(2.48 \times 10^{15} \text{ Hz})(41-1)^2}$

$$\Rightarrow \lambda = 7.6 \times 10^{-11} \text{ m}$$

43-48: (a) The radius is inversely proportional to Z, so the classical turning radius is $2a/Z$. (b) The normalized wave function is

$$\psi_{1s}(r) = \frac{1}{\sqrt{\pi a^3/Z^3}} e^{-Zr/a}$$

and the probability of the electron being found outside the classical turning point is

$$P = \int_{2a/Z}^{\infty} |\psi_{1s}|^2 4\pi r^2 dr = \frac{4}{a^3/Z^3} \int_{2a/Z}^{\infty} e^{-2Zr/a} r^2 dr .$$

Making the change of variable $u = Zr/a$, $dr = (a/Z)\, du$ changes the integral to

$$P = 4\int_{2}^{\infty} e^{-2u} u^2 du ,$$

which is independent of Z. The probability is that found in Problem 43-35, 0.238, independent of Z.

Chapter 44: Molecules and Condensed Matter

44-1: (a) $K = \frac{3}{2}kT \Rightarrow T = \frac{2K}{3k} = \frac{2(7.9\times10^{-4}\ \text{eV})(1.60\times10^{-19}\ \text{J/eV})}{3(1.38\times10^{-23}\ \text{J/K})}$

$\Rightarrow T = 6.1\ \text{K}$

(b) $T = \frac{2(4.48\ \text{eV})(1.60\times10^{-19}\ \text{J/eV})}{3(1.38\times10^{-23}\ \text{J/K})} = 34{,}600\ \text{K}.$

44-2: (a) $U = -\frac{1}{4\pi\varepsilon_0}\frac{e^2}{r} = -5.0\ \text{eV}$. (b) $-5.0\ \text{eV} + (4.3\ \text{eV} - 3.5\ \text{eV}) = -4.2\ \text{eV}$.

44-3: (a) $E_l = \frac{l(l+1)\hbar^2}{2I}$

$E_{l-1} = \frac{l(l-1)\hbar^2}{2I}$

$\Rightarrow \Delta E = \frac{\hbar^2}{2I}(l^2 + l - l^2 + l) = \frac{l\hbar^2}{I}$

(b) $f = \frac{\Delta E}{h} = \frac{\Delta E}{2\pi\hbar} = \frac{l\hbar}{2\pi I}$

44-4: Each atom has a mass m and is at a distance $L/2$ from the center, so the moment of inertia is $2(m)(L/2)^2 = mL^2/2 = 1.38\times10^{-46}\ \text{kg}\cdot\text{m}^2$.

44-5: (a) $I = m_r r^2 = \left(\frac{m_{\text{Li}}m_{\text{H}}}{m_{\text{Li}} + m_{\text{H}}}\right)r^2 = \frac{(1.17\times10^{-26}\ \text{kg})(1.67\times10^{-27}\ \text{kg})(1.59\times10^{-10}\ \text{m})^2}{(1.17\times10^{-26}\ \text{kg} + 1.67\times10^{-27}\ \text{kg})}$

$= 3.69\times10^{-47}\ \text{kg}\cdot\text{m}^2$

$\Delta E = E_3 - E_2 = \frac{\hbar^2}{2I}(3(3+1) - (2)(2+1)) = \frac{3\hbar^2}{I}$

$\Rightarrow \Delta E = \frac{3(1.054\times10^{-34}\ \text{J}\cdot\text{s})^2}{3.69\times10^{-47}\ \text{kg}\cdot\text{m}^2}$

$\Rightarrow \Delta E = 9.02\times10^{-22}\ \text{J} = 5.64\times10^{-3}\ \text{eV}$

(b) $\lambda = \frac{hc}{\Delta E} = \frac{(6.63\times10^{-34}\ \text{J}\cdot\text{s})(3.00\times10^{8}\ \text{m/s})}{9.02\times10^{-22}\ \text{J}}$

$\Rightarrow \lambda = 2.20\times10^{-4}\ \text{m}$

44-6: $\hbar\omega = \hbar\sqrt{k'/m_r} = \hbar\sqrt{2k'/m} = 3.14\times10^{-20}\ \text{J} = 0.196\ \text{eV}$, where $m_r = m/2$ has been used.

44-7: (a) $f = \dfrac{\omega}{2\pi} = \dfrac{1}{2\pi}\sqrt{\dfrac{k'}{m_r}}$

$$\Rightarrow k' = m_r(2\pi f)^2 = \frac{m_1 m_2 (2\pi f)^2}{m_1 + m_2}$$

$$\Rightarrow k' = \frac{(1.67\times10^{-27}\ \text{kg})(3.15\times10^{-26}\ \text{kg})[2\pi(1.24\times10^{14}\ \text{Hz})]^2}{(1.67\times10^{-27}\ \text{kg} + 3.15\times10^{-26}\ \text{kg})}$$

$$\Rightarrow k' = 963\ \text{N/m}$$

(b) $\Delta E = \left(n + \dfrac{3}{2}\right)\hbar\omega - \left(n + \dfrac{1}{2}\right)\hbar\omega = \hbar\omega = hf$

$$\Rightarrow \Delta E = (6.63\times10^{-34}\ \text{J}\cdot\text{s})(1.24\times10^{14}\ \text{Hz}) = 8.22\times10^{-20}\ \text{J} = 0.513\ \text{eV}$$

(c) $\lambda = \dfrac{c}{f} = \dfrac{3.00\times10^{8}\ \text{m/s}}{1.24\times10^{14}\ \text{Hz}} = 2.42\times10^{-6}\ \text{m}$ (infrared)

44-8: (a) $\Delta E = \dfrac{hc}{\lambda} = \hbar\sqrt{k'/m_r}$, and solving for k',

$$k' = \left(\frac{2\pi c}{\lambda}\right)^2 m_r = 205\ \text{N/m}.$$

(b) $\Delta l = \pm1$, and so such a transition, with $\Delta l = 0$, is forbidden.

44-9: Energy levels are $E = E_n + E_l = \left(n + \dfrac{1}{2}\right)\hbar\omega + l(l+1)\dfrac{\hbar^2}{2I}$

$$= (n + \frac{1}{2})(0.269\ \text{eV}) + l(l+1)(2.395\times10^{-4}\ \text{eV})$$

where the values are from Example 44-2, 44-3.

(a) $n = 0,\ l = 1 \rightarrow n = 1,\ l = 2$:

$$\Rightarrow \Delta E = E_f - E_i = (1)(0.2690\ \text{eV}) + (4)(2.395\times10^{-4}\ \text{eV}) = 0.2700\ \text{eV}$$

$$\Rightarrow \lambda = \frac{hc}{\Delta E} = \frac{(6.626\times10^{-34}\ \text{J}\cdot\text{s})(2.998\times10^{8}\ \text{m/s})}{(0.2700\ \text{eV})(1.602\times10^{-19}\ \text{J/eV})} = 4.592\times10^{-6}\ \text{m}.$$

(b) $n = 0,\ l = 2 \rightarrow n = 1,\ l = 1$:

$$\Rightarrow \Delta E = E_f - E_i = (1)(0.2690\ \text{eV}) + (-4)(2.395\times10^{-4}\ \text{eV}) = 0.2680\ \text{eV}$$

$$\Rightarrow \lambda = \frac{hc}{\Delta E} = \frac{(6.626\times10^{-34}\ \text{J}\cdot\text{s})(2.998\times10^{8}\ \text{m/s})}{(0.2680\ \text{eV})(1.602\times10^{-19}\ \text{J/eV})}$$

$$\Rightarrow \lambda = 4.627\times10^{-6}\ \text{m}$$

(c) $n = 0,\ l = 3 \rightarrow n = 1,\ l = 2$:

$$\Delta E = E_f - E_i = (1)(0.2690\ \text{eV}) + (-6)(2.395\times10^{-4}\ \text{eV}) = 0.2676\ \text{eV}$$

$$\Rightarrow \lambda = \frac{hc}{\Delta E} = \frac{(6.626\times10^{-34}\ \text{J}\cdot\text{s})(2.998\times10^{8}\ \text{m/s})}{(0.2676\ \text{eV})(1.602\times10^{-19}\ \text{J/eV})} = 4.634\times10^{-6}\ \text{m}$$

44-10: (a) $\lambda = \dfrac{hc}{E} = 0.100\text{ nm}$, (b) $\lambda = \dfrac{h}{p} = \dfrac{h}{\sqrt{2mE}} = 0.100\text{ nm}$ and

(c) $\lambda = \dfrac{h}{\sqrt{2mE}} = 0.100\text{ nm}$.

44-11: The volume enclosing a single sodium and chlorine atom $= 2(2.82\times10^{-10}\text{ m})^3 = 4.49\times10^{-29}\text{ m}^3$. So the density

$$\rho = \frac{m_{\text{Na}} + m_{\text{Cl}}}{V} = \frac{3.82\times10^{-26}\text{ kg} + 5.89\times10^{-26}\text{ kg}}{4.49\times10^{-29}\text{ m}^3}$$

$$\Rightarrow \rho = 2.16\times10^3\text{ kg/m}^3$$

44-12: For an average spacing a, the density is $\rho = m/a^3$, where m is the average of the ionic masses, and so

$$a^3 = \frac{m}{\rho} = \frac{(6.49\times10^{-26}\text{ kg} + 1.33\times10^{-25}\text{ kg})/2}{(2.75\times10^3\text{ kg/m}^3)} = 3.60\times10^{-29}\text{ m}^3,$$

and $a = 3.30\times10^{-10}\text{ m} = 0.330\text{ nm}$. b) The larger (higher atomic number) atoms have the larger spacing.

44-13: To be detected the photon must have enough energy to bridge the gap width $\Delta E = 1.12\text{ eV}$

$$\Rightarrow \lambda = \frac{hc}{\Delta E} = \frac{(6.63\times10^{-34}\text{ J}\cdot\text{s})(3.00\times10^8\text{ m/s})}{(1.12\text{ eV})(1.60\times10^{-19}\text{ J/eV})} = 1.11\times10^{-6}\text{ m}$$

44-14: $\dfrac{hc}{\Delta E} = 2.27\times10^{-7}\text{ m} = 227\text{ nm}$, in the ultraviolet.

44-15: $\Delta E = \dfrac{hc}{\lambda} = \dfrac{(6.63\times10^{-34}\text{ J}\cdot\text{s})(3.00\times10^8\text{ m/s})}{6.67\times10^{-12}\text{ m}} = 2.98\times10^{-14}\text{ J}$

$= 1.86\times10^5\text{ eV}$

So the number of electrons that can be excited to the conduction band is

$$n = \frac{1.86\times10^5\text{ eV}}{1.12\text{ eV}} = 1.66\times10^5\text{ electrons}$$

44-16: $1 = \int |\psi|^2 dV$

$$= A^2\left(\int_0^L \sin^2\left(\frac{n_x\pi x}{L}\right)dx\right)\left(\int_0^L \sin^2\left(\frac{n_y\pi y}{L}\right)dy\right)\left(\int_0^L \sin^2\left(\frac{n_z\pi z}{L}\right)dz\right)$$

$$= A^2\left(\frac{L}{2}\right)^3,$$

so $A = (2/L)^{3/2}$ (assuming A to be real and positive).

44-17: Density of states:

$$g(E) = \frac{(2m)^{3/2}V}{2\pi^2\hbar^3}E^{1/2}$$

$$\Rightarrow g(E) = \frac{(2(9.11\times10^{-31}\text{ kg}))^{3/2}(1.0\times10^{-6}\text{ m}^3)(7.0\text{ eV})^{1/2}(1.60\times10^{-19}\text{ J/eV})^{1/2}}{2\pi^2(1.054\times10^{-34}\text{ J}\cdot\text{s})^3}$$

$$= (1.13\times10^{41}\text{ states/J})(1.60\times10^{-19}\text{ J/eV})$$

$$= 1.80\times10^{22}\text{ states/eV})$$

44-18: $v_{\text{rms}} = \sqrt{3kT/m} = 1.17\times10^5$ m/s, as found in Example 44-9. The equipartition theorem does not hold for the electrons at the Fermi energy. Although these electrons are very energetic, they cannot lose energy, unlike electrons in a free electron gas.

44-19: (a) $\frac{-\hbar^2}{2m}\left(\frac{\partial^2\psi}{\partial x^2}+\frac{\partial^2\psi}{\partial y^2}+\frac{\partial^2\psi}{\partial z^2}\right) = E\psi$

where $\psi = A\sin\frac{n_x\pi x}{L}\sin\frac{n_y\pi y}{L}\sin\frac{n_z\pi z}{L}$

$\frac{\partial^2\psi}{\partial x^2} = -\left(\frac{n_x\pi}{L}\right)^2\psi$, and similarly for $\frac{\partial^2\psi}{\partial y^2}$ and $\frac{\partial^2\psi}{\partial z^2}$.

$$\Rightarrow -\frac{\hbar^2}{2m}\left(-\left(\frac{n_x\pi}{L}\right)^2-\left(\frac{n_y\pi}{L}\right)^2-\left(\frac{n_z\pi}{L}\right)^2\right)\psi = E\psi$$

$$\Rightarrow E = \frac{(n_x{}^2+n_y{}^2+n_z{}^2)\pi^2\hbar^2}{2mL^2}$$

(b) Ground state $\Rightarrow n_x = n_y = n_z = 1 \Rightarrow E = \frac{3\pi^2\hbar^2}{2mL^2}$,

The only degeneracy is from the two spin states.

The first excited state $\Rightarrow$ (2, 1, 1) or (1, 2, 1) or (1, 1, 2)

$$\Rightarrow E = \frac{3\pi^2\hbar^2}{mL^2}$$

and the degeneracy is $(2)\times(3) = 6$.

2nd excited state $\Rightarrow$ (2, 2, 1) or (2, 1, 2) or (1, 2, 2)

$$\Rightarrow E = \frac{9\pi^2\hbar^2}{2mL^2}$$

and the degeneracy is $(2)\times(3) = 6$.

44-20: Equation (44-13) may be solved for $n_{\text{rs}} = (2mE)^{1/2}(L/\hbar\pi)$, and substituting this into Eq. (44-12), using $L^3 = V$, gives Eq. (44-14).

44-21: Eq (44-13): $E = \frac{n_{rs}^2 \pi^2 \hbar^2}{2mL^2}$

$\Rightarrow n_{rs} = \frac{L}{\pi\hbar}\sqrt{2mE}$

$= \frac{0.010\ \text{m}}{\pi(1.054\times10^{-34}\ \text{J}\cdot\text{s})}\sqrt{2(9.11\times10^{-31}\ \text{kg})(5.0\ \text{eV})(1.60\times10^{-19}\ \text{J/eV})}$

$\Rightarrow n_{rs} = 3.6\times10^{7}.$

44-22: (a) From Eq. (44-22), $E_{av} = \frac{3}{5}E_F = 3.29\ \text{eV}$. (b) $\sqrt{2E/m} = 1.08\times10^{6}\ \text{m/s}$.

(c) $\frac{E_F}{k} = 6.36\times10^{4}\ \text{K}$.

44-23: $C = \left(\frac{\pi^2 kT}{2E_F}\right)R = \left(\frac{\pi^2(1.38\times10^{-23}\ \text{J/K})(300\ \text{K})}{2(3.23\ \text{eV})(1.60\times10^{-19}\ \text{J/eV})}\right)R$

$\Rightarrow C = 0.0395\,R$

44-24: (a) See Example 44-10: The probabilities are 1.78×10^{-7}, 2.37×10^{-6} and 1.51×10^{-5}. b) The Fermi distribution, Eq. (44-17), has the property that $f(E_F - E) = 1 - f(E)$ (see Problem 44-42), and so the probability that a state at the top of the valence band is occupied is the same as the probability that a state of the bottom of the conduction band is filled (this result depends on having the Fermi energy in the middle of the gap).

44-25: $f(E) = \frac{1}{e^{(E-E_F)/kT}+1}$

$\Rightarrow E_F = E - kT\ \ln\left(\frac{1}{f(E)} - 1\right)$

$E_F = E - (1.38\times10^{-23}\ \text{J/K})(300\ \text{K})\ln\left(\frac{1}{4.4\times10^{-4}} - 1\right)$

$E_F = E - (3.20\times10^{-20}\ \text{J}) = E - 0.20\ \text{eV}$

So the Fermi level is 0.20 eV below the conduction band.

44-26: (a) Solving Eq. (44-23) for the voltage as a function of current,

$$V = \frac{kT}{e}\ln\left(\frac{I}{I_S} + 1\right) = \frac{kT}{e}\ln\left(\frac{40.0\ \text{mA}}{3.60\ \text{mA}} + 1\right) = 0.0645\ \text{V}.$$

(b) From part (a), the quantity $e^{eV/kT} = 12.11$, so for a reverse-bias voltage of the same magnitude,

$$I = I_S\left(e^{-eV/kT} - 1\right) = I_S\left(\frac{1}{12.11} - 1\right) = -3.30\ \text{mA}.$$

44-27: $I = I_S(e^{eV/kT} - 1) \Rightarrow I_S = \dfrac{I}{e^{eV/kT} - 1}$

(a) $\dfrac{eV}{kT} = \dfrac{(1.60\times10^{-19}\ \text{C})(1.50\times10^{-2}\ \text{V})}{(1.38\times10^{-23}\ \text{J/K})(300\ \text{K})} = 0.580$

$\Rightarrow I_S = \dfrac{9.25\times10^{-3}\ \text{A}}{e^{0.580} - 1} = 0.0118\ \text{A}$

Now for $V = 0.0100\ \text{V}$, $eV/kT = 0.387$

$\Rightarrow I = (0.0118\ \text{A})(e^{+0.387} - 1) = 5.56\times10^{-3}\ \text{A} = 5.56\ \text{mA}$

(b) Now with $V = -15.0\ \text{mV}$, $\dfrac{eV}{kT} = -0.580$

$\Rightarrow I = (0.0118\ \text{A})(e^{-0.580} - 1) = -5.18\times10^{-3}\ \text{A}$

If $V = -10.0\ \text{mV} \Rightarrow \dfrac{eV}{kT} = -0.387$

$\Rightarrow I = (0.0118\ \text{A})(e^{-0.387} - 1) = -3.77\times10^{-3}\ \text{A}.$

44-28: The electrical potential energy is $U = -5.13\ \text{eV}$, and

$r = -\dfrac{1}{4\pi\varepsilon_0}\dfrac{e^2}{U} = 2.8\times10^{-10}\ \text{m}.$

44-29: (a) For maximum separation of Na^+ and Cl^- for stability:

$U = \dfrac{-e^2}{4\pi\varepsilon_0 r} = -(5.1\ \text{eV} - 3.6\ \text{eV})(1.60\times10^{-19}\ \text{J/eV}) = -2.40\times10^{-19}\ \text{J}$

$\Rightarrow r = \dfrac{(1.60\times10^{-19}\ \text{C})^2}{4\pi\varepsilon_0(2.40\times10^{-19}\ \text{J})} = 9.6\times10^{-10}\ \text{m}.$

(b) For K^+ and Br^-:

$U = -\dfrac{e^2}{4\pi\varepsilon_0 r} = -(4.3\ \text{eV} - 3.5\ \text{eV})(1.60\times10^{-19}\ \text{J/eV}) = -1.28\times10^{-19}\ \text{J}$

$\Rightarrow r = \dfrac{(1.60\times10^{-19}\ \text{C})^2}{4\pi\varepsilon_0(1.28\times10^{-19}\ \text{J})} = 1.8\times10^{-9}\ \text{m}.$

44-30: See Problem (44-3): $I = \dfrac{2\hbar^2}{\Delta E} = \dfrac{h\lambda}{2\pi^2 c} = 7.14\times10^{-48}\ \text{kg}\cdot\text{m}^2.$

44-31: (a) $p = qd = (1.60\times10^{-19}\ \text{C})(2.4\times10^{-10}\ \text{m}) = 3.8\times10^{-19}\ \text{C}\cdot\text{m}$

(b) $q = \dfrac{p}{d} = \dfrac{3.0\times10^{-29}\ \text{C}\cdot\text{m}}{2.4\times10^{-10}\ \text{m}} = 1.3\times10^{-19}\ \text{C}$

(c) $q/e = 0.78$

(d) $q = \dfrac{p}{d} = \dfrac{1.5\times10^{-30}\ \text{C}\cdot\text{m}}{1.6\times10^{-10}\ \text{m}} = 9.4\times10^{-21}\ \text{C}$

$\Rightarrow q/e = 0.059$

This is much less than for sodium chloride (part (c)). Therefore the bond for hydrogen iodide is more covalent in nature than ionic.

44-32: From the result of Problem 44-3, the moment of inertia of the molecule is

$$I = \frac{\hbar^2 l}{\Delta E} = \frac{hl\lambda}{4\pi^2 c} = 6.43 \times 10^{-46}\ \text{kg}\cdot\text{m}^2,$$

and from Eq. (44-6) the separation is

$$r_0 = \sqrt{\frac{I}{m_r}} = 0.193\ \text{nm}.$$

44-33: $E_n = \left(n + \frac{1}{2}\right)\hbar\sqrt{\frac{k'}{m_r}} \Rightarrow E_0 = \frac{1}{2}\hbar\sqrt{\frac{2k'}{m_H}}$

$$\Rightarrow E_0 = \frac{1}{2}(1.054 \times 10^{-34}\ \text{J}\cdot\text{s})\sqrt{\frac{2(576\ \text{N/m})}{1.67 \times 10^{-27}\ \text{kg}}} = 4.38 \times 10^{-20}\ \text{J} = 0.274\ \text{eV}$$

This is much less than the H_2 bond energy.

44-34: The vibration frequency is, from Eq. (44-8), $f = \frac{\Delta E}{h} = 1.12 \times 10^{14}$ Hz. The force constant is

$$k' = (2\pi f)^2 m_r = 777\ \text{N/m}.$$

44-35: (a)

$$I = m_r r^2 = \frac{m_{Na} m_{Cl}}{m_{Na} + m_{Cl}} r^2 = \frac{(3.8176 \times 10^{-26}\ \text{kg})(5.8068 \times 10^{-26}\ \text{kg})(2.361 \times 10^{-10}\ \text{m})^2}{(3.8176 \times 10^{-26}\ \text{kg} + 5.8068 \times 10^{-26}\ \text{kg})}$$

$$\Rightarrow I = 1.284 \times 10^{-45}\ \text{kg}\cdot\text{m}^2$$

For $l = 2 \to l = 1$:

$$\Rightarrow \Delta E = E_2 - E_1 = (6-2)\frac{\hbar^2}{2I} = \frac{2(1.054 \times 10^{-34}\ \text{J}\cdot\text{s})^2}{1.284 \times 10^{-45}\ \text{kg}\cdot\text{m}^2} = 1.730 \times 10^{-23}\ \text{J}$$

$$\Rightarrow \lambda = \frac{hc}{\Delta E} = 0.01148\ \text{m} = 1.148\ \text{cm}.$$

For $l = 1 \to l = 0$: $\Delta E = E_1 - E_0 = (2-0)\frac{\hbar^2}{2I} = \frac{1}{2}(1.730 \times 10^{-23}\ \text{J})$

$$= 8.650 \times 10^{-24}\ \text{J}$$

$$\lambda = \frac{hc}{\Delta E} = 2.297\ \text{cm}.$$

(b) Carrying out exactly the same calculation for $Na^{37}Cl$, where $m_r(37) = 2.354 \times 10^{-26}$ kg and $I(37) = 1.312 \times 10^{-45}\ \text{kg}\cdot\text{m}^2$ we find for $l = 2 \to l = 1$: $\Delta E = 1.693 \times 10^{-23}$ J and $\lambda = 1.173$ cm. For $l = 1 \to l = 0$: $\Delta E = 8.465 \times 10^{-24}$ J and $\lambda = 2.347$ cm.

So the difference in wavelength are:

$l = 2 \to l = 1$: $\Delta\lambda = 1.173\ \text{cm} - 1.148\ \text{cm} = 0.025\ \text{cm}.$

$l = 1 \to l = 0$: $\Delta\lambda = 2.347\ \text{cm} - 2.297\ \text{cm} = 0.050\ \text{cm}.$

44-36: The energies corresponding to the observed wavelengths are 3.29×10^{-21} J, 2.87×10^{-21} J, 2.47×10^{-21} J, 2.06×10^{-21} J and 1.65×10^{-21} J. The average spacing of these energies is 0.410×10^{-21} J, and using the result of Problem 44-4, these are seen to correspond to transition from levels 8, 7, 6, 5 and 4 to the respective next lower levels. Then, $\frac{\hbar^2}{I}=0.410\times10^{-21}$ J, from which $I=2.71\times10^{-47}\ \text{kg}\cdot\text{m}^2$.

44-37: (a) Pr. (44-36) yields $I=2.71\times10^{-47}\ \text{kg}\cdot\text{m}^2$, and so $r=\sqrt{\frac{I}{m_r}}=\sqrt{\frac{I(m_H+m_{Cl})}{m_H m_{Cl}}}$

$$\Rightarrow r=\sqrt{\frac{(2.71\times10^{-47}\ \text{kg}\cdot\text{m}^2)(1.67\times10^{-27}\ \text{kg}+5.81\times10^{-26}\ \text{kg})}{(1.67\times10^{-27}\ \text{kg}+5.81\times10^{-26}\ \text{kg})}}$$

$\Rightarrow r=1.29\times10^{-10}$ m

(b) From $l\rightarrow l-1$: $\Delta E=\frac{\hbar^2}{2I}[l(l+1)-(l-1)l]=\frac{l\hbar^2}{I}$

But $\Delta E=\frac{hc}{\lambda}\Rightarrow l=\frac{2\pi cI}{\hbar\lambda}=\frac{4.84\times10^{-4}\ \text{m}}{\lambda}$

So the l-values that lead to the wavelength of Pr(44-36) are:

$$\lambda=6.04\times10^{-5}\ \text{m}:\ l=\frac{4.84\times10^{-4}\ \text{m}}{6.04\times10^{-5}\ \text{m}}=8$$

Similarly for:

$\lambda=6.90\times10^{-5}$ m: $l=7$; $\lambda=8.04\times10^{-5}$ m: $l=6$

$\lambda=9.64\times10^{-5}$ m: $l=5$; $\lambda=1.204\times10^{-4}$ m: $l=4$.

(c) The longest wavelength mean the least transition energy ($l=1\rightarrow l=0$)

$$\Rightarrow\Delta E=\frac{(1)(1.054\times10^{-34}\ \text{J}\cdot\text{s})^2}{2.71\times10^{-47}\ \text{kg}\cdot\text{m}^2}=4.10\times10^{-22}\ \text{J}$$

$$\Rightarrow\lambda=\frac{hc}{\Delta E}=4.85\times10^{-4}\ \text{m}.$$

(d) If the hydrogen atom is replaced by deuterium, then the reduced mass changes to $m_r'=3.16\times10^{-27}$ kg. Now $\Delta E=\frac{l\hbar^2}{I'}=\frac{hc}{\lambda'}\Rightarrow\lambda'=\frac{2\pi cI'}{l\hbar}=\frac{2\pi cm_r'r^2}{l\hbar}$

$$\Rightarrow\lambda'=\left(\frac{m_r'}{m_r}\right)\lambda=\left(\frac{3.16\times10^{-27}\ \text{kg}}{1.62\times10^{-27}\ \text{kg}}\right)\lambda=(1.95)\lambda$$

So for $l=8\rightarrow l=7$: $\lambda=(60.4\mu\text{m})(1.95)=118\mu\text{m}$.

$l=7\rightarrow l=6$: $\lambda=(69.0\mu\text{m})(1.95)=134\mu\text{m}$.

$l=6\rightarrow l=5$: $\lambda=(80.4\mu\text{m})(1.95)=156\mu\text{m}$.

$l=5\rightarrow l=4$: $\lambda=(96.4\mu\text{m})(1.95)=188\mu\text{m}$.

$l=4\rightarrow l=3$: $\lambda=(120.4\mu\text{m})(1.95)=234\mu\text{m}$.

44-38: (a) The frequency is proportional to the reciprocal of the square root of the reduced mass, and in terms of the atomic masses, the frequency of the isotope with the deuterium atom is

$$f = f_0\left(\frac{m_F m_H/(m_H + m_F)}{m_F m_D/(m_D + m_F)}\right)^{1/2} = f_0\left(\frac{1+(m_F/m_D)}{1+(m_F/m_H)}\right)^{1/2}.$$

Using f_0 from Exercise 44-7 and the given masses, $f = 8.99\times10^{13}$ Hz.

44-39: (a) $I = m_r r^2 = \dfrac{m_H m_I r^2}{m_H + m_I} = (1.657\times10^{-27}\ \text{kg})(0.160\times10^{-9}\ \text{m})^2$

$$= 4.24\times10^{-47}\ \text{kg}\cdot\text{m}^2$$

(b) Vibration-rotation energy levels are:

$$E_l = l(l+1)\frac{\hbar^2}{2I} + \left(h + \frac{1}{2}\right)\hbar\sqrt{\frac{k'}{m_r}}$$

$$= l(l+1)\frac{\hbar^2}{2I} + \left(n + \frac{1}{2}\right)hf \quad \left(\text{since: } \omega = 2\pi f = \sqrt{\frac{k'}{m_r}}\right)$$

(i) $n = 1, l = 1 \to n = 0, l = 0$:

$$\Delta E = (2-0)\frac{\hbar^2}{2I} + \left(\frac{3}{2} - \frac{1}{2}\right)hf = \frac{\hbar^2}{I} + hf.$$

$$\Rightarrow \lambda = \frac{hc}{\Delta E} = \frac{hc}{\frac{\hbar^2}{I} + hf} = \frac{c}{\frac{\hbar}{2\pi I} + f} = \frac{3.00\times10^8\ \text{m/s}}{3.96\times10^{11}\ \text{Hz} + 6.93\times10^{13}\ \mu}.$$

$$\Rightarrow \lambda = 4.30\times10^{-6}\ \text{m}.$$

(ii) $n = 1, l = 2 \to n = 0, l = 1$:

$$\Delta E = (6-2)\frac{\hbar^2}{2I} + hf$$

$$\Rightarrow \lambda = \frac{c}{2\left(\frac{\hbar^2}{2\pi I}\right) + f} = \frac{3.00\times10^8\ \text{m/s}}{2(3.96\times10^{11}\ \text{Hz}) + 6.93\times10^{13}\ \text{Hz}} = 4.28\times10^{-6}\ \text{m}$$

(iii) $n = 2, l = 2 \to n = 1, l = 3$

$$\Delta E = (6-12)\frac{\hbar^2}{2I} + hf$$

$$\Rightarrow \lambda = \frac{c}{-3\left(\frac{\hbar^2}{2\pi I}\right) + f} = \frac{3.00\times10^8\ \text{m/s}}{-3(3.96\times10^{11}\ \text{Hz}) + 6.93\times10^{13}\ \text{Hz}} = 4.40\times10^{-6}\ \text{m}$$

44-40: (a) $$\frac{d}{dr}U_{\text{tot}} = \frac{\alpha e^2}{4\pi\varepsilon_0}\frac{1}{r^2} - 8A\frac{1}{r^9}.$$

Setting this equal to zero when $r = r_0$ gives

$$r_0^7 = \frac{8A4\pi\varepsilon_0}{\alpha e^2}$$

and so

$$U_{\text{tot}} = \frac{\alpha e^2}{4\pi\varepsilon_0}\left(-\frac{1}{r} + \frac{r_0^7}{8r^8}\right).$$

At $r = r_0$,

$$U_{\text{tot}} = -\frac{7\alpha e^2}{32\pi\varepsilon_0 r_0} = -1.26\times10^{-18}\ \text{J} = -7.85\ \text{eV}.$$

(b) To remove a Na^+Cl^- ion pair from the crystal requires 7.85 eV. When neutral Na and Cl atoms are formed from the Na^+ and Cl^- atoms there is a net release of energy $-5.14\ \text{eV} + 3.61\ \text{eV} = -1.53\ \text{eV}$, so the net energy required to remove a neutral Na, Cl pair from the crystal is $7.85\ \text{eV} - 1.53\ \text{eV} = 6.32\ \text{eV}$.

44-41: Since potassium is a metal we approximate $E_F = E_{F0}$.

$$\Rightarrow E_F = \frac{3^{2/3}\pi^{4/3}\hbar^2 n^{2/3}}{2m}$$

But the electron concentration $n = \dfrac{\rho}{m}$

$$\Rightarrow n = \frac{851\ \text{kg/m}^3}{6.49\times10^{-26}\ \text{kg}} = 1.31\times10^{28}\ \text{electron/m}^3$$

$$\Rightarrow E_F = \frac{3^{2/3}\pi^{4/3}(1.054\times10^{-34}\ \text{J}\cdot\text{s})^2(1.31\times10^{28}\ /\text{m}^3)^{2/3}}{2(9.11\times10^{-31}\ \text{kg})} = 3.24\times10^{-19}\ \text{J}$$

$$= 2.03\ \text{eV}.$$

44-42: The sum of the probabilities is

$$f(E_F+\Delta E) + f(E_F-\Delta E) = \frac{1}{e^{-\Delta E/kT}+1} + \frac{1}{e^{\Delta E/kT}+1}$$

$$= \frac{1}{e^{-\Delta E/kT}+1} + \frac{e^{-\Delta E/kT}}{1+e^{-\Delta E/kT}}$$

$$= 1.$$

44-43: (a) $$U = \frac{1}{4\pi\varepsilon_0}\sum_{i<j}\frac{q_i q_j}{r_{ij}} = \frac{q^2}{4\pi\varepsilon_0}\left(\frac{-1}{d} + \frac{1}{r} - \frac{1}{r+d} - \frac{1}{r-d} + \frac{1}{r} - \frac{1}{d}\right)$$

$$= \frac{q^2}{4\pi\varepsilon_0}\left(\frac{2}{r} - \frac{2}{d} - \frac{1}{r+d} - \frac{1}{r-d}\right)$$

but $$\frac{1}{r+d}+\frac{1}{r-d}=\frac{1}{r}\left(\frac{1}{1+\frac{d}{r}}+\frac{1}{1-\frac{d}{r}}\right)\approx\frac{1}{r}\left(1-\frac{d}{r}+\frac{d^2}{r^2}+\ldots+1+\frac{d}{r}+\frac{d^2}{r^2}\right)$$

$$\approx\frac{2}{r}+\frac{2d^2}{r^3}$$

$$\Rightarrow U=\frac{-2q^2}{4\pi\varepsilon_0}\left(\frac{1}{d}+\frac{d^2}{r^3}\right)=\frac{-2p^2}{4\pi\varepsilon_0 r^3}-\frac{2p^2}{4\pi\varepsilon_0 d^3}$$

(b) $$U=\frac{1}{4\pi\varepsilon_0}\sum_{i<j}\frac{q_i q_j}{r_{ij}}$$

$$=\frac{q^2}{4\pi\varepsilon_0}\left(\frac{-1}{d}-\frac{1}{r}+\frac{1}{r+d}+\frac{1}{r-d}-\frac{1}{r}-\frac{1}{d}\right)$$

$$=\frac{q^2}{4\pi\varepsilon_0}\left(\frac{-2}{d}-\frac{2}{r}+\frac{2}{r}+\frac{2d^2}{r^3}\right)=\frac{-2q^2}{4\pi\varepsilon_0}\left(\frac{1}{d}-\frac{d^2}{r^3}\right)$$

$$\Rightarrow U=\frac{-2p^2}{4\pi\varepsilon_0 d^3}+\frac{2p^2}{4\pi\varepsilon_0 r^3}.$$

If we ignore the potential energy involved in forming each individual molecules, which just involves a different choice for the zero of potential energy, then the answers are: (a) $U=\frac{-2p^2}{4\pi\varepsilon_0 r^3}$. (b) $U=\frac{+2p^2}{4\pi\varepsilon_0 r^3}$.

44-44: Following the hint,

$$k'\,dr=-d\left(\frac{1}{4\pi\varepsilon_0}\frac{e^2}{r^2}\right)_{r=r_0}=\frac{1}{2\pi\varepsilon_0}\frac{e^2}{r_0^3}dr \quad \text{and}$$

$$\hbar\omega=\hbar\sqrt{2k'/m}=\hbar\sqrt{\frac{1}{\pi\varepsilon_0}\frac{e^2}{mr_0^3}}=1.23\times10^{-19}\ \text{J}=0.77\ \text{eV},$$

where $(m/2)$ has been used for the reduced mass. b) The reduced mass is doubled, and the energy is reduced by a factor of $\sqrt{2}$ to 0.54 eV.

Chapter 45: Nuclear Physics

45-1: (a) ${}^{20}_{10}\text{Ne}$ has 10 protons and 20 neutrons
(b) ${}^{59}_{27}\text{Co}$ has 27 protons and 32 neutrons
(c) ${}^{209}_{83}\text{Bi}$ has 83 protons and 126 neutrons.

45-2: (a) From Eq. (45-1), the radii are roughly 3.3 fm, 4.7 fm and 7.1 fm.
(b) Using $4\pi R^2$ for each of the radii in part (a), the areas are 130 fm^2, 270 fm^2 and 640 fm^2 (c) $\frac{4}{3}\pi R^3$ gives 140 fm^3, 420 fm^3 and 1510 fm^3. (d) The density is the same, since the volume and the mass are both proportional to A^3: $2.3\times10^{17}\ \text{kg/m}^3$ (see Example 45-1). (e) Dividing the result of part (d) by the mass of a nucleon, the number density is $0.14/\text{fm}^3 = 1.40\times10^{44}/\text{m}^3$.

45-3: $\Delta E = \mu_z B - (-\mu_z B) = 2\mu_z B$

But $\Delta E = hf$, so $B = \dfrac{hf}{2\mu_z}$

$$\Rightarrow B = \frac{(6.63\times10^{-34}\ \text{J}\cdot\text{s})(3.69\times10^{7}\ \text{Hz})}{2(2.7928)(5.051\times10^{-27}\ \text{J/T})} = 0.867\ \text{T}$$

45-4: (a) As in Example 45-2,

$$\Delta E = 2(1.9130)(3.15245\times10^{-8}\ \text{eV/T})(2.30\ \text{T}) = 2.77\times10^{-7}\ \text{eV},$$

smaller than but comparable to that found in the example. (b) $f = \dfrac{\Delta E}{h} =$ 67.1 MHz, $\lambda = \dfrac{c}{f} = 4.47\ \text{m}$.

45-5: (a) $U = \vec{\mu}\cdot\vec{B} = -\mu_z B$. So if the spin magnetic moment of the proton is parallel to the magnetic field, $U < 0$, and if they are antiparallel, $U > 0$. So the parallel case has lower energy.
The frequency of an emitted photon has a transition of the protons between the two states is given by:

$$f = \frac{\Delta E}{h} = \frac{E_+ - E_-}{h} = \frac{2\mu_z B}{h}$$

$$= \frac{2(2.7928)(5.051\times10^{-27}\ \text{J/T})(1.80\ \text{T})}{(6.63\times10^{-34}\ \text{J}\cdot\text{s})} = 7.66\times10^{7}\ \text{Hz}.$$

$$\Rightarrow \lambda = \frac{c}{f} = \frac{3.00\times10^{+8}\ \text{m/s}}{7.66\times10^{7}\ \text{Hz}} = 3.92\ \text{m}.$$ This is a radio wave.

(b) For electrons, the negative charge means that the argument from part (a) leads to the $m_s = -\frac{1}{2}$ state (antiparallel) having the lowest energy. So an emitted photon in a transition from one state to the other has a frequency

$$f = \frac{\Delta E}{h} = \frac{E_{-1/2} - E_{+1/2}}{h} = \frac{-2\mu_z B}{h}$$

But from eq. (43-22), $\mu_z = -(2.00232)\frac{e}{2m_e}S_z = \frac{-(2.00232)e\hbar}{4m_e}$

$$\Rightarrow f = \frac{(2.00232)\,eB}{4\pi m_e} = \frac{(2.00232)(1.60\times10^{-19}\,\mathrm{C})(1.80\,\mathrm{T})}{4\pi(9.11\times10^{-31}\,\mathrm{kg})}$$

$$\Rightarrow f = 5.04\times10^{10}\,\mathrm{Hz}$$

so $\lambda = \frac{c}{f} = \frac{3.00\times10^{8}\,\mathrm{m/s}}{5.04\times10^{10}\,\mathrm{Hz}} = 5.95\times10^{-3}\,\mathrm{m}$

This is a microwave.

45-6: (a) $7(m_\mathrm{n} + m_\mathrm{H}) - m_\mathrm{N} = 0.112\,\mathrm{u}$, which is 105 MeV, or 7.48 MeV per nucleon.
(b) Similarly, $2(m_\mathrm{H} + m_\mathrm{n}) - m_\mathrm{He} = 0.03038\,\mathrm{u} = 28.3\,\mathrm{MeV}$, or 7.07 MeV per nucleon, slightly lower (compare to Fig. (45-2)).

45-7: Z is a magic number for the elements helium(Z = 2), oxygen (Z = 8), calcium (Z = 20), nickel (Z = 28), tin (Z = 50) and lead (Z = 82). The elements are especially stable, with large energy jumps to the next allowed energy level. The binding energy for these elements is also large. The protons' net magnetic moments are zero.

45-8: (a) $(13.6\,\mathrm{eV})/(0.511\times10^{6}\,\mathrm{eV}) = 2.66\times10^{-5} = 0.0027\%$.
(b) $(8.795\,\mathrm{MeV})/(938.3\,\mathrm{MeV}) = 9.37\times10^{-3} = 0.937\%$.

45-9: The binding energy of a deuteron is $2.224\times10^{6}\,\mathrm{eV}$. The photon with this energy has wavelength equal to

$$\lambda = \frac{hc}{E} = \frac{(6.626\times10^{-34}\,\mathrm{J\cdot s})(2.998\times10^{8}\,\mathrm{m/s})}{(2.224\times10^{6}\,\mathrm{eV})(1.602\times10^{-19}\,\mathrm{J/eV})} = 5.576\times10^{-13}\,\mathrm{m}.$$

45-10: (a) $146m_\mathrm{n} + 92m_\mathrm{H} - m_\mathrm{U} = 1.93\,\mathrm{u}$, which is (b) 1.80×10^{3} MeV, or (c) 7.57 MeV per nucleon.

45-11: (a) for ${}^{11}_{5}B$ the mass detect is:

$$\begin{aligned}\Delta m &= 5m_p + 6m_n + 5m_e - M({}^{11}_{5}B)\\ &= 5(1.007277\,\mathrm{u}) + 6(1.008665\,\mathrm{u}) + 5(0.000549\,\mathrm{u}) - 11.009305\,\mathrm{u}\\ &= 0.081815\,\mathrm{u}\end{aligned}$$

$$\Rightarrow \text{The binding energy } E_B = (0.081815\,\mathrm{u})(931.5\,\mathrm{MeV/u}) = 76.21\,\mathrm{MeV}$$

(b) From Eq. (45-9): $E_B = C_1A - C_2A^{2/3} - C_3\dfrac{Z(Z-1)}{A^{1/3}} - C_4\dfrac{(A-2Z)^2}{A}$
and there is no fifth term since Z is odd and A is even.

$$\Rightarrow E_B = (15.75\ \text{MeV})11 - (17.80\ \text{MeV})(11)^{2/3} - (0.7100\ \text{MeV})\frac{5(4)}{11^{1/3}}$$

$$-(23.69\ \text{MeV})\frac{(11-10)^2}{11}$$

$$\Rightarrow E_B = 76.68\ \text{MeV}$$

So the percentage difference is $\dfrac{76.68\ \text{MeV} - 76.21\ \text{MeV}}{76.21\ \text{MeV}} \times 100 = 0.62\%$

45-12: (a) $34m_{\text{n}} + 29m_{\text{H}} - m_{\text{Cu}} = 0.592\ \text{u}$, which is 551 MeV, or 8.75 MeV per nucleon.
(b) In Eq. (45-9), $Z = 29$ and $N = 34$, so the fifth term is zero. The predicted binding energy is

$$\begin{aligned} E_{\text{B}} &= (15.75\ \text{MeV})(63) \\ &- (17.80\ \text{MeV})(63)^{2/3} \\ &- (0.7100\ \text{MeV})\frac{(29)(28)}{(63)^{1/3}} \\ &- (23.69\ \text{MeV})\frac{(5)^2}{(63)} \\ &= 556\ \text{MeV}, \end{aligned}$$

which differs from the binding energy found from the mass deficit by 0.86%, a very good agreement comparable to that found in Example 45-4.

45-13: $\Delta m = 2M({}^{4}_{2}\text{He}) - M({}^{8}_{4}\text{Be})$

$$= 2(4.002603\ \text{u}) - 8.005305\ \text{u}$$

$$\Rightarrow \Delta m = -9.9 \times 10^{-5}\ \text{u}$$

45-14: (a) A proton changes to a neutron, so the emitted particle is a positron (β^+).
(b) The number of nucleons in the nucleus decreases by 4 and the number of protons by 2, so the emitted particel is an alpha-particle. (c) A neutron changes to a proton, so the emitted particle is an electron (β^-).

45-15: (a) α decay: Z decreases by 2, A decreases by 4
$\Rightarrow {}^{239}_{94}\text{Pb} \rightarrow {}^{235}_{92}\text{U} + \alpha$
(b) β^- decay: Z increases by 1, A remains the same:
$\Rightarrow {}^{24}_{11}\text{Na} \rightarrow {}^{24}_{12}\text{Mg} + \beta^-$
(c) β^+ decay: Z decreases by 1 and A remains the same:
$\Rightarrow {}^{15}_{8}\text{O} \rightarrow {}^{15}_{7}\text{N} + \beta^+$

45-16: (a) The energy released is the energy equivalent of $m_{\rm n} - m_{\rm p} - m_{\rm e} = 8.40\times10^{-4}\,{\rm u}$, or 783 keV (b) $m_{\rm n} > m_{\rm p}$, and the decay is not possible.

45-17: (a) If tritium is to be unstable with respect to β^- decay, then the mass of the products of the decay must be less than the parent nucleus.

$M({}^3_1{\rm H}^+) = 3.016049\ {\rm u} - 0.00054858\ {\rm u} = 3.015500\ {\rm u}$

$M({}^3_2{\rm He}^{2+}) = 3.016029\ {\rm u} - 2(0.00054858\ {\rm u}) = 3.017932\ {\rm u}$

$\Rightarrow \Delta m = M({}^3_1{\rm H}^+) - M({}^3_2{\rm He}^-) - m_e = 2.0\times10^{-5}\,{\rm u}$,

so the decay is possible.

(b) The energy of the products is just $E = (2.0\times10^{-5}\,{\rm u})(931.5\ {\rm MeV/u})$

$= 0.019\ {\rm MeV}$

$= 19\ {\rm keV}$

45-18: (a) As in the example, $(0.000898\,{\rm u})(931.5\ {\rm MeV/u}) = 0.836\ {\rm MeV}$. (b) 0.836 MeV– 0.122 MeV – 0.014 MeV = 0.700 MeV.

45-19: If β^- decay of ${}^{14}C$ is possible, then we are considering the decay ${}^{14}_6{\rm C}\rightarrow{}^{14}_7{\rm N}+\beta^-$.

$\Delta m = M({}^{14}_6C) - M({}^{14}_7N) - m_e$

$= (14.003242\ {\rm u} - 6(0.000549\ {\rm u})) - (14.003074\ {\rm u} - 7(0.000549\ {\rm u})) - 0.0005491$

$= +1.68\times10^{-4}\ {\rm u}: {\rm So}\ E = (1.68\times10^{-4}\ {\rm u})(931.5\ {\rm MeV/u}) = 0.156\ {\rm MeV}$

$= 156\ {\rm keV}$

45-20: $\dfrac{360\times10^6\ {\rm decays}}{86{,}400\ {\rm s}} = 4.17\times10^3\ {\rm Bq} = 1.13\times10^{-7}\ {\rm Ci} = 0.113\ \mu{\rm Ci}.$

45-21: (a) $\lambda = \dfrac{0.693}{T_{1/2}} = \dfrac{0.693}{(4.47\times10^9\ {\rm y})(3.156\times10^7\ {\rm s/y})} = 4.91\times10^{-18}\,{\rm s}^{-1}$

(b) $6.00\times10^{-6}\,{\rm Ci} = (6.00\times10^{-6}\,{\rm Ci})\left(3.70\times10^{10}\dfrac{\rm decays}{\rm s\cdot Ci}\right) = 2.22\times10^5\,{\rm decays/s}.$

But $\left|\dfrac{dN}{dt}\right| = \lambda N \Rightarrow N = \dfrac{1}{\lambda}\left|\dfrac{dN}{dt}\right| = \dfrac{2.22\times10^5\,{\rm decays/s}}{4.91\times10^{-18}\,{\rm s}^{-1}}$

$\Rightarrow N = 4.52\times10^{22}\,{\rm nuclei} \Rightarrow m = (238\ {\rm u})N$

$\Rightarrow m = (238)(1.66\times10^{-27}\ {\rm kg})(4.52\times10^{22}) = 0.0179\ {\rm kg}.$

(c) Each decay emits one alpha particle. If 30.0 g of uranium there are

$N = \dfrac{0.0300\ {\rm kg}}{238(1.66\times10^{-27}\ {\rm kg})} = 7.59\times10^{22}\,{\rm nuclei}$

$\Rightarrow \left|\dfrac{dN}{dt}\right| = \lambda N = (4.91\times10^{-18}\,{\rm s}^{-1})(7.59\times10^{22}\,{\rm nuclei}) = 3.73\times10^5\,{\rm alpha\ particles}$

emitted each second.

45-22: (a) Solving Eq. (45-17) for $\lambda, \lambda = \dfrac{\ln 2}{T_{1/2}} = 4.17\times10^{-9}\,\text{s}^{-1}$. (b) $N = \dfrac{m}{Au} = 4.26\times10^{17}$.

(c) $\dfrac{dN}{dt} = \lambda N = 1.78\times10^{9}$ Bq, which is (d) 0.0480 Ci. The same calculation for radium, with larger A and longer half-life (lower λ) gives 4.2×10^{-5} Ci.

45-23: (a) $\left|\dfrac{dN(0)}{dt}\right| = 9.75\times10^{10}\,\text{Bq} = 9.75\times10^{10}\,\text{decays/s}.$

and $\lambda = \dfrac{0.693}{T_{1/2}} = \dfrac{0.693}{(30.8\,\text{min})(60\,\text{s/min})} = 3.75\times10^{-4}\,\text{s}^{-1}$

So $N(0) = \dfrac{1}{\lambda}\left|\dfrac{dN(0)}{dt}\right| = \dfrac{9.75\times10^{10}\,\text{decays/s}}{3.75\times10^{-4}\,\text{s}^{-1}} = 2.60\times10^{14}\,\text{nuclei}.$

(b) The number of nuclei left after one half-life is $\dfrac{N(0)}{2} = 1.30\times10^{14}$ nuclei, and the activity is half: $\left|\dfrac{dN}{dt}\right| = 4.88\times10^{10}\,\text{decays/s}.$

(c) After two half lives (61.6 minutes) there is a quarter of the original amount $N = 6.5\times10^{13}$ nuclei, and a quarter of the activity:

$\left(\dfrac{dN}{dt}\right) = 2.44\times10^{10}\,\text{decays/s}.$

45-24: The activity of the sample is $\dfrac{4030\,\text{decays/min}}{0.500\,\text{kg}} = 134$ Bq/kg, while the activity of atmospheric carbon is 255 Bq/kg (see Example 45-8). The age of the sample is then

$$t = \frac{\ln(134/255)}{\lambda} = -\frac{\ln(134/255)}{1.21\times10^{-4}/\text{y}} = 5030\,\text{y}.$$

45-25: (a) $\lambda = \dfrac{0.693}{T_{1/2}} = \dfrac{0.693}{(1.28\times10^{9}\,\text{y})(3.156\times10^{7}\,\text{s/y})}$

$\Rightarrow \lambda = 1.72\times10^{-17}\,\text{s}^{-1}$

In $m = 3.26\times10^{-6}$ g of ^{40}K there are

$$N = \frac{3.26\times10^{-9}\,\text{kg}}{40(1.66\times10^{-27}\,\text{kg})} = 4.91\times10^{16}\,\text{nuclei}.$$

So $\left|\dfrac{dN}{dt}\right| = \lambda N = (1.72\times10^{-17}\,\text{s}^{-1})(4.91\times10^{16}\,\text{nuclei}) = 0.842\,\text{decays/s}$

(b) $\left|\dfrac{dN}{dt}\right| = \dfrac{0.842\,\text{Bq}}{3.70\times10^{10}\,\text{Bq/Ci}} = 2.28\times10^{-11}\,\text{Ci}$

45-26: (a) From Table (45-3), the absorbed dose is 0.0900 rad. (b) The energy absorbed is $(9.00\times10^{-4}\text{ J/kg})(0.150\text{ kg}) = 1.35\times10^{-4}\text{ J}$, each proton has energy $1.282\times10^{-13}\text{ J}$, so the number absorbed is 1.05×10^{9}. (c) The RBE for alpha particles is twice that for protons, so only half as many, 5.27×10^{8}, would be absorbed.

45-27: (a) $E_{\text{total}} = NE_{\gamma} = \dfrac{Nhc}{\lambda} = \dfrac{(6.50\times10^{10})(6.63\times10^{-34}\text{ J}\cdot\text{s})(3.00\times10^{8}\text{ m/s})}{2.00\times10^{-11}\text{ m}}$

$\Rightarrow E_{\text{total}} = 6.46\times10^{-4}\text{ J}$

(b) The absorbed dose is the energy divided by tissue mass:

$$\text{dose} = \frac{6.46\times10^{-4}\text{ J}}{0.600\text{ kg}} = \left(1.08\times10^{-3}\frac{\text{J}}{\text{kg}}\right)\left(\frac{100\text{ rad}}{\text{J/kg}}\right) = 0.108\text{ rad}.$$

The rem dose for x rays (RBE = 1) is just 0.108 rad.

45-28: $(0.72\times10^{-6}\text{ Ci})(3.7\times10^{10}\text{ Bq/Ci})(3.156\times10^{7}\text{ s}) = 8.41\times10^{11}\ \alpha$ particles. The absorbed dose is

$$\frac{(8.41\times10^{11})(4.0\times10^{6}\text{ eV})(1.602\times10^{-19}\text{ J/eV})}{(0.50\text{ kg})} = 1.08\text{ Gy} = 108\text{ rad}.$$

The equivalent dose is $(20)(108\text{ rad}) = 2160$ rem.

45-29: (a) We need to know how many decays per second occur.

$$\lambda = \frac{0.693}{T_{1/2}} = \frac{0.693}{(12.3\text{ y})(3.156\times10^{7}\text{ s/y})} = 1.79\times10^{-9}\text{s}^{-1}$$

The number of tritium atoms is $N(0) = \dfrac{1}{\lambda}\left|\dfrac{dN}{dt}\right| = \dfrac{(0.35\text{ Ci})(3.70\times10^{10}\text{ Bq/Ci})}{1.79\times10^{-9}\text{s}^{-1}}$

$\Rightarrow N(0) = 7.2540\times10^{18}$ nuclei.

The number of remaining nuclei after one week is just

$N(1\text{ week}) = N(0)e^{-\lambda t} = (7.25\times10^{18})e^{-(1.79\times10^{-9}\text{s}^{-1})(7)(24)(3600\text{s})}$

$\Rightarrow N(1\text{ week}) = 7.2462\times10^{18}$ nuclei

$\Rightarrow \Delta N = N(0) - N(1\text{ week}) = 7.8\times10^{15}$ decays.

So the energy absorbed is

$E_{\text{total}} = \Delta N\cdot E_{\gamma} = (7.8\times10^{15})(5000\text{ eV})(1.60\times10^{-19}\text{ J/eV}) = 6.24\text{ J}$

So the absorbed dose is $\dfrac{(6.24\text{ J})}{(50\text{ kg})} = 0.125\text{ J/kg} = 12.5$ rad.

Since RBE = 1, then the equivalent dose is 12.5 rem.

(b) In the decay, antineutrino's are also emitted. These are not absorbed by the body, and so some of the energy of the decay is lost (about 12 keV).

45-30: (a) $5.4\text{ Sv}(100\text{ rem/Sv}) = 540$ rem. (b) The RBE of 1 gives an absorbed dose of 540 rad (c) The absorbed dose is 5.4 Gy, so the total energy absorbed is

$(5.4\ \text{Gy})(65\ \text{kg}) = 351\ \text{J}$. The energy required to raise the body temperature is $(65\text{kg})(4190\ \text{J/kg}\cdot\text{K})(0.01\ ^\circ\text{C}) = 3\text{kJ}$.

45-31: The mass defect is $\Delta m = M({}^{235}_{92}\text{U}) + m_\text{n} - M({}^{236}_{92}\text{U}^*)$
$\Rightarrow \Delta m = 235.043924\ \text{u} + 1.008665\ \text{u} - 236.045563\ \text{u}$
$= 0.007026\ \text{u}$
So the internal excitation of the nucleus is:
$Q = (\Delta m)c^2 = (0.007026\ \text{u})(931.5\ \text{MeV/u})$
$= 6.545\ \text{MeV}$.

45-32: $(200\times10^6\ \text{eV})(1.602\times10^{-19}\ \text{J/eV})(6.023\times10^{23}\ \text{molecules/mol})$
$= 1.93\times10^{13}\ \text{J/mol}$,
which is far higher than typical heats of combustion.

45-33: (a) ${}^2_1\text{H} + {}^9_4\text{Be} \to {}^A_Z\text{X} + {}^4_2\text{He}$
So $2+9 = A+4 \Rightarrow A = 7$
and $1+4 = Z+2 \Rightarrow Z = 3$, so X is ${}^7_3\text{Li}$.
(b) $\Delta m = M({}^2_1\text{H}) + M({}^9_4\text{Be}) - M({}^7_3\text{Li}) - M({}^4_2\text{He})$
$= 2.014102\ \text{u} + 9.012182\ \text{u} - 7.016003\ \text{u} - 4.002603\ \text{u}$
$= 7.678\times10^{-3}\ \text{u}$
So $E = (\Delta m)c^2 = (7.678\times10^{-3}\ \text{u})(931.5\ \text{MeV/u}) = 7.152\ \text{MeV}$
(c) The threshold energy is taken to be the potential energy of the two reactants when they just "touch." So we need to know their radii:
$r_{{}^2\text{H}} = (1.2\times10^{-15}\ \text{m})(2)^{1/3} = 1.5\times10^{-15}\ \text{m}$
$r_{{}^9\text{Be}} = (1.2\times10^{-15}\ \text{m})(9)^{1/3} = 2.5\times10^{-15}\ \text{m}$
So the centers' separation is $r = 4.0\times10^{-15}\ \text{m}$.
Thus $U = \dfrac{1}{4\pi\varepsilon_0}\cdot\dfrac{q_\text{H}\,q_\text{Be}}{r} = \dfrac{4(1.60\times10^{-19}\ \text{C})^2}{4\pi\varepsilon_0(4.0\times10^{-15}\ \text{m})} = 2.3\times10^{-13}\ \text{J}$
$\Rightarrow U = 1.4\times10^6\ \text{eV} = 1.4\ \text{MeV}$.

45-34: (a) $Z = 3+2-0 = 5$ and $A = 4+7-1 = 10$ (b) The nuclide is a boron nucleus, and $m_\text{He} + m_\text{Li} - m_\text{n} - m_\text{B} = -3.00\times10^{-3}$ u, and so 2.79 MeV of energy is absorbed.

45-35: (a) As in Ex. (45-33a), $2+14 = A+10 \Rightarrow A = 6$, and $1+7 = Z \to 5 \Rightarrow Z = 3$, so X $= {}^6_3\text{Li}$
(b) As in Ex. (45-33b), using $M({}^2_1\text{H}) = 2.014102\ \text{u}$, $M({}^{14}_7\text{N}) = 14.003074\ \text{u}$, $M({}^6_3\text{Li}) = 6.015121\ \text{u}$, *and* $M({}^{10}_5\text{B}) = 10.012937\ \text{u}$, $\Delta m = -0.010882\ \text{u}$, so energy is absorbed in the reaction. $\Rightarrow Q = (-0.010882\ \text{u})(931.5\ \text{MeV/u}) = -10.14\ \text{MeV}$

(c) From Eq. (45-22): $K_{cm} = \dfrac{M}{M+m}K$

so $K = \left(\dfrac{M+m}{M}\right)K_{cm} = \dfrac{14.0\text{ u} + 2.01\text{ u}}{14.0\text{ u}}(10.14\text{ MeV}) = 11.6\text{ MeV}.$

45-36: $m_{^3_2\text{He}} + m_{^2_1\text{H}} - m_{^4_2\text{He}} - m_{^1_1\text{H}} = 1.97 \times 10^{-2}$ u,
so the energy released is 18.4 MeV.

45-37: Nuclei: $^A_Z\text{X}^{Z+} \rightarrow ^{A-4}_{Z-2}\text{Y}^{(Z-2)+} + ^4_2\text{He}^{2+}$
Add the mass of Z electrons to each side and we find: $\Delta m = M(^A_Z\text{X}) - M(^{A-4}_{Z-2}\text{Y}) - M(^4_2\text{He})$, where now we have the mass of the neutral atoms. So as long as the mass of the original neutral atom is greater than the sum of the neutral products masses, the decay can happen.

45-38: Denote the reaction as

$$^A_Z\text{X} \rightarrow ^A_{Z+1}\text{Y} + e^-.$$

The mass defect is related to the change in the neutral atomic masses by

$$[m_\text{X} - Zm_\text{e}] - [m_\text{Y} - (Z+1)m_\text{e}] - m_\text{e} = (m_\text{X} - m_\text{Y}),$$

where m_X and m_Y are the masses as tabulated in, for instance, Table (45-2).

45-39: $^A_Z\text{X}^{Z+} \rightarrow ^A_{Z-1}\text{Y}^{(Z-1)+} + \beta^+$
Adding (Z–1) electron to both sides yields
$^A_Z\text{X}^+ \rightarrow ^A_{Z-1}\text{Y} + \beta^+$
So in terms of masses:

$$\begin{aligned}\Delta m &= M(^A_Z\text{X}^+) - M(^A_{Z-1}\text{Y}) - m_e \\ &= (M(^A_Z\text{X}) - m_e) - M(^A_{Z-1}\text{Y}) - m_e \\ &= M(^A_Z\text{X}) - M(^A_{Z-1}\text{Y}) - 2m_e.\end{aligned}$$

So the decay will occur as long as the original neutral mass is greater than the sum of the neutral product mass and two electron masses.

45-40: Denote the reaction as

$$^A_Z\text{X} + e^- \rightarrow ^A_{Z-1}\text{Y}.$$

The mass defect is related to the change in the neutral atomic masses by

$$[m_\text{X} - Zm_\text{e}] + m_\text{e} - [m_\text{Y} - (Z-1)m_\text{e}] = (m_\text{X} - m_\text{Y}),$$

where m_X and m_Y are the masses as tabulated in, for instance, Table (45-2).

45-41: (a) Only the heavier one $(^{25}_{13}\text{Al})$ can decay into the lighter one $(^{25}_{12}\text{Mg})$.
(b) $(^{25}_{13}\text{Al}) \rightarrow (^{25}_{12}\text{Mg}) + ^A_Z\text{X} \Rightarrow A = 0, \quad Z = +1 \Rightarrow X$ is a positron
$\Rightarrow \beta^+$decay
or $(^{25}_{13}\text{Al}) + ^A_Z\text{X}' \rightarrow ^{25}_{12}\text{Mg} \Rightarrow A = 0, Z = -1 \Rightarrow X'$ is an electron
$\Rightarrow$ electron capture
(c) Using the nuclear masses, we calculate the mass defect for β^+ decay:

$$\Delta m = \left(M\left({}^{25}_{13}\text{Al}\right) - 13m_e\right) - \left(M\left({}^{25}_{12}\text{Mg}\right) - 12m_e\right) - m_e$$
$$= 24.990429\text{ u} - 24.985837\text{ u} - 2(0.00054858\text{ u})$$
$$= 3.495 \times 10^{-3}\text{ u}$$
$$\Rightarrow Q = (\Delta m)c^2 = \left(3.495\times10^{-3}\text{ u}\right)(931.5\text{ MeV/u}) = 3.255\text{ MeV}.$$

For electron capture:

$$\Delta m = M\left({}^{25}_{13}\text{Al}\right) - M\left({}^{25}_{12}\text{Mg}\right) = 24.990429\text{ u} - 24.985837\text{ u}$$
$$= 4.592 \times 10^{-3}\text{ u}$$
$$\Rightarrow Q = (\Delta m)c^2 = \left(4.592\times10^{-3}\text{ u}\right)(931.5\text{ MeV/u}) = 4.277\text{ MeV}.$$

45-42: (a) $m_{{}^{210}_{84}\text{Po}} - m_{{}^{206}_{82}\text{Pb}} - m_{{}^{4}_{2}\text{He}} = 5.81\times10^{-3}$ u, or Q = 5.41 MeV. The energy of the alpha particle is (206/210) times this, or 5.30 MeV (see Example 45-5).
(b) $m_{{}^{210}_{84}\text{Po}} - m_{{}^{209}_{82}\text{Pb}} - m_{{}^{1}_{1}\text{H}} = -5.35\times10^{-3}$ u < 0, so the decay is not possible.
(c) $m_{{}^{210}_{84}\text{Po}} - m_{{}^{209}_{82}\text{Pb}} - m_{\text{n}} = -8.22\times10^{-3}$ u < 0, so the decay is not possible.
(d) $m_{{}^{210}_{85}\text{At}} > m_{{}^{210}_{84}\text{Po}}$, so the decay is not possible (see Problem 45-38).
(e) $m_{{}^{210}_{83}\text{Bi}} + 2m_e > m_{{}^{210}_{84}\text{Po}}$ so the decay is not possible (see Problem 45-39).

45-43: Using Eq. (45-10): ${}^{A}_{Z}M = ZM_{\text{H}} + Nm_{\text{n}} - E_{\text{B}}/c^2$

$$\Rightarrow M\left({}^{24}_{11}\text{Na}\right) = 11M_{\text{H}} + 13m_{\text{n}} - E_{\text{B}}/c^2$$

But $E_B = 15.75\text{ MeV}(24) - (17.80\text{ MeV})(24)^{2/3} - (0.7100\text{ MeV})\dfrac{(11)(10)}{(24)^{1/3}}$

$$-(23.69\text{ MeV})\frac{(24-2(11))^2}{24} - (39\text{ MeV})(24)^{-4/3}$$
$$= 198.31\text{ MeV}.$$

$$\Rightarrow M\left({}^{24}_{11}\text{Na}\right) = 11(1.007825\text{ u}) + 13(1.008665\text{ u}) - \frac{(198.31\text{ MeV})}{931.5\text{ MeV/u}} = 23.9858\text{ u}$$

$$\%\,\text{error} = \frac{23.990961 - 23.9858}{23.990961}\times 100 = 0.021\,\%$$

If the binding energy term is neglected, $M\left({}^{24}_{11}\text{Na}\right) = 24.1987$ u and so the percentage error would be $\dfrac{24.1987 - 23.99096}{23.99096}\times 100 = 0.87\,\%$.

45-44: The α-particle will have $\dfrac{226}{230}$ of the mass energy,

$$\frac{226}{230}\left(m_{\text{Th}} - m_{\text{Ra}} - m_{\alpha}\right) = 5.036\times10^{-3}\text{ u} \quad \text{or} \quad 4.69\text{ MeV}.$$

45-45: ${}^{198}_{79}\text{Au} \rightarrow {}^{198}_{80}\text{Hg} + \beta^-$

$$\Delta m = M\left({}^{198}_{79}\text{Au}\right) - M\left({}^{198}_{80}\text{Hg}\right) = 197.968217\text{ u} - 197.966743\text{ u}$$
$$= 1.473\times10^{-3}\text{ u}$$

And the total energy available was $Q = (\Delta m)c^2$

$\Rightarrow Q = (1.473\times10^{-3}\,\text{u})(931.5\,\text{MeV/u}) = 1.372\,\text{MeV}$

The emitted photon has energy 0.412 MeV, so the emitted electron must have kinetic energy equal to 1.372 MeV – 0.412 MeV = 0.960 MeV.

45-46: (See Problem 45-39) $m_{^{11}_{6}\text{C}} - m_{^{11}_{5}\text{B}} - 2m_e = 1.03\times10^{-3}$ u. Decay is energetically possible.

45-47: $^{13}_{7}\text{N}\rightarrow{}^{13}_{6}\text{C}+\beta^{+}$

As in Pr. (45-41(c)), β^{+} decay has a mass defect in terms of neutral atoms of

$$\Delta m = M(^{13}_{7}\text{N}) - M(^{13}_{6}\text{C}) - 2m_e$$

$$= 13.005739\,\text{u} - 13.003355 - 2(0.00054858\,\text{u})$$

$$= 1.287\times10^{-3}\,\text{u}$$

Therefore the decay is possible because the initial mass is greater than the final mass.

45-48: (a) A least-squares fit to log of the activity *vs.* time gives a slope of $\lambda = 0.5995\,\text{hr}^{-1}$, for a half-life of $\dfrac{\ln 2}{\lambda} = 1.16\,\text{hr}$. (b) The initial activity is $N_0\lambda$, so

$$N_0 = \frac{(2.00\times10^{4}\,\text{Bq})}{(0.5995\,\text{hr}^{-1})(1\,\text{hr}/3600\,\text{s})} = 1.20\times10^{8}.$$

c) $N_0 e^{-\lambda t} = 1.81\times10^{6}$.

45-49: $\lambda = \dfrac{0.693}{T_{1/2}} = \dfrac{0.693}{(4.89\times10^{10}\,\text{y})(3.156\times10^{7}\,\text{s/y})} = 4.49\times10^{-19}\,\text{s}^{-1}$

$$N(87) = N_0(87)e^{-\lambda t} \Rightarrow N_0(87) = N(87)e^{+\lambda t}$$

$$\Rightarrow N_0(87) = N(87)e^{(4.49\times10^{-19}\,\text{s}^{-1})(4.6\times10^{9}\,\text{y})(3.156\times10^{7}\,\text{s/y})}$$

$$\Rightarrow N_0(87) = 1.0674\,N(87)$$

But we also know that $\dfrac{N(87)}{N(85)+N(87)} = 0.2783$

$$\Rightarrow N(87) = \frac{0.2783\,N(85)}{(1-0.2783)} = 0.3856\,N(85)$$

$$= 0.3856\,N_0(85)$$

So $\dfrac{N_0(87)}{N_0(85)+N_0(87)} = \dfrac{1.0674(0.3856)}{(1+1.0674(0.3856))} = 0.2916$

So the original percentage of ^{87}Rb is 29 %.

45-50: (a) $(6.25\times10^{12})(4.77\times10^{6}\,\text{MeV})(1.602\times10^{-19}\,\text{J/eV})/(70.0\,\text{kg}) = 0.0682\,\text{Gy}$ $= 6.82\,\text{rad}$. (b) $(20)(6.82\,\text{rad}) = 136\,\text{rem}$. (c) $N\lambda = \dfrac{m}{Au}\dfrac{\ln(2)}{T_{1/2}} =$

1.17×10^{9} Bq = 31.6 mCi. (d) $\dfrac{6.25\times10^{12}}{1.17\times10^{9}\ \text{Bq}} = 5.34\times10^{3}$ s, about an hour and a half. Note that this time is so small in comparison with the half-life that the decrease in activity of the source may be neglected.

45-51: (a) $\left|\dfrac{dN}{dt}\right| = 2.60\times10^{-5}\,\text{Ci}(3.70\times10^{10}\,\text{decays/s}\cdot\text{Ci}) = 9.62\times10^{5}\,\text{decays/s}$

so in one second there is an energy delivered of

$$E = \frac{1}{2}\left(\frac{dN}{dt}\right)\cdot t\cdot E_\gamma = \frac{1}{2}(9.62\times10^{5}\,\text{s}^{-1})(1.00\,\text{s})(1.25\times10^{6}\,\text{eV})(1.60\times10^{-19}\,\text{J/eV})$$
$$= 9.62\times10^{-8}\,\text{J/s}$$

(b) Absorbed dose $= \dfrac{E}{m} = \dfrac{9.62\times10^{-8}\,\text{J/s}}{0.500\,\text{kg}}$

$$= 1.92\times10^{-7}\,\text{J/kg}\cdot\text{s}\left(100\,\frac{\text{rad}}{\text{J/kg}\cdot\text{s}}\right) = 1.92\times10^{-5}\,\text{rad}$$

(c) Equivalent dose $= 0.7(1.92\times10^{-5})\,\text{rad} = 1.34\times10^{-5}\,\text{rad}$.

(d) $\dfrac{200\ \text{rem}}{1.34\times10^{-5}\,\text{rem/s}} = 1.49\times10^{7}\,\text{s} = 172\ \text{days}$.

45-52: (a) After 4.0 min = 240 s, the ratio of the number of nuclei is

$$\frac{2^{-240/122.2}}{2^{-240/26.9}} = 2^{(240)\left(\frac{1}{26.9}-\frac{1}{122.2}\right)} = 124.$$

(b) After 15.0 min = 900 s, the ratio is 7.15×10^{7}.

45-53: $\dfrac{N}{N_0} = 0.21 = e^{-\lambda t}$

$$\Rightarrow t = \frac{-\ln(0.21)}{\lambda} = -\ln(0.21)\cdot\frac{5730\ \text{y}}{0.693} = 13000\ \text{y}.$$

45-54: The activity of the sample will have decreased by a factor of

$$\frac{(4.2\times10^{-6}\,\text{Ci})(3.70\times10^{10}\,\text{Bq/Ci})}{(8.5\,\text{counts/min})(1\,\text{min}/60\,\text{s})} = 1.097\times10^{6} = 2^{20.06};$$

this corresponds to 20.06 half-lifes, and the elapsed time is 40.1 h. Note the retention of extra figures in the exponent to avoid roundoff error. To the given two figures the time is 40 h.

45-55: For deuterium:

$$\text{(a) } U = \frac{1}{4\pi\varepsilon_0}\frac{e^2}{r} = \frac{(1.60\times10^{-19}\,\text{C})^2}{4\pi\varepsilon_0\left[2(1.2\times10^{-15}\,\text{m})(2)^{1/3}\right]} = 7.61\times10^{-14}\,\text{J}$$
$$= 0.48\ \text{MeV}$$

(b) $\Delta m = 2M\left({}_1^2\text{H}\right) - M\left({}_2^3\text{He}\right) - m_\text{n}$

$$= 2(2.014102\text{ u}) - 3.016029\text{ u} - 1.008665\text{ u}$$

$$= 3.51\times10^{-3}\text{ u}$$

$$\Rightarrow E = (\Delta m)c^2 = \left(3.51\times10^{-3}\text{ u}\right)(931.5\text{ MeV/u}) = 3.270\text{ MeV}$$

$$= 5.231\times10^{-13}\text{ J}$$

(c) A mole of deuterium has 6.022×10^{23} molecules, so the energy per mole is $\left(6.022\times10^{23}\right)\left(5.231\times10^{-13}\text{ J}\right) = 3.150\times10^{11}\text{ J}$. This is over a million times more than the heat of combustion.

45-56: (a) $m_{{}^{16}_{8}\text{O}} - m_{{}^{15}_{7}\text{N}} - m_{{}^{1}_{1}\text{H}} = -1.30\times10^{-2}$ u, so the proton separation energy is 12.1 MeV. (b) $m_{{}^{16}_{8}\text{O}} - m_{{}^{15}_{8}\text{O}} - m_\text{n} = -1.68\times10^{-2}$ u, so the neutron separation energy is 15.7 MeV. (c) It takes less energy to remove a proton.

45-57: Mass of ^{40}K atoms in 1.00 kg is $\left(2.1\times10^{-3}\right)\left(1.2\times10^{-4}\right)\text{kg} = 2.52\times10^{-7}\text{ kg}$. Number of atoms $N = \dfrac{2.52\times10^{-7}\text{ kg}}{40\text{ u}\left(1.661\times10^{-27}\text{ kg/u}\right)} = 3.793\times10^{18}$

$$\frac{dN}{dt} = \lambda N = \frac{(0.693)\left(3.793\times10^{18}\right)}{1.25\times10^{9}\text{ y}} = 2.103\times10^{9}\text{ decays/y}$$

So in 50 years the energy absorbed is:

$$E = (0.50\text{ MeV/decay})(50\text{ y})\left(2.103\times10^{9}\text{ decay/y}\right) = 5.26\times10^{10}\text{ MeV}$$

$$= 8.41\times10^{-3}\text{ J}$$

So the absorbed dose is $\left(8.41\times10^{-3}\text{ J}\right)(100\text{J/rad}) = 0.84$ rad and since the RBE = 1.0, the equivalent dose is 0.84 rem.

45-58: In terms of the number N of cesium atoms that decay in one week and the mass m = 1.0 kg, the equivalent dose is

$$3.5\text{ Sv} = \frac{N}{m}\left((\text{RBE})_\gamma E_\gamma + (\text{RBE})_\text{e} E_\text{e}\right)$$

$$= \frac{N}{m}\left((1)(0.66\text{ MeV}) + (1.5)(0.51\text{ MeV})\right)$$

$$= \frac{N}{m}\left(2.283\times10^{-13}\text{ J}\right),\quad \text{so}$$

$$N = \frac{(1.0\text{ kg})(3.5\text{ Sv})}{\left(2.283\times10^{-13}\text{ J}\right)} = 1.456\times10^{13}.$$

The number N_0 of atoms present is related to N by $N_0 = N/\lambda t$, so

$$N_0 = \frac{N}{\lambda t} = \frac{NT_{1/2}}{\ln(2)t} = \frac{\left(1.456\times10^{13}\right)(30.17\text{ y})\left(3.156\times10^{7}\text{ s/yr}\right)}{\ln(2)(7\text{ d})(86{,}400\text{ s/d})} = 3.31\times10^{16}.$$

45-59: $v_{cm} = v \cdot \frac{m}{m+M}$

$$v'_m = v - v\frac{m}{m+M} = \left(\frac{M}{m+M}\right)v \qquad v'_m = \frac{vm}{m+M}$$

$$K' = \frac{1}{2}mv'^2_m + \frac{1}{2}Mv'^2_M = \frac{1}{2}\frac{mM^2}{(m+M)^2}v^2 + \frac{1}{2}\frac{Mm^2}{(m+M)^2}v^2$$

$$= \frac{1}{2}\frac{M}{(m+M)}\left(\frac{mM}{m+M} + \frac{m^2}{m+M}\right)v^2$$

$$= \frac{M}{m+M}\left(\frac{1}{2}mv^2\right) \qquad \Rightarrow K' = \frac{M}{m+M}K$$

45-60: $K = \frac{M_\alpha}{M_\alpha + M}K_\infty$, where K_∞ is the energy that the α-particle would have if the nucleus were infinitely massive. Then,

$$M = M_{\text{Os}} - M_\alpha - K_\infty = M_{\text{Os}} - M_\alpha - \frac{186}{182}(2.76\ \text{MeV}/c^2) = 181.94820\ \text{u}.$$

45-61: $\Delta m = M({}^{235}_{92}\text{U}) - M({}^{140}_{54}\text{Xe}) - M({}^{94}_{38}\text{Sr}) - m_\text{n}$

$$= 234.043924\ \text{u} - 139.921620\ \text{u} - 93.915367\ \text{u} - 1.008665\ \text{u}$$

$$= 0.1983\ \text{u}$$

$$\Rightarrow E = (\Delta m)c^2 = (0.1983\ \text{u})(931.5\ \text{MeV/u}) = 185\ \text{MeV}$$

45-62: (a) A least-squares fit of the log of the activity *vs.* time for the times later than 4.0 hr gives a fit with correlation $-(1-2\times10^{-6})$ and decay constant of $0.361\ \text{hr}^{-1}$, corresponding to a half-life of 1.92 hr. Extrapolating this back to time 0 gives a contribution to the rate of about 2500/s for this longer-liver species. A least-squares fit of the log of the activity *vs.* time for times earlier than 2.0 hr gives a fit with correlation = 0.994, indicating the presence of only two species.
(b) By trial and error, the data is fit by a decay rate modeled by

$$R = (5000\ \text{Bq})e^{-t(1.733/\text{hr})} + (2500\ \text{Bq})e^{-t(0.361/\text{hr})}.$$

This would correspond to half-lives of 0.400 hr and 1.92 hr. (c) In this model, there are 1.04×10^7 of the shorter-lived species and 2.49×10^7 of the longer-lived species. (d) After 5.0 hr, there would be 1.80×10^3 of the shorter-lived species and 4.10×10^6 of the longer-lived species.

45-63: (a) There are two processes occurring: the creation of ${}^{128}\text{I}$ by the neutron irradiation, and the decay of the newly produced ${}^{128}\text{I}$. So $\frac{dN}{dt} = K - \lambda N$ where K is the rate of production by the neutron irradiation.

Then $\int_0^N \frac{dN'}{K - \lambda N'} = \int_0^t dt$

$$\Rightarrow \left[\ln(K - \lambda N')\right]_0^N = -\lambda t$$

$$\Rightarrow \ln(K - \lambda N) = \ln K - \lambda t$$

$$\Rightarrow N(t) = \frac{K(1 - e^{-\lambda t})}{\lambda}$$

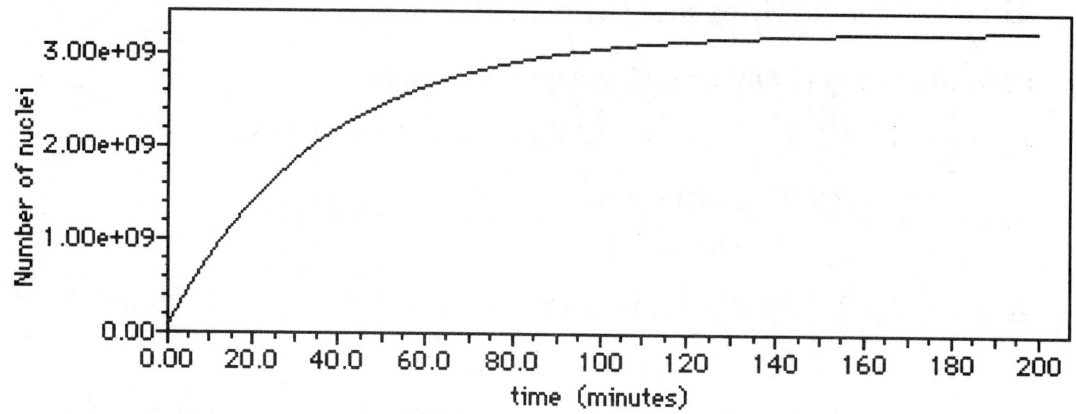

(b) The activity of the sample is $\lambda N(t) = K(1 - e^{-\lambda t})$

$$= (1.5\times10^6\,\text{decays/s})\left(1 - e^{-\left(\frac{0.693}{25\,\text{min}}\right)t}\right)$$

So the activity is $(1.5\times10^6\,\text{decays/s})(1 - e^{-0.02772\,t})$, with t in minutes.

So the activity $\left(\frac{-dN'}{dt}\right)$ at various times is:

$$\frac{-dN'}{dt}(t = 1\,\text{min}) = 4.1\times10^4\,\text{Bq}\,; \quad \frac{-dN'(t = 10\,\text{min})}{dt} = 3.6\times10^4\,\text{Bq}\,;$$

$$\frac{-dN'}{dt}(t = 25\,\text{min}) = 7.5\times10^5\,\text{Bq}\,; \quad \frac{-dN'(t = 50\,\text{min})}{dt} = 1.1\times10^6\,\text{Bq}\,;$$

$$\frac{-dN'}{dt}(t = 75\,\text{min}) = 1.3\times10^6\,\text{Bq}\,; \quad \frac{-dN'(t = 180\,\text{min})}{dt} = 1.5\times10^6\,\text{Bq}\,;$$

(c) $N_{\text{max}} = \frac{K}{\lambda} = \frac{(1.5\times10^6)(60)}{(0.02772)} = 3.2\times10^9\,\text{atoms}$

(d) The maximum activity is at saturation, when the rate being produced equals that decaying and so it equals $1.5\times10^6\,\text{decays/s}$.

45-64: The activity of the original iron, after 1000 hours of operation, would be

$$(9.4\times10^{-6}\ \text{Ci})(3.7\times10^{10}\ \text{Bq/Ci})2^{-(1000\,\text{hr})/(45\text{d}\times24\text{hr/d})} = 1.8306\times10^5\ \text{Bq}.$$

The activity of the oil is 84 Bq, or 4.5886×10^{-4} of the total iron activity, and this must be the fraction of the mass worn, or a mass of 4.59×10^{-2} g. The rate at which the piston rings lost their mass is then 4.59×10^{-5} g/hr.

Chapter 46: Particle Physics and Cosmology

46-1: $E=(\Delta m)c^2=(300\text{ kg}+300\text{ kg})(3.00\times10^8\text{ m/s})^2=5.40\times10^{19}\text{ J}.$

46-2: (a) The energy will be the proton rest energy, 938.3 MeV, corresponding to a frequency of 2.27×10^{23} Hz and a wavelength of 1.32×10^{-15} m. (b) The energy of each photon will be 938.3 MeV + 830 MeV = 1768 MeV, with frequency 42.8×10^{22} Hz and wavelength 7.02×10^{-16} m.

46-3: Each photon gets half of the energy of the pion

$$E_\gamma=\frac{1}{2}m_\pi c^2=\frac{1}{2}(270\,m_e)c^2=\frac{1}{2}(270)(0.511\text{ MeV})=69\text{ MeV}$$

$$\Rightarrow f=\frac{E}{h}=\frac{(6.9\times10^7\text{ eV})(1.6\times10^{-19}\text{ J/eV})}{(6.63\times10^{-34}\text{ J}\cdot\text{s})}=1.7\times10^{22}\text{ Hz}$$

$$\Rightarrow \lambda=\frac{c}{f}=\frac{3.00\times10^8\text{ m/s}}{1.7\times10^{22}\text{ Hz}}=1.8\times10^{-14}\text{ m}$$

46-4: (a) $\lambda=\dfrac{hc}{E}=\dfrac{hc}{m_\mu c^2}=\dfrac{h}{m_\mu c}=\dfrac{(6.626\times10^{-34}\text{ J}\cdot\text{s})}{(207)(9.11\times10^{-31}\text{ kg})(3.00\times10^8\text{ m/s})}$

$=1.17\times10^{-14}\text{ m}=0.0117\text{ pm}.$

In this case, the muons are created at rest (no kinetic energy). (b) Shorter wavelengths would mean highter photon energy, and the muons would be created with non-zero kinetic energy.

46-5: (a) $\Delta m=m_{\pi^+}-m_{\mu^+}=270\,m_e-207\,m_e=63\,m_e$

$\Rightarrow E=63(0.511\text{ MeV})=32\text{ MeV}.$

(b) A positive muon has less mass than a positive pion, so if the decay from muon to pion was to happen, you could always find a frame where energy was not conserved. This cannot occur.

46-6: (a) $2f=\dfrac{\omega}{\pi}=\dfrac{eB}{m\pi}=3.97\times10^7\text{ /s}.$ (b) $\omega R=\dfrac{eBR}{m}=3.12\times10^7\text{ m/s}.$ (c) For three-figure precision, the relativistic form of the kinetic energy must be used, $eV=(\gamma-1)mc^2$, so $eV=(\gamma-1)mc^2$, so $V=\dfrac{(\gamma-1)mc^2}{e}=5.11\times10^6\text{ V}.$

46-7: (a) $\omega=\dfrac{|q|B}{m}\Rightarrow B=\dfrac{m\omega}{|q|}=\dfrac{2\pi mf}{|q|}$

$$\Rightarrow B=\frac{2\pi(201\text{u})(1.66\times10^{-27}\text{ kg/u})(1.10\times10^7\text{ Hz})}{1.60\times10^{-19}\text{ C}}$$

$\Rightarrow B=1.44\text{T}$

(b) $K = \frac{q^2B^2R^2}{2m} = \frac{(1.60\times10^{-19}\text{C})^2(1.44\text{T})^2(0.32\text{m})^2}{2(2.01\text{u})(1.66\times10^{-27}\text{kg/u})} = 8.16\times10^{-13}\text{J}$

$= 5.10\times10^6 \text{ eV} = 5.10 \text{ MeV}$

and $v = \sqrt{\frac{2K}{m}} = \sqrt{\frac{2(8.16\times10^{-13}\text{J})}{(2.01\text{u})(1.66\times10^{-27}\text{kg/u})}} = 2.21\times10^7 \text{m/s}.$

46-8: (a) The energy is so high that the total energy of each particle is half of the available energy, 50 GeV. (b) Equation 46-11 is applicable, and $E_a = 226$ MeV.

46-9: (a) With $E_m >> mc^2, E_m = \frac{E_a^2}{2mc^2}$ from Eq. (46-11)

So $E_m = \frac{[2(38.7 \text{ GeV})]^2}{2(0.938 \text{ GeV})} = 3190 \text{ GeV}$

(b) For colliding beams the available energy E_a is that of both beams). So two proton beams colliding would each need energy of 38.7 GeV to give a total of 77.4 GeV.

46-10: The available energy E_a must be $(m_{\eta^0} + 2m_p)c^2$, so Eq. (46-10) becomes

$$(m_{\eta^0} + 2m_p)^2c^4 = 2m_pc^2(E_t + 2m_pc^2), \text{ or}$$

$$E_t = \frac{(m_{\eta^0} + 2m_p)^2c^2}{2m_p} - 2m_pc^2 = 1258 \text{ MeV},$$

or 1260 MeV to three significant figures.

46-11: (a) $E_a^2 = 2mc^2(E_m + mc^2)$

$\Rightarrow E_m = \frac{E_a^2}{2mc^2} - mc^2$

The mass of the alpha particle is that of a ${}^4_2\text{He}$ atomic mass, minus two electron masses. But to 3 significant figures this is just $M({}^4_2\text{He}) = 4.00$ u

$= (4.00 \text{ u})(0.9315 \text{ GeV/u})$

$= 3.73 \text{ GeV}$

So $E_m = \frac{(16.0 \text{ GeV})^2}{2(3.73 \text{ GeV})} - 3.73 \text{ GeV} = 30.6 \text{ GeV}.$

(b) For colliding beams of equal mass, each has half the available energy, so each has 8.0 GeV.

46-12: (a) $\gamma = \frac{1000\times10^3 \text{ MeV}}{938.3 \text{ MeV}} = 1065.8$, so $v = 0.999999559c$. (b) Non-relativistic: $\omega = \frac{eB}{m} = 3.83\times10^8$ rad/s. Relativistic: $\omega = \frac{eB}{m}\frac{1}{\gamma} = 3.59\times10^5$ rad/s.

46-13: $\left[\frac{f^2}{\hbar c}\right] = \frac{(\text{J}\cdot\text{m})}{(\text{J}\cdot\text{s})(\text{m}\cdot\text{s}^{-1})} = 1$

and thus $\frac{f^2}{\hbar c}$ is dimensionless. (Recall f^2 has with units of energy times distance.)

46-14: Using the values of the constants from Appendix F,

$$\frac{e^2}{4\pi\varepsilon_0\hbar c} = 7.29660475\times10^{-3} = \frac{1}{137.050044},$$

or 1/137 to three figures.

46-15: $\Delta m = M(\Sigma^+) - m_{\text{p}} - m_{\pi^0}$

Using table 46-3:

$$\Rightarrow E = (\Delta m)c^2 = 1189\text{ MeV } - 938.3\text{ MeV } - 135.0\text{ MeV} = 116\text{ MeV}$$

46-16: From Table (46-46-2), $(m_\mu - m_{\text{e}} - 2m_\nu)c^2 = 105.2$ MeV.

46-17: Conservation of lepton number.

(a) $\mu^- \to e^- + \nu_e + \bar{\nu}_\mu \Rightarrow L_\mu : +1 \neq -1,\ L_e : 0 \neq +1+1$

so lepton numbers are not conserved

(b) $\tau^- \to e^- + \bar{\nu}_e + \nu_\tau \Rightarrow L_e : 0 = +1-1$

$L_\tau : +1 = +1$

so lepton numbers are conserved.

(c) $\pi^+ \to e^+ + \gamma$. Lepton numbers are not conserved since just one lepton is produced from zero original leptons.

(d) $\text{n} \to p + e^- + \bar{\gamma}_e \Rightarrow L_e : 0 = +1-1$, so the lepton numbers are conserved.

46-18: (a) Conserved: Both the neutron and proton have baryon number 1, and the electron and neutrino have baryon number 0. (b) Not conserved: The proton has baryon number 1, and the pions have baryon number 0. (c) Not conserved: The initial baryon number is 1 + 1 = 2 and the final baryon number is 1. (d) Conserved: The initial and final baryon numbers are 1 + 1 = 1 + 1 + 0.

46-19: Conservation of strangeness:

(a) $K^+ \to \mu^+ + \nu_\mu$. Strangeness is not conserved since there is just one strange particle, in the initial states.

(b) $\text{n} + K^+ \to p + \pi^0$. Again there is just one strange particle so strangeness cannot be conserved.

(c) $K^+ + K^- \to \pi^0 \Rightarrow S : +1-1 = 0$, so strangeness is conserved.

(d) $p + K^- \to \Lambda^0 + \pi^0 \Rightarrow S : 0-1 = -1+0$, so strangeness is conserved.

46-20: (a) $S = 1$ indicates the presence of one $\bar{s}$ antiquark and no s quark. To have baryon number 0 there can be only one other quark, and to have net charge $+e$ that quark must be a u, and the quark content is $u\bar{s}$. (b) The particle has an $\bar{s}$ antiquark, and for a baryon number of –1 the particle must consist of three antiquarks. For a net charge of $-e$, the quark content must be $\bar{d}\bar{d}\bar{s}$. (c) $S = -2$ means that there are two s quarks, and for baryon number 1 there must be one more quark. For a charge of 0 the third quark must be a u quark and the quark content is uss.

46-21: (a) The antiparticle must consist of the anti-quarks so:
$\bar{\mathrm{n}} = \bar{u}\bar{d}\bar{d}$
(b) So $\mathrm{n} = udd$ is not its own antiparticle
(c) $\pi^0 = u\bar{u}$ or $d\bar{d}$, so $\bar{\pi}^0 = \bar{u}u = \pi^0$
or $\bar{\pi}^0 = \bar{d}d = \pi^0$, so the neutral pion is its own antiparticle.

46-22: $(m_\gamma - 2m_\tau)c^2 = 5906$ MeV.

46-23: (a) uds:

$$Q/e = \frac{2}{3} + \left(-\frac{1}{3}\right) + \left(-\frac{1}{3}\right) = 0;$$
$$B = \frac{1}{3} + \frac{1}{3} + \frac{1}{3} = 1;$$
$$(S = 0 + 0 + (-1) = -1)$$
$$C = 0 + 0 + 0 = 0$$

(b) $c\bar{u}$:

$$\frac{Q}{e} = \frac{2}{3} + \frac{-2}{3} = 0;$$
$$B = \frac{1}{3} + \left(\frac{-1}{3}\right) = 0;$$
$$S = 0 + 0 = 0;$$
$$C = 1 + 0 = 1.$$

(c) ddd:

$$\frac{Q}{e} = 3\left(\frac{-1}{3}\right) = -1; \; B = 3\left(\frac{1}{3}\right) = 1;$$
$$S = 3(0) = 0; \quad C = 3(0) = 0.$$

(d) $d\bar{c}$:

$$\frac{Q}{e} = \frac{-1}{3} + \left(\frac{-2}{3}\right) = -1; \quad B = \frac{1}{3} + \left(\frac{-1}{3}\right) = 0;$$
$$S = 0 + 0 = 0; \quad C = 0 + (-1) = -1.$$

46-24: (a) The Ω^- particle has $Q = -1$ (as its label suggests) and $S = -3$.

46-25: (a)
$$v = \left[\frac{(\lambda_0/\lambda_s)^2 - 1}{(\lambda_0/\lambda_s)^2 + 1}\right] c = \left[\frac{(629.2/590.0)^2 - 1}{(629.2/590.0)^2 + 1}\right](2.998 \times 10^8 \text{ m/s})$$
$$= 1.92 \times 10^7 \text{ m/s}.$$

(b)
$$r = \frac{v}{H_0} = \frac{1.926 \times 10^7 \text{ m/s}}{2.3 \times 10^4 \text{ m/Mly}} = 840 \text{ Mly}$$

46-26: Squaring both sides of Eq. (46-13) and multiplying by $c-v$ gives $\lambda_0^2(c-v)=\lambda_S^2(c+v)$, and solving this for v gives Eq. (46-14).

46-27: (a) $v = H_0 r = (23\ (\text{km/s})/\text{Mly})\ (1440\ \text{Mly}) = 3.3\times10^4\ \text{km/s}$

(b) $\dfrac{\lambda_0}{\lambda_s} = \sqrt{\dfrac{c+v}{c-v}} = \sqrt{\dfrac{3.0\times10^5\ \text{km/s}+3.3\times10^4\text{km/s}}{3.0\times10^5\ \text{km/s}-3.3\times10^4\text{km/s}}} = 1.1$

46-28: From Eq. (46-15), $r = \dfrac{c}{H_0} = \dfrac{3.00\times10^8\ \text{m/s}}{23\ (\text{km/s})/\text{Mly}} = 1.3\times10^4\ \text{Myl}$. (b) This distance represents looking back in time so far that the light has not been able to reach us (see the discussion on Page 1439).

46-29: For blackbody radiation $\lambda_m T = 2.90\times10^{-3}\,\text{m}\cdot\text{K}$,
so $\lambda_m(1)T(1) = \lambda_m(2)T(2)$

$\Rightarrow \lambda_m(1) = (1.063\times10^{-3}\ \text{m})\dfrac{2.726\text{K}}{3000\text{K}} = 9.66\times10^{-7}\ \text{m}$

46-30: (a) The dimensions of $\hbar$ are energy times time, the dimensions of G are energy times time per mass squared, and so the dimensions of $\sqrt{\hbar G/c^3}$ are

$$\left[\frac{(\text{E}\cdot\text{T})(\text{E}\cdot\text{L}/\text{M}^2)}{(\text{L}/\text{T})^3}\right]^{1/2} = \left[\frac{\text{E}}{\text{M}}\right]\left[\frac{\text{T}^2}{\text{L}}\right] = \left[\frac{\text{L}}{\text{T}}\right]^2\left[\frac{\text{T}^2}{\text{L}}\right] = \text{L}.$$

(b) $$\left(\frac{\hbar G}{c^3}\right)^{1/2} = \left(\frac{(6.626\times10^{-34}\ \text{J}\cdot\text{s})(6.673\times10^{-11}\text{N}\cdot\text{m}^2/\text{kg}^2)}{2\pi(3.00\times10^8\ \text{m/s})^3}\right)^{1/2}$$
$$= 1.616\times10^{-35}\,\text{m}.$$

46-31: (a) $\Delta m = M({}^1_1H) + M({}^2_1H) - M({}^3_2He)$ where atomic masses are used to balance electron masses.

$\Rightarrow \Delta m = 1.007825\text{u} + 2.014102\text{u} - 3.016029\text{u}$
$= 5.898\times10^{-3}\text{u}$
$\Rightarrow E = (\Delta m)c^2 = (5.898\times10^{-3}\text{u})(931.5\ \text{MeV/u}) = 5.494\ \text{MeV}.$

(b) $\Delta m = m_n + M({}^3_2He) - M({}^4_2He) =$
$= 1.0086649\text{u} + 3.016029\text{u} - 4.002603\text{u}$
$= 0.022091\text{u}$
$\Rightarrow E = (\Delta m)c^2 = (0.022091\text{u})(931.5\ \text{MeV/u}) = 20.58\ \text{MeV}.$

46-32: $3m({}^4\text{He}) - m({}^{12}\text{C}) = 7.80\times10^{-3}\ \text{u}$, or $7.27\ \text{MeV}$.

46-33: $\Delta m = m_e + m_p - m_n - m_{\nu_e}$ so assuming $m_{\nu_e} \approx 0$,
$\Delta m = 0.0005486\text{u} + 1.007276\text{u} - 1.008665\text{u} = -8.40\times10^{-4}\text{u}$
$\Rightarrow E = (\Delta m)c^2 = (-8.40\times10^{-4}\text{u})(931.5\ \text{MeV/u}) = -0.783\ \text{MeV}$ and is endoergic.

46-34: $m_{{}^{12}_6\text{C}} + m_{{}^4_2\text{He}} - m_{{}^{16}_8\text{O}} = 7.69\times10^{-3}\ \text{u}$, or $7.16\ \text{MeV}$, an exoergic reaction.

46-35: (a) $\Delta m = M({}^4_2He) + M({}^9_4Be) - M({}^{12}_6C) - m_n$
$= 4.002603u + 9.012182u - 12.000000u - 1.008665u$
$= 6.120 \times 10^{-3}u$
$\Rightarrow E = (\Delta m)c^2 = (6.120 \times 10^{-3}u)(931.5\ \text{MeV/u}) = 5.701\ \text{MeV}$
(b) Threshold kinetic energy must be:

$$K = U = \frac{1}{4\pi\epsilon_0}\frac{q_1 q_2}{r} = \frac{1}{4\pi\epsilon_0}\frac{(2e)(4e)(1.60\times10^{-19}\ \text{C})^2}{(1.2\times10^{-15}\ \text{m})(4^{1/3}+9^{1/3})}$$

$= 4.19 \times 10^{-13}\ \text{J} = 2.6\ \text{MeV}.$

46-36: $m_n + m_{{}^{10}_5B} - m_{{}^7_3Li} - m_{{}^4_2He} = 3.00 \times 10^{-3}$ u, and so Q = 2.79 MeV. (b) The incident neutron will not be repelled by the nucleus, and there is no threshold neutron kinetic energy.

46-37: (a) $E_a = 2(20\ \text{TeV}) = 40\ \text{TeV}$
(b) Fixed target; equal mass particles,

$$E_m = \frac{E_a^2}{2mc^2} - mc^2 = \frac{(4.0\times10^7\ \text{MeV})^2}{2(938.3\ \text{MeV})^2} - 938.3\ \text{MeV}$$

$= 8.53 \times 10^{11}\ \text{MeV} = 8.5 \times 10^5\ \text{TeV}$

46-38: $K + m_p c^2 = \frac{hc}{\lambda}, K = \frac{hc}{\lambda} - m_p c^2 = 652\ \text{MeV}.$

46-39: The available energy must be the sum of the final rest masses: (at least)
$E_a = 2m_e c^2 + m_{\pi^0} c^2$
$= 2(0.511\ \text{MeV}) + 13.50\ \text{MeV}$
$= 136.0\ \text{MeV}.$
For alike target and beam particles:

$$E_{m_e} = \frac{E_a^2}{2m_e c^2} - m_e c^2 = \frac{(136.0\ \text{MeV})^2}{2(0.511\ \text{MeV})} - 0.511\ \text{MeV} = 1.81\times10^4\ \text{MeV}$$

So $K = (1.81 \times 10^4\ \text{MeV}) - m_e c^2 = 1.81 \times 10^4\ \text{MeV}.$

46-40: In Eq. (46-9), $E_a = (m_{\Sigma^0} + m_{K^0})c^2$, and with $M = m_p, m = m_{\pi^-}$ and $E_m = (m_{\pi^-})c^2 + K$,

$$K = \frac{E_a^2 - (m_\pi - c^2)^2 - (m_p c^2)^2}{2m_p c^2} - (m_{\pi^-})c^2$$

$$= \frac{(1193\ \text{MeV} + 497.7\ \text{MeV})^2 - (139.6\ \text{MeV})^2 - (938.3\ \text{MeV})^2}{2(938.3\ \text{MeV})} - 139.6\ \text{MeV}$$

$= 904\ \text{MeV}.$

46-41: The available energy must be at least the sum of the final rest masses.
$E_a = (m_{\Lambda^0})c^2 + (m_{K^+})c^2 + (m_{K^-})c^2 = 1116\ \text{MeV} + 2(493.7\ \text{MeV}) = 2103\ \text{MeV}.$
$E_a^2 = 2(m_p)c^2 E_{K^-} + ((m_p)c^2)^2 + ((m_{K^-})c^2)^2$

So $E_{K^-} = \dfrac{E_a^2 - ((m_p)c^2)^2 - ((m_{K^-})c^2)^2}{2(m_p)c^2} = \dfrac{(2103)^2 - (938.3)^2 - (493.7)^2}{2(938.3)^2}$ MeV

$\Rightarrow E_{K^-} = 1759 \text{ MeV} = (m_{K^-})c^2 + K$

So the threshold energy $K = 1759 \text{ MeV} - 493.7 \text{ MeV} = 1265 \text{ MeV}$

46-42: (a) The decay products must be neutral, so the only possible combinations are $\pi^0\pi^0\pi^0$ or $\pi^0\pi^+\pi^-$ (b) $m_{\eta_0} - 3m_{\pi^0} = 143 \text{ MeV}/c^2$, so the kinetic energy of the mesons is 143 MeV. $(m_{\eta_0} - m_{\pi^0} - m_{\pi^+} - m_{\pi^-})c^2 = 133$ MeV.

46-43: (a) If the π^- decays, it must end in an electron and neutrinos. The rest energy of π^- (139.6 MeV) is shared between the electron rest energy (0.511 MeV) and kinetic energy (assuming the neutrino masses are negligible).
So the energy released is $139.6 \text{ MeV} - 0.511 \text{ MeV} = 139.1 \text{ MeV}$
(b) Conservation of momentum leads to the neutrinos carrying away most of the energy.

46-44: $\dfrac{\hbar}{\Delta E} = 1.5 \times 10^{-22}$ s.

46-45: (a) $E = (\Delta m)c^2 = (m_p)c^2 - (m_{K^+})c^2 - (m_{K^-})c^2$

$= 1019.4 \text{ MeV} - 2(493.7 \text{ MeV})$

$= 32.0 \text{ MeV}$

Each kaon gets half the energy so the kinetic energy of the K^+ is 16.0 MeV.
(b) Since the π^0 mass is greater than the energy left over in part(a), it could not have been produced in addition to the kaons.
(c) Conservation of strangeness will not allow $\phi \to K^+ + \pi^-$ or $\phi \to K^+ + \pi^+$

46-46: (a) The baryon number is 0, the charge is $+e$, the strangeness is 1, all lepton numbers are zero, and the particle is K^+. (b) The baryon number is 0, the charge is $-e$, the strangeness is 0, all lepton numbers are zero, and the particle is π^-. (c) The baryon number is –1, the charge is 0, the strangeness is zero, all lepton numbers are 0, and the particle is an antineutron. (d) The baryon number is 0, the charge is $+e$, the strangeness is 0, the muonic lepton number is –1, all other lepton numbers are 0, and the particle is μ^+.

46-47: $\Delta t = 7.5 \times 10^{-21} \text{s}. \Rightarrow \Delta E = \dfrac{\hbar}{\Delta t} = \dfrac{1.05 \times 10^{-34} \text{ J}\cdot\text{s}}{7.5 \times 10^{-21} \text{s}} = 1.41 \times 10^{-14} \text{J} = 88 \text{ keV}$

$\dfrac{\Delta E}{m_\psi c^2} = \dfrac{0.088 \text{ MeV}}{3097 \text{ MeV}} = 2.8 \times 10^{-5}$

46-48: (a) The number of protons in a kilogram is

$(1.00 \text{ kg})\left(\dfrac{6.023 \times 10^{23} \text{ molecules/mol}}{18.0 \times 10^{-3} \text{ kg/mol}}\right)(2 \text{ protons/molecule}) = 6.7 \times 10^{25}.$

Note that only the protons in the hydrogen atoms are considered as possible sources of proton decay. The energy per decay is $m_p c^2 = 938.3\text{ MeV} = 1.503\times10^{-10}$ J, and so the energy deposited in a year, per kilogram, is $(6.7\times10^{25})\left(\dfrac{\ln(2)}{1.0\times10^{18}\text{ y}}\right)(1\text{ y})(1.50\times10^{-10}\text{ J}) = 7.0\times10^{-3}\text{ Gy} = 0.70\text{ rad}$.

(b) For an RBE of unity, the equivalent dose is (1)(0.70 rad) = 0.70 rem.

46-49: (a) $E = (\Delta m)c^2 = (m_{\Xi^-})c^2 - (m_{\Lambda^0})c^2 - (m_{\pi^-})c^2$
$= 1321\text{ MeV} - 1116\text{ MeV} - 139.6\text{ MeV}$
$\Rightarrow E = 65\text{ MeV}$

(b) Using (non-relativistic) conservation of momentum and energy:
$P_{\Lambda^0} = 0 = P_f = m_{\Lambda^0} v_{\Lambda^0} - m_{\pi^-} - v_{\pi^-}$

$$\Rightarrow v_{\pi^-} = \left(\frac{m_{\Lambda^0}}{m_{\pi^-}}\right) v_{\Lambda^0}$$

Also $K_{\Lambda^0} + K_{\pi^-} = E$ from part (a)

$$\text{So } K_{\Lambda^0} + \frac{1}{2} m_{\pi^-} v_{\pi^-}{}^2 = K_{\Lambda^0} + \frac{1}{2}\left(\frac{m_{\Lambda^0}}{m_{\pi^-}}\right) m_{\Lambda^0} v_{\Lambda^0}{}^2 = K_{\Lambda^0}\left(1 + \frac{m_{\Lambda^0}}{m_{\pi^-}}\right)$$

$$\Rightarrow K_{\Lambda^0} = \frac{E}{1 + \frac{m_{\Lambda^0}}{m_{\pi^-}}} = \frac{65\text{ MeV}}{1 + \frac{1116\text{ MeV}}{139.6\text{ MeV}}} = 7.2\text{ MeV}$$

$$\Rightarrow K_{\pi^-} = 65 - 7.2\text{ MeV} = 57.8\text{ MeV}$$

So the fractions of energy carried off by the particles is $\dfrac{7.2}{65} = 0.11$ for the Λ^0 and 0.89 for the π^-.

46-50: (a) For this model, $\dfrac{dR}{dt} = HR$, so $\dfrac{dR/dt}{R} = \dfrac{HR}{R} = H$, presumed to be the same for all points on the surface. (b) For constant θ, $\dfrac{dr}{dt} = \dfrac{dR}{dt}\theta = HR\theta = Hr$. (c) See part (a), $H_0 = \dfrac{dR/dt}{R}$. (d) The equation $\dfrac{dR}{dt} = H_0 R$ is a differential equation, the solution to which, for constant H_0 is $R(t) = R_0 e^{H_0 t}$, where R_0 is the value of R at t = 0. This equation may be solved by separation of variables, as

$$\frac{dR/dt}{R} = \frac{d}{dt}\ln(R) = H_0$$

and integrating both sides with respect to time. (e) A constant H_0 would mean a constant critical density , which is inconsistent with uniform expansion.

46-51:

From Pr. (46-50): $r = R\theta \Rightarrow R = \dfrac{r}{\theta}$

So $\dfrac{dR}{dt} = \dfrac{1}{\theta}\dfrac{dr}{dt} - \dfrac{r}{\theta^2}\dfrac{d\theta}{dt} = \dfrac{1}{\theta}\dfrac{dr}{dt}$ since $\dfrac{d\theta}{dt} = 0.$

So $\frac{1}{R}\frac{dR}{dt} = \frac{1}{R\theta}\frac{dr}{dt} = \frac{1}{r}\frac{dr}{dt} \Rightarrow v = \frac{dr}{dt} = \left(\frac{1}{R}\frac{dR}{dt}\right)r = H_0 r$

Now $\frac{dv}{d\theta} = 0 = \frac{d}{d\theta}\left(\frac{r}{R}\frac{dR}{dt}\right) = \frac{d}{d\theta}\left(\theta\frac{dR}{dt}\right)$

$\Rightarrow \theta\frac{dR}{dt} = K$ where K is a constant.

$\Rightarrow \frac{dR}{dt} = \frac{K}{\theta} \Rightarrow R = \left(\frac{K}{\theta}\right)t$ since $\frac{d\theta}{dt} = 0.$

$\Rightarrow H_0 = \frac{1}{R}\frac{dR}{dt} = \frac{\theta}{Kt}\frac{K}{\theta} = \frac{1}{t}$

So the current value of the Hubble constant is $\frac{1}{T}$ where T is the present age of the universe.

46-52: (a) For mass m, in Eq. (39-24) $u = -v_{\text{cm}}, v' = v_0$, and so $v_m = \frac{v_0 - v_{cm}}{1 - v_0 v_{cm}/c^2}$. For mass M, $u = -v_{\text{cm}}, v' = 0$, so $v_M = -v_{\text{cm}}$. (b) The condition for no net momentum in the center of mass frame is $m\gamma_m v_m + M\gamma_M v_M = 0$, where γ_m and γ_M correspond to the velocities found in part (a). The algebra reduces to $\beta_m\gamma_m = (\beta_0 - \beta')\gamma_0\gamma_M$, where $\beta_0 = \frac{v_0}{c}, \beta' = \frac{v_{\text{cm}}}{c}$, and the condition for no net momentum becomes

$$m(\beta_0 - \beta')\gamma_0\gamma_M = M\beta'\gamma_M, \quad \text{or}$$

$$\beta' = \frac{\beta_0}{1 + \frac{M}{m\gamma_0}} = \beta_0 \frac{m}{m + M\sqrt{1 - \beta_0^2}}, \quad \text{and}$$

$$v_{\text{cm}} = \frac{mv_0}{m + M\sqrt{1 - (v_0/c)^2}}.$$

(c) Substitution of the above expression for into the expressions for the velocities found in part (a) gives the relatively simple forms

$$v_m = v_0\gamma_0 \frac{M}{m + M\gamma_0}, \quad v_M = -v_0\gamma_0 \frac{m}{m\gamma_0 + M}.$$

After some more algebra,

$$\gamma_m = \frac{m + M\gamma_0}{\sqrt{m^2 + M^2 + 2mM\gamma_0}}, \quad \gamma_M = \frac{M + m\gamma_0}{\sqrt{m^2 + M^2 + 2mM\gamma_0}},$$

from which

$$m\gamma_m + M\gamma_M = \sqrt{m^2 + M^2 + 2mM\gamma_0}.$$

This last expression, multiplied by c^2, is the available energy E_a in the center or mass frame, so that

$$\begin{aligned} E_a^2 &= (m^2 + M^2 + 2mM\gamma_0)c^4 \\ &= (mc^2)^2 + (Mc^2)^2 + (2Mc^2)(m\gamma_0 c^2) \\ &= (mc^2)^2 + (Mc^2)^2 + 2Mc^2 E_m, \end{aligned}$$

which is Eq. (46-9).

46-53: $\Lambda^0 \to n + \pi^0$

(a) $E = (\Delta m)c^2 = (m_{\Lambda^0})c^2 - (m_n)c^2(m_{\pi^0})c^2$

$= 1116\text{ MeV} - 939.6\text{ MeV} - 135.0\text{ MeV}$

$= 41.4\text{ MeV}$

(b) Using conservation of momentum and kinetic energy, we know that the momentum of the neutron and pion must have the same magnitude, $p_n = p_\pi$

$K_n = E_n - m_n c^2 = \sqrt{(m_n c^2)^2 + (p_n c)^2} - m_n c^2$

$= \sqrt{(m_n c^2)^2 + (p_\pi c)^2} - m_n c^2$

$= \sqrt{(m_n c^2)^2 + K_\pi{}^2 + 2m_\pi c^2 K_\pi} - m_n c^2$

$\Rightarrow K_\pi + K_n = K_\pi + \sqrt{(m_n c^2)^2 + K_\pi{}^2 + 2m_\pi c^2 K_\pi} - m_n c^2 = E$

$(m_n c^2)^2 + K_\pi{}^2 + 2m_\pi c^2 K_\pi = E^2 + (m_n c^2)^2 + K_\pi{}^2 + 2Em_\pi c^2 - 2EK_\pi - 2m_n c^2 K_\pi$.

Collecting terms we find:

$K_\pi(2m_\pi c^2 + 2E + 2m_n c^2) = E^2 + 2Em_n c^2$

$$\Rightarrow K_\pi = \frac{(41.4\text{ MeV})^2 + 2(41.4\text{ MeV})(939.6\text{ MeV})}{2(135.0\text{ MeV}) + 2(41.4\text{ MeV}) + 2(939.6\text{ MeV})}$$

$\Rightarrow K_\pi = 35.62\text{ MeV}$

So the fractional energy carried by the pion is $\frac{35.62}{41.4} = 0.86$, and that of the neutron is 0.14.

CPSIA information can be obtained
at www.ICGtesting.com
Printed in the USA
BVHW060709280521
608295BV00002B/833

9 781802 973556